T0214865

Communications in Computer and Information Science 553

Commenced Publication in 2007
Founding and Former Series Editors:
Alfredo Cuzzocrea, Dominik Ślęzak, and Xiaokang Yang

More information about this series at http://www.springer.com/series/7899

Ana Fred · Jan L.G. Dietz
David Aveiro · Kecheng Liu
Joaquim Filipe (Eds.)

Knowledge Discovery, Knowledge Engineering and Knowledge Management

6th International Joint Conference, IC3K 2014
Rome, Italy, October 21–24, 2014
Revised Selected Papers

 Springer

Editors

Ana Fred
Instituto de Telecomunicações
Lisboa
Portugal

Jan L.G. Dietz
Delft University of Technology
Delft, Zuid-Holland
The Netherlands

David Aveiro
University of Madeira
Funchal
Portugal

Kecheng Liu
Henley Business School
University of Reading
Reading
UK

Joaquim Filipe
INSTICC
Setubal
Portugal

ISSN 1865-0929   ISSN 1865-0937 (electronic)
Communications in Computer and Information Science
ISBN 978-3-319-25839-3   ISBN 978-3-319-25840-9 (eBook)
DOI 10.1007/978-3-319-25840-9

Library of Congress Control Number: 2015952528

Springer International Publishing AG Switzerland is part of Springer Science+Business Media
(www.springer.com)

Preface

The present book includes extended and revised versions of a set of selected papers from the 6th International Joint Conference on Knowledge Discovery, Knowledge Engineering and Knowledge Management (IC3K 2014), held in Rome, Italy, during October 21–24, 2014. IC3K was sponsored by the Institute for Systems and Technologies of Information, Control and Communication (INSTICC) and was organized in cooperation with the Association for the Advancement of Artificial Intelligence (AAAI), Information and Knowledge Management Society (iKMS), ACM Special Interest Group on Management Information Systems (ACM SIGMIS), ACM Special Interest Group on Artificial Intelligence (ACM SIGART), European Research Consortium for Informatics and Mathematics (ERCIM) and Associazione Italiana per l'Intelligenza Artificiale (AI*IA).

The main objective of IC3K is to provide a point of contact for scientists, engineers, and practitioners interested in the areas of knowledge discovery, knowledge engineering, and knowledge management.

IC3K is composed of three co-located complementary conferences, each specialized in one of the aforementioned main knowledge areas, namely:

- International Conference on Knowledge Discovery and Information Retrieval (KDIR)
- International Conference on Knowledge Engineering and Ontology Development (KEOD)
- International Conference on Knowledge Management and Information Sharing (KMIS)

The International Conference on Knowledge Discovery and Information Retrieval (KDIR) aims to provide a major forum for the scientific and technical advancement of knowledge discovery and information retrieval. Knowledge discovery is an interdisciplinary area focusing on methodologies for identifying valid, novel, potentially useful, and meaningful patterns from data, often based on underlying large data sets. A major aspect of knowledge discovery is data mining, i.e., applying data analysis and discovery algorithms that produce a particular enumeration of patterns (or models) over the data. Knowledge discovery also includes the evaluation of patterns and identification of which add to knowledge.

Information retrieval (IR) is concerned with gathering relevant information from unstructured and semantically fuzzy data in texts and other media, searching for information within documents and for metadata about documents, as well as searching relational databases and the Web. Automation of IR enables the reduction of what has been called "information overload." IR can be combined with knowledge discovery to create software tools that empower users of decision support systems to better understand and use the knowledge underlying large data sets.

The purpose of the International Conference on Knowledge Engineering and Ontology Development (KEOD) is to provide a major meeting point for researchers and practitioners interested in the study and development of methodologies and technologies for knowledge engineering and ontology development. Knowledge engineering (KE) refers to all technical, scientific and social aspects involved in building, maintaining, and using knowledge-based systems. KE is a multidisciplinary field, bringing in concepts and methods from several computer science domains such as artificial intelligence, databases, expert systems, decision support systems, and geographic information systems.

Ontology development (OD) aims at building reusable semantic structures that can be informal vocabularies, catalogs, glossaries as well as more complex finite formal structures representing the entities within a domain and the relationships between those entities. Ontologies have been gaining interest and acceptance in computational audiences: formal ontologies are a form of software, thus software development methodologies can be adapted to serve OD. A wide range of applications are emerging, especially given the current Web emphasis, including library science, ontology-enhanced search, e-commerce, and business process design.

The goal of the International Conference on Knowledge Management and Information Sharing (KMIS) is to provide a major meeting point for researchers and practitioners interested in the study and application of all perspectives of knowledge management and information sharing. Knowledge management (KM) is a discipline concerned with the analysis and technical support of practices used in an organization to identify, create, represent, distribute, and enable the adoption and leveraging of good practices embedded in collaborative settings and, in particular, in organizational processes. Effective KM is an increasingly important source of competitive advantage, and a key to the success of contemporary organizations, bolstering the collective expertise of its employees and partners. Information sharing (IS) is a term that has been used for a long time in the information technology (IT) lexicon, related to data exchange, communication protocols, and technological infrastructures.

The joint conference IC3K received 287 paper submissions from 56 countries, which demonstrates the success and global dimension of this conference. From these, 39 papers were accepted as full papers, 89 were accepted for short presentation, and another 54 for poster presentation. These numbers suggest a "full-paper" acceptance ratio of about 14 %.

IC3K was complemented by the 5th International Workshop on Software Knowledge (SKY; chaired by Iaakov Exman, Juan Llorens, Anabel Fraga and Juan Miguel Gómez) and the Special Session on Text Mining (SSTM; chaired by Ana Fred).

On behalf of the conference Organizing Committee, we would like to thank all participants. First of all the authors, whose quality work is the essence of the conference, and the members of the Program Committee, who helped us with their expertise and diligence in reviewing the papers. As we all know, holding a conference requires the effort of many individuals. We wish to thank also all the members of our Organizing Committee, whose work and commitment were invaluable.

April 2015

Ana Fred
Jan L.G. Dietz
David Aveiro
Kecheng Liu
Joaquim Filipe

Organization

Conference Chair

Joaquim Filipe   Polytechnic Institute of Setúbal/INSTICC, Portugal

Program Co-chairs

KDIR

Ana Fred   Instituto de Telecomunicações, IST - University
                                  of Lisbon, Portugal

KEOD

Jan Dietz   Delft University of Technology, The Netherlands
David Aveiro   University of Madeira, Portugal

KMIS

Kecheng Liu   University of Reading, UK

Organizing Committee

Helder Coelhas	INSTICC, Portugal
Vera Coelho	INSTICC, Portugal
Lucia Gomes	INSTICC, Portugal
Ana Guerreiro	INSTICC, Portugal
André Lista	INSTICC, Portugal
Andreia Moita	INSTICC, Portugal
Vitor Pedrosa	INSTICC, Portugal
Andreia Pereira	INSTICC, Portugal
João Ribeiro	INSTICC, Portugal
Rui Rodrigues	INSTICC, Portugal
Susana Ribeiro	INSTICC, Portugal
Sara Santiago	INSTICC, Portugal
Mara Silva	INSTICC, Portugal
José Varela	INSTICC, Portugal
Pedro Varela	INSTICC, Portugal

KDIR Program Committee

Sherief Abdallah	British University in Dubai, UAE
Muhammad Abulaish	Jamia Millia Islamia, India

Samad Ahmadi	De Montfort University, UK
Mayer Aladjem	Ben-Gurion University of the Negev, Israel
Eva Armengol	IIIA CSIC, Spain
Zeyar Aung	Masdar Institute of Science and Technology, UAE
Vladan Babovic	National University of Singapore, Singapore
Márcio Basgalupp	Universidade Federal de São Paulo, Brazil
Alejandro Bellogin	Universidad Autonoma de Madrid, Spain
Jiang Bian	Microsoft Research, China
Alberto Del Bimbo	Università degli Studi di Firenze, Italy
Pavel Brazdil	University of Porto, Portugal
Fabricio Breve	State University of São Paulo, Brazil
Emanuele Di Buccio	University of Padua, Italy
Maria Jose Aramburu Cabo	Jaume I University, Spain
Yi Cai	South China University of Technology, China
Rui Camacho	Universidade do Porto, Portugal
Luis M. de Campos	University of Granada, Spain
Chien-Chung Chan	University of Akron, USA
Keith C.C. Chan	The Hong Kong Polytechnic University, Hong Kong, SAR China
Meng Chang Chen	Academia Sinica, Taiwan
Shu-Ching Chen	Florida International University, USA
Philipp Cimiano	University of Bielefeld, Germany
Thanh-Nghi Do	College of Information Technology, Can Tho University, Vietnam
Dejing Dou	University of Oregon, USA
Antoine Doucet	University of Caen, France
Tapio Elomaa	Tampere University of Technology, Finland
Iaakov Exman	JCE - The Jerusalem College of Engineering, Israel
Katti Facelli	UFSCAR, Brazil
Nastaran Fatemi	HEIG-VD, University of Applied Sciences of Western Switzerland, Switzerland
Elisabetta Fersini	University of Milano-Bicocca Consorzio Milano Ricerche, Italy
Philippe Fournier-Viger	University of Moncton, Canada
Fabrício Olivetti de França	Universidade Federal do ABC, Brazil
Ana Fred	Instituto de Telecomunicações/IST, Portugal
Panorea Gaitanou	Ionian University, Greece
Susan Gauch	University of Arkansas, USA
Rosario Girardi	UFMA, Brazil
Rosalba Giugno	University of Catania, Italy
Manuel Montes y Gómez	INAOE, Mexico
Nuno Pina Gonçalves	EST-Setúbal/IPS, Portugal
Antonella Guzzo	University of Calabria, Italy
Yaakov Hacohen-Kerner	Jerusalem College of Technology (Machon Lev), Israel
Greg Hamerly	Baylor University, USA
Karin Harbusch	Universität Koblenz-Landau, Germany

Jennifer Harding	Loughborough University, UK
José Hernández-Orallo	Universitat Politècnica de València, Spain
Beatriz de la Iglesia	University of East Anglia, UK
Szymon Jaroszewicz	Polish Academy of Sciences, Poland
Liu Jing	Xidian University, China
Mouna Kamel	IRIT, France
Mehmed Kantardzic	University of Louisville, USA
Ron Kenett	KPA Ltd., Israel
Claus-Peter Klas	GESIS Leibniz Institute for the Social Sciences, Germany
Natallia Kokash	Leiden University, The Netherlands
Steven Kraines	The University of Tokyo, Japan
Ralf Krestel	Hasso-Plattner-Institut, Germany
Hagen Langer	University of Bremen, Germany
Mark Last	Ben-Gurion University of the Negev, Israel
Anne Laurent	Lirmm, University of Montpellier 2, France
Carson K. Leung	University of Manitoba, Canada
Chun Hung Li	Hong Kong Baptist University, Hong Kong, SAR China
Lei Li	Florida International University, USA
Xia Lin	Drexel University, USA
Jun Liu	University of Ulster, UK
Rafael Berlanga Llavori	Jaume I University, Spain
Alicia Troncoso Lora	Pablo de Olavide University of Seville, Spain
Misael Mongiovi	Università di Catania, Italy
Stefania Montani	Piemonte Orientale University, Italy
Eduardo F. Morales	INAOE, Mexico
Gianluca Moro	Università di Bologna, Italy
Henning Müller	University of Applied Sciences Western Switzerland, Switzerland
Wolfgang Nejdl	L3S and University of Hannover, Germany
Engelbert Mephu Nguifo	LIMOS, Université Blaise Pascal, France
Giorgio Maria Di Nunzio	Università degli Studi di Padova, Italy
Mitsunori Ogihara	University of Miami, USA
Elias Oliveira	Universidade Federal do Espirito Santo, Brazil
Nicola Orio	Università degli Studi di Padova, Italy
Rui Pedro Paiva	University of Coimbra, Portugal
Krzysztof Pancerz	University of Information Technology and Management in Rzeszow, Poland
Panos M. Pardalos	University of Florida, USA
Luigi Pontieri	National Research Council (CNR), Italy
Alfredo Pulvirenti	University of Catania, Italy
Marcos Gonçalves Quiles	Federal University of Sao Paulo - UNIFESP, Brazil
Bijan Raahemi	University of Ottawa, Canada
Maria RIFQI	University of Panthéon-Assas, France
Fabio Rinaldi	University of Zurich, Switzerland

Carolina Ruiz	WPI, USA
Henryk Rybinski	Warsaw University of Technology, Poland
Ovidio Salvetti	National Research Council of Italy - CNR, Italy
Filippo Sciarrone	Open Informatica srl, Italy
Fabricio Silva	FIOCRUZ - Fundação Oswaldo Cruz, Brazil
Dan A. Simovici	University of Massachusetts Boston, USA
Minseok Song	UNIST (Ulsan National Institue of Science and Technology), Korea, Republic of
Marcin Sydow	IPI PAN (and PJIIT), Warsaw, Poland
Andrea Tagarelli	University of Calabria, Italy
Kosuke Takano	Kanagawa Institute of Technology, Japan
Ulrich Thiel	Fraunhofer Gesellschaft, Germany
Kar Ann Toh	Yonsei University, Republic of Korea
Vicenc Torra	IIIA-CSIC, Spain
Evelyne Tzoukermann	The MITRE Corporation, USA
Domenico Ursino	University "Mediterranea" of Reggio Calabria, Italy
Slobodan Vucetic	Temple University, USA
Jiabing Wang	South China University of Technology, China
Harco Leslie Hendric Spits Warnars	Surya University, Indonesia
Leandro Krug Wives	Universidade Federal do Rio Grande do Sul, Brazil
Yinghui Wu	UCSB, USA
Yang Xiang	The Ohio State University, USA

KDIR Auxiliary Reviewers

Lamberto Ballan	Università degli Studi di Firenze, Italy
Arlene Casey	De Montfort University, UK
Arnaud Castelltort	LIRMM, France
Giacomo Domeniconi	University of Bologna, Italy
Dario Di Fina	MICC - University of Florence, Italy
Roberto Interdonato	DIMES - Unversità della Calabria, Italy
Mohamed Nader Jelassi	Université Blaise Pascal Clermont Ferrand, Tunisia
Myoung-Ah KANG	Blaise Pascal University, France
Giuseppe Lisanti	MICC-UNIFI, Italy
Elio Masciari	ICAR-CNR, Italy
Victoria Nebot	Jaume I University, Spain
Roberto Pasolini	University of Bologna, Italy
Pietro Pinoli	Politecnico di Milano, Italy
Salvatore Romeo	UNICAL, Italy
Tiberio Uricchio	Università degli Studi di Firenze, Italy
Peiyuan Zhou	The Hong Kong Polytechnic University, Hong Kong, SAR China

KEOD Program Committee

Alia Abdelmoty	Cardiff University, UK
Alessandro Agostini	University of Trento, Italy
Salah Ait-Mokhtar	Xerox Research Centre Europe, France
Raian Ali	Bournemouth University, UK
Francisco Antunes	Institute of Computer and Systems Engineering of Coimbra and Beira Interior University, Portugal
Panos Balatsoukas	University of Manchester, UK
Claudio de Souza Baptista	Universidade Federal de Campina Grande, Brazil
Jean-Paul Barthes	Université de Technologie de Compiègne, France
Teresa M.A. Basile	Università degli Studi di Bari, Italy
Punam Bedi	University of Delhi, India
Sonia Bergamaschi	University of Modena and Reggio Emilia, Italy
Joost Breuker	University of Amsterdam, The Netherlands
Patrick Brezillon	University Pierre and Marie Curie (UPMC), France
Giacomo Bucci	Università degli Studi di Firenze, Italy
Gerhard Budin	University of Vienna, Austria
Vladimír Bureš	University of Hradec Kralove, Czech Republic
Ismael Caballero	Universidad de Castilla-La Mancha (UCLM), Spain
Doina Caragea	Kansas State University, USA
Jin Chen	Michigan State University, USA
Ricardo Colomo-Palacios	Østfold University College, Norway
João Paulo Costa	Institute of Computer and Systems Engineering of Coimbra, Portugal
Constantina Costopoulou	Agricultural University of Athens, Greece
Erdogan Dogdu	TOBB University of Economics and Technology, Turkey
John Edwards	Aston University, UK
Nicola Fanizzi	Università degli studi di Bari Aldo Moro, Italy
Anna Fensel	STI Innsbruck, University of Innsbruck, Austria
Dieter A. Fensel	University of Innsbruck, Austria
Jesualdo Tomás Fernández-Breis	University of Murcia, Spain
Raul Garcia-Castro	Universidad Politécnica de Madrid, Spain
Faiez Gargouri	ISIMS, Tunisia
Manolis Gergatsoulis	Ionian University, Greece
George Giannakopoulos	SKEL Lab – NCSR Demokritos, Greece
Rosario Girardi	UFMA, Brazil
Matteo Golfarelli	University of Bologna, Italy
Sergio Greco	University of Calabria, Italy
Sven Groppe	University of Lübeck, Germany
Ourania Hatzi	Harokopio University of Athens, Greece
Christopher Hogger	Imperial College London, UK

Mari Carmen Suárez-Figueroa	Ontology Engineering Group, UPM, Spain
Domenico Talia	University of Calabria and ICAR-CNR, Italy
George Tsatsaronis	Technische Universität Dresden, Germany
Shengru Tu	University of New Orleans, USA
Manolis Tzagarakis	University of Patras, Greece
Rafael Valencia-Garcia	Universidad de Murcia, Spain
Iraklis Varlamis	Harokopio University of Athens, Greece
Cristina Vicente-Chicote	Universidad de Extremadura, Spain
Bruno Volckaert	Ghent University, Belgium
Yue Xu	Queensland University of Technology, Australia
Gian Piero Zarri	Sorbonne University, France
Jinglan Zhang	Queensland University of Technology, Australia
Catherine Faron Zucker	I3S, Université Nice Sophia Antipolis, CNRS, France

KEOD Auxiliary Reviewers

Lorenzo Albano	University of Modena and Reggio Emilia, Italy
Alessia Bardi	Italian National Research Council, Italy
Fabio Benedetti	Unimore, Italy
Domenico Beneventano	DII - Università di Modena e Reggio Emilia, Italy
Alban Gaignard	CNRS, France
José María García	University of Innsbruck, Austria
Francesco Guerra	University of Modena and Reggio Emilia, Italy
Nelia Lasierra	University of Innsbruck, Austria
Cristian Molinaro	University of Calabria, Italy
Laura Po	University of Modena and Reggio Emilia, Italy
Alvaro E. Prieto	Univesity of Extremadura, Spain
Andrea Pugliese	University of Calabria, Italy
Giovanni Simonini	Università di Modena e Reggio Emilia, Italy
Maurizio Vincini	DIEF - Università di Modena e Reggio Emilia, Italy

KMIS Program Committee

Marie-Helene Abel	HEUDIASYC CNRS UMR, University of Compiègne, France
Shamsuddin Ahmed	University of Malaya, Malaysia
Miriam C. Bergue Alves	Institute of Aeronautics and Space, Brazil
Rangachari Anand	IBM T.J. Watson Research Center, USA
Silvana Castano	Università degli Studi di Milano, Italy
Marcello Castellano	Politecnico di Bari, Italy
Xiaoyu Chen	SKLSDE, Beihang University, China
Roger Chiang	University of Cincinnati, USA
Dickson K.W. Chiu	Dickson Computer Systems, Hong Kong, SAR China
Byron Choi	Hong Kong Baptist University, Hong Kong, SAR China

Paolo Spagnoletti   LUISS Guido Carli University, Italy
Esaú Villatoro Tello   Universidad Autonoma Metropolitana (UAM), Mexico
Wendy Hui Wang   Stevens Institute of Technology, USA
Robert Warren   Carleton University, Canada
Rosina Weber   iSchool at Drexel, USA
Martin Wessner   Fraunhofer IESE, Germany
Uffe K. Wiil   University of Southern Denmark, Denmark
Leandro Krug Wives   Universidade Federal do Rio Grande do Sul, Brazil

KMIS Auxiliary Reviewers

Chulaka Gunasekara   IBM, USA
Rutilio López   University of Alcala, Spain

Invited Speakers

Domenico Talia   University of Calabria and ICAR-CNR, Italy
Sonia Bergamaschi   DIEF - University of Modena and Reggio Emilia, Italy
Michele Missikoff   CNR and Polytechnic University of Marche, Italy
Wil Van Der Aalst   Technische Universiteit Eindhoven, The Netherlands
Wim Van Grembergen   University of Antwerp, Belgium
Marie-Jeanne Lesot   Université Pierre et Marie Curie - LIP6, France

Contents

Knowledge Engineering and Ontology Development

Knowledge Discovery
and Information Retrieval

Efficient Discovery of Episode Rules with a Minimal Antecedent and a Distant Consequent

Lina Fahed[✉], Armelle Brun, and Anne Boyer

Lorraine University, LORIA - KIWI Team, Campus Scientifique, BP 239,
54506 Vandoeuvre-lès-Nancy Cedex, France
{lina.fahed,armelle.brun,anne.boyer}@loria.fr

Abstract. This paper focuses on event prediction in an event sequence, particularly on distant event prediction. We aim at mining episode rules with a consequent temporally distant from the antecedent and with a minimal antecedent. To reach this goal, we propose an algorithm that determines the consequent of an episode rule at an early stage in the mining process, and that applies a span constraint on the antecedent and a gap constraint between the antecedent and the consequent. This algorithm has a complexity lower than that of state of the art algorithms, as it is independent of the gap between the antecedent and the consequent. In addition, the determination of the consequent at an early stage allows to filter out many non relevant rules early in the process, which results in an additional significant decrease of the running time. A new confidence measure is proposed, the temporal confidence, which evaluates the confidence of a rule in relation to the predefined gap. The temporal confidence is used to mine rules with a consequent that occurs mainly at a given distance. The algorithm is evaluated on an event sequence of social networks messages. We show that our algorithm mines minimal rules with a distant consequent, while requiring a small computation time. We also show that these rules can be used to accurately predict distant events.

Keywords: Data mining · Episode rules mining · Minimal rules · Distant event prediction

1 Introduction

The flow of messages posted in blogs and social networks is an important and valuable source of information that can be analyzed, modeled (through the extraction of hidden relationships) and from which information can be predicted. This last aspect is the focus of our work. Prediction of information or events in a flow of messages is of the highest importance for companies, which may be interested in what will be said about them in social networks. Prediction can also be viewed as a way to recommend items, for example, when the flow is made up of items (customer consultations or purchases, etc.).

© Springer International Publishing Switzerland 2015
A. Fred et al. (Eds.): IC3K 2014, CCIS 553, pp. 3–18, 2015.
DOI: 10.1007/978-3-319-25840-9_1

We consider that the sooner an event is predicted, the more useful this prediction is. When the data is a flow of messages from blogs or social networks, predicting early a negative event (a criticism of a company) allows to have enough time to act before the occurrence of thus event, or to prevent its occurrence. Predicting distant events is the focus of our work.

Temporal data mining is related to the mining of sequential patterns ordered by a given criterion such as time or position [6]. Episode mining is the appropriate pattern discovery task related to the case the data is made up of a single long sequence. An episode is a temporal pattern made up of "relatively close" partially ordered items (or events), which often appears throughout the sequence or in a part of it [9]. When the order of items is total, the episode is said to be serial.

Similarly to the extraction of association rules from itemsets, episode rules can be extracted from episodes, and used to predict events [3]. The rule mining task, whether they are association or episode rules, is usually decomposed into two sub-problems. The first one is the discovery of frequent itemsets or episodes that have a support higher than a predefined threshold. The second one is the generation of rules from those frequent itemsets or episodes, with the constraint of a confidence threshold [2]. In general, rules are generated by considering some items in the itemsets (or the last items in the case of episodes) as the consequent of the rule, and the rest of the items as the antecedent. Since the second sub-problem is quite straightforward, most of the researches focus on the first one: the extraction of itemsets or episodes. Episode and episode rules mining are used in many areas, such as telecommunication alarm management [9], intrusion detection [8], discovery of relation between financial events [12], etc.

Not only predicting distant events is more useful than predicting close events, but also predicting these events as soon as possible is of higher usefulness. Therefore, we aim at mining serial episode rules with a consequent distant from the antecedent. Due to complexity reasons, traditional episode rules mining algorithms form episode rules with a consequent close to the antecedent, through the use of a predefined span. To mine rules with a distant consequent, these algorithms have to use a larger span and perform a post-processing step: the rules are filtered to keep only rules with a consequent that may occur far from the antecedent. This post-processing is time consuming.

We propose a new algorithm that serial episode rules with a consequent temporally distant from the antecedent, and with a small antecedent (in number of events and in time), to be able to predict events as early as possible. This algorithm has a complexity lower than that of traditional algorithms.

An example of rules we focus on is presented below (from a sequence of annotated messages of blogs about finance issues, where each event includes a sentiment polarity): *R: (interest rate, neutral) (credit, negative) (waiting, neutral) → (concurrence, negative); the antecedent occurs within 5 days, the gap between the antecedent and the consequent is 15 days.*

The rest of this paper is organized as follows: Sect. 2 presents related works about episode rules mining. Our algorithm is introduced in Sect. 3, followed by experimental results in Sect. 4. We conclude and provide some perspectives in Sect. 5.

2 Related Works

We first start by introducing few concepts. Let $I = \{i_1, i_2, ..., i_m\}$ be a finite set of items. I_t is the set of items that occur at a timestamp t, referred to as an **event**. An **event sequence** S is an ordered list of events, $S =< (t_1, I_{t_1}), (t_2, I_{t_2}), ...,$ $(t_n, I_{t_n}) >$ with $t_1 < t_2 < ... < t_n$, (see Fig. 1). The **serial episode** $P =<$ $p_1, p_2, ..., p_k >$ on I^k is an ordered list of events. Its support, denoted by $supp(P)$, represents the number of occurrences of P, according to a frequency measure. P is said to be a frequent episode if $supp(P) \geq minsupp$ where $minsupp$ is the predefined minimal threshold. An **occurrence window** of the episode P is a segment $< I_{t_s}, ..., I_{t_e} >$ of the sequence, denoted as $OW(S, t_s, t_e)$ that starts at timestamp t_s and ends at timestamp t_e, where $P \subseteq < I_{t_s}, ..., I_{t_e} >$, $p_1 \subseteq I_{t_s}$ and $p_k \subseteq I_{t_e}$. It represents the interval that bounds the episode. Let P and Q be two episodes. An **episode rule** $R : P \rightarrow Q$ means that Q appears after P. The **confidence** of this episode rule is the probability to find Q after P: $conf$ $(P \rightarrow Q) = supp(P \cdot Q)/supp(P)$. The rule is said to be confident if its confidence exceeds a predefined threshold $minconf$.

Winepi and *Minepi* are seminal episode mining algorithms [9], and are the basis of many recent algorithms [5, 7]. To extract episodes, both algorithms start by extracting 1-tuple episodes (made up of one item/event), then iteratively seek larger ones by merging items/events on their right side.

In the frequent pattern mining task (which includes episode mining), anti-monotonicity is a common property that has to be respected by any frequency measure [2]. Several frequency measures for episode mining have been proposed. In [9], window-based and minimal occurrence-based frequency measures are introduced in both *Winepi* and *Minepi*. *Winepi* evaluates the frequency of an episode as the number of windows of length w that contain the episode. *Minepi* evaluates the frequency of an episode as the number of minimal windows that contain the episode. A minimal window is a window such that no sub-window contains the episode. [7] has introduced the non-overlapped occurrence-based frequency measure. Two occurrences of an episode are non-overlapped if no item of one occurrence appears in between items of the other. It is shown that the non-overlapped occurrence-based algorithms are much more efficient in terms of space and time needed.

When mining serial episodes, additional constraints on the episodes may be imposed, mainly to reduce the complexity of the algorithms. The span constraint [1] imposes an upper bound (of distance or time) between the first and last event of the occurrence of an episode. The gap constraint [10] imposes an upper bound between successive events in the occurrence of an episode. If the extracted serial episodes have to represent causative chains, such constraints are important.

Traditional episode rules mining algorithms construct episode rules with a large antecedent (made up of many events) [13].

Discovering rules with a small antecedent was introduced for association rules, called "minimal association rules discovery" [14], where it is assumed that no new knowledge is given by larger antecedents. This algorithm has the constraint that the consequents are fixed in advance by the user. Minimal rules have also been studied with the aim to reduce time and space complexity of the mining task, as well as to avoid redundancy in the resulting set of rules [11]. These works focus on association rules; recall we want to form episode rules.

Mining episodes in an event sequence is a task which has received much attention [4]. In an event sequence each data element may contain several items (an event). In [5], the algorithm *Emma* encodes the event sequence with frequent itemsets then serial episodes are mined. In [4], episodes are first extracted, then non-derivable episodes rules are formed (where no rule can be derived from another).

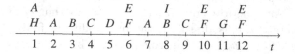

Fig. 1. Example of an event sequence S.

3 The Proposed Algorithm

3.1 Principle

Our goal is to mine episode rules that can be used to efficiently predict distant events. To achieve this goal, the episode rules formed have to hold the characteristic that the consequent is temporally distant from the antecedent. Traditional algorithms are not designed to form such rules. Recall that they first form episodes from left to right by iteratively appending events to the episode being formed. Due to the limited of the predefined span, these events are close to the episode. Second, the episode rules are built by considering the last element(s) of the episodes as consequent of the rules. When forming these episodes, it is impossible to know if the event being appended will be part of the consequent or not. So, it is impossible to know if a distance to other events has to be imposed, while forming the episode. The only way to mine rules with a consequent distant from the antecedent is to use a larger span, mine all rules and then filter out the occurrences that do not respect this distance. Both modifications make the algorithm more time consuming.

We propose to mine episode rules without any episode mining phase. To be able to constrain the distance between the antecedent and the consequent, the consequent is determined early in the mining process. We think that, by determining the consequent at an early stage, the occurrence windows of an episode rule will be filtered early, thus the search space will be pruned, and no post-processing is required. We also aim at predicting events as early as possible. We assume that the more the antecedent of a rule is small in number of events and in time, the earliest it ends, and the earliest the consequent can be predicted. Therefore, we propose to extract episode rules that have an additional characteristic: an antecedent as small as possible (in number of events and in time), which we call "minimal episode rules". For that, we apply the traditional minimal occurrence-based frequency measure. Determining the consequence early in the mining process allows to get such rules, without relying on a post-processing phase. Indeed, knowing the consequence makes the confidence of a rule computable. The construction of the episode rule can be stopped as yet as the confidence threshold is reached.

Before presenting our algorithm in details, we introduce the new concepts of **Sub-windows of** $Win(S, t_s, w)$. Let $Win(S, t_s, w)$ be a window in the sequence S of length w that starts at t_s, with its first element containing the prefix of an episode rule (the first event) to be built. In order to mine episode rules with a distant consequent and a minimal antecedent, we split this window into three sub-windows as follows (see Fig. 2):

- Win_{begin} is a segment of $Win(S, t_s, w)$ of length $w_{begin} < w$, starting at t_s. Win_{begin} can be viewed as an expiry time for the antecedent of an episode rule. It represents the span of the antecedent of an episode rule to guarantee that the antecedent occurs within a determined time.
- Win_{end} is a segment of $Win(S, t_s, w)$ of length $w_{end} < w$, that ends at $t_s + w$. Win_{end} represents the time window of occurrence of the consequent.
- $Win_{between}$ is the remaining sub-window of length $w_{between}$, in which neither the antecedent nor the consequent can appear. $Win_{between}$ guarantees the temporal distance between the antecedent and the consequent of an episode rule. It represents a minimal gap between the antecedent and the consequent to guarantee that the consequent is far from the antecedent.

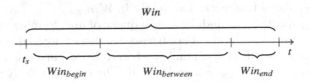

Fig. 2. Sub-windows of $Win(S, t_s, w)$.

3.2   Steps of the Algorithm

We present now the different steps of our algorithm.

Initialization. The algorithm we propose starts by an initialization phase, which reads the event sequence to extract all frequent events and their associated occurrence timestamps. An event represents a 1-tuple episode and will be denoted by P. Table 1 presents the list of 1-tuple episodes of the sequence S (Fig. 1) and their associated occurrence windows when $minsupp = 2$ (see Algorithm line 2).

Table 1. 1-tuple episodes of S.

1-tuple episode p	List of occurrence windows
A	[1,1], [2,2], [7,7]
B	[3,3], [8,8]
C	[4,4], [9,9]
E	[6,6], [10,10], [12,12]
F	[6,6], [10,10], [12,12]
EF	[6,6], [10,10], [12,12]

Prefix Identification. Episode rules are built iteratively by first fixing the prefix (the first event of the antecedent). The antecedent is denoted by ant. Each 1-tuple episode p is viewed as a prefix of the antecedent of an episode rule R to be built. Once the prefix R is fixed, the set of its occurrence windows $OW(S, t_s, t_e)$ is known. For example, let $minsupp = 2$, A can be considered as a prefix of an episode rule. The list of occurrence windows of $A = ([1, 1], [2, 2], [7, 7])$ (see Table 1).

Consequent Identification. A candidate consequent of an antecedent ant (at this step of the algorithm, ant corresponds to a single element, the prefix) is chosen in the windows $Win(S, t_s, w)$ where $[t_s, t_s]$ is in the set of occurrence windows of ant. Recall we want to form episode rules with a consequent distant from the antecedent. Thus, the candidate consequents are not searched in the entire windows of size w, they are searched only in the last part of these windows: Win_{end}, where the farthest candidates are. We construct $\mathcal{P}_{end}(ant)$, the ordered list of 1-tuple episodes that occur frequently in Win_{end}.

Let $p_j \in \mathcal{P}_{end}(ant)$ be a candidate consequent of ant, forming the candidate episode rule $R : ant \rightarrow p_j$. Its set of occurrence windows is formed. The frequency of this rule is computed as the number of minimal occurrence windows. For example, let $w_{begin} = 2$, $w_{end} = 2$ and $w = 6$. The episode rule $R : A \rightarrow E$ has three occurrence windows: $([1, 6], [2, 6], [7, 12])$. The first occurrence window $[1, 6]$ is not minimal. So, the support of $R : A \rightarrow E$ is 2. Notice that all occurrence windows are kept in memory, they will be used to complete the antecedent. This allows to guarantee to discover all interesting episode rules, which could be missed if we kept only the minimal occurrences in memory.

If $R : ant \rightarrow p_j$ is not frequent, p_j cannot be a consequent of ant. This iteration is stopped and the rule is discarded. There is no need to complete ant as whatever are the events that may be appended to the antecedent, the resulting rule will not be frequent. The algorithm will iterate on another consequent. If $R : ant \rightarrow p_j$ is frequent, its confidence is computed.

If the rule $R : ant \rightarrow p_j$ is confident, this rule is added to the set of rules formed by the algorithm. It is minimal and has a consequent far from the antecedent; it fulfills our goal. If the rule is frequent but not confident, the antecedent of the rule $R : ant \rightarrow p_j$ is further completed (see next subsection).

For example, let $w = 6$, $w_{begin} = 2$ and $w_{end} = 2$. For the episode rule R with prefix A, $\mathcal{P}_{end}(A) = [E, F, EF, A]$. We first construct the episode rule R with the consequent E. Thus, for $R : A \rightarrow E$, $supp(R) = 2$ and $conf(R) = 2/3 = 0.67$. For $minsupp = 2$ and $minconf = 0.7$, R is frequent but not confident, so its antecedent has to be completed. If $minconf = 0.5$, $R : A \rightarrow E$ is confident, its construction is stopped; it is a frequent, confident and minimal episode rule.

Antecedent Completion. In this step, the antecedent ant is iteratively completed with 1-tuple episodes from the initialization step, placed on its right side in the limit of the predefined sub-window Win_{begin}. At the first iteration, ant is a unique element (the prefix) (see Algorithm 1, line 15). Recall that we aim at forming rules having the last event of the antecedent as far as possible from the consequent, so as close as possible of the prefix. We construct $\mathcal{P}_{begin}(ant)$, the ordered list of 1-tuple episodes that occur frequently after ant in the windows of R, in Win_{begin}. The new candidate rule $R : ant, p_i \rightarrow p_j$ is formed. Similarly to the consequent identification step, the occurrence windows of R are computed, as well as its support. We apply the same support and confidence verifications, similarly to the previous step.

To speed up the episode rules mining process we use a heuristic. We sort the list of candidates $\mathcal{P}_{begin}(ant)$ in descending order of the number of windows in which each candidate appears. We assume that this number is highly correlated with the support of the corresponding episode rules. So, in the traversal of this list, when we observe that candidates tend to form infrequent episode rules (several consecutive 1-tuple episodes lead to infrequent episode rules), we stop the traversal. We consider that the remaining candidates in this list will lead to infrequent episode rules. This heuristic is used to reduce the number of iterations. Although it may discard interesting rules, it allows to reduce the number of iterations thus allows to increase the size of the span of the rule (Win).

For example, let $minsupp = 2$ and $minconf = 0.7$, for $R : A \rightarrow E$, $\mathcal{P}_{begin}(A) = [B, C, A]$. The antecedent of R is completed with B and forms the episode rule $R : A, B \rightarrow E$. Thus, $supp(R) = 2$ and $conf(R) = 2/2 = 1$. The episode rule R is now confident, the phase of completing its antecedent is stopped.

3.3 Temporal Confidence

$Win_{between}$ has been introduced to guarantee that the consequent of an episode rule occurs in Win_{end}, after at least $w_{begin} + w_{between}$ events from the prefix of the episode rule. However, the consequent of a rule may also occur closer to the antecedent, in the window $Win_{between}$, which should affect the confidence of the rule. This information may be important in some applications. In the previous example about social networks, where the prediction of a negative event allows the company to prevent its occurrence, it is important to mine rules with a consequent that never occurs in $Win_{between}$. Indeed, predicting a consequent at a given distance, which may appear closer is useless, even dangerous. Consequently, the occurrence of the consequent in $Win_{between}$ has to be considered. We introduce a new confidence measure, the temporal confidence, which represents the probability that the consequent occurs in Win_{end} and only in it. For an episode rule $R : P \rightarrow Q$, the temporal confidence $conf_t(R)$ is equal to the ratio between the support of $P \cdot Q$ when Q occurs only in Win_{end} (and not in $Win_{between}$) and the support of $P \cdot Q$. $conf_t(R) = 1$ if no occurrence of the consequent is found in $Win_{between}$. The rules with a temporal confidence above $minconf_t$ are kept. For example, let $w = 6$, $w_{begin} = 2$ and $w_{end} = 2$, the temporal confidence of the frequent confident episode rule $R : A \rightarrow E$ depends on the occurrences of E in $Win_{between}$ which is equal to 1 (E appears in the timestamp t_{10} in $Win_{between}$). Thus, $conf_t(R : A \rightarrow E) = 1/2 = 0.5$. For $minconf_t = 0.5$, R is temporally confident and is a rule formed by our algorithm.

Algorithm 1. Episode rules mining

input : S: event sequence, $minsupp$, $minconf$,
$minconf_t$, w, w_{begin}, $w_{between}$, w_{end}
output: ER : List of episode rules

1 **Procedure** *Episode rules mining*
2   extract frequent 1-tuple episodes;
3   **foreach** $p_i \in$ *1-tuple episodes* **do**
4          $ant \leftarrow p_i$;
5   Construct ordered lists
            $\mathcal{P}_{end}(ant), \mathcal{P}_{begin}(ant)$;
6   Consequent $(ant, \mathcal{P}_{end}(ant), \mathcal{P}_{begin}(ant))$

7 **Procedure** *Consequent* $(ant, \mathcal{P}_{end}(ant), \mathcal{P}_{begin}(ant))$
8   **foreach** $p_j \in \mathcal{P}_{end}(ant)$ **do**
9   **if** $ant \rightarrow p_j$ *is frequent* **then**
10   **if** $ant \rightarrow p_j$ *is confident* **then**
11   **if** $ant \rightarrow p_j$ *is temporally confident* **then**
12   Add $ant \rightarrow p_j$ to ER
13   **else**
14   Antecedent $(ant, p_j, \mathcal{P}_{begin}(ant))$

15 **Procedure** *Antecedent* $(ant, p_j, \mathcal{P}_{begin}(ant))$
16   **foreach** $p_k \in \mathcal{P}_{begin}(ant)$ **do**
17   **if** $ant, p_k \rightarrow p_j$ *is frequent* **then**
18   **if** $ant, p_k \rightarrow p_j$ *is confident* **then**
19   **if** $ant, p_k \rightarrow p_j$ *is temporally confident* **then**
20   Add $ant, p_k \rightarrow p_j$ to ER
21   **else**
22                 $ant \leftarrow < ant, p_k >$;
23   Antecedent $(ant, p_j, \mathcal{P}_{begin}(ant))$

---

4   Experimental Results

We now evaluate the algorithm through the study of the characteristics of the episode rules formed, as well as its performance in a prediction task and its running time in comparison to traditional algorithm.

4.1   Dataset

The dataset we use is made up of 27, 612 messages extracted from blogs about finance. Messages are annotated using the $Temis$[1] software. Each message is represented by its corresponding set of annotations (items). For example, the message: *"Not only my bank propose the best, but also online banks, so how to optimize my savings? accounts in other banks? life insurance? i need more info."* is annotated with the following items, each one being associated with an opinion degree: *{(online banks, negative), (savings, neutral), (life insurance, neutral), (needs, neutral)}* .

In this dataset, the messages are annotated with 4.8 items on average, ranging from 1 to 50 and the median is equal to 4. There are about 4, 000 distinct annotations (items), with an average frequency of 88.5. 1, 981 of these items have a frequency equal to 1. These items will be automatically filtered out in the initialization phase of our algorithm.

4.2   Characteristics of the Resulting Rules

Initialization Phase. In this phase, frequent 1-tuple episodes are extracted (a 1-tuple episode is made up of one or more items). We fix $minsupp$ to 30. This phase results in 652 frequent 1-tuple episodes. Table 2 shows that the 1-tuple episodes are made up of one to three items only. They are mainly made up of one item (76 % of them), which also have a high support (on average 149.7).

Table 2. 1-tuple episodes : length and support.

1-tuple episode length (#items)	Number(%)	Support			
		Min	Max	Mean	Median
1	498(76.4)	30	797	149.7	89.5
2	147(22, 5)	30	376	71.5	55
3	7(1)	32	54	44.4	46

In order to study the episode rules formed, we make vary $minsupp$, $minconf$ and $w_{between}$ one at a time, while fixing others.

Making Vary $minsupp$**.** We make vary $minsupp$ from 10 to 35 to study the number of rules, represented in Fig. 3. Other parameters remain fixed: $minconf = 0.4$, $w = 40$ (with $w_{begin} = 20$, $w_{between} = 10$ and $w_{end} = 10$). As expected, the smaller $minsupp$, the higher the number of rules. When $minsupp$ is fixed to 10, the number of rules is high (about 10^4). Recall that the number of 1-tuple episodes is only 652. However, the number of rules is dramatically decreased when $minsupp$ is increased: only 1, 200 when $minsupp = 15$ and 250 when $minsupp = 20$. These low values are due to the small average frequency of 1-tuple episodes (about 88).

---

[1] http://www.temis.com.

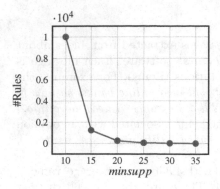

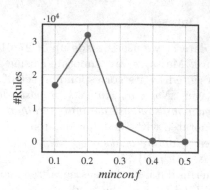

Fig. 3. Number of rules (10^4) vs. *minsupp*.

Fig. 4. Number of rules (10^4) vs. *minconf*.

In addition, these rules represent a temporal dependence between the antecedent and the consequent which, in these experiments, is at least $w_{between} = 10$. This may explain the low number of rules. A thorough study shows that their average confidence increases with *minsupp*.

Making Vary *minconf*. We make vary *minconf* from 0.1 to 0.5. Figure 4 presents the number of episode rules according to the value of *minconf* (*minsupp* = 20, $w = 40$ ($w_{begin} = 20$, $w_{between} = 10$ and $w_{end} = 10$)). The number of rules is particularly high with *minconf* = 0.2. This is explained by the way our rules are formed: when the confidence of a rule does not exceed *minconf*, its antecedent is extended. Given an antecedent *ant*, events in $\mathcal{P}_{begin}(ant)$ (up to 652) are appended to it, resulting in a large number of candidate rules. Some of them are confident, which explains the increase of the number of rules.

Table 3 presents the length of the antecedent of the rules (in number of events), according to *minconf*. The maximum length of an antecedent is three. Thus, our algorithm, forms rules with a small antecedent, which was one of its goals. The average length of the antecedent increases with *minconf*: when *minconf* = 0.1, most of the rules have an antecedent of length 1, whereas when *minconf* = 0.3, most of the rules have an antecedent of length 2. This was expected, as minimal antecedents are searched. Indeed, when a frequent rule has a confidence below *minconf*, its antecedent is extended, till it is confident or not frequent. So, the higher *minconf*, the larger the antecedents. A thorough study shows that the average length of occurrence windows of the antecedents is 8 timestamps (for antecedents of length 2 and 3), which is smaller than the span of the antecedent (w_{begin}). We conclude that our algorithm succeeds in forming rules with a distant consequent, a small antecedent (in length and time) and a relatively high confidence.

Making Vary $w_{between}$. We now focus on the number of rules formed, according to $w_{between}$ (w and w_{begin} remain fixed), presented in Fig. 5 where *minsupp* = 20 and *minconf* = 0.4. Two values of w are studied: $w = 40$ (with $w_{begin} = 20$)

Table 3. Antecedent length when making vary $minconf$.

$minconf$	#Rules	%Rules		
		Ant 1	Ant 2	Ant 3
0.1	16,850	57.2	42.84	0
0.2	31,972	4.2	95.6	0.2
0.3	5,072	0.9	97.5	1.6
0.4	251	0.8	90.9	8.3
0.5	9	0	100	0

and $w = 100$ ($w_{begin} = 20$). Notice that the cases $w_{between} = 0$ represent similar cases than state of the art. We note that the larger $w_{between}$, the smaller the number of rules. Two reasons may explain this decrease. First, when $w_{between}$ increases, w_{end} (the window in which the consequent is searched) decreases, as well as the number of consequents studied. Second, the larger $w_{between}$, the more distant the consequent, thus the lower the probability of having a dependence between the antecedent and the consequent. However, even with a large value of $w_{between}$, some rules are formed: 210 rules when $w_{between} = 70$. We conclude that there is actually a temporal dependence between messages in blogs. When the minimal distance between the antecedent and the consequence is 50, more than 140k confident rules are formed: there is a strong dependence between messages with such a distance. When $w = 100$, an episode rule: *(price, positive), (information, positive)* → *(buy, positive)*, means that when someone talks about the price of an article, then asks for information, he/she will buy this article after some time. Thus, we have time to recommend him similar articles or to propose to him a credit to buy it. **Influence of** $minconf_t$**:** In this section we study the temporal confidence of the resulting rules. Table 4 presents the evolution of the temporal confidence according to $w_{between}$. We remark that the smaller $w_{between}$, the higher the temporal confidence: the consequent does rarely occur between the antecedent and the consequent when the gap between them is small, which was expected. When $w_{between} = 30$, the average temporal confidence is 0.6, which is quite high. A thorough study shows that among $2.7 \cdot 10^5$ rules (see Fig. 5), about 40 ones have a temporal confidence equal to 1 (the consequent never occurs in $Win_{between}$) and 1,400 rules have a temporal confidence higher than 0.9 (the consequent appears in $Win_{between}$ in less than 10 % of the cases). This shows that in this dataset there is a strong temporal dependence between events, and that some events are interdependent at a distance of 30. So, when exploiting the temporal confidence as a filter, a great number of rules remain.

As mentioned before, one may be interested in the extraction of rules with a consequent that occurs in Win_{end} most of the time (with a high temporal confidence). Therefore, a minimal threshold ($minconf_t$) is fixed to keep only rules having a temporal confidence that exceeds this threshold. In Fig. 6, we make vary $minconf_t$ when varying $w_{between}$ with the same parameters as in Fig. 5. It is intuitive that the higher $minconf_t$, the lower the number of rules. We remark

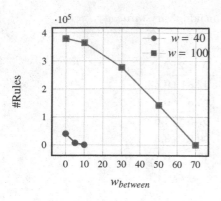

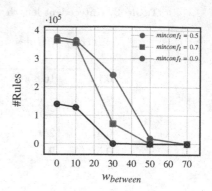

Fig. 5. Number of rules (10^5) vs. $w_{between}$.

Fig. 6. Number of rules (10^5) vs. $w_{between}$, $minconf_t$.

also that the larger $Win_{between}$, the lower the number of temporally confident rules. This decrease is intuitive since the farthest the consequents are searched (for large $Win_{between}$), the lower the dependence between events. Despite this strong filter of rules, a considerable number of temporally confident rules is extracted. For example, when $w_{between} = 30$ (which represents a quite large distance), about 1,400 rules are kept when $minconf_t = 0.9$. 70,800 rules are kept when $minconf_t = 0.7$ and about 242,900 rules when $minconf_t = 0.5$, which is quite high. Once again, this confirms the existence of a strong temporal dependence between events in these experiments.

Table 4. $w_{between}$ vs. Temporal confidence $(conf_t)$.

$w_{between}$	min-$conf_t$	max-$conf_t$	mean-$conf_t$	median-$conf_t$
70	0	0.5	0.2	0.2
50	0.2	0.9	0.4	0.4
30	0.3	1	0.6	0.6
10	0.2	1	0.9	0.9

4.3 Performance

In this section we focus on the accuracy of the rules formed, when they are used to predict events. we also perform a comparison of these rules with those of a traditional algorithm.

We evaluate the accuracy of our algorithm with the traditional recall and precision measures. The episode rules are trained on the first 75 % messages (in the temporal order of publication) and are tested on the 25 % of messages left. Figure 7 presents the resulting precision and recall at 20. We fix $minsupp = 20$,

$minconf = 0.4$, $w = 100$ and $w_{begin} = 20$. We make vary $w_{between}$ from 10 to 70. First of all, mention that two precision and recall values with two different values of $w_{between}$ are not directly comparable as they are not computed on the same data (the windows Win_{end}, on which they are computed vary in size). Both precision and recall curves decrease as $w_{between}$ increases. This was expected as the number of rules decreases. When $w_{between} = 70$ (and $w_{end} = 10$), both precision and recall values are quite low. This was expected as the rules aim at predicting events distant to at least 70, in an occurrence window of length $w_{end}=10$. The prior probability of predicting events accurately is low. Let us now consider $w_{between} = 30$, as in the previous section. We can see that both precision and recall values are quite high. When an event is predicted, in 37 % of the cases, it actually occurs and events that occur in the sequence are predicted by our rules in 70 % of the cases.

Comparison of the Rules Formed with Traditional Algorithms. Contrary to traditional algorithms, our algorithm forces a minimum gap of length $w_{between}$ between the antecedent and the consequent of an episode rule. Our algorithm can be compared to traditional algorithms when $w_{between} = 0$. Since we apply the minimal occurrence-based frequency, we choose to compare it to the well-known $Minepi$ [9].

Table 5 compares the number of rules formed by our algorithm and by $Minepi$, for $minsupp = 20$, $minconf = 0.4$, $w = 40$, $w_{begin} = 20$ $w_{end} = 20$ and $w_{between} = 0$ (the only parameter used by $Minepi$ is $w = 40$).

$Minepi$ forms more than $136,000$ episode rules, whereas our algorithm (when $w_{between} = 0$) extracts about $40,000$ episode rules (70 % less). This decrease is due to two reasons. First, the constraint about the position of consequent of the episode rules from our algorithm (in this case the distance between the antecedent and the consequent is at least 20, even if $w_{between} = 0$), makes the number of rules resulting from our algorithm lower (also their support is lower). Second, our algorithm aims at forming minimal rules, thus few rules with a large antecedent are formed. We remark that 25 % of the rules extracted by $Minepi$ have an antecedent larger or equal to 3, whereas this rate is only 1.8 % for our algorithm.

Here is an example of an episode rule extracted by both our algorithm ($w_{between} = 0$) and by $Minepi$: *(credit, positive), (consultant, positive)* → *(loan subscription, positive)*.

Here is a rule that has not been extracted by our algorithm, as it does not satisfy the desired characteristics of episode rule (minimal antecedent): *(consultant, neutral), (interest rate, positive)* → *(request interest rate 0, positive)*, where the antecedent occurs in 5 timetstamps and consequent occurs in the 7th timestamp. This rule is useful in traditional cases of event prediction (prediction of close events). However, it does not fit our objective of early prediction of distant events, as the antecedent is so long both in time and in number and the consequent is too close to the antecedent.

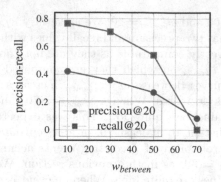

Fig. 7. Precision, recall vs. $w_{between}$.

Table 5. *Our algorithm* vs. *Minepi*.

Algorithm	#Rules	#R/ant 1	#R/ant 2	#R/ant 3	#R/ant 4
This paper	40,061	578	38,742	741	0
Minepi	136,225	9,300	93,389	33,505	31

4.4 Running Time

In this last section, we are interested in the running time of our algorithm according to the minimal gap between the antecedent and the consequent ($w_{between}$), in comparison to the running time of *Minepi*. We consider the two previously studied values of w: 40 and 100. The running time is presented in Fig. 8, in terms of relative running time compared to that of *Minepi*. First, we can see that our algorithm runs faster than *Minepi*. The most comparable points are those corresponding to $w_{between} = 0$: no minimal gap between the antecedent and the consequent, where our algorithm runs 5 times faster than *Minepi* when $w = 100$, and 4 times faster when $w = 40$. This decrease is due to the same two factors than presented below. The first one is related to the consequent, which is fixed at an early stage of the algorithm and which allows to filter infrequent rules. The second one is due to the fact that the proposed algorithm mines rules with a minimal antecedent, which avoids some iterations once a confident rule is found.

We can also see that the larger $w_{between}$, the faster the algorithm. This trend is intuitive since large $w_{between}$ values imply small w_{end} values (as w remains fixed), so a smaller number of consequent are studied in Win_{end} for each potential rule. For example, when $w = 100$, the running time is divided by almost 2 when $w_{between}$ ranges from 10 to 50.

Last, when comparing running times when $w = 40$ and $w = 100$, in comparable configurations: $w_{begin} = 20$ and $w_{end} = 10$ ($w_{between}$ is equal to respectively 10 and 70), we remark that they have nearly the same running time (these figures are not presented in Fig. 8 as it shows relative time compared to *Minepi*). This confirms that the running time of our proposed algorithm is independent of w.

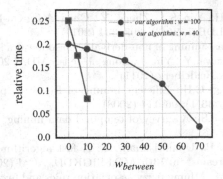

Fig. 8. Relative time according to $w_{between}$ (*relative time* = 1 represents running time of *Minepi*)

It only dependents on w_{begin} and w_{end}, in which antecedents and consequents are searched.

5 Conclusion

In this paper, we have proposed an algorithm that mines episode rules, in order to predict distant events. To achieve our goal, the algorithm mines serial episode rules with a distant consequent. We determine two main characteristics of the episode rules formed: minimal antecedent and a consequent temporally distant from the antecedent. A new confidence measure, the temporal confidence, is proposed to evaluate the confidence on distant consequents. Our algorithm is evaluated on an event sequence of annotated social networks messages. We show that our algorithm is efficient in extracting episode rules with the desired characteristics and in predicting distant events.

Since we use data from social networks, we aim to use multi-thread sequences. This means that we intend to construct a sequence for each thread of messages: user messages thread, topic messages thread and discussion thread, etc. and the algorithm is run on each one. Using multi-thread sequences allows to build more diverse episode rules which are all together more significant. The presence of a rule in several threads will increase its confidence.

Acknowledgements. This research is supported by Crédit Agricole S.A.

References

1. Achar, A., Sastry, P., et al.: Pattern-growth based frequent serial episode discovery. Data Knowl. Eng. **87**, 91–108 (2013)
2. Agrawal, R., Imieliński, T., Swami, A.: Mining association rules between sets of items in large databases. In: ACM SIGMOD Record, vol. 22, pp. 207–216. ACM (1993)

3. Cho, C.W., Wu, Y.H., Yen, S.J., Zheng, Y., Chen, A.L.: On-line rule matching for event prediction. VLDB J. **20**(3), 303–334 (2011)
4. Gan, M., Dai, H.: Fast mining of non-derivable episode rules in complex sequences. In: Torra, V., Narakawa, Y., Yin, J., Long, J. (eds.) MDAI 2011. LNCS, vol. 6820, pp. 67–78. Springer, Heidelberg (2011)
5. Huang, K.Y., Chang, C.H.: Efficient mining of frequent episodes from complex sequences. Inf. Syst. **33**(1), 96–114 (2008)
6. Laxman, S., Sastry, P.S.: A survey of temporal data mining. Sadhana **31**(2), 173–198 (2006)
7. Laxman, S., Sastry, P., Unnikrishnan, K.: A fast algorithm for finding frequent episodes in event streams. In: 13th ACM SIGKDD. ACM (2007)
8. Luo, J., Bridges, S.M.: Mining fuzzy association rules and fuzzy frequency episodes for intrusion detection. Int. J. Intell. Syst. **15**(8), 687–703 (2000)
9. Mannila, H., Toivonen, H., Verkamo, A.I.: Discovery of frequent episodes in event sequences. Data Min. Knowl. Discov. **1**(3), 259–289 (1997)
10. Méger, N., Rigotti, C.: Constraint-based mining of episode rules and optimal window sizes. In: Boulicaut, J.-F., Esposito, F., Giannotti, F., Pedreschi, D. (eds.) PKDD 2004. LNCS (LNAI), vol. 3202, pp. 313–324. Springer, Heidelberg (2004)
11. Neeraj, S., Swati, L.S.: Overview of non-redundant association rule mining. Res. J. Recent Sci. **1**(2), 108–112 (2012). ISSN 2277-2502
12. Ng, A., Fu, A.W.: Mining frequent episodes for relating financial events and stock trends. In: Whang, K.-Y., Jeon, J., Shim, K., Srivastava, J. (eds.) PAKDD 2003. LNCS, vol. 2637, pp. 27–39. Springer, Heidelberg (2003)
13. Pasquier, N., Bastide, Y., Taouil, R., Lakhal, L.: Discovering frequent closed itemsets for association rules. In: Beeri, C., Bruneman, P. (eds.) ICDT 1999. LNCS, vol. 1540, pp. 398–416. Springer, Heidelberg (1998)
14. Rahal, I., Ren, D., Wu, W., Perrizo, W.: Mining confident minimal rules with fixed-consequents. In: 16th IEEE ICTAI 2004 (2004)

URL-Based Web Page Classification: With n-Gram Language Models

Tarek Amr Abdallah and Beatriz de La Iglesia[✉]

School of Computing Sciences, University of East Anglia, Norwich Research Park,
Norwich NR4 7TJ, UK
b.Iglesia@uea.ac.uk

Abstract. There are some situations these days in which it is important to have an efficient and reliable classification of a web-page from
the information contained in the Uniform Resource Locator (URL) only,
without the need to visit the page itself. For example, a social media
website may need to quickly identify status updates linking to malicious
websites to block them. The URL is very concise, and may be composed
of concatenated words so classification with only this information is a
very challenging task. Methods proposed for this task, for example, the
all-grams approach which extracts all possible sub-strings as features,
provide reasonable accuracy but do not scale well to large datasets.

We have recently proposed a new method for URL-based web page
classification. We have introduced an n-gram language model for this
task as a method that provides competitive accuracy and scalability to
larger datasets. Our method allows for the classification of new URLs
with unseen sub-sequences. In this paper we extend our presentation
and include additional results to validate the proposed approach. We
explain the parameters associated with the n-gram language model and
test their impact on the models produced. Our results show that our
method is competitive in terms of accuracy with the best known methods but also scales well for larger datasets.

Keywords: Language models · Information retrieval · Web classification · Web mining · Machine learning

1 Introduction

During 2010 twitter users sent about 90 million updates every day, as reported
by Thomas et al. [1]. It is estimated that 25 % of those updates contain web-
links. Similarly, a huge number of links are carried by the millions of email
messages and Facebook updates sent every day. In such context, it is crucial to
be able to classify web-pages in real-time using their URLs only, without the
need to visit the pages themselves, even if some accuracy is sacrificed for the
sake of greater speed of classification. Also, search engines depend mainly on
textual data to retrieve on-line resources. However, they are often faced with
multimedia content such as videos and images with scarce descriptive tags or

© Springer International Publishing Switzerland 2015
A. Fred et al. (Eds.): IC3K 2014, CCIS 553, pp. 19–33, 2015.
DOI: 10.1007/978-3-319-25840-9_2

surrounding text. Thus, in this context, URL-based classification can be used to decide the categories of such content enhancing the retrieval performance.

Additionally, the classification approach presented here is not limited to URL-based classification tasks only. It can also be adapted for similar problems where there is a need to classify very concise documents with no obvious boundaries between words, e.g. social networks folksonomies.

Unlike documents, URLs are very concise as they are composed of very few words. Usually, words are also concatenated without intermediate punctuations or spaces; for example: carsales.com and vouchercodes.co.uk. They also contain various abbreviations and domain-specific terms. Therefore, classification requires specific approaches that can deal with the special characteristics of the data under consideration.

2   Related Work

Previous researchers have focused on how to extract features from URLs. Early approaches segmented URLs based on punctuation marks using the resulting terms as the classifier's feature-set [2]. Later on, researchers used either statistical or brute-force approaches to further segment URLs beyond the punctuation marks. The non-brute-force approaches used *information content* [2], *dictionary based tokenizes* [3] and *symmetric/non-symmetric sliding windows* [4]. The brute-force approach, on the other hand, tends to extract all possible sub-strings, *all-grams*, to use them as the classifier's feature-set [5–8]. To our knowledge, this is the most successful so far, however, it is obvious that it does not scale very well. In our experiments reported here, the resulting datasets from applying the all-grams approach can be very large, going beyond our computational resources and therefore it becomes difficult to store, classify or even to select subsets of features. For example, in a dataset of 43,223 URLs, when extracting all-grams between 4 and 8 characters-long, we ended up with 1,681,223 n-grams as our feature-set.

The aforementioned classification algorithms are sometimes called batch algorithms, as opposed to online algorithms. In recent research, online learners have been used in URL-based classifications [9,10]. Nevertheless, they incorporate meta-features, such as those obtained from WHOIS and geographic information, in addition to the URLs' lexical features. We prefer to limit ourselves here to features found in the URLs only.

Our proposed approach tries to classify URLs without the need to segment them. We borrow the concept of language models from the information retrieval and automatic speech recognition field. We apply a similar approach to that used by Peng et al. to classify Japanese and Chinese documents [11]. They used an *n-Gram Language Model (LM)* in order to classify textual data without the need for segmenting into separate terms. These two East Asian languages are similar to URLs in the sense that spaces between words are absent so we hypothesise that a similar approach can work for the URL classification problem. We have adapted the model used by Peng et al to be used with URLs, given their format and punctuations. Furthermore, we made use of the *Linked Dependence Assumption* to relax the model's independence assumption and to improve its performance. We further expand on this in Sect. 3.2.

In the next section we are going to explain the n-gram Language Model and its use for document classification. In Sect. 4 we give more details on the dataset used, and the experiments done. Then, we present our results in Sect. 5. Finally, we conclude our findings and offer suggestions for future researchers in the last section.

3   The n-Gram Language Model

Let us assume we have a set of documents $D = \{d_1, d_2, ..., d_m\}$, and a set of classes $C = \{c_1, c_2, ..., c_k\}$, where each document is classified as member of one of these classes. For any document, d_i, the probability that it belongs to class c_j, can be represented as $Pr(c_j/d_i)$ and using Bayes rules [11,12], this probability is calculated by:

$$Pr(c_j/d_i) = \frac{Pr(d_i/c_j) * Pr(c_j)}{Pr(d_i)} \qquad (1)$$

The term $Pr(d_i)$ is constant for all documents. The term $Pr(c_j)$ can represent the distribution of class j in the training set. A uniform class distribution can also be assumed, so we end up with the term $Pr(d_i/c_j)$ only [13]. For a document d_i, that is composed of a sequence of words $w_1, w_2, ..., w_L$, $Pr(d_i/c_j)$ it is expressed as follows: $Pr(w_1, w_2, ...w_L/c_j)$. We are going to write it as $Pr_{c_j}(w_1, w_2, ...w_L)$ for simplicity.

$Pr_{c_j}(w_1, w_2, ...w_L)$ is the likelihood that $w_1, w_2, ..., w_L$ occurs in c_j. This can be calculated as shown in Eq. 2.

$$Pr_{c_j}(w_1, w_2, .., w_{L-1}, w_L) \qquad (2)$$
$$= Pr_{c_j}(w_1) * Pr_{c_j}(w_2/w_1)$$
$$... * Pr_{c_j}(w_L/w_{L-1}, .., w_1)$$
$$= \Pi_{i=1}^{L} Pr_{c_j}(w_i/w_{i-1}, w_{i-2}, ..., w_1)$$

Nevertheless, in practice, the above dependency is relaxed and it is assumed that each word w_i is only dependent on the previous $n - 1$ words [11]. Hence, Eq. 2 is transformed to the following equation:

$$Pr_{c_j}(w_1, w_2, ...w_L) \qquad (3)$$
$$= \Pi_{i=1}^{L} Pr_{c_j}(w_i/w_{i-1}, w_{i-2}, ..., w_{i-n+1})$$

The n-gram model is the probability distribution of sequences of length n, given the training data [14]. Therefore, $Pr_{c_j}(w_1, w_2, ...w_L)$ is referred to as the n-gram language model approximation for class c_j. Now, from the training set and for each class, the n-gram probabilities are calculated using the maximum likelihood estimation (MLE) shown in Eq. 4 [15]:

$$Pr_{c_j}(wi/w_{i-n+1}^{i-1}) = \frac{Pr(w_{i-n+1}^i)}{Pr(w_{i-n+1}^{i-1})} \tag{4}$$

$$= \frac{count(w_{i-n+1}^i)/N_w}{count(w_{i-n+1}^{i-1})/N_w}$$

$$= \frac{count(w_{i-n+1}^i)}{count(w_{i-n+1}^{i-1})}$$

where N_w is the total number of words, and w_{i-n+1}^i is the string formed of the 'n' consecutive words between w_{i-n+1} and w_i. We are proposing to use the n-Gram Language model for URL-based classification. However, in our case, we will use characters instead of words as a basis of the language model. We construct a separate LM for each class of URLs as follows. The above probabilities are calculated for each class in the training set by counting the number of times all sub-strings of lengths n and $n-1$ occur in the member URLs of that class. For example, suppose we have the following strings as members of class c_j, {'ABCDE','ABC','CDE'}. In a 3-gram LM, for class c_j we will store all sub-strings of length 3 and those of length 2, along with their counts, as shown in Table 1.

Table 1. Sample data-structure for 3-gram LM counts.

3-grams	('ABC': 2), ('BCD': 1), ('CDE': 2)
2-grams	('AB': 2), ('BC': 2), ('CD': 2), ('DE': 2)

Counts in Table 1 are acquired during the training phase. Then in the testing phase, URLs are converted into n-grams, and for each n-gram, its probability is calculated using Eq. 4. A new URL, URL_i, is classified as member of class c_j, if the language model of c_j maximizes Eq. 1, i.e. maximizes $Pr(c_j/URL_i)$.

3.1 Dealing with Unseen n-Grams

The maximum likelihood in Eq. 4 can be zero for n-grams not seen in the training set. Therefore, smoothing is used to deal with the problem by assigning non-zero counts to unseen n-grams. Laplace smoothing is one of the simplest approaches [15], calculated as follows:

$$Pr_{c_j}(wi/w_{i-n+1}^{i-1}) = \frac{count(w_{i-n+1}^i) + 1}{count(w_{i-n+1}^{i-1}) + V} \tag{5}$$

In Eq. 5, the count is increased by 1 in the numerator, and by V in the denominator, where V represents the number of unique sequences of length $n-1$ found in the training set. By using this, we are effectively lowering the count of the non-zero sequences and assigning a discounted value to the unseen sequences [16]. Both 1 and V can be multiplied by a coefficient γ in order to control the amount of the probability mass to be re-assigned to the unseen sequences. There are other more sophisticated smoothing techniques that could be applied including Witten-Bell discounting [17] and Good Turing discounting [18].

3.2   Linked Dependence Assumption

In the n-gram LM, in order to move from Eqs. 2 to 3, we need to assume that the probability of w_i depends only on that of the previous $n-1$ terms. Similarly, in the uni-gram LM, all terms are assumed to be totally independent, i.e. it is equivalent to a bag of words approach. Although, increasing the value of n relaxes the *independence assumption*, the assumption is still strong. Cooper [19], points out the *linked dependence assumption* (LDA) as a weaker alternative assumption. Lavrenko [20] explained the *linked dependence* as follows. Consider the case of a two words vocabulary, $V = \{a, b\}$. In the case of two classes, c_1 and c_2, and under the *independence assumption*, $Pr_{c1}(a, b) = Pr_{c1}(a) * Pr_{c1}(b)$. Similarly $Pr_{c2}(a, b)$ is the product of $Pr_{c2}(a)$ and $Pr_{c2}(b)$. Otherwise, when terms are assumed to be dependent, $Pr_{c1}(a, b)$ and $Pr_{c2}(a, b)$ can be expressed as follows:

$$Pr_{c_j}(a, b) = K_{c_j} * Pr_{c_j}(a) * Pr_{c_j}(b) \tag{6}$$

where K_{c_j} measures the dependence of the terms in class c_j. Terms are positively correlated if $K_{c_j} > 1$, and they are negatively correlated if $K_{c_j} < 1$. As mentioned earlier, with the independence assumption, K_{c_j} is equal to 1. Now, in Cooper's LDA, K_{c_j} is not assumed to be equal to 1, however it is assumed to be the same for all classes, i.e. $K_{c_1} = K_{c_2} = K_{c_j} = K$

Accordingly, the value of K might not be needed if we try to maximize the log-likelihood ratio of relevance of $Pr(c_j/d_i)$ divided by $Pr(\bar{c}_j/d_i)$, rather than $Pr(c_j/d_i)$ as in Eq. 1. $Pr(\bar{c}_j/d_i)$ is the posterior probability of all other classes except c_j. This is similar to the approach used in the *binary independence model* (BIM) [21,22]. Similarly, when using Language Models for spam detection, Terra created two models for ham and spam messages [23], and a message was considered to be spam if its log-likelihood odds ratio exceeded a certain ratio. Hence, the equation of our proposed classifier will look as follows.

$$logLL_{c_j} = log\left(\frac{Pr(c_j/d_i)}{Pr(\bar{c}_j/d_i)}\right) \tag{7}$$

$$= log\left(\frac{Pr(d_i/c_j) * Pr(c_j)}{Pr(d_i/\bar{c}_j) * Pr(\bar{c}_j)}\right)$$

$$= \Sigma_{i=1}^{L} log\left(\frac{Pr_{c_j}(w_{i-n+1}^i)}{Pr_{\bar{c}_j}(w_{i-n+1}^i)}\right) + log\left(\frac{Pr(c_j)}{Pr(\bar{c}_j)}\right)$$

A new URL, URL_i, is classified as member of class c_j, if the language model of c_j maximizes Eq. 7, i.e. maximizes the $logLL_{c_j}$. Hereafter, we refer to this variation of the n-gram LM as *Log-likelihood Odds* (LLO) model. It is worth mentioning that the use of logarithmic scale also helps in preventing decimal point overflow during the implementation.

4   Experiments and Datasets

After some preliminary results in [24], here we extend the experimentation to 3 datasets with different classification objectives. Our first dataset, WebKB corpus, is commonly used for web classification (e.g. [25]). It contains pages collected from the computer science departments in 4 universities. Pages are labelled according to their function in the university websites. In total, there are 7 classes-labels: course, faculty, student, project, staff, department and other. We employed the same subset of the dataset used in previous research, to be able to compare our results to them [2,5,26]. The subset contains 4,167 pages. Following previous researches, we used the same training and test-sets and a *leave-one-university-out cross-validation* for the WebKB URLs [2].

In addition to WebKB, we also used the categorized web pages from DMOZ, which was historically known as the Open Directory Project (ODP). This represents a problem of topic classification. Baykan et al. [5] selected 15 topics from DMOZ categories; 1,000 URLs were put aside for testing, and the remaining URLs were used to create 15 balanced training sets for their 15 binary classifiers. For the sake of useful comparison, we calculated the precision, recall and F-measure for this dataset in the same fashion as explained in [27].

Besides functional and topic classification, we also wanted to apply our approach to a different type of classification problem, namely one of classification based on language. Global Voices Online (GVO) is a website that publishes social and political articles in different languages. Thus, we created our third dataset by extracting the URLs of the most recent articles published there in 5 languages. We choose articles in 2 Latin languages (Spanish and Italian), 2 Germanic languages (Deutsch and Dutch) and articles in English. For the first 4 languages we got the URLs of the most recent 100 articles in each of them. For English articles, we got the URLs of the most recent 150 articles, in order to also test the effect of having imbalanced classes. In total we have 550 URLs. The URLs were equally split into training and test sets.

An example URL looks as follows:

"es.globalvoicesonline.org/2013/07/08/edward-snowden-divide-a-los-rusos"

The host part of the URL reflects the article's language. For example, in the above URL, *es* stands for Spanish. It is then followed by the website's domain globalvoicesonline.org, then the article's data in the form of year, month and day, with forward slashes in between. The rest of the URL comes after another slash.

The final part of the URL is normally constructed from the article's headline, however, this is not always the case. The presence of an identifier to the article's language in the host part makes our classification problem trivial, hence we removed that part of the URLs while constructing our dataset. We also removed the domain part since it is constant in all our URLs, as well as the date part.

Hence, the example UR mentioned earlier, is being saved as follows in our dataset:

"*edward-snowden-divide-a-los-rusos*"

We are going to call the resulting dataset GVO.

The core functionality of the code used for the experiments is implemented in IRLib (Information Retrieval Library). IRLib[1] is written in Python and is available as Free and Open Source Software on-line.

5   Results

5.1   Results for the Primary Dataset

Kan [2] achieved an average $F_1 - measure$ of 22.1 % for the WebKB dataset using punctuation-based (terms) approach. The next step was to use *information content* (IC) reduction and *title token-based finite state transducer* (FST) to further segment URL terms and expand abbreviations. This achieved an average $F_1-measure$ of 33.2 % and 34 % respectively. For the same dataset, the proposed n-gram LM classifier achieved an average $F_1 - measure$ of 51.4 %, where $n = 4$ and $\gamma = 0.0062$. The *log-likelihood odds* (LLO) variation of the same LM increased the average $F_1 - measure$ to 59.25 %. Detailed results are shown in Table 2.

Table 2. Comparing $F_1 - measure$ for the WebKB dataset. Results in first 3 rows are from [2] using SVM^{light}, the last two rows are using the proposed n-gram Language Model ($\gamma = 0.0062$). IC, FST and LLO stand for information content reduction, title token-based finite state transducer, and Log-likelihood Odds respectively. All F_1 values are multiplied by 100.

Classifier	Course	Faculty	Project	Student	Macro avg
Terms	13.5	23.4	35.6	15.8	22.1
IC	50.2	31.8	35.0	15.7	33.2
FST	52.7	31.5	36.3	15.6	34.0
All-Grams	78	**75**	50	**63**	**66.5**
4-gram LM/LLO	**83.6**	40.2	**53.7**	59.4	59.25

---

[1] https://github.com/gr33ndata/irlib.

In later research, Kan [26] tried additional feature extraction methods, achieving the highest $F_1 - measure$ of 52.5 %. For the same dataset, Baykan et al. [5,6] reported $F_1 - measure$ of 66.5 % using the all-gram approach. It is clear that the classification performance of the n-gram LM for this dataset is better than all previous approaches except for all-grams. Nevertheless, the difference between results for all-grams and that of the n-Gram LM are not statistically significant (p=0.5) applying a pairwise t-test with Bonferroni-Holm adjustment. Furthermore, it is worth noting that the n-gram LM uses only 4-grams and requires about 0.04 % of the storage and memory needed for the all-grams approach. More discussion on the scalability of the n-gram LM is included in Sect. 6.

5.2   Results for the Secondary Dataset

The results for DMOZ dataset are shown in Table 3. The best results for the n-gram LM were achieved using 7-grams and $\gamma = 0.004$. The results for the previous research using SVM and all-gram features (all 4,5,6,7 and 8-grams) [5], are also shown in Table 3. The performance of the n-gram LM is marginally better, however the statistical analysis of the results confirms that there is no statistical significance between the accuracy of the two approaches. Again, for some classes, the n-Gram LM requires less than 0.001 % of the memory and storage needed by the all-gram approach.

Table 3. Comparing the F-measure of the n-Gram LM and SVM (all-gram features) classifiers for DMOZ dataset. All F_1 values are multiplied by 100.

Topic	SVM all-gram	n-Gram LM/LLO
Adult	**87.6%**	87.58 %
Arts	81.9 %	**82.03%**
Business	**82.9%**	82.71 %
Computers	82.5 %	**82.79%**
Games	**86.7%**	86.43 %
Health	82.4 %	**82.49%**
Home	81 %	**81.13%**
Kids	80 %	**81.09%**
News	**80.1%**	79.01 %
Recreation	79.7 %	**80.22%**
Reference	**84.4%**	83.37 %
Science	80.1 %	**82.52%**
Shopping	**83.1%**	82.48 %
Society	80.2 %	**81.66%**
Sports	84 %	**85.30%**
Average	82.44 %	**82.72%**

5.3   Results for the Tertiary Dataset

Figure 1 shows the variation of the F-measure with the value of n the in n-gram LM. It is clear that the best results were achieved for bi-grams. Thus, we compared our proposed 2-gram LM with SVM and Naive Bayes classifiers using bi-grams as their feature-sets. Table 4 shows the classification performance for the different classifiers. Applying a pairwise t-test with Bonferroni-Holm adjustment shows that the performance of the LLO variation of the 2-gram LM is significantly better than SVM ($p = 0.004$) and is marginally better than NB ($p = 0.054$).

We believe the reason the classification performance was much better for the GVO dataset compared to the other 2 datasets, even for a small training-set, is due to the nature of the classification problem under scrutiny. Usually, the most common n-grams in a document are the ones correlated with the language of the document. Table 5 shows the top 20 terms seen in the GVO dataset. It is clear

Table 4. Comparing $F_1 - measure$ for 2-gram LM and the Log-likelihood Odds (LLO) variation of the 2-gram LM with SVM and Naive Bayes (NB). $\gamma = 0.5$ for all n-gram LMs. F_1 values are multiplied by 100 (GVO Dataset).

Classifier	DE	EN	ES	IT	NL	Macro avg.
2-gram SVM (Poly. Kernel)	52.5	60.6	51.7	69.1	70.2	60.7
2-gram NB (Multinomial)	84.6	78.4	75.5	73.6	91.3	80.4
2-gram LM	87.5	88.2	92.2	92.9	88.9	89.93
2-gram LM/LLO	90.3	89.0	91.3	92.9	89.8	90.67

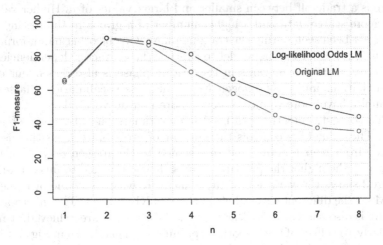

Fig. 1. The variation of the classification performance for GVO dataset, for different values of n. The value of γ is set to 1.

Table 5. The top 20 most frequent terms in GVO dataset, compared to the top 20 bi-grams there. Terms which have parts of them appearing in the top 20 bi-grams are listed in bold letters.

Terms	'die', 'il', 'di', 'of', 'per', 'china', 'i', 'op', 'de', 'en', 'und', 'van', 'the', 'to', 'a', 'video', 'del', 'in', 'y', 'la'
2-grams	'an', 'al', 'on', 'la', 'ti', 'de', 're', 'ta', 'nt', 'or', 'in', 'si', 'di', 'ra', 'te', 'en', 'nd', 'st', 'er', 'es'

that most of them are stop words used in the 5 different languages. Stop words are normally removed in topic classification tasks since they are not correlated with specific topics. For language classification, however, as each language has its own set of stop words they are very useful. In their use of n-grams for text categorization, Cavnar et al. [28] noted that the top 300 n-grams are highly correlated with the documents' languages. Then, the less common n-grams are correlated with the documents' topics. In other words, the majority of the top 300 n-grams are common in documents of the same language, even when the documents' topics are different. We had similar findings to [28]. In combination with that, the low order of n enables the small training-set to cover a high percentage of the total bi-gram vocabulary.

5.4   n-Gram LM Parameter Experimentation

Two main parameters play an important role in our n-gram LM results:

1. The order of n in the n-gram LM.
2. The value of γ in Laplace smoothing.

There is a trade-off between smaller and larger values of n. Higher values of n imply more scarce data and a higher number of n-grams in the testing phase that have not been seen during the training phase. On the other hand, for a lower value of n, it is harder for the model to capture the character dependencies [11]. The quantity of unseen n-grams in the testing phase is also dependent on the class distributions and the homogeneity of the class vocabularies. Classes with more samples have more chance to cover more n-gram vocabulary.

In this context, smoothing is needed to estimate the likelihood of unseen n-grams. The value of γ controls the amount of probability mass that is to be discounted from seen n-grams and re-assigned to the unseen ones. The higher the value of γ the higher the probability mass being assigned to unseen n-grams.

Figure 2 shows the variation of the F-measure with the value of n the in n-gram LM for the different class labels in the WebKB dataset. The macro-average F-measure is also shown. It is clear that the best results are achieved at n = 4.

Similarly, the effect of the smoothing parameter (γ) is shown in Fig. 3. Figure 3 also shows that relaxing the model's *independence assumption*, by using the *Log-likelihood Odds* model, results in better performance, and more immunity to the variations of the smoothing parameter.

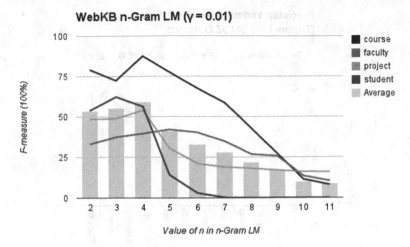

Fig. 2. Variations of F-measures with n.

When the model encounters a high percentage of n-grams that were never seen during the training phase, the precision of the model is affected. Smoothing, on the other hand, tries to compensate this effect by moving some of the probability mass to the unseen n-grams. As stated earlier, the amount of the probability mass assigned to the unseen n-grams is controlled by the value of γ.

In Fig. 4, we can see the correlation between the precision and the percentage of seen n-grams for the different classes. It is also clear that the correlation gets stronger with lower values of γ. For the shown models, the Pearson correlation

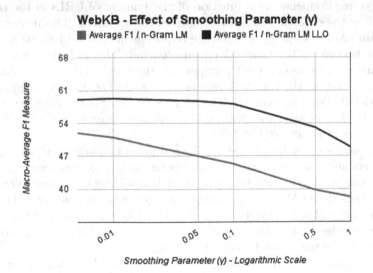

Fig. 3. Variations of F-measures with γ.

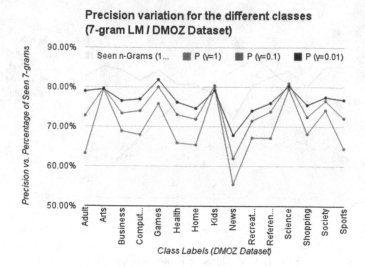

Fig. 4. Variations of Precision with classes and with the smoothing parameter, γ.

coefficients for the precision values with the percentages of seen n-grams are 0.51, 0.65 and 0.74 for $\gamma = 1$, 0.1 and 0.01 respectively.

6   n-Gram LM Scalability

The storage size needed for the n-gram LM is a function of the number of n-grams and classes we have, while for the all-grams approach used by Baykan et al. [5], the storage requirements are a function of the number of URLs in the training set as well as the different orders of 'n' used in the all-grams. This means that in the n-gram LM the memory and storage requirements can be 100,000 times less than that needed by the conventional approaches. This reduction was shown, during our tests, to also have a big impact on the classification processing time.

Let us use any of the binary-classifiers used in DMOZ dataset to explain this in more detail. We have about 100,000 URLs in the 'Sports' category, thus as shown in Baykan et al. [5], we will build a balanced training-set of positive and negative cases of about 200,000 URLs.

As we have seen in Eqs. 4 and 5, for an n-gram language model we need to store the counts of n-grams and (n-1)-grams for each class. Since we can achieve slightly better results than Baykan et al. [5] with 'n = 7', we will do our calculations based on the 7-gram LM here. The number of 7-grams in the positive and negative classes are 746,024 and 1,037,419 respectively, while the number of 6-grams for the same two classes are 568,162 and 795,192. Thus the total storage needed is the summation of the above 4 values, i.e. 3,146,797

For the approach used by Baykan et al. [5], we need to construct a matrix of all features and training-data records. The features in this case will be the all-grams, i.e. 4, 5, 6, 7 and 8-grams, and the training-data records are the 200,000

URLs in the training-set. This matrix is to be used by a Naive Bayes or SVM classifiers later on. The counts for the 4, 5, 6, 7 and 8-grams are 222,649, 684,432, 1,198,689, 1,628,422, 2,008,153 respectively. Thus, the total number of features is the summation of the above 5 values, i.e. 5,742,345. Given that there are about 200,000 URLs in the training-set, the total size of the matrix will be the product of the above 2 numbers, $5,742,345 * 200,000$, which is 1,148,469,000,000.

As we can see in the above example, the memory and storage requirements for the n-gram LM is 1:364,964 ($\approx 0.0003\%$) of that needed for the conventional approaches. Similarly, even for a small datasets such as WebKB, the memory needed for n-gram LM is about 1:2600 ($\approx 0.04\%$) of that needed for the all-grams approach.

As we have discussed earlier, such reduction in storage and processing requirements for the n-Gram LM, does not impact negatively on its classification performance compared the the previous classification approaches.

7 Conclusions

Here we have presented a new LM approach for URL classification that cuts down on the number of features, and therefore, the storage and processing requirements, and still manages to achieve comparable levels of performance. We have tested our approach on 3 different web page classification settings: based on function, topic and language. Our experiments show that the n-gram LM approach with very basic smoothing is offering some significant improvements for classification performance in some cases or at least equal performance over other methods such as *terms* or *all-grams* used with NB and SVM classifiers.

The n-gram LM requires less processing power compared to all-gram. For some cases the proposed model required less that 0.001% of the storage and processing power needed by the previous methods.

Our method has application to real world URL classification, an important emerging problem. We have tested it on a large dataset (some classes of DMOZ dataset have more that 200,000 URLs) as well as on the WebKB dataset. We have also performed parameter experimentation to establish the importance of parameters in the new LM.

As further work, we believe that more sophisticated smoothing methods and interpolating multiple n-gram models, with different values of n, could improve the performance of the LM model. Thus, we propose to continue our research in that direction.

References

1. Thomas, K., Grier, C., Ma, J., Paxson, V., Song, D.: Design and evaluation of a real-time URL spam filtering service. In: 2011 IEEE Symposium on Security and Privacy (SP), pp. 447–462. IEEE (2011)
2. Kan, M.: Web page classification without the web page. In: Proceedings of the 13th International World Wide Web Conference on Alternate Track Papers & Posters, pp. 262–263. ACM (2004)

3. Vonitsanou, M., Kozanidis, L., Stamou, S.: Keywords identification within greek URLs. Polibits **43**, 75–80 (2011)
4. Nicolov, N., Salvetti, F.: Efficient spam analysis for Weblogs through URL segmentation. Amsterdam Stud. Theory History Linguist. Sci. Ser. 4 **292**, 125 (2007)
5. Baykan, E., Henzinger, M., Marian, L., Weber, I.: Purely URL-based topic classification. In: Proceedings of the 18th International Conference on World Wide Web, pp. 1109–1110. ACM (2009)
6. Baykan, E., Marian, L., Henzinger, M., Weber, I.: A comprehensive study of features and algorithms for URL-based topic classification. ACM Trans. Web (TWEB) **5**, 15 (2011)
7. Baykan, E., Henzinger, M., Weber, I.: A comprehensive study of techniques for URL-based web page language classification. ACM Trans. Web (TWEB) **7**, 3 (2013)
8. Chung, Y., Toyoda, M., Kitsugeregawa, M.: Topic classification of spam host based on URLs. In: Proceedings of the Forum on Data Engineering and Information Management (DEIM) (2010)
9. Ma, J., Saul, L.K., Savage, S., Voelker, G.M.: Identifying suspicious URLs: an application of large-scale online learning. In: Proceedings of the 26th Annual International Conference on Machine Learning, pp. 681–688. ACM (2009)
10. Zhao, P., Hoi, S.C.: Cost-sensitive online active learning with application to malicious URL detection. In: Proceedings of the 19th ACM SIGKDD International Conference on Knowledge Discovery and Data Mining, pp. 919–927. ACM (2013)
11. Peng, F., Huang, X., Schuurmans, D., Wang, S.: Text classification in asian languages without word segmentation. In: Proceedings of the Sixth International Workshop on Information Retrieval with Asian Languages, vol. 11, pp. 41–48. Association for Computational Linguistics (2003)
12. Zhai, C., Lafferty, J.: A study of smoothing methods for language models applied to Ad Hoc information retrieval. In: Proceedings of the 24th Annual International ACM SIGIR Conference on Research and Development in Information Retrieval, pp. 334–342. ACM (2001)
13. Grau, S., Sanchis, E., Castro, M.J., Vilar, D.: Dialogue act classification using a Bayesian approach. In: 9th Conference Speech and Computer (2004)
14. Manning, C.D., Schütze, H.: Foundations of Statistical Natural Language Processing, vol. 999. MIT Press, Cambridge (1999)
15. Chen, S.F., Goodman, J.: An empirical study of smoothing techniques for language modeling. In: Proceedings of the 34th annual meeting on Association for Computational Linguistics, pp. 310–318. Association for Computational Linguistics (1996)
16. Jurafsky, D., Martin, J.: Speech & Language Processing. Pearson Education India, New Delhi (2000)
17. Witten, I.H., Bell, T.C.: The zero-frequency problem: estimating the probabilities of novel events in adaptive text compression. IEEE Trans. Inf. Theory **37**, 1085–1094 (1991)
18. Good, I.J.: The population frequencies of species and the estimation of population parameters. Biometrika **40**, 237–264 (1953)
19. Cooper, W.S.: Some inconsistencies and misidentified modeling assumptions in probabilistic information retrieval. ACM Trans. Inf. Syst. (TOIS) **13**, 100–111 (1995)
20. Lavrenko, V.: A Generative Theory of Relevance, vol. 26. Springer, New York (2009)
21. Robertson, S.E., Jones, K.S.: Relevance weighting of search terms. J. Am. Soc. Inf. Sci. **27**, 129–146 (1976)

22. Jones, Sk, Walker, S., Robertson, S.E.: A probabilistic model of information retrieval: development and comparative experiments: Part 1. Inf. Process. Manag. **36**, 779–808 (2000)

23. Terra, E.: Simple language models for spam detection. In: TREC (2005)

24. Abdallah, T.A., De la Iglesia, B.: URL-based web page classification - a new method for URL-based web page classification using n-gram language models. In: International Conference on Knowledge Discovery and Information Retrieval (KDIR 2014) (2014)

25. Slattery, S., Craven, M.: Combining statistical and relational methods for learning in hypertext domains. In: Page, David L. (ed.) ILP 1998. LNCS, vol. 1446, pp. 38–52. Springer, Heidelberg (1998)

26. Kan, M., Thi, H.: Fast webpage classification using URL features. In: Proceedings of the 14th ACM International Conference on Information and Knowledge Management, pp. 325–326. ACM (2005)

27. Baykan, E., Henzinger, M., Weber, I.: Web page language identification based on URLs. Proc. VLDB Endowment **1**, 176–187 (2008)

28. Cavnar, W.B., Trenkle, J.M., et al.: N-gram-based text categorization, pp. 161–175. Ann Arbor MI 48113 (1994)

Exploiting Guest Preferences with Aspect-Based Sentiment Analysis for Hotel Recommendation

Fumiyo Fukumoto[1]([⊠]), Hiroki Sugiyama[2], Yoshimi Suzuki[1],
and Suguru Matsuyoshi[1]

[1] Interdisciplinary Graduate School of Medicine and Engineering
University of Yamanashi, Yamanashi 400-8511, Japan
{fukumoto,ysuzuki,sugurum}@yamanashi.ac.jp
[2] Universal Computer Co., Ltd., Osaka 540-6126, Japan

Abstract. This paper presents a collaborative filtering method for hotel recommendation incorporating guest preferences. We used the results of aspect-based sentiment analysis to recommend hotels because whether or not the hotel can be recommended depends on the guest preferences related to the aspects of a hotel. For each aspect of a hotel, we identified the guest preference by using dependency triples extracted from the guest reviews. The triples represent the relationship between aspect and its preference. We calculated transitive association between hotels by using the positive/negative preference on some aspect. Finally, we scored hotels by Markov Random Walk model to explore transitive associations between the hotels. The empirical evaluation showed that aspect-based sentiment analysis improves overall performance. Moreover, we found that it is effective for finding hotels that have never been stayed at but share the same neighborhoods.

Keywords: Collaborative filtering · Markov random walk model · Aspect-based sentiment analysis

1 Introduction

Collaborative filtering (CF) identifies the potential preference of a consumer/ guest for a new product/hotel by using only the information collected from other consumers/guests with similar products/hotels in the database. It is a simple technique as it is not necessary to apply more complicated content analysis compared to the content-based filtering framework [1]. CF has been very successful in both research and practical systems. It has been widely studied [2–6,11], and many practical systems such as Amazon for book recommendation and Expedia for hotel recommendation have been developed.

Item-based collaborative filtering is one of the major recommendation techniques [6,7]. It assumes that the consumers/guests are likely to prefer product/hotel that are similar to what they have bought/stayed before. Unfortunately, most of them only consider star ratings and leave consumers/guests textual reviews. Several authors focused on the problem, and attempted to

© Springer International Publishing Switzerland 2015
A. Fred et al. (Eds.): IC3K 2014, CCIS 553, pp. 34–49, 2015.
DOI: 10.1007/978-3-319-25840-9_3

improve recommendation results using the techniques on text analysis, *e.g.*, sentiment analysis, opinion mining, or information extraction [8–10]. However, major approaches aim at finding the positive/negative opinions for the product/hotel, and do not take users preferences related to the *aspects* of a product/hotel into account. For instance, one guest is interested in a hotel with nice restaurants for enjoying her/his vacation, while another guest, *e.g.*, a businessman prefers to the hotel which is close to the station. In this case, the aspect of the former is different from the latter.

This paper presents a collaborative filtering method for hotel recommendation incorporating guest preferences. We rank hotels according to scores. The score is obtained using the analysis of different aspects of guest preferences. The method utilizes a large amount of guest reviews which make it possible to solve the item-based filtering problem of data sparseness, *i.e.*, some items were not assigned a label of users preferences. We used the results of aspect-based sentiment analysis to recommend hotels because whether or not the hotel can be recommended depends on the guest preferences related to the aspects of a hotel. For instance, if one guest stays at hotels for her/his vacation, a room with nice views may be an important factor to select hotels, whereas another guest stayed at hotels for business, may select hotels near to the station. We parsed all reviews by using syntactic analyzer, and extracted dependency triples which represent the relationship between aspect and its preference. For each aspect of a hotel, we identified the guest preference related to the aspect to good or not, based on the dependency triples in the guest reviews. The positive/negative opinion on some aspect is used to calculate transitive association between hotels. Finally, we scored hotels by Markov Random Walk (MRW) model, *i.e.*, we used MRW based recommendation technique to explore transitive associations between the hotels. Random Walk based recommendation overcomes the item-based CF problem that the inability to explore transitive associations between the hotels that have never been stayed but share the same neighborhoods [11].

2   Related Work

Sarwar *et al.* mentioned that CF mainly consists of two procedures, *prediction* and *recommendation* [7]. Prediction refers to a numerical value expressing the predicted likeliness of item for user, and recommendation is a list of items that the user will like the most. As the volume of online reviews has drastically increased, sentiment analysis, opinion mining, and information extraction for the process of prediction are a practical problem attracting more and more attention. Several efforts have been made to utilize these techniques to recommend products [9,12]. Cane *et al.* have attempted to elicit user preferences expressed in textual reviews, and map such preferences onto some rating scales that can be understood by existing CF algorithms [8]. They identified sentiment orientations of opinions by using a relative-frequency-based technique that estimates the strength of a word with respect to a certain sentiment class as the relative frequency of its occurrence in the class. The results using movie reviews from the Internet Movie Database (IMDb) for the MovieLens 100k dataset showed the effectiveness of the method. However, the sentiment analysis they used is limited, *i.e.*, they used only adjectives or verbs.

Niklas *et al.* have attempted to improve the accuracy of movie recommendations by using the results of opinion extraction from free-text reviews [9]. They presented three approaches: (i) manual clustering, (ii) semi-automatic clustering by Explicit Semantic Analysis (ESA), and (iii) fully automatic clustering by Latent Dirichlet Allocation (LDA) [13] to extract movie aspects as opinion targets, and used them as features for the collaborative filtering. The results using 100 random users from the IMDb showed that the LDA-based movie aspect extraction yields the best results. Our work is similar to Niklas *et al.* method in the use of LDA. The difference is that our approach applied LDA to the dependency triples, although Niklas applied LDA to single words. Raghavan *et al.* have attempted to improve the performance of collaborative filtering in recommender systems by incorporating quality scores to ratings [10]. They estimated the quality scores of ratings using the review and user data set, and ranked according to the scores. They adapted the Probabilistic Matrix Factorization (PMF) framework. The PMF aims at inferring latent factors of users and items from the available ratings. The experimental evaluation on two product categories of a benchmark data set, *i.e.*, *Book* and *Audio CDs* from Amazon.com showed the efficacy of the method.

In the context of recommendation, several authors have attempted to rank items by using graph-based ranking algorithms [14,15]. Wijaya *et al.* have attempted to rank items directly from the text of their reviews [16]. They constructed a sentiment graph by using simple contextual relationships such as collocation, negative collocation and coordination by pivot words such as conjunctions and adverbs. They applied PageRank algorithm to the graph to rank items. Li *et al.* proposed a basket-sensitive random walk model for personalized recommendation in the grocery shopping domain [11]. The method extends the basic random walk model by calculating the product similarities through a weighted bi-partite network which allows the current shopping behaviors to influence the product ranking. Empirical results using three real-world data sets, LeShop, TaFeng and an anonymous Belgium retailer showed that a performance improvement of the method over other existing collaborative filtering models, the cosine, conditional probability and the bi-partite network based similarities. However, the transition probability from one product node to another is computed based on a user's purchase frequency of a product with regardless of the users' positive or negative opinions concerning to the product.

There are three novel aspects in our method. Firstly, we propose a method to incorporate different aspect of a hotel into users preferences/criteria to improve quality of recommendation. Secondly, from a ranking perspective, the MRW model we used is calculated based on the polarities of reviews. Finally, from the opinion mining perspective, we propose overcoming with the unknown polarized words by utilizing LDA.

3   Framework of the System

Figure 1 illustrates an overview of the method. It consists of four steps: (1) Aspect analysis, (2) Positive/negative opinion detection based on aspect analysis, (3) Positive/negative review identification, and (4) Scoring hotels by MRW model.

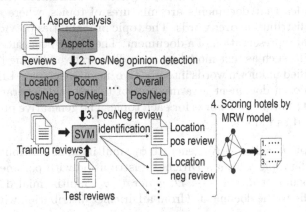

Fig. 1. Overview of the method.

3.1   Aspect Analysis

The first step to recommend hotels based on guest preferences is to extract aspects for each hotel from a guest review corpus. All reviews were parsed by the syntactic analyzer CaboCha [17], and all the dependency triples (rel, x, y) are extracted. Here, x refers to a noun/compound noun word related to the aspect. y shows verb or adjective word related to the preference for the aspect. rel denotes a grammatical relationship between x and y. We classified rel into 9 types of Japanese particle, "$ga(ha)$", "wo", "ni", "he", "to", "de", "$yori$", "$kara$" and "$made$". For instance, from the sentence "$Cyousyoku$ (breakfast) ga totemo (very) yokatta (was delicious)." (The breakfast was very delicious.), we can obtain the dependency triplet, (ga, $cyousyoku$, $yokatta$). The triplet represents positive opinion, "$yokatta$"(was $delicious$) concerning to the aspect, "$Cyousyoku$"($breakfast/meal$).

3.2   Positive/Negative Opinion Detection

The second step is to identify positive/negative opinion related to the aspects of a hotel. We classified aspects into seven types: "Location", "Room", "Meal", "Spa", "Service", "Amenity", and "Overall". These types are used in the Rakuten travel data[1] which we used in the experiments. Basically, the identification of positive/negative opinion is done using Japanese sentiment polarity dictionary [18]. More precisely, if y in the triplet (rel, x, y) is classified into positive/negative classes in the dictionary, we regarded the extracted dependency

---

[1] http://rit.rakuten.co.jp/opendata.html.

triplet as positive/negative opinion. However, the dictionary makes it nearly impossible to cover all of the words in the review corpus.

For unknown verb or adjective words that were extracted from the review corpus, but did not occur in any of the dictionary classes, we classified them into positive or negative class using a topic model. Topic models such as probabilistic latent semantic indexing [19] and Latent Dirichlet Allocation (LDA) [13] are based on the idea that documents are mixtures of topics, where each topic is captured by a distribution over words. The topic probabilities provide an explicit low-dimensional representation of a document. They have been successfully used in many domains such as text modeling and collaborative filtering [20]. We used LDA and classified unknown words into positive/negative classes. LDA presented by [13] models each document as a mixture of topics, and generates a discrete probability distribution over words for each topic. The generative process of LDA can be described as follows:

1. For each topic $k = 1, \cdots, K$, generate ϕ_k, multinomial distribution of words specific to the topic k from a Dirichlet distribution with parameter β;
2. For each document $d = 1, \cdots, D$, generate θ_d, multinomial distribution of topics specific to the document d from a Dirichlet distribution with parameter α;
3. For each word $n = 1, \cdots, N_d$ in document d;
   a. Generate a topic z_{dn} of the nth word in the document d from the multinomial distribution θ_d
   b. Generate a word w_{dn}, the word associated with the nth word in document d from multinomial $\phi_{z_{dn}}$

Like much previous work on LDA, we used Gibbs sampling to estimate ϕ and θ. The sampling probability for topic z_i in document d is given by:

$$P(z_i \mid z_{\backslash i}, W) = \frac{(n^v_{\backslash i,j} + \beta)(n^d_{\backslash i,j} + \alpha)}{(n_{\backslash i,j} + W\beta)(n^d_{\backslash i,.} + T\alpha)}. \tag{1}$$

$z_{\backslash i}$ refers to a topic set Z, not including the current assignment z_i. $n^v_{\backslash i,j}$ is the count of word v in topic j that does not include the current assignment z_i, and $n_{\backslash i,j}$ indicates a summation over that dimension. W refers to a set of documents, and T denotes the total number of unique topics. After a sufficient number of sampling iterations, the approximated posterior can be used to estimate ϕ and θ by examining the counts of word assignments to topics and topic occurrences in documents. The approximated probability of topic k in the document d, $\hat{\theta}^k_d$, and the assignments word w to topic k, $\hat{\phi}^w_k$ are given by:

$$\hat{\theta}^k_d = \frac{N_{dk} + \alpha}{N_d + \alpha K}. \tag{2}$$

$$\hat{\phi}^w_k = \frac{N_{kw} + \beta}{N_k + \beta V}. \tag{3}$$

For each aspect, we manually collected reviews and created a review set. We applied LDA to each set of reviews consisted of triples. We need to estimate two parameters, *i.e.*, the number of reviews, and the number of topics k for the result obtained by LDA. We note that the result can be regarded as a clustering result: each cluster is positive/negative opinion, and each element of the cluster is positive/negative opinion according to the sentiment polarity dictionary, or unknown words. For each number of reviews, we applied LDA, and as a result, we used Entropy measure which is widely used to evaluate clustering techniques to estimate the number of topics (clusters) k. The Entropy measure is given by:

$$E = -\frac{1}{\log k} \sum_j \frac{N_j}{N} \sum_i P(A_i, C_j) \log P(A_i, C_j). \qquad (4)$$

k refers to the number of clusters. $P(A_i, C_j)$ is a probability that the elements of the cluster C_j assigned to the correct class A_i. N denotes the total number of elements and N_j shows the total number of elements assigned to the cluster C_j. The value of E ranges from 0 to 1, and the smaller value of E indicates better result. We chose the parameter k whose value of E is smallest. For each cluster, if the number of positive opinion is larger than those of negative ones, we regarded a triplet including unknown word in the cluster as positive and vice versa.

3.3   Positive/Negative Review Identification

We used the result of positive/negative opinion detection to classify guest reviews into positive or negative related to the aspect. Like much previous work on sentiment analysis based on supervised machine learning techniques [21] or corpus-based statistics, we used Support Vector Machine (SVMs) to annotate automatically [22]. For each aspect, we collected positive/negative opinion (triples) from the results of LDA[2]. Each review in the test data is represented as a vector where each dimension of a vector is positive/negative triplet appeared in the review, and the value of each dimension is a frequency count of the triplet. For each aspect, the classification of each review can be regarded as a two-class problem: positive or negative.

3.4   Scoring Hotels by MRW Model

The final procedure for recommendation is to rank hotels. We used a ranking algorithm, the MRW model that has been successfully used in Web-link analysis, social networks [23], and recommendation [11,14,15]. Given a set of hotels H, $Gr = (H, E)$ is a graph reflecting the relationships between hotels in the set. H is the set of nodes, and each node h_i in H refers to the hotel. E is a set of edges, which is a subset of $H \times H$. Each edge e_{ij} in E is associated with an affinity

---

[2] We used the clusters that the number of positive and negative words is not equal.

weight $f(i \rightarrow j)$ between hotels h_i and h_j $(i \neq j)$. The weight of each edge is a value of transition probability $P(h_j \mid h_i)$ between h_i and h_j, and defined by:

$$P(h_j \mid h_i) = \sum_{k=1}^{|Gr|} \frac{c(g_k, h_j)}{(\sum c(g_k, \cdot))} \cdot \frac{c(g_k, h_i)}{(\sum c(\cdot, h_i))}. \tag{5}$$

Equation (5) shows the preference voting for target hotel h_j from all the guests in Gr stayed at h_i. We note that we classified reviews into positive/negative. We used the results to improve the quality of score. More precisely, we used only the positive review counts to calculate transition probability. $c(g_k, h_j)$ and $c(g_k, h_i)$ in Eq. (5) refer to the lodging count that the guest g_k reviewed the hotel $h_j(h_i)$ as *positive*. $P(h_j \mid h_i)$ in Eq. (5) is the marginal probability distribution over all the guests. The transition probability obtained by Eq. (5) shows a weight assigned to the edge between hotels h_i and h_j.

We used the row-normalized matrix $U_{ij} = (U_{ij})_{|H| \times |H|}$ to describe Gr with each entry corresponding to the transition probability, where $U_{ij} = p(h_j \mid h_i)$. To make U a stochastic matrix, the rows with all zero elements are replaced by a smoothing vector with all elements set to $\frac{1}{|H|}$. The matrix form of the recommendation score $Score(h_i)$ can be formulated in a recursive form as in the MRW model: $\lambda = \mu U^T \lambda + \frac{(1-\mu)}{|H|} e$, where $\lambda = [Score(h_i)]_{|H| \times 1}$ is a vector of saliency scores for the hotels. e is a column vector with all elements equal to 1. μ is a damping factor. We set μ to 0.85, as in the PageRank [24]. The final transition matrix is given by:

$$M = \mu U^T + \frac{(1-\mu)}{|H|} ee^T. \tag{6}$$

Each score is obtained by the principal eigenvector of the new transition matrix M. We applied the algorithm to the graph. The higher score based on transition probability the hotel has, the more suitable the hotel is recommended. For each aspect, we chose the topmost k hotels according to rank score. For each selected hotel, if the negative review is not included in the hotel reviews, we regarded the hotel as a recommendation hotel.

4 Experiments

4.1 Data

We used Rakuten travel data[3]. It consists of 11,468 hotels, 348,564 reviews submitted from 157,729 guests. We used plda[4] to assign positive/negative tag to the aspects. For each aspect, we estimated the number of reviews, and the number of topics (clusters) by searching in steps of 100 from 200 to 1,000. Table 1 shows

---

[3] http://rit.rakuten.co.jp/opendata.html.
[4] http://code.google.com/p/plda.

Table 1. The minimum entropy value and the # of topics.

Aspect	Entropy	Reviews	Topics
Location	0.209	600	700
Room	0.460	700	600
Meal	0.194	500	700
Spa	0.232	400	500
Service	0.226	500	700
Amenity	0.413	600	600
Overall	0.202	500	700

Table 2. Data used in the experiments.

Hotels	30,358
Different hotels	6,387
Guests	23,042
Reviews	116,033

the minimum entropy value, the number of reviews, and the number of topics for each aspect. Table 1 shows that the number of reviews ranges from 400 to 700, and the number of topics are from 500 to 700. For each of the seven aspects, we used these numbers of reviews and topics in the experiments. We used linear kernel of SVM-Light [22] and set all parameters to their default values. All reviews were parsed by the syntactic analyzer CaboCha [17], and 633,634 dependency triples are extracted. We used them in the experiments.

We had an experiment to classify reviews into positive or negative. For each aspect, we chose the topmost 300 hotels whose number of reviews are large. We manually annotated these reviews. The evaluation is made by two humans. The classification is determined to be correct if two human judges agree. We obtained 400 reviews consisting 200 positive and 200 negative reviews. 400 reviews are trained by using SVMs for each aspect, and classifiers are obtained. We randomly selected another 100 test reviews from the topmost 300 hotels, and used them as test data. Each of the test data was classified into positive or negative by SVMs classifiers. The process is repeated five times. As a result, the macro-averaged F-score concerning to positive across seven aspects was 0.922, and the F-score for negative was 0.720. For each aspect, we added the reviews classified by SVMs to the original 400 training reviews, and used them as a training data to classify test reviews.

We created the data which is used to test our recommendation method. More precisely, we used the topmost 100 guests staying at a large number of different hotels as recommendation. For each of the 100 guests, we sorted hotels in chronological order. We used these with the latest five hotels as test data. To score hotels by MRW model, we used guest data staying at more than three times. The data is shown in Table 2. "Hotels" and "Different hotels" in Table 2 refer to the total number of hotels, and the number of different hotels that the guests stayed at more than three times, respectively. "Guests" shows the total number of guests who stayed at one of the "Different hotels". "Reviews" shows the number of reviews with these hotels.

We used MAP (Mean-Averaged Precision) as an evaluation measure [25]. For a given set of guests $G = \{g_1, \cdots, g_n\}$, and $H = \{h_1, \cdots, h_{m_j}\}$ be a set of hotels that should be recommended for a guest g_j, the MAP of G is given by:

$$\text{MAP}(G) = \frac{1}{|G|} \sum_{j=1}^{|G|} \frac{1}{m_j} \sum_{k=1}^{m_j} Precision(R_{jk}). \tag{7}$$

R_{jk} in Eq. (7) refers to the set of ranked retrieval results from the top result until we get hotel h_k. $Precision$ indicates a ratio of correct recommendation hotels by the system divided by the total number of recommendation hotels.

4.2 Basic Results

The results across seven aspects are shown in Table 3. As shown in Table 3, there are no significant difference among seven aspects, and the averaged MAP obtained by our method, aspect-based sentiment analysis (ASA) was 0.392. Table 4 shows sample clusters regarded as positive for three aspects, "location", "room", and "meal" obtained by LDA. Each cluster shows the top 5 triples and content words. We observed that the extracted triples show positive opinion for each aspect. This indicates that aspect extraction contributes to improve overall performance. In contrast, some words such as *yoi* (be good) and *manzoku* (satisfy) in content word based clusters appear across aspects. Similarly, some words such as *ricchi* (location) and *cyousyoku* (breakfast) which appeared in negative cluster are an obstacle to identify positive/negative reviews in SVMs classification.

We recall that we classified aspects into seven types according to the guest preferences. There are other aspects for the hotels such as hotel types and area. We used three types of the hotels, *i.e.*, "Japanese style inn at a hot spring", "Pension", and "Business hotel". Similarly, we used two area, *i.e.*, "Tokyo" and "Nagano prefecture". We had an experiment to examine how these aspects affect the overall performance of recommendation. The data and the results are shown in Tables 5 and 6. We can see from Table 5 that there are no significant difference among hotel types as the averaged MAP against the hotel types are from 0.384 to 0.394. However, the Map of "Tokyo" related to "Amenity" was 0.376 while

Table 3. Basic results.

	Location	Room	Meal	Spa	Service	Amenity	Overall	Avg
MAP	0.391	0.373	0.403	0.392	0.382	0.391	0.414	0.392

Table 4. Top 5 triples and content words.

Rank	Aspects		
	Location	Room	Meal
1	*(ni, eki, chikai)*	*(ga, heya, yoi)*	*(ga, shokuji, yoi)*
	be near to the station	room was nice	breakfast was nice
2	*(ha, hotel, chikai)*	*(ha, heya, hiroi)*	*(ha, shokuji, yoi)*
	the hotel is close	the room is wide	meal was nice
3	*(ni, hotel, chikai)*	*(ga, heya, kirei)*	*(ha, restaurant, good)*
	be near to the hotel	A room is clean	a restaurant is good
4	*(ni, parking, chikai)*	*(de , sugoseru, heya)*	*(ha, restaurant, yoi)*
	be near to the parking	can spend in the room	restaurant is nice
5	*(ga, konbini, aru)*	*(ha, heya, jyuubun)*	*(ha, buffet, yoi)*
	be near to the convenience store	a room is enough goo	Buffet is delicious
Rank	Content Words		
	Location	Room	Meal
1	*ricchi*	*heya*	*syokuji*
	location	room	meal
2	*eki*	*hiroi*	**yoi**
	station	be wide	be good
3	**yoi**	*kirei*	*cyousyoku*
	be good	be clean	breakfast
4	*mise*	**manzoku**	*oishii*
	store	satisfy	be delicious
5	*subarashii*	**yoi**	**manzoku**
	be great	be good	satisfy

Table 5. Data and results (hotel types).

	Data		Results							
	Hotels	Reviews	Location	Room	Meal	Spa	Service	Amenity	Overall	Avg
Hot spring	3,073	52,798	0.393	0.375	0.411	0.394	0.381	0.392	0.411	0.394
Pension	1,845	30,275	0.383	0.364	0.388	0.379	0.376	0.386	0.401	0.384
Business	2,759	38,569	0.394	0.370	0.394	0.386	0.379	0.390	0.408	0.389

that of "Nagano Pref." was only 0.329, and the difference was 0.047. One reason behind this lies the small number of reviews as the "Amenity" of "Nagano Pref."consisted of only 47 reviews. For future work, we should be extend our method for further efficacy by overcoming the lack of sufficient reviews in data sets.

Table 6. Data and results (area).

	Data		Results							
	Hotels	Reviews	Location	Room	Meal	Spa	Service	Amenity	Overall	Avg
Tokyo	982	2,902	0.369	0.359	0.387	0.381	0.376	0.382	0.399	0.394
Nagano Pref.	897	2,093	0.361	0.353	0.378	0.369	0.329	0.371	0.380	0.367

4.3 Comparative Experiments

We compared the results obtained by our method, ASA with the following four approaches to examine how the results of each method affect the overall performance.

1. Transition Probabilities without Review (TPWoR). The probability $P(h_j \mid h_i)$ used in the method is the preference voting for the target hotel h_j from all the guests in a set G who stayed at h_i, regardless of positive or negative review of G.
2. Content Words (CW). The difference between content words method and our method, ASA is that the former applies LDA to the content words.
3. Without Reviews Classified by SVMs (WoR). SVMs used in this method classifies test data by using only the original 400 training reviews.
4. Without Negative Review Filtering (WoNRF). The method selected the topmost k hotels according to the MRW model, and the method dose not use negative reviews as a filtering.

Table 7 shows averaged MAP across seven aspects. As we can see from Table 7 that aspect-based sentiment analysis was the best among four baselines, and MAP score attained at 0.392. The result obtained by transition probability without review was worse than any other results. This shows that the use of guest review information is effective for recommendation. Table 7 shows that the result obtained by content words method was worse than the result obtained by aspect-based sentiment analysis, and even worse than the results without reviews classified by SVMs (WoR) and without negative review filtering (WoNRF). Furthermore, we can see from Table 7 that negative review filtering was a small contribution, *i.e.*, the improvement was 0.014 as the result without negative review filtering was 0.378 and aspect-based SA was 0.392. One reason is that the accuracy of negative review identification. The macro-averaged F-score concerning to negative across seven aspects was 0.720, while the F-score for positive was 0.922. Negative review filtering depends on the performance of negative review identification. Therefore, it will be necessary to examine features other than word triples to improve negative review identification.

It is very important to compare the results of our method with four baselines against each aspect. Figure 2 shows MAP against each aspect. The results obtained by aspect-based sentiment analysis were statistically significant compared to other methods except for the aspects "spa" and "overall in "without negative review filtering" method.

Table 7. Recommendation results.

Method	MAP
Trans. pro. without review	0.257
Content Words	0.304
Without reviews by SVMs	0.356
Without neg review filtering	0.378
Aspect-based SA	0.392

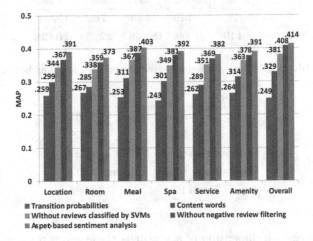

Fig. 2. The results against each aspect.

Table 8 shows a ranked list of the hotels for one guest (guest ID: 2037) obtained by using each method. The aspect is "meal", and each number shows hotel ID. Bold font in Table 8 refers to the correct hotel, *i.e.*, the latest five hotels that the guest stayed at. As can be seen clearly from Table 8, the result obtained by our method includes all of the five correct hotels within the topmost eight hotels, while without negative review filtering (WoNRF) was four. TPWoR, CW, and WoR did not work well as the number of correct hotel was no more than three, and these were ranked seventh and eighth.

It is interesting to note that some recommended hotels are very similar to the correct hotels, while most of the eight hotels did not exactly match these correct hotels except for the result obtained by aspect-based sentiment analysis method. If these hotels were similar to the correct hotels, the method is effective for finding transitive associations between the hotels that have never been stayed but share the same neighborhoods. Therefore, we examined how these hotels are similar to the correct hotels. To this end, we calculated distance between correct hotels and other hotels within the rank for each method by using seven preferences. The preferences have star rating, *i.e.*, each has been scored from 1 to 5, where 1(bad) is lowest, and 5(good) is the best score. We represented each ranked hotel

Table 8. Recommendation list for user ID 2037.

Rank	TPWoR	CW	WoR	WoNRF	ASA
1	2349	2203	2614	3022	**449**
2	604	2349	554	604	30142
3	12869	30142	30142	**449**	**18848**
4	90	604	604	30142	531
5	666	12869	3022	**18848**	**769**
6	2149	39502	531	531	**2223**
7	38126	**449**	**449**	**769**	15204
8	**449**	31209	**18848**	**2223**	**20428**

Table 9. Distance between correct hotel and another hotel.

Method	Dis
Trans. pro. without review	3.067
Content Words	2.859
Without reviews by SVMs	2.721
Without neg review filtering	2.532
Aspect-based SA	2.396

as a vector where each dimension of a vector is these seven preferences and the value of each dimension is its score value. The distance between correct hotel and other hotels within the rank for each method X is defined as:

$$\text{Dis}(X) = \frac{1}{|G|} \sum_{i=1}^{|G|} \operatorname*{argmin}_{j,k} \; d(R\_h_{ij}, C\_h_{ik}). \qquad (8)$$

$|G|$ refers to the number of guests. R_h_{ij} refers to a vector of the j-th ranked hotels except for the correct hotels. Similarly, C_h_{ik} stands for a vector representation of the k-th correct hotel. d refers to Euclidean distance. Equation (8) shows that for each guest, we obtained the minimum value of Euclidean distance between R_h_{ij} and C_h_{ik}. We calculated the averaged summation of the 100 guests. The results are shown in Table 9. The value of "Dis" in Table 9 shows that the smaller value indicates a better result. We can see from Table 9 that the hotels except for the correct hotels obtained by our method are more similar to the correct hotels than those obtained by four baselines. The results show that our method is effective for finding hotels that have never been stayed at but share the same neighborhoods.

5   Conclusions

We proposed a method for hotel recommendation by incorporating preferences related to the different aspects of a hotel to improve quality of the score. We used the results of aspect-based sentiment analysis to detect guest preferences. We parsed all reviews by the syntactic analyzer, and extracted dependency triples. For each aspect, we identified the guest opinion to positive or negative using dependency triples in the guest review. We calculated transitive association between hotels based on the positive/negative opinion. Finally, we scored hotels by Markov Random Walk model. The comparative results using Rakuten travel data showed that aspect analysis of guest preferences improves overall performance and especially, it is effective for finding hotels that have never been stayed at but share the same neighborhoods.

There are a number of directions for future work. In the aspect-based sentiment analysis for guest preferences, we should be able to obtain further advantages in efficacy by overcoming the lack of sufficient reviews in data sets by incorporating transfer learning approaches [26, 27]. We used only surface information of terms (words) and ignore their senses in the aspect-based sentiment analysis. A number of methodologies have been developed for identifying semantic related words in natural language processing research field. This is a rich space for further exploration. We used Rakuten Japanese travel data in the experiments, while the method is applicable to other textual reviews. To evaluate the robustness of the method, experimental evaluation by using other data such as grocery stores: LeShop[5] and movie data: movieLens[6] can be explored in future. Finally, comparison to other recommendation methods, *e.g.*, matrix factorization methods (MF) [28] and combination of MF and the topic modeling [29] will also be considered in the future.

References

1. Balabanovic, M., Shoham, Y.: Fab content-based collaborative recommendation. Commun. ACM **40**, 66–72 (1997)
2. Park, S.T., Pennock, D.M., Madani, O., Good, N., DeCoste, D.: Naive filterbots for robust cold-start recommendations. In: Proceeding of the 12th ACM SIGKDD Conference on Knowledge Discovery and Data Mining, pp. 699–705 (2006)
3. Yildirim, H., Krishnamoorthy, M.S.: A random walk method for alleviating the sparsity problem in collaborative filtering. In: Proceeding of the 3rd ACM Conference on Recommender Systems, pp. 131–138 (2008)
4. Liu, N.N., Yang, Q.: A ranking-oriented approach to collaborative filtering. In: Proceeding of the 31st Annual International ACM SIGIR Conference on Research and Development in Information Retrieval, pp. 83–90 (2008)
5. Lathia, N., Hailes, S., Capra, L., Amatriain, X.: Temporal diversity in recommender systems. In: Proceeding of the 33rd ACM SIGIR Conference on Research and Development in Information Retrieval, pp. 210–217 (2010)
6. Zhao, X., Zhang, W., Wang, J.: Interactive collaborative filtering. In: Proceeding of the 22nd ACM Conference on Information and Knowledge Management, pp. 1411–1420 (2013)

---

[5] www.Leshop.ch.

[6] http://www.grouplens.org/node/73.

7. Sarwar, B., Karypis, G., Konstan, J., Reidl, J.: Automatic multimedia Cross-Model correlation discovery. In: Proceeding of the 10th ACM SIGKDD Conference on Knowledge Discovery and Data Mining, pp. 653–658 (2001)
8. Cane, W.L., Stephen, C.C., Fu-lai, C.: Integrating collaborative filtering and sentiment analysis. In: Proceeding of the ECAI 2006 Workshop on Recommender Systems, pp. 62–66 (2006)
9. Niklas, J., Stefan, H.W., Mark, C.M., Iryna, G.: Beyond the Stars: exploiting free-text user reviews to improve the accuracy of movie recommendations. In: Proceeding of the 1st International CIKM workshop on Topic-Sentiment Analysis for Mass Opinion, pp. 57–64 (2009)
10. Raghavan, S., Gunasekar, S., Ghosh, J.: Review quality aware collaborative filtering. In: Proceeding of the 6th ACM Conference on Recommender Systems, pp. 123–130 (2012)
11. Li, M., Dias, B., Jarman, I.: Grocery shopping recommendations based on basket-sensitive random walk. In: Proceeding of the 15th ACM SIGKDD Conference on Knowledge Discovery and Data Mining, pp. 1215–1223 (2009)
12. Faridani, S.: Using canonical correlation analysis for generalized sentiment analysis product recommendation and search. In: Proceeding of 5th ACM Conference on Recommender Systems, pp. 23–27 (2011)
13. Blei, D.M., Ng, A.Y., Jordan, M.I.: Latent dirichlet allocation. Mach. Learn. **3**, 993–1022 (2003)
14. Yin, Z., Gupta, M., Weninger, T., Han, J.: A unified framework for link recommendation using random walks. In: Proceeding of the Advances in Social Networks Analysis and Mining, pp. 152–159 (2010)
15. Li, L., Zheng, L., Fan, Y., Li, T.: Modeling and broadening temporal user interest in personalized news recommendation. Expert Syst. Appl. **41**(7), 3163–3177 (2014)
16. Wijaya, D.T., Bressan, S.: A random walk on the red carpet: rating movies with user reviews and pagerank. In: Proceeding of the ACM International Conference on Information and Knowledge Management CIKM 2008, pp. 951–960 (2008)
17. Kudo, T., Matsumoto, Y.: Fast method for kernel-based text analysis. In: Proceeding of the 41st Annual Meeting of the Association for Computational Linguistics, pp. 24–31 (2003)
18. Kobayashi, N., Inui, K., Matsumoto, Y., Tateishi, K., Fukushima, S.: Collecting evaluative expressions for opinion extraction. J. Nat. Lang. Process. **12**(3), 203–222 (2005)
19. Hofmann, T.: Probabilistic latent semantic indexing. In: Proceeding of the 22nd Annual International ACM SIGIR Conference on Research and Development in Information Retrieval, pp. 50–57 (1999)
20. Li, Y., Yang, M., Zhang, Z.: Scientific articles recommendation. In: Proceeding of the ACM International Conference on Information and Knowledge Management CIKM 2013, pp. 1147–1156 (2013)
21. Turney, P.D.: Thumbs Up or Thumbs Down? semantic orientation applied to unsupervised classification of reviews. In: Proceeding of the 40th Annual Meeting of the Association for Computational Linguistics, pp. 417–424 (2002)
22. Joachims, T.: SVM light support vector machine. Dept. of Computer Science Cornell University (1998)
23. Xue, G.R., Yang, Q., Zeng, H.J., Yu, Y., Chen, Z.: Exploiting the hierarchical structure for link analysis. In: Proceeding of the 28th ACM SIGIR Conference on Research and Development in Information Retrieval, pp. 186–193 (2005)
24. Brin, S., Page, L.: The anatomy of a large-scale hypertextual web search engine. Comput. Networks **30**(1–7), 107–117 (1998)

25. Yates, B., Neto, R.: Modern Information Retrieval. Addison Wesley, Boston (1999)
26. Blitzer, J., Dredze, M., Pereira, F.: Biographies, Bollywood, Boom-boxes and Blenders: domain adaptation for sentiment classification. In: Proceeding of the 45th Annual Meeting of the Association for Computational Linguistics, pp. 187–295 (2007)
27. Dai, W., Yang, Q., Xue, G., Yu, Y.: Boosting for transfer learning. In: Proceeding of the 24th International Conference on Machine Learning, pp. 193–200 (2007)
28. Koren, Y., Bell, R.M., Volinsky, C.: Matrix factorization techniques for recommender systems. IEEE Comput. **42**(8), 30–37 (2009)
29. Wang, C., Blei, D.M.: Collaborative topic modeling for recommending scientific articles. In: Proceeding of the 17th ACM SIGKDD Conference on Knowledge Discovery and Data Mining, pp. 448–456 (2011)

Iterative Refining of Category Profiles for Nearest Centroid Cross-Domain Text Classification

Giacomo Domeniconi, Gianluca Moro, Roberto Pasolini[✉],
and Claudio Sartori

Università di Bologna, Bologna (BO), Italy
{giacomo.domeniconi,gianluca.moro,roberto.pasolini,
claudio.sartori}@unibo.it

Abstract. In cross-domain text classification, topic labels for documents of a target domain are predicted by leveraging knowledge of labeled documents of a source domain, having equal or similar topics with possibly different words. Existing methods either adapt documents of the source domain to the target or represent both domains in a common space. These methods are mostly based on advanced statistical techniques and often require tuning of parameters in order to obtain optimal performances. We propose a more straightforward approach based on nearest centroid classification: profiles of topic categories are extracted from the source domain and are then adapted by iterative refining steps using most similar documents in the target domain. Experiments on common benchmark datasets show that this approach, despite its simplicity, obtains accuracy measures better or comparable to other methods, obtained with fixed empirical values for its few parameters.

1 Introduction

Text categorization (or *classification*), widely addressed in literature, is the general problem of automatically organizing large sets of text documents into meaningful categories, usually according to the discussed topics. This organization can be applied to mail messages, news stories and in many other contexts. Many works propose a machine learning-based approach, where a training set of documents pre-labeled with the correct categories is used to infer a knowledge model, used to classify subsequent documents [23].

The practical usefulness of this approach, which usually proves to be effective, can be hindered by the necessity of an adequately sized training set of documents, which must contain documents discussing exactly the same topics to be identified with the same words, in order to get a representative model. Denoting the set of known categories and the distribution of words across them as a *domain*, the training set should be constituted of documents of the same domain of those which are to be classified. Such set may be unavailable and may require to manually label many documents, which usually is a time-consuming task.

© Springer International Publishing Switzerland 2015
A. Fred et al. (Eds.): IC3K 2014, CCIS 553, pp. 50–67, 2015.
DOI: 10.1007/978-3-319-25840-9_4

Under some circumstances, however, might be available a set of documents pre-classified under topics which are similar to those of interest but use some different words, as may happen for example between documents of different time periods. Theoretically, categories are considered to be the same and to be equally conditioned by the words, but the words themselves have different distributions.

Cross-Domain Text Categorization is the task of automatically classifying into categories a set of given documents belonging to a *target domain* by leveraging the knowledge of a set of pre-labeled documents from a similar but not identical *source domain*. The set of categories is considered to be shared across the two domains, but the topics represented by the same category are slightly different in the two or are discussed with some different words. Cross-domain categorization allows, in spite of the absence of labels in the target domain, to apply supervised learning methods, which are generally more accurate than unsupervised approaches in practically any context [7].

Several effective methods have been proposed to tackle cross-domain text classification: some based on adapting documents of the source domain to the target one, others based on creating a new, common feature space where both domains are represented. In both cases, these methods are usually based on advanced statistical techniques, thus requiring significant effort and expertise to be correctly implemented [21].

Another limit of some methods is the presence of a number of parameters to be tuned in order to obtain optimal performances. In the experimental evaluation of these methods, is sometimes necessary to go through a trial-and-error process to find good combinations of values for the parameters for each dataset. This limits their usability in real use cases, where categories of documents of the target domain must be predicted without knowing whether they are correct.

In this paper is proposed a simple method for cross-domain text categorization based on nearest centroid classification, where explicit profiles of categories are created and compared to documents.

In our method, such profiles are created from labeled documents of the source domain as an approximation of how they would be in the target domain; then, through an iterative process, new refined profiles are progressively extracted from documents of the target domain which are most similar to the current ones. In the end, target documents are compared to the obtained final profiles, which are ought to be sufficiently refined to accurately represent categories of the target domain. A base method following this scheme, along with a couple of variants to obtain minor improvements, is presented.

This method has the advantage of a simple implementation and few parameters to be set. Experiments on benchmark datasets commonly used in the related literature show that, even by equally setting parameters for all tests, results better or comparable to other works are obtained, also with good running times.

This paper generalizes the method proposed in [9], where is entailed the extraction of a regression model for correct measurement of similarity. Here this specific step is considered to be optional and its effect is investigated in the experiments. In addition, alternative term weighting schemes have been tested, leading to an overall slight improvement of the previously presented results.

The rest of the paper is organized as follows. Section 2 gives an overview of existing literature about cross-domain classification, with focus on text categorization. Section 3 formally defines the problem and exposes the base method, whose variants are then discussed in Sect. 4. Section 5 discusses experimental evaluation of the method, with comparison to results reported in other works and analysis of the effects of variations of parameters. In the end, Sect. 6 recaps the work and proposes some future developments.

2 Related Work

The problem of categorization of text documents by topic has been widely addressed in literature, with the machine learning-based approach, where a classifier is extracted from a set of pre-labeled documents, being a very common solution. Each document is commonly represented as a *bag of words*, a vector of weights indicating the presence of a set of relevant words in it. [23] As possible ways to improve the representation of documents and consequently improve the final classification accuracy, different *term weighting* schemes have been devised: deriving from research on information retrieval [22], these are employed in many text analysis tasks [16] and even in different domains, such as prediction of gene function annotations in biology [8].

In most text classification methods, after representing documents as bags of words, standard machine learning algorithms are used: among these works, some recurring learning schemes are probabilistic classifiers [19] and support vector machines [15]. Nearest centroid classification, also known as *Rocchio classification*, where explicit profiles of categories are built [14], is a somewhat distinct approach which will be described in the next section to introduce our method.

Text categorization is perhaps the most recurring application for cross-domain classification (or *domain adaptation*), a case of *transductive transfer learning* where knowledge must be transferred across two domains which are generally different while having the same set of class labels $\mathcal{Y}$ [21]. In many cases, including text classification, the two domains are represented in a common feature space $\mathcal{X}$. It is generally assumed that labels in source and target domains are equally conditioned by the input data, which is however distributed differently between the two. Denoting with X_S and Y_S data and labels of source domain and with X_T and Y_T those of target domain, is assumed $P(Y_S|X_S) = P(Y_T|X_S)$, but $P(X_S) \neq P(X_T)$: this condition is known as *covariate shift* [24].

Two major approaches to transductive transfer learning are generally distinguished: *instance transfer* and *feature representation transfer* [21].

Instance transfer-based approaches generally work by re-weighting instances (data samples) from the source domain to adapt them to the target domain, in order to compensate the discrepancy between $P(X_S)$ and $P(X_T)$: this generally involves estimating an *importance* $\frac{P(x_S)}{P(x_T)}$ for each source instance x_S to reuse it as a training instance x_T under the target domain. This is roughly the group where our method falls in, although source documents are in practice transferred to the target domain in aggregate form, as category profiles.

Some works mainly address the related problem of sample selection bias, where a classifier must be learned from a training set with a biased data distribution. In [29] is analyzed the bias impact on various learning methods and a correction method using knowledge of selection probabilities is proposed.

The *kernel mean matching* method [13] learns re-weighting factors by matching the means between the domains data in a reproducing kernel Hilbert space (RKHS); this is done without estimating $P(X_S)$ and $P(X_T)$ from a possibly limited quantity of samples. Among other works operating under this restriction there is the *Kullback-Liebler importance estimation procedure* [25], a model to estimate importance based on minimization of the Kullback-Liebler divergence between real and expected $P(X_T)$. Among works specifically considering text classification, [5] trains a Naïve Bayes classifier on the source domain and transfers it to the target domain through an iterative Expectation-Maximization algorithm. In [10] multiple classifiers are trained on possibly multiple source domains and combined in a *locally weighted ensemble* based on similarity to a clustering of the target documents to classify them.

On the other side, feature representation transfer-based approaches generally work by mapping instances of both domains to a new feature space, where their differences are reduced and standard learning methods can be applied.

Structural Correspondence Learning [1] introduces *pivot* features: weights from linear predictors for them are transformed through Singular Value Decomposition and used to augment training data instances. In [6] is presented a simple method to augment instances with features differentiating the two domains, possibly improvable through a nonlinear kernel. In [18] a spectral classification-based framework is introduced, with an objective function balancing source domain supervision and target domain structure. Among works focused on text classification, [4] proposes co-clustering of words and documents: word clusters act as a bridge across domains. *Topic-bridged PLSA* [28] extends Probabilistic Latent Semantic Analysis to accept unlabeled data. In [30] is proposed a framework for joint non-negative matrix tri-factorization of both domains. *Topic correlation analysis* [17] extracts both shared and domain-specific latent features and groups them, to support higher distribution gaps between domains.

As in traditional text categorization, some methods leverage external knowledge bases, which can be helpful to link knowledge across domains. The cited co-clustering-based approach [4] is improved in [26] by representing documents with concepts extracted from Wikipedia. *Bridging information gap* [27] exploits instead an auxiliary domain acting as a bridge between source and target, using Wikipedia articles as a practical example. These methods usually offer very high performances, but need a suitable knowledge base for the context of the analyzed documents, which might not be easily available for overly specialized domains, thus we do not compare them to our method, which uses no external knowledge.

Other than topic classification, a related application of interest of domain adaptation is *sentiment analysis*, where positive and negative opinions about specific objects (products, brands, etc.) must be distinguished. It can be useful to extract knowledge from labeled reviews for some products to classify reviews for products of a different type, with possibly different terminologies. *Spectral feature alignment* [21] clusters domain-specific words leveraging the more general terms. In [2] a sentiment-sensitive thesaurus is built from possibly multiple source domains. In [3] a Naïve Bayes classifier on syntax trees-based features is used.

3   Base Cross-Domain Method

Here we present our general method for cross-domain text classification: a couple of variants are addressed in detail in the subsequent section.

3.1   Rationale

Basically, the method is based on *nearest centroid* classification, where explicit *representations* or *profiles* are built for each category in the same form of bags of words for documents, so that each new document is assigned to the category having the profile most similar to its bag of words. These profiles are often simply constituted by the mean point (centroid) of bags for known documents belonging to respective categories. As a measure to compare vector-like representations of documents and categories, *cosine similarity* is usually employed.

$$\forall \mathbf{a}, \mathbf{b} \in \mathbb{R}^n : \cos(\mathbf{a}, \mathbf{b}) = \frac{\sum_{i=1}^n a_i \cdot b_i}{\sqrt{\sum_{i=1}^n a_i^2} \cdot \sqrt{\sum_{i=1}^n b_i^2}} \tag{1}$$

In standard text categorization, these profiles can be created from documents of the training set and this potentially leads to optimal representations of categories. In the cross-domain case, profiles can be created from documents of the source domain, whose labels for each are known, but these are presumably not optimal to classify documents under the target domain, so some sort of adaptation is needed to use them.

The proposed idea is to use profiles extracted from the source domain as a starting point, expecting that at least some documents of the target domain will be significantly similar to them. Considering these documents as correctly classified, updated profiles for categories can be computed by averaging them, assuming them to constitute better representations of the categories in the target domain. By iteratively repeating this step, somewhat similarly to what happens in the *k-means* clustering algorithm, category profiles can be furtherly improved, as they are progressively extracted from more documents. After a number of iterations, the final profiles are used as a reference to classify documents.

3.2   Problem Formalization

The proposed algorithm has as input a set $\mathcal{D}_S$ of *source* documents which constitute the *source domain* and a disjoint set $\mathcal{D}_T$ of *target* documents making up the *target domain*; we denote with $\mathcal{D} = \mathcal{D}_S \cup \mathcal{D}_T$ their union. Each document is presumed to be related to one and only one category of a set $\mathcal{C}$: this set is shared between the two domains, although categories are generally represented differently across the two. For source documents, a labeling $C_S : \mathcal{D}_S \rightarrow \mathcal{C}$ mapping each of them to one of the categories is given.

The output of the algorithm is a predicted mapping $\hat{C}_T : \mathcal{D}_T \to \mathcal{C}$ of each target document to one of the categories in $\mathcal{C}$. Within experimental evaluation, where a correct mapping $C_T : \mathcal{D}_T \to \mathcal{C}$ is known, the performance of the method is evaluated by measuring how much $\hat{C}_T$ is similar to C_T.

3.3 Document Pre-processing and Term Weighting

As a first step, as typically happens in text categorization, a pre-processing phase is run where all documents in $\mathcal{D}$ are reduced to bags of words.

For each document, single words are extracted, then those shorter than three letters or found in a list of *stopwords* (articles, prepositions, etc.) are removed. Among all distinct words found within documents, likely to many other cross-domain methods, only those appearing in at least three of them are used as features or *terms*: the global set of selected features is denoted with $\mathcal{W}$. Each document d is then represented with a vector $\mathbf{w}_d \in \mathbb{R}^{|\mathcal{W}|}$, containing for each term t a weight $w_{t,d}$ of its relevance within the document.

Each weight $w_{t,d}$ is generally obtained by product of a *local* factor denoting the relevance of t in d itself and a *global* factor equal for all documents, indicating how important is t across them. The most basic local factor is *term frequency* (tf): $\mathrm{tf}(t, d)$ denotes the number of occurrences of a term t in a document d. A common variation is logarithmic term frequency.

$$\mathrm{logtf}(t, d) = \log(1 + \mathrm{tf}(t, d)) \tag{2}$$

Regarding global term weights, the most common scheme is *inverse document frequency* (idf), which gives less importance to terms appearing in too many documents, which generally are not indicative of the category. Denoting with $\mathcal{D}_t$ the set of documents where t appears at least once, formulas below define both idf and its *probabilistic* variant for a term t.

$$\mathrm{idf}(t) = \log\left(\frac{|\mathcal{D}|}{|\mathcal{D}_t|}\right) \qquad \mathrm{idf}_p(t) = \log\left(\frac{|\mathcal{D}| - |\mathcal{D}_t|}{|\mathcal{D}_t|}\right) \tag{3}$$

Effective term weights are given by a local factor optionally multiplied by a global factor: for example, the common "tf.idf" scheme is the product of tf and idf. Moreover, *cosine normalization* can be applied to any local or composite scheme by dividing all the weights of a vector for its length: this is useful to obtain unitary vectors regardless of the length of the documents.

$$w_n(t, d) = \frac{w(t, d)}{\sqrt{\sum_{\tau \in \mathcal{W}} w^2(\tau, d)}} \tag{4}$$

These schemes can be combined in different ways: Sect. 5 will initially present various composite schemes before establishing one of them to be used.

3.4 Cross-Domain Learning Algorithm

After processing documents, a profile $\mathbf{w}_c^0$ for each category $c \in \mathcal{C}$ is computed as the centroid of source documents labeled with c, whose set is denoted with R_c^0.

$$R_c^0 = \{d \in \mathcal{D}_S : C_S(d) = c\} \tag{5}$$

$$\mathbf{w}_c^0 = \frac{1}{|R_c^0|} \sum_{d \in R_c^0} \mathbf{w}_d \tag{6}$$

The "0" index denotes that these are *initial* profiles, which constitute the starting point for the iterative phase, which is explained in the following. The i index in the following is the iterations counter, which starts from 0.

Firstly the similarity score $s^i(d, c)$ between each target document $d \in \mathcal{D}_T$ and each category $c \in \mathcal{C}$ is computed. In the base method, it is simply the cosine similarity between the bag of words for d and the current profile for c.

$$s^i(d, c) = \cos(\mathbf{w}_d, \mathbf{w}_c^i) \tag{7}$$

This similarity is considered as the *absolute* likelihood of c being related to d, i.e. the probability that it is the correct category for d. In order to be confident in the assignment of a category c to a document d, the relevant score $s^i(d, c)$ must be significantly higher than those for the same document and other categories. To evaluate where this is the case, *relative* scores are computed by normalizing those of each document for all categories: this makes them constitute in practice a probability distribution among categories.

$$p^i(d, c) = \frac{s^i(d, c)}{\sum_{\gamma \in C} s^i(d, \gamma)} \tag{8}$$

The value of $p^i(d, c)$ indicates the estimated probability of c being the correct category for d, considering similarities between d and all categories. For each document, the most likely category is obviously that with the highest score: we denote with $A_c^i \subseteq \mathcal{D}_T$ the set of target documents for which c is the predicted category. However, the score could indicate more or less certainty about the prediction. Setting a threshold ρ, we can define for each category $c \in \mathcal{C}$ a set $R_c^{i+1} \subseteq A_c^i$ of documents for which the assignment of the c label is "sure enough".

$$A_c^i = \{d \in \mathcal{D}_T : c = \underset{\gamma \in C}{\operatorname{argmax}}\, p^i(d, \gamma)\} \tag{9}$$

$$R_c^{i+1} = \{d \in A_c^i : p^i(d, c) > \rho\} \tag{10}$$

R_c^{i+1} is a set of *representative* documents for the category c in the target domain. A new profile for the category is built by averaging these documents.

$$\mathbf{w}_c^{i+1} = \frac{1}{|R_c^{i+1}|} \sum_{d \in R_c^{i+1}} \mathbf{w}_d \tag{11}$$

At this point, conditions for the termination of the iterative phase are checked. A maximum number M_I of iterations is set to ensure that the algorithm does not run for an excessively long time. However, after a limited number of iterations, category profiles usually tend to cease to change from one iteration to another.

At a certain iteration, if category profiles are identical to those of the previous one, the same representative documents as before will be selected and the same profiles will keep to be computed, so in this case the iterative phase can be safely terminated. This leads to the following termination condition.

$$\forall c \in \mathcal{C} : \mathbf{w}_c^{i+1} = \mathbf{w}_c^i \tag{12}$$

If this condition does not hold and the number of finished iterations $(i+1)$ is below M_I, all steps from (7) up to here are repeated with the iteration counter i incremented by 1. Otherwise, the iterative phase terminates with an iteration count $n_I = i+1$. When this happens, the final predicted category for each target document d is computed as the one whose latest computed profile is most similar to the bag of words for d.

$$\hat{C}_T(d) = \underset{c \in \mathcal{C}}{\operatorname{argmax}} \ \cos(\mathbf{w}_d, \mathbf{w}_c^{n_I}) \tag{13}$$

Other than to documents of the target domain known and used in the iterative phase, this formula can be applied to any previously unseen document of the same domain, comparing its bag of words to final category profiles.

3.5   Computational Complexity

The process performs many operations on vectors of length $|\mathcal{W}|$: using suitable data structures, both storage space and computation time can be bound linearly w.r.t. the mean number of non-zero elements, which will be denoted with l_D and l_C for bags of words for documents and categories, respectively. By definition, we have $l_D \leq |\mathcal{W}|$ and $l_C \leq |\mathcal{W}|$; from our experiments we also generally observed $l_D \ll l_C < |\mathcal{W}|$.

Initial profiles for categories are built in $O(|\mathcal{D}_S| \cdot l_D)$ time, as all values of all bags of words for documents must be summed up. Cosine similarity between vectors with l_D and l_C non-zero elements respectively can be computed in $O(l_D + l_C)$ time, which can be written as $O(l_C)$ given that $l_D < l_C$.

In each iteration of the refining phase, the method computes cosine similarity for $N_T = |\mathcal{D}_T| \cdot |\mathcal{C}|$ document-category pairs and normalizes them to obtain distribution probabilities in $O(N_T \cdot l_C)$ time; then, to build new bags of words for categories, up to $|\mathcal{D}_T|$ document bags must be summed up, which is done in $O(|\mathcal{D}_T| \cdot l_D)$ time. The sum of these two steps, always considering $l_D < l_C$, is $O(|\mathcal{D}_T| \cdot |\mathcal{C}| \cdot l_C)$, which must be multiplied by the final number n_I of iterations.

Summing up, the overall complexity of the method is $O(|\mathcal{D}_S| \cdot l_D + n_I \cdot |\mathcal{D}_T| \cdot |\mathcal{C}| \cdot l_C)$, which can be simplified to $O(n_I \cdot |\mathcal{D}| \cdot |\mathcal{C}| \cdot l_C)$, with $l_C \leq |\mathcal{W}|$. The complexity is therefore linear in the number $|\mathcal{D}|$ of documents, the number $|\mathcal{C}|$ of top categories (usually very small), the mean number l_C of mean terms per category (having $|\mathcal{W}|$ as an upper bound) and the number n_I of iterations in the final phase, which in our experiments is almost always within 20. This complexity is comparable to many other cross-domain classification methods.

4   Variants

The section above describes the base method which, as will be shown in the experiments, already ensures optimal accuracy with correct setting of parameters. In the following, two variants are proposed to make the method more robust to variation of parameters and to set a tradeoff between accuracy and running times. The two proposed modifications are effective in two different points of the base algorithm and can therefore also be applied at the same time.

4.1   Logistic Regression on Similarity

Normally, the cosine similarity between vectors of a document d and a category c directly constitutes the absolute likelihood for them of being related, i.e. of c being the correct label for d. However, in practice, such cosine similarity is generally largely below 1 even if d and c are related and slightly above 0 otherwise. In an ideal situation, the similarity score $s^x(d, c)$ should be very close to 1 if d and c are related, so that also the relative score $p^x(d, c)$ is significantly high.

A solution is to apply a correction function $[0, 1] \rightarrow [0, 1]$ to the "raw" cosine similarity, in order to boost values generally indicating high relatedness to be close to 1. The function to be used must be somehow infered from known data about similarity and relatedness of documents and categories: such data can be obtained from the source domain, whose labeling of documents is known. To extract a function from this data, univariate logistic regression is employed, where a function π with the following form is obtained [12].

$$\pi(x) = \frac{1}{1 + e^{-(\beta_0 + \beta_1 x)}} \tag{14}$$

In practice, after extracting initial categories representations from the source domain, for each couple $(d, c) \in \mathcal{D}_S \times \mathcal{C}$, an observation $(x_{d,c}, y_{d,c})$ is extracted.

$$x_{d,c} = \cos(\mathbf{w}_d, \mathbf{w}_c^0) \qquad y_{d,c} = \begin{cases} 1 & \text{if } C_S(d) = c \\ 0 & \text{if } C_S(d) \neq c \end{cases} \tag{15}$$

All these observations are used for logistic regression, which computes optimal parameters β_0 and β_1 of function π in order to maximize the following objective function.

$$\prod_{(d,c) \in \mathcal{D}_S \times \mathcal{C}} \pi(x_{d,c})^{y_{d,c}} (1 - \pi(x_{d,c}))^{1-y_{d,c}} \tag{16}$$

Following this, the iterative phase is executed as in the base method, but integrating the function π in the calculation of the absolute similarity score between documents and categories. In practice, (7) is substituted with the following.

$$s^i(d, c) = \pi(\cos(\mathbf{w}_d, \mathbf{w}_c^i)) \tag{17}$$

Regarding the computational complexity to fit the logistic regression model, the cosine similarity for $N_S = |\mathcal{D}_S| \cdot |\mathcal{C}|$ pairs must be computed to acquire input data, which requires $O(l_c \cdot N_S)$ time; then the model can be fit with one of various optimization methods, which are generally linear in the number N_S of data samples [20].

4.2 Relaxed Termination Condition

To terminate the iterative phase before the maximum number M_I of iterations is reached, is normally required that the last computed profile of each category is identical to the one computed in the previous iteration. However, changes of these profiles across later iterations are in practice very small; earlier iterations are by far more important to reach a nearly optimal configuration of the profiles.

Using a substantially low maximum number of iterations could improve the running time of the algorithm with limited losses of classification accuracy, but the number of iterations necessary to obtain a fair accuracy can be significantly different according to the specific data under analysis.

The proposed solution is, rather than testing strict equality between current and previous profiles, measuring the mutual cosine similarities and checking that they are all close enough to 1. Formally, fixed a threshold parameter $\beta \in [0, 1]$, the termination condition given in (12) is here substituted with the following.

$$\forall c \in \mathcal{C} : \cos(\mathbf{w}_c^{i+1}, \mathbf{w}_c^i) \geq \beta \qquad (18)$$

This modification ensures a number of iterations not higher than those obtained with the base method, which can be tuned by using different values of β, which should be very close to 1 (such as 0.999). In the experiments section, the effectiveness of this variant on benchmark datasets will be shown.

5 Experiments

Experiments have been conducted to assess the performances of the proposed method, to compare them to the state of the art and to test how they vary for different values of the parameters.

The method has been implemented in Java within a software framework. For logistic regression, we relied upon the WEKA machine learning software [11].

5.1 Benchmark Datasets

For our experiments, we considered three text collections commonly used in cross-domain classification, allowing to compare our results with those reported by other works for the same collections. Their classes taxonomy exhibits a shallow hierarchical structure, allowing to isolate a small set $\mathcal{C}$ of *top categories*, each including a number of *sub-categories* in which documents are organized. Each input dataset is obtained by picking a small set of top categories of a collection and splitting documents of these categories between a source and a target domain, which contain documents of different branches of the top categories.

By labeling each document of the two domains with the correct top category, we obtain suitable benchmark datasets.

The **20 Newsgroups** collection[1] (*20NG* for short) is a set of posts from 20 different Usenet discussion groups arranged in a hierarchy, each represented by almost 1,000 posts. We consider the 4 most frequent top categories *comp*, *rec*, *sci* and *talk*, each represented by 4 sub-categories (5 for *comp*). Each test involves two or more top categories, with disjoint source and target domains each composed by documents of 2 or 3 sub-categories for each of them. We performed tests with two, three and all four top categories, considered the sub-categories splits used in [4] and other works for two top categories and those suggested in [26] for less common tests with three or four top categories.

The **SRAA** text collection[2] is also drawn from Usenet: it consists of posts from discussion groups about simulated autos, simulated aviation, real autos and real aviation. Tests are performed using two different sets of top categories: {*real, simulated*} (with documents about aviation as source and about autos as target) and {*auto, aviation*} (simulated as source, real as target). Likely to other works, 4,000 documents are considered for each of the four groups.

The **Reuters-21578** collection[3] contains 21,578 newswire stories from year 1992 about economy and finance. Documents are tagged with 5 types of labels, among which *orgs*, *people* and *places* are used as top categories: we considered the three possible pairs of them, using the same split between source and target employed by other works where sub-categories are evenly divided.

5.2   Results

In each test run, to evaluate the goodness of the predicted labeling $\hat{C}_T$ with respect to the correct one C_T, likely to other works, we measure the *accuracy* as the ratio of documents in the target domain for which the correct label was predicted: as almost all target domains have evenly distributed documents across categories, this is a fairly valid measure.

$$\text{Accuracy}(C_T, \hat{C}_T) = \frac{|\{d \in \mathcal{D}_T : \hat{C}_T(d) = C_T(d)\}|}{|\mathcal{D}_T|} \tag{19}$$

Firstly, using the base method, we tested on all the datasets different term weighting schemes and different values of the ρ parameter to check how accuracy varies with them. Plots in Fig. 1 show results for these tests on some of the considered test datasets with some of the best performing weighting schemes, namely cosine normalized raw or logarithmic term frequency multiplied by standard or probabilistic idf. The maximum number of iterations (also for subsequent tests) has been set to $M_I = 50$, which has been rarely reached.

From the plots, it can be noticed that optimal accuracy is generally reached with values of ρ near to the minimum possible value, which is $1/|\mathcal{C}|$ (e.g. 0.5 with

---

[1] http://qwone.com/~jason/20Newsgroups/ (we used the `bydate` distribution).
[2] http://people.cs.umass.edu/~mccallum/data/sraa.tar.gz.
[3] http://www.cse.ust.hk/TL/dataset/Reuters.zip.

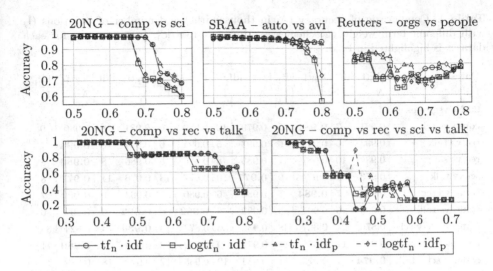

Fig. 1. Classification accuracy (Y axis) of the base cross-domain method for different datasets (plot titles), term weighting schemes (data series, see legend) and values of the threshold parameter ρ (X axis).

2 top categories). This suggests that the selection of representative documents must be large enough in order to obtain good accuracies. By considering all target documents at each iteration to rebulid profiles (i.e. setting $\rho = 1/|\mathcal{C}|$) accuracy is very good, but usually not at its highest possible value.

Given the results, we set $\rho = 1.08/|\mathcal{C}|$ as the default threshold, which is then 0.54, 0.36 and 0.27 for problems with respectively 2, 3 and 4 top-categories: these values are among those performing best over all datasets. Table 1 compares results with this setting for a selection of tested term weighting schemes. From now on, given the overall good results it yields, we choose $\text{logtf}_n \cdot \text{idf}$ (cosine normalized logarithmic tf by standard idf) as the default scheme.

Table 2 reports the accuracy measures with this configuration for each considered dataset, along with the number of iterations, results reported in other works and also two *baseline* results, which represent expected lower and upper bounds for the accuracy of our method. The "min" accuracy is obtained by classifying target documents directly using the initial category profiles, thus suppressing the iterative phase. The "max" accuracy is instead obtained by classifying target documents using profiles extracted from the target domain itself. Regarding other works, we reported the available results from the following ones, also cited in Sect. 2: (**CoC**) co-clustering [4], (**TbP**) topic-bridged PLSA [28], (**SC**) spectral classification [18], (**MTr**) matrix trifactorization [30] and (**TCA**) topic correlation analysis [17].

We can see from the table that our approach performs better than reported methods in most cases. Also, the effective accuracy is usually fairly close to the "max" baseline, suggesting that the final category profiles obtained by iterative

Table 1. Comparison of accuracy (A, in thousandths) and number of iterations (I) with different term weighting schemes, setting $\rho = 1.08/|\mathcal{C}|$; best accuracy for each dataset is highlighted in bold.

Dataset	$\text{logtf} \cdot \text{idf}$		$\text{logtf} \cdot \text{idf}_p$		$\text{tf}_n \cdot \text{idf}$		$\text{tf}_n \cdot \text{idf}_p$		$\text{logtf}_n \cdot \text{idf}$		$\text{logtf}_n \cdot \text{idf}_p$	
	A	I	A	I	A	I	A	I	A	I	A	I
20 Newsgroups												
comp vs sci	0.981	7	**0.981**	7	0.979	10	0.980	10	0.981	8	0.980	9
rec vs talk	0.993	6	0.993	6	0.993	5	0.994	6	0.994	6	**0.994**	7
rec vs sci	0.980	10	0.981	7	0.985	8	0.985	8	0.986	8	**0.986**	8
sci vs talk	0.973	9	0.974	9	**0.977**	10	0.977	13	0.976	11	**0.977**	8
comp vs rec	0.983	7	**0.984**	7	0.980	6	0.980	7	0.983	9	**0.984**	7
comp vs talk	0.990	6	0.991	5	0.992	6	**0.992**	6	**0.992**	6	0.992	6
comp v rec v sci	0.876	25	0.827	48	0.932	36	0.873	13	**0.934**	34	0.824	33
rec v sci v talk	0.978	11	0.978	11	0.977	9	0.979	12	0.979	11	**0.980**	11
comp v sci v talk	**0.974**	11	0.973	11	0.964	12	0.965	12	0.972	18	0.972	17
comp v rec v talk	0.977	5	0.978	6	0.980	7	**0.980**	7	0.979	6	0.980	6
comp rec sci talk	0.969	8	0.969	9	0.964	11	0.965	12	0.970	18	**0.970**	17
SRAA												
real v simulated	0.940	10	0.942	10	0.934	10	0.937	14	0.950	16	**0.950**	16
auto v aviation	0.963	10	0.962	12	**0.966**	16	0.965	15	0.964	14	0.965	12
Reuters-21578												
orgs vs places	0.729	7	**0.730**	6	0.722	8	0.722	8	0.726	7	**0.730**	7
orgs vs people	0.768	37	0.778	40	0.841	12	**0.873**	9	0.826	8	0.861	13
people vs places	0.513	48	0.413	44	**0.741**	35	0.640	22	0.722	23	0.652	24

refining are similar enough to the ideal ones which would be extracted directly from the target domain. Within our environment, the running times for single test runs have always ranged between about 10 s for tests on Reuters datasets (the smallest ones) and one minute when considering 20 Newsgroups with all four top categories.

We now analyze the effect of the variants, starting from the application of logistic regression. Plots in Fig. 2 compare the accuracy obtained either applying or not this variant for some datasets and for different values of the ρ threshold. While with regression the method reaches roughly the same levels of accuracy as does without it, these levels are generally reached for a wider range of values of the confidence threshold. We then suppose that the general effect of logistic regression is to make the algorithm more robust with respect to variations of the ρ parameters and then possibly more likely to perform well.

Regarding instead the use of the relaxed termination condition for the iterative phase, Table 3 compares the results obtained with the base method (already reported in Table 2) with those obtained from this variant with the β parameter set either to 0.9999 or 0.999. It is shown that in the latter cases, the number of iterations is generally sensibly lowered with negligible or acceptable losses of

Table 2. Results of our method (on rightmost columns) on selected test datasets using logtf$_n$ · idf weighting and $\rho = 1.08/|\mathcal{C}|$, compared with those reported by other works: the results in bold are the best for each dataset (excluding baselines).

Dataset	Other methods					$\rho = 1.08/|\mathcal{C}|$		Baselines	
	CoC	TbP	SC	MTr	TCA	Acc	Iters	min	max
20 Newsgroups									
comp vs sci	0.870	**0.989**	0.902	-	0.891	0.981	8	0.803	0.986
rec vs talk	0.965	0.977	0.908	*0.950*	0.962	**0.994**	6	0.646	0.997
rec vs sci	0.945	0.951	0.876	*0.955*	0.879	**0.986**	8	0.837	0.990
sci vs talk	0.946	0.962	0.956	*0.937*	0.940	**0.976**	11	0.781	0.989
comp vs rec	0.958	0.951	0.958	-	0.940	**0.983**	9	0.896	0.990
comp vs talk	0.980	0.977	0.976	-	0.967	**0.992**	6	0.970	0.995
comp v rec v sci	-	-	-	*0.932*	-	**0.934**	34	0.687	0.970
rec v sci v talk	-	-	-	*0.936*	-	**0.979**	11	0.503	0.988
comp v sci v talk	-	-	-	*0.921*	-	**0.972**	18	0.719	0.985
comp v rec v talk	-	-	-	*0.955*	-	**0.979**	6	0.922	0.988
comp rec sci talk	-	-	-	-	-	**0.970**	18	0.618	0.981
SRAA									
real v simulated	0.880	0.889	0.812	-	-	**0.950**	16	0.618	0.969
auto v aviation	0.932	0.947	0.880	-	-	**0.964**	14	0.807	0.979
Reuters-21578									
orgs vs places	0.680	0.653	0.682	**0.768**	0.730	0.726	7	0.732	0.915
orgs vs people	0.764	0.763	0.768	0.808	0.792	**0.826**	8	0.772	0.930
people vs places	**0.826**	0.805	0.798	0.690	0.626	0.722	23	0.632	0.920

Values of **MTr** for 20NG (in italic) are averages of multiple runs with equal top categories where a baseline classifier trained on source domain and tested on target got 65 % or higher accuracy

accuracy (with only one exception on "comp vs rec vs sci") or, in some cases, even with slight improvements of it. Interestingly, the final number of iterations with this variant is still different across datasets, but its effects in terms of variation of accuracy is similar in many of them. This suggests that the β parameter could be set to impose a roughly predictable tradeoff between the classification accuracy and the running time, which depends from the number of iterations.

Finally, we show results obtained by simulating the situation where not all documents of the target domain are known in advance. Specifically, we run tests where only a set ratio of randomly sampled target documents are assumed to be known in the iterative phase and so used to build refined category profiles, but final accuracy evaluation is carried out on all target documents as always. Plots in Fig. 3 show the results for some datasets, namely those with the best and the worst accuracy measures among those with 2 or 3 top categories of 20 Newsgroups: results for other datasets of the same groups lie between the

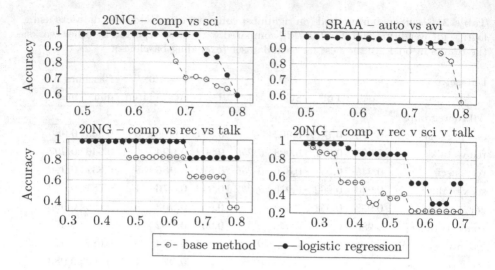

Fig. 2. Comparison of classification accuracy (Y axis) with and without application of logistic regression to similarity on different datasets (plot titles) and for different values of the threshold parameter ρ (X axis).

two. It is shown that, even if only 10 % of the target documents are known (leftmost points in the plots), the obtained accuracy would often be only few percentage points below the one obtained with all documents. This indicates that the method is fairly accurate in classifying even documents of the target domain which are not known while computing category profiles, assuming anyway that a representative enough set of them is known during the iterative phase.

Table 3. Accuracy (A, in thousandths) and number of iterations (I) for all dataset with $\text{logtf}_n \cdot \text{idf}$ as weighting, $\rho = 1/|\mathcal{C}|$ and different settings for termination.

$\beta \rightarrow$	none		0.9999		0.999		$\beta \rightarrow$	none		0.9999		0.999	
Dataset	A	I	A	I	A	I	Dataset	A	I	A	I	A	I
20 Newsgroups – 2 top-categories							20 Newsgroups – 3 and 4 top-categories						
comp vs sci	981	8	980	6	979	4	comp rec sci	934	34	885	11	878	6
rec vs talk	994	6	994	5	994	4	rec sci talk	979	11	979	7	978	6
rec vs sci	986	8	985	5	985	4	comp sci talk	972	18	958	10	955	7
sci vs talk	976	11	977	6	977	5	comp rec talk	979	6	979	4	979	3
comp vs rec	983	9	983	5	983	4	all 4 cats	970	18	961	11	958	8
comp vs talk	992	6	992	3	992	3	Reuters-21578						
SRAA							orgs places	726	7	726	7	727	5
real vs sim	950	16	950	8	951	5	orgs people	826	8	826	8	830	4
auto vs avi	964	14	966	6	969	4	people places	722	23	731	16	739	12

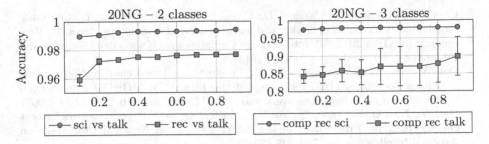

Fig. 3. Classification accuracy (Y axis) of the base cross-domain method for different datasets (data series, see legend) where only a given ratio (X axis) of documents of the target domain is known; each point is an average on 5 runs with different random selections of documents, error bars indicate standard deviation.

6 Conclusions and Future Work

We presented a simple method for cross-domain text categorization based on nearest centroid classification. Profiles for categories are initially built from the source domain, to be then refined by iteratively selecting for each category a set of target documents with high enough classification confidence and using it to build a new profile.

Experiments shown that this method, despite its conceptual simplicity and relative ease of implementation, achieves accuracy results better or comparable to many of those reported throughout the literature for common benchmark datasets. This is obtained with fixed values for the few parameters to be set and with fast running times. By measuring how much category profiles change across successive iterations, it is possible to set a tradeoff between accuracy and running times in a way which guarantees an optimal number of iterations for most datasets.

Among possible future developments of the method, we contemplate the possibility to improve the computation of similarity between document and category profiles, for example by the integration of semantic knowledge. Also, we are planning to test the efficacy of the method, other than on categorization by topic, on other important applications such as sentiment analysis, possibly applying suitable adjustments to the method.

References

1. Blitzer, J., McDonald, R., Pereira, F.: Domain adaptation with structural correspondence learning. In: Proceedings of the 2006 Conference on Empirical Methods in Natural Language Processing, pp. 120–128. Association for Computational Linguistics (2006)
2. Bollegala, D., Weir, D., Carroll, J.: Cross-domain sentiment classification using a sentiment sensitive thesaurus. IEEE Trans. Knowl. Data Eng. **25**(8), 1719–1731 (2013)

3. Cheeti, S., Stanescu, A., Caragea, D.: Cross-domain sentiment classification using an adapted naive bayes approach and features derived from syntax trees. In: Proceedings of KDIR 2013, 5th International Conference on Knowledge Discovery and Information Retrieval, pp. 169–176 (2013)
4. Dai, W., Xue, G.-R., Yang, Q., Yu, Y.: Co-clustering based classification for out-of-domain documents. In: Proceedings of the 13th ACM SIGKDD International Conference on Knowledge Discovery and Data Mining, pp. 210–219. ACM (2007)
5. Dai, W., Xue, G.-R., Yang, Q., Yu, Y.: Transferring naive bayes classifiers for text classification. In: Proceedings of the AAAI 2007, 22nd National Conference on Artificial Intelligence, pp. 540–545 (2007)
6. Hal Daumé III. Frustratingly easy domain adaptation. In: Proceedings of the 45th Annual Meeting of the Association of Computational Linguistics, pp. 256–263 (2007)
7. Domeniconi, G., Masseroli, M., Moro, G., Pinoli, M.: Discovering new gene functionalities from random perturbations of known gene ontological annotations. In: Proceedings of the 6th International Conference on Knowledge Discovery and Information Retrieval (2014)
8. Domeniconi, G., Masseroli, M., Moro, G., Pinoli, M.: Random perturbations and term weighting of gene ontology annotations for unknown gene function discovering. In: Fred, A. et al. (eds.) IC3K 2014. CCIS, vol. 553, pp. xx–yy. Springer, Heidelberg (2015)
9. Domeniconi, G., Moro, G., Pasolini, R., Sartori, C.: Cross-domain text classification through iterative refining of target categories representations. In: Proceedings of the 6th International Conference on Knowledge Discovery and Information Retrieval (2014)
10. Gao, J., Fan, W., Jiang, J., Han, J.: Knowledge transfer via multiple model local structure mapping. In: Proceedings of the 14th ACM SIGKDD International Conference on Knowledge Discovery and Data Mining, pp. 283–291. ACM (2008)
11. Hall, M., Frank, E., Holmes, G., Pfahringer, B., Reutemann, P., Witten, I.H.: The WEKA data mining software: an update. ACM SIGKDD Explor. Newsl. 11(1), 10–18 (2009)
12. Hosmer Jr., D.W., Lemeshow, S.: Applied Logistic Regression. Wiley, New York (2004)
13. Huang, J., Smola, A.J., Gretton, A., Borgwardt, K.M., Schölkopf, B.: Correcting sample selection bias by unlabeled data. Adv. Neural Inf. Process. Syst. 19, 601–608 (2007)
14. Joachims, T.: A probabilistic analysis of the Rocchio algorithm with TFIDF for text categorization. In: Proceedings of ICML 1997, 14th International Conference on Machine Learning, pp. 143–151 (1997)
15. Joachims, T.: Text categorization with support vector machines: learning with many relevant features. In: Nédellec, C., Rouveirol, C. (eds.) ECML 1998. LNCS, vol. 1398, pp. 137–142. Springer, Heidelberg (1998)
16. Lan, M., Tan, C.L., Su, J., Lu, Y.: Supervised and traditional term weighting methods for automatic text categorization. IEEE Trans. Pattern Anal. Mach. Intell. 31(4), 721–735 (2009)
17. Li, L., Jin, X., Long, M.: Topic correlation analysis for cross-domain text classification. In: Proceedings of the Twenty-Sixth AAAI Conference on Artificial Intelligence (2012)
18. Ling, X., Dai, W., Xue, G.-R., Yang, Q., Yu, Y.: Spectral domain-transfer learning. In: Proceedings of the 14th ACM SIGKDD International Conference on Knowledge Discovery and Data Mining, pp. 488–496. ACM (2008)

19. McCallum, A., Nigam, K., et al.: A comparison of event models for naive bayes text classification. In: AAAI 1998 Workshop on Learning for Text Categorization, vol. 752, pp. 41–48. Citeseer (1998)
20. Minka, T.P.: A comparison of numerical optimizers for logistic regression. http://research.microsoft.com/en-us/um/people/minka/papers/logreg/ (2003)
21. Sinno Jialin Pan and Qiang Yang: A survey on transfer learning. IEEE Trans. Knowl. Data Eng. **22**(10), 1345–1359 (2010)
22. Salton, G., Buckley, C.: Term-weighting approaches in automatic text retrieval. Inf. Process. Manag. **24**(5), 513–523 (1988)
23. Sebastiani, F.: Machine learning in automated text categorization. ACM Comput. Surv. (CSUR) **34**(1), 1–47 (2002)
24. Shimodaira, H.: Improving predictive inference under covariate shift by weighting the log-likelihood function. J. Stat. Plann. Infer. **90**(2), 227–244 (2000)
25. Sugiyama, M., Nakajima, S., Kashima, H., Von Buenau, P., Kawanabe, M.: Direct importance estimation with model selection and its application to covariate shift adaptation. Advances in Neural Information Processing Systems 2007, vol. 20, pp. 1433–1440 (2007)
26. Wang, P., Domeniconi, C., Hu, J.: Using Wikipedia for co-clustering based cross-domain text classification. In: ICDM 2008, 8th IEEE International Conference on Data Mining, pp. 1085–1090. IEEE (2008)
27. Xiang, E.W., Cao, B., Hu, D.H., Yang, Q.: Bridging domains using world wide knowledge for transfer learning. IEEE Trans. Knowl. Data Eng. **22**(6), 770–783 (2010)
28. Xue, G.-R., Dai, W., Yang, Q., Yu, Y.: Topic-bridged PLSA for cross-domain text classification. In: Proceedings of the 31st Annual International ACM SIGIR Conference on Research and Development in Information Retrieval, pp. 627–634. ACM (2008)
29. Zadrozny, B.: Learning and evaluating classifiers under sample selection bias. In: Proceedings of the 21st International Conference on Machine Learning, pp. 114. ACM (2004)
30. Zhuang, F., Luo, P., Xiong, H., He, Q., Xiong, Y., Shi, Z.: Exploiting associations between word clusters and document classes for cross-domain text categorization. Stat. Anal. Data Min. **4**(1), 100–114 (2011)

Cluster Summarization with Dense Region Detection

Elnaz Bigdeli[1]([envelope]), Mahdi Mohammadi[2], Bijan Raahemi[2],
and Stan Matwin[3,4]

[1] School of Electrical Engineering and Computer Science,
University of Ottawa, Ottawa, Canada
ebigd008@uottawa.ca
[2] Knowledge Discovery and Data Mining Lab,
Telfer School of Management, University of Ottawa, Ottawa, Canada
{mmohamm6, braahemi}@uottawa.ca
[3] Department of Computing, Dalhousie University, Halifax, Canada
stan@cs.dal.ca
[4] Polish Academy of Sciences, Warsaw, Poland

Abstract. This paper introduces a new approach to summarize clusters by finding dense regions, and representing each cluster as a Gaussian Mixture Model (GMM). The GMM summarization allows us to summarize a cluster efficiently, then regenerate the original data with high accuracy. Unlike the classical representation of a cluster using a radius and a center, the proposed approach keeps information of the shape, as well as distributions of the samples in the clusters. Considering the GMM as a parametric model (number of Gaussian mixtures in each GMM), we propose a method to find number of Gaussian mixtures automatically. Each GMM is able to summarize a cluster generated by any kind of clustering algorithms and regenerate the original data with high accuracy. Moreover, when a new sample is presented to the GMMs of clusters, a membership value is calculated for each cluster. Then, using the membership values, the new incoming sample is assigned to the closest cluster. Employing the GMMs to summarize clusters offers several advantages with regards to accuracy, detection rate, memory efficiency and time complexity. We evaluate the proposed method on a variety of datasets, both synthetic dataset and real datasets from the UCI repository. We examine the quality of the summarized clusters generated by the proposed method in terms of DUNN, DB, SD and SSD indexes, and compare them with that of the well-known ABACUS method. We also employ the proposed algorithm in anomaly detection applications, and study the performance of the proposed method in terms of false alarm and detection rates, and compare them with Negative Selection, Naïve models, and ABACUS. Furthermore, we evaluate the memory usage and processing time of the proposed algorithms with other algorithms. The results illustrate that our algorithm outperforms other well-known anomaly detection algorithms in terms of accuracy, detection rate, as well as memory usage and processing time.

© Springer International Publishing Switzerland 2015
A. Fred et al. (Eds.): IC3K 2014, CCIS 553, pp. 68–83, 2015.
DOI: 10.1007/978-3-319-25840-9_5

1   Introduction

Nowadays, a large volume of data is being generated in variety of applications throughout the Internet and computer networks. The large amount of data is mostly unlabeled and consequently cannot be modeled by supervised machine learning models like classification approaches. Therefore, the attention is drawn to clustering approaches. There are different clustering methods that can be applied on various kinds of data. Each clustering approach has its own characteristics and limitations. K-means clustering is one of the well-known and mostly used algorithms which is mainly used for clustering dataset in which the number of clusters is known. On the other side, the other clustering approaches like DBSCAN [1] and STING [2] are used for data with arbitrary shape clusters. In some applications in which the hierarchy of data is important, hierarchical clustering is a better choice [8].

Clusters represent characteristics of data. They are also employed to label new incoming samples, and consequently, they might change over time with new coming samples. However, keeping the samples of entire clusters which are growing over time is not feasible or memory efficient. Therefore, one important problem for clustering algorithm is how to represent clusters that is small enough to be preserved in memory for a long time. Most of the clustering algorithms need to keep the entire samples to represent the shape of the clusters. Moreover, in applications that the size of data is growing very fast, new coming samples have to be assigned to right cluster and the whole cluster has to be summarized. In this paper, a new summarization approach is presented to summarize clusters. The proposed summarization approach represents the clusters as a Gaussian Mixture Model (GMM). GMM is a good representative of each cluster because not only it preserves the statistical information of clusters it also creates a straightforward way to find a membership value for new coming samples. The membership value is then used to find the most similar cluster to the new sample. However, finding a proper GMM that can be a good representative of clusters is a challenging task. The proposed algorithm finds the number of GMM components and assigns a proper GMM for clusters and creates a good summary for clusters.

The structure of the paper is as follows: in Sect. 2, related works in the area of clustering and cluster summarization are presented. Section 3 presents a review of different approaches for labelling new samples in different clustering techniques. In Sect. 4 the proposed summarization approaches is introduced. This approach has three steps which are explained in more detail in Sect. 4. In Sect. 5, the analysis of algorithm is presented. Section 6 presents the experimental results to verify the performance of the proposed algorithm. Finally, Sect. 7 presents conclusions and future works.

2   Related Works

There are various algorithms available for clustering which can be categorized into four groups; partition-based, hierarchical, density-based and spectral-based clustering [3]. K-means is one of the popular algorithms in the area of partition-based clustering. K-means gets number of clusters as an input then finds the centers of clusters in iterative algorithm and then assigns points to each center based the distance of points to the center of clusters. The centers of k-means are artificial which are not parts of input data [4]. K-median is another version of k-means that finds the centers from available data [5, 6]. The center-based approaches are sensitive to noise, and ignore many interesting information such as shape and distribution of clusters. Hieratical clustering methods present data clusters in a hierarchy [7]. Hieratical clustering methods can be divisive or agglomerative [8] depending on the bottom-up or top-down strategy that is chosen. Considering clusters in hierarchy saves time in finding a proper cluster for a point. However, there are some issues related to this kind of presentation for clusters. The basic problem in this algorithm is to find a measure to divide or combine clusters and also how to preserve the cluster members and the clusters hierarchy. The main approach to create and store clusters hierarchy is similar to the idea of preserving a center and radius. Spectral-based clustering or grid-based methods divide the space of objects to cells, and then use the cells to cluster data. STING [2] and CLUIQE [9] are two well-known algorithms in this area. Based on some parameters such as distribution, density and distance of cells, the cells are merged and the clusters are created. These algorithms are able to create arbitrary shape clusters in which preserving all cluster members is the only way to present the accurate shape of the clusters. In arbitrary shape clustering, DBSCAN [1] and DENCLUE [10] are the most famous ones. In these methods, clusters are created using the concept of connecting dense regions to find arbitrary shape clusters. Density-based clustering algorithms find arbitrary-shape clusters without prior knowledge of the number of clusters. However, the major concern for these algorithms is the time and memory complexity. Based on prevalence of real time applications, there is more interest to make these algorithms fast for streaming applications [11–13].

In the mentioned arbitrary-shape clustering methods, cluster representation is the major concern. Preserving either the whole data of each cluster or centers are among the most prevalent approaches to represent the clusters. Obviously, these approaches do not reflect the cluster properties. Summarization is the solution to alleviate the complexity of clustering methods. There are different ways to summarize a cluster [14–16]. These algorithms employ the general idea behind the density based clustering methods. In the area of summarization, the idea is to detect dense regions and summarize the regions using some core points. Then, a set of proper features is considered to summarize the dense regions and their connectivity. In [14], a grid is created for each cluster and based on the idea of connecting dense regions, the core or dense cells with their connections and their related features are kept. In summarization approaches, these features play crucial role. In [14], location and range of values and status connection vector are kept that has some drawbacks. First, creating grids on each cluster is time consuming. Second, considering all grids requires significant processing time and memory space which is impractical in many cases. Cao et al. [15] use the idea of finding core points to generate the cluster summery. A major drawback of this work is that the number of core points is large and in some cases, it is equal to the number of input samples. Moreover, a fixed radius specifies the neighborhood that does not represent the distribution of objects in each cluster [15]. Chaoji et al. represent a density-based clustering algorithm named ABACUS for creating arbitrary shape clusters [16]. The summarization part in their approach is based on finding the core points and the relative variance around the objects. Most of clustering methods require two parameters; number of neighbors and a radius. In [16], the only input parameter is the number of neighbors and they estimate the radius using data distribution automatically. A drawback in their work is that the algorithm generates many core points.

In the above mentioned summarization approaches, the focus are on preserving the cluster members. They do not usually consider the classification application of clustering techniques. Clustering approaches are created to summarize data and categorized the input data into some groups, but it can also be used as a pre-processing phase of classification task [1, 10]. Each cluster has a label and for each new object, the closest cluster is found, then the object gets the label of that cluster. Anomaly detection is one of the applications of clustering in classification [17–19]. In this area, those objects which are outside of the cluster boundaries are considered as anomaly. In this paper, we present an approach to summarize clusters using Gaussian Mixture Models which is an efficient representation of a cluster, and also, can be employed for classification purpose.

3 Application of Cluster Summarization in Classification

For classification purpose, clusters are employed to label new incoming samples. Each new sample is compared to the clusters, and the closest one is chosen as a cluster that the new sample belongs to. There are different options to find out to which cluster a new sample belongs. One way to find an appropriate cluster for an object is to create a boundary around each cluster. If the new sample is inside the boundary of a cluster, then the new sample belongs to the cluster. Finding the boundary of arbitrary shape clusters, especially in high dimensional problems is complex and time consuming. Moreover, it is necessary to consider too many faces to just keep the borders of cluster created by convex in higher dimensions and the number of faces grows exponentially with dimension [20, 21]. As an example, the boundary of a cluster is represented in Fig. 1a. This figure shows that even for a simple cluster in two dimensional space, there is a need to find many vectors to separate inside and outside of the cluster.

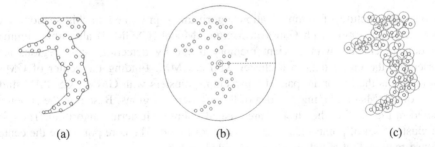

(a)   (b)   (c)

Fig. 1. Cluster representation with different approaches. (a) Internal area of cluster is separated from outside using boundary vectors. (b) K-means cluster with a center specified by red circle and radius r. (c) A cluster represented by set of circles with same radius (Color figure online).

Another approach to label a new object is to use a cluster notion. In this case, the distance of a new sample to the centre of a cluster is calculated, then, if the distance is less than the radius of a cluster (Fig. 1b), the new sample is assigned to that cluster [22, 23]. Since the boundary is inaccurate, the labelling could be faulty. Another approach to preserve the shape and specify border of a cluster is to consider a circle around each sample in a cluster. First, some points are selected as the centers, and a radius r is chosen. Then, all the samples close to the center (within a distance less than radius r) are removed from the dataset, making the center a representative of the removed samples. With these circle-based (hyper spherical-based in higher dimension) methods, centers and the radius are kept instead of the samples. To label a new sample, its distances form the centers are calculated, and if the new sample is inside the boundary of one of the centers, it is assigned to that cluster. Otherwise, it is outside of the area of the clusters, and consequently, is considered an anomaly [15]. This structure is represented in Fig. 1c, in which to preserve the accuracy, the radius clusters is small. There are some drawbacks with this type of summarization. To keep an accurate boundary, a small radius should be considered, which means many points need to be kept like what depicted in Fig. 1c. In this Figure, the boundary is well preserved, but for this, we reduced the radius such that all points in the cluster are preserved. Moreover, since we just categorize a new sample by comparing it with all circles inside each cluster, the decision is binary, i.e. there is no probabilistic value to label a new sample. To address these issues, we propose a new approach for cluster representation. The crucial point to keep in mind for cluster representation is to preserve the cluster features while keeping the generated summery as small as possible. Our proposed approach is to represent each cluster as a Gaussian Mixture Model. Further-more, unlike the above mentioned methods, in GMM-based model, there is a probabilistic representation for clusters, and consequently, the assignment (prediction) of a new sample to a cluster is calculated based on a probabilistic membership value. In the following, first some basic concepts are described and then the algorithm that finds a proper GMM for a cluster is explained.

4   Cluster Summarization with Gaussian Mixture Model (CSGMM)

The overall structure of summarization algorithm is presented in this section. The Cluster Summarization with Gaussian Mixture Model (CSGMM) algorithm summa-rizes data in clusters with efficient memory usage by detecting dense regions and representing the entire data in a cluster with a GMM. Finding the center of GMM components is the important part of representing clusters with GMMs. CSGMM finds centers of GMMs employing the idea of finding dense regions. Based on the structure depicted in Fig. 2, first the clusters are created using a clustering algorithm. Then for each cluster a set of points called *core points* are found. The core points are the center of dense regions that are shown by red circled in Fig. 2.

Core points are representative of their neighborhoods specified by circles. To summarize cluster with descriptive information each core point and its neighborhood are represented as a normal distribution and consequently the whole cluster is repre-sented as a GMM. Based on procedure depicted in Fig. 2, the CSGMM algorithm presented in this paper has three major steps: finding core points, absorption, and GMM representation. In the following sections each step is described in more details.

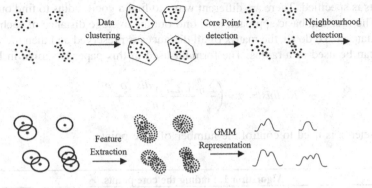

Fig. 2. Cluster summarization using CSGMM algorithm in three steps (Color figure online).

4.1 Finding Core Points

Algorithm starts with core point detection phase. In this phase a radius is considered as an input parameter for the algorithm. The radius is employed to find dense regions. Based on the input radius, we find the number of neighbors for each object to find out core points.

Definition 1. Let consider r as a radius for point o_i. $N(o_i)$ specifies the neighbors of point o_i.

$$N(o_i) = \{o_j \in D | dist(o_i, o_j) < r\}$$

Where $dist(o_i, o_j)$ is Euclidian distance of points o_i and o_j and D is dataset.

Definition 2. Core point is a point O_i that has more number of neighbors in comparison with all its neighbors O_j.

$$\{O_i \in C | O_j \in N(O_i), \|N(O_i)\| > \|N(O_j)\|\}$$

Where the operator $\|O\|$ indicates the count of samples within radius r of the sample O. There is a temporary list of possible core points. At first, all points are in this list. There is another list which has the final core points. Every point with more number of neighbors in comparison with other points is a good candidate to be a core point. Algorithm starts with a point with maximum number of neighbors and puts it on the list of core points. Since, this point is a representative of all its neighbors; we remove all of its neighbors from temporary list of core points. After removing the first core point and its neighbors, we find the next point with maximum number of neighbors in the temporary list. This point is added to the list of core points and then all its neighbors are deleted from the temporary list. We do this for the rest of points in the temporary list and find all the possible core points. The proposed method is explained in the Algorithm 1. The algorithm is recursive and it terminates when it processes all points in the temporary list. Therefore, by setting a proper radius and after all iterations, number of core points is specified. There are different ways to find a good radius to find out dense regions. The first and basic approach is to find out the average distance of each pair of points in dataset. To do so, the matrix of all distances is preserved and then the average distance can be used as a radius. The formula used in this paper is shown in Eq. 1.

$$Radius = \alpha * \left(\frac{\sum_{j=1}^{n} \sum_{i=1}^{n} dist(o_i, o_j)}{n} \right) \tag{1}$$

Parameter α is used to control the number of core points.

Algorithm 1. Finding the core points.

```
Input:    ObjectList, all objects in cluster C
Output:   CorePointList, core points of cluster C
For i=1:ClusterSize
          [O_i, Number of O_i Neighbors]=Find Neighbors(O_i)
end
ObjectLists=SortObjects (O_i,Number of O_i Memebers)
While  ObjectList<>empty
          CorePointList =FindObjectwithMaxNeighbors(ObjectList);
          ObjectList=Remove Neighbors of  O_i from  CorePointList ;
End
```

4.2 Absorption and Cluster Feature Extraction

The goal of summarization is to find a good representative of the clusters. Core points are the only objects preserved in each cluster while the rest of the objects in the cluster are removed. After finding all core points in each cluster, the next step is to define a cluster using core points. It is obvious that considering only core points, cannot represent cluster distribution. Therefore, core points have to be accompanied with a set of features related to the cluster to represent its characteristics. Accordingly, we define a set of features for each core point which are good representative for distribution around each core point.

Definition 3. (Core point Feature) (CF) Each core point is represented by a triple $CF_i = c_i, \Sigma_i, \omega_i$.

In this definition c_i is the core point and Σ_i is the covariance calculated using the core point and all objects in its neighborhood, $\omega_i = n/CS$ is the weight of core point, n is the number of objects in the neighborhood of core point c_i, and CS is cluster size. ω_i shows the proportion of objects in the neighbourhood of the core point c_i. Using the features for each core point, we estimate samples scattering around each core point without keeping the entire samples in their neighborhoods.

4.3 GMM Representation of Clusters

After finding the core points and the entire characteristic for each cluster, we generate a Gaussian Mixture Model for each cluster.

Definition 4. A Gaussian Mixture Model is a combination of a set of normal distributions. Given feature space $f \subset R^d$, a Gaussian Mixture Model $g : f \rightarrow R$ with n component is defined as:

$$g(x) = \sum_{i=1}^{n} w_i N_{\mu_i \Sigma_i}(x) \qquad N_{\mu_i \Sigma_i}(x) = \frac{1}{\sqrt{(2\pi)^d |\Sigma_i|}} e^{-\frac{1}{2}(x-\mu_i)\Sigma_i^{-1}(x-\mu_i)^T} \qquad (2)$$

Based on all $CF_i, i = 1 \cdots m$, a GMM is defined over a cluster. Each component for GMM is created using a core point, a covariance and a weight. In the formula in Eq. 2, μ_i is the centre of the i_{th} GMM component, which is set to the coordination of core point and therefore, $\mu_i = c_i$. The covariance is set to covariance of i_{th} core point covariance, that is Σ_i. The weight for each component is the weight of core point and as a result w_i in the Eq. 2 is set to the weight of core point c_i which is ω_i. To use a cluster as a class in classification problem, we need to find the closest cluster to a new incoming sample. In an arbitrary shape cluster, all objects represent a cluster and to find a right cluster for a new sample, it has to be compared with all objects in a cluster. Whereas, the GMM-based presentation finds the closest cluster by introducing the new sample to the GMM formula, which then returns a membership value.

5 Analysis of the CSGMM Algorithm

As each GMM consists of a set of Gaussian distributions, finding the number of GMM components is one of the important issues of the proposed method. In this paper, we proposed a solution to calculate number of components which is one of the major and unique innovations in the proposed method. In our approach, the number of components for GMM is set to the number of core points in a cluster. The summarization techniques should preserve the original shape and the distribution of the data. The proposed summarization of data using GMM has both characteristics. In CSGMM, not only finding the core points summarizes the clusters but also the shape of the summarized cluster still follows the shape of the original cluster. Since the CSGMM starts from dense regions in the cluster, it reduces the number of final core points which improves the performance of the proposed method in terms of memory complexity. In addition, the CSGMM is able to regenerate the original data with an acceptable accuracy. That is to say, while the CSGMM reduces the number of samples in each cluster, it is able to preserve the main characteristics of the cluster such as shape of the cluster as well as samples scattering. Based on the samples placed in neighborhood of a core point, the mean, variance and weight of a normal distribution are calculated. The normal distribution is the representative of the core point and samples scattering around the point. By combining the normal distributions a GMM is estimated which makes the proposed algorithm to preserve the shape and the sample scattering of the cluster without keeping the entire samples in memory.

In terms of time complexity, the proposed method competes with other fast and well-known algorithms such as ABACUS which has the time complexity of $O(N^2)$. In the proposed method, first we find the neighbors of each sample; with N as the number of objects it takes $O(N^2)$ to find neighbors. Then, we need to sort the objects and absorb neighbors of the core points that takes O(NlogN). In the absorbing step, we find the number of neighbors for each core point and its related variance. We consider O(NlogN) for the second part of the proposed method. As such, the final complexity of the proposed algorithm is $O(N^2 + NlogN)$ that is equal to $O(N^2)$. In the grid-based method a considerable amount of time is spent on creating the grid and further investigation to find core grids and connecting them. In ABACUS there are iterations to find core points and to reallocate them which are time consuming. In the next section, we develop a comparison between the required time of our method and the one for ABACUS. The results illustrate that the proposed method outperforms ABACUS in terms of time.

6 Experimental Results

CSGMM algorithm is a more accurate model in comparison with other models that summarize clusters. In this section, we compare the CSGMM algorithm with similar approaches to study the efficiency of the proposed method. We set up different experiments to find the efficiency of the method from different aspects. In some applications like network security, there is a need to regenerate data in periodic fashions to find out the changes that happens during the time. CSGMM is able to regenerate the original data using the related GMMs with good accuracy. The experiments in Sect. 6.1 verify the efficiency of CSGMM in data regeneration. The experiments in Sect. 6.2 are set to show the accuracy of CSGMM model in labeling new samples. In this sub-section, we focus on anomaly detection application using clustering approach. In Sect. 6.3, the second set of experiments is set to compare the complexity of CSGMM with other algorithms.

To achieve these three objectives, three sets of experiments were set up on different datasets. We consider both synthetic and the real datasets in our experiments. All experimental results are generated in Matlab running on a machine with Intel CPU 3.4 GHz and 4 GB memory. Selected UCI datasets are considered in our experiments to evaluate the accuracy of our algorithm on real datasets. These datasets and their features are presented in Table 1. The dataset are chosen in a way that different varieties of dataset with different features are included in the experiments. The Synthetic dataset is a 2-Dimensional dataset that helps us to evaluate the algorithms and visualize the results.

6.1 Clustering Goodness

To show the efficiency of the CSGMM method, we set up experiments to summarize the dataset using core points, then we use these core points and their related variances

Table 1. Dataset features.

Dataset/features	Dataset size	Number of features	Number of classes
MagicGamma	19020	10	2
Wave	5000	40	3
Shuttle	43500	9	7
Segment	2310	19	7
P2P	32767	4	2
KDD	10001	41	5
Diabet	768	8	2
CMC	1472	9	3
Synthetic	6000	2	4

and weights to regenerate the dataset. The difference between the original dataset and the regenerated one will verify the performance of the proposed summarization algorithm. The goodness of clusters for original data and the regenerated data are measured by four different measure; Dunn index [24], DB index [25], SD index [26] and purity [13]. Table 2 shows the description of these indexes.

Table 2. Clustering goodness indexes definition.

Index	Definition
Dunn	$D_{nc} = \min\limits_{i=1,...,nc}\left\{ \min\limits_{j=i+1,...,nc}\left(\dfrac{dist(c_i,c_j)}{\max\limits_{k=1,...,nc} diam(c_k)} \right) \right\}$
DB	$D_{nc} = \min\limits_{i=1,...,nc}\left\{ \min\limits_{j=i+1,...,nc}\left(\dfrac{dist(c_i,c_j)}{\max\limits_{k=1,...,nc} diam(c_k)} \right) \right\}$
DS	$D_{nc} = \min\limits_{i=1,...,nc}\left\{ \min\limits_{j=i+1,...,nc}\left(\dfrac{dist(c_i,c_j)}{\max\limits_{k=1,...,nc} diam(c_k)} \right) \right\}$
Purity	$purity = \sum\limits_{i=1}^{k} \dfrac{\|T_i\|}{\|C_i\|}, T_i = \left\{ s_i \in C_j | label(s_i) = label(C_i) \right\}$

The values in Table 3 show the differences between clustering indexes value for original and regenerated data when the dataset is summarized by CSGMM and ABACUS. The smaller these values, the more similar the original data and the regenerated one. The results in Table 3 indicate that the CSGMM algorithm regenerates the data with high accuracy in comparison with ABACUS algorithm.

6.2   Clustering Accuracy

To study the accuracy of the CSGMM method in labelling new samples, we applied it to anomaly detection. In cluster-based anomaly detection, each cluster is representative of a normal class. A new coming sample is introduced to each cluster, and if the new sample is in the boundary of cluster, it is considered a *normal* sample; otherwise it is an abnormal (*anomaly*) sample. In our approach each cluster is represented by a GMM.

Table 3. The difference of clustering goodness between original and regenerated dataset when summarized by CSGMM and ABACUS.

Dataset/Index	Dunn (CSGMM, ABACUS)	DB (CSGMM, ABACUS)	SD (CSGMM, ABACUS)	SSQ (CSGMM, ABACUS)
MagicGamma	(0.0010,0.0032)	(1.04, 1.09)	(0.48, 0.64)	(169,1235)
Wave	(0.020, 0.024)	(0.11, 0.29)	(0.62,0.83)	(125, 2243)
Shuttle	(4.6e-05, 2.74e-05)	(0.49, 0.49)	(0.32,0.45)	(179, 7999)
Segment	(0.0045,0.014)	(0.33,0.44)	(0.12,0.41)	(7790, 9314)
P2P	(6.5e-06, 0.00014)	(0.46, 0.67)	(0.55,0.63)	(207,1021)
KDD	(2.04-05, 0.0038)	(1.04,1.05)	(0.69, 1.11)	(1463,7788)
Diabet	(0.001, 0.008)	(1.65, 1.98)	(0.29, 0.31)	(135, 192)
CMC	(0.09, 0.17)	(2.93, 3.11)	(0.09, 0.17)	(116,1497)
Synthetic	(0.22, 0.33)	(1.82, 2.15)	(0.05, 0.19)	(275,1823)

When a new sample is introduced to the GMM, a membership value is calculated. If the membership value of a cluster is less than a threshold, it is considered as normal; otherwise it is abnormal. To set up this experiment in the dataset, some classes are considered as normal and the rest as abnormal. The abnormal samples are considered only in testing dataset. Then, we divide the normal classes into two parts, one of which is the training set and the rest is added to the testing set. This means that the training dataset just consists of normal samples but the testing dataset is a combination of normal and abnormal samples.

To study the accuracy of the CSGMM method, we measure the false alarm and detection rates which are two parameters for measuring accuracy in anomaly detection systems. In our experiments, we compare the CSGMM with Naïve method [15], ABACUS and Negative Selection [27]. The Naïve method classifies new samples using the idea of considering circle around each sample inside the cluster that is presented in Fig. 1c. ABACUS method is clustering algorithm and it is not for classification, However, we have used the core points of this algorithm to find out whether the core points generated by CSGMM is better than ABACUS or not. The Negative Selection is another well-known algorithm in the area of anomaly detention that we consider to compare CSGMM with that.

The ROC curve based on detection rate and false alarm rate for the KDD dataset is depicted in Fig. 3. The results show that the CSGMM method is more accurate than the other methods in terms of false alarm and detection rate. The results also show that the false alarm rate of Negative Selection is more than the CSGMM and other methods (that is the basic problem with Negative Selection method). The result for Magic-Gamma dataset (Fig. 4) confirms the accuracy of CSGMM in comparison with other algorithms. Figure 5 shows the result for shuttle dataset in which we have 6 classes and among those, 4 classes are considered as normal, and the rest is abnormal. This dataset is a proper dataset for anomaly detection purposes because it has more than two classes,

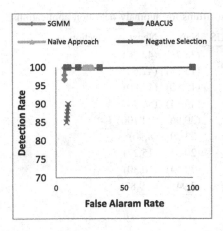

Fig. 3. False alarm and detection rates on the KDD dataset.

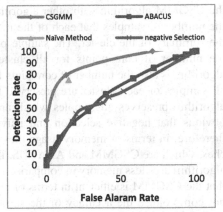

Fig. 4. False alarm and detection rates on of the MagicGamma dataset.

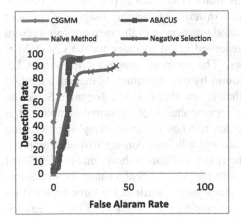

Fig. 5. False alarm and detection rates on the Shuttle dataset.

Fig. 6. False alarm and detection rates on of the Waves dataset.

part of classes can be considered as normal and the rest as anomaly. Therefore, we are able to test our algorithm in finding different types of anomalies.

The result of Fig. 6 shows that the CSGMM is better than other algorithms on the Waves dataset. In the Waves dataset the data dimension is 40, which is high. In spite of the complexity that is added to the data in higher dimensions, the CSGMM algorithm still works better than other algorithms.

Table 4 shows the results of applying CSGMM, ABACUS, Naïve, and Negative Selection algorithm on different datasets. The values of (false alarm rate, detection rate) listed in this table are the best ones recorded for each algorithm.

Table 4. (false alarm rate, detection rate) of algorithms for anomaly detection applications.

Dataset/(FA,DR)	CSGMM	ABACUS	Naïve	Negative selection
MagicGamma	(30,80)	(32,73)	(38,73)	(42,78)
Wave	(13,79)	(40,62)	(15,75)	(42,60)
Shuttle	(5,96)	(1,93)	(15,80)	(15,80)
Segment	(33,94)	(32,94)	(45,94)	(50,94)
P2P	(5,90)	(3,84)	(30,96)	(24,100)
KDD	(3,100)	(6,100)	(23,99)	(8,90)
Diabet	(31,67)	(35,53)	(20,57)	(15,27)
CMC	(16,57)	(20,40)	(20,44)	(10,30)
Synthetic	(2,94)	(1.9,87)	(2,94)	(1.9,87)

6.3 Memory Complexity

One of the important factors in measuring the performance of summarization approaches is the memory usage. A good summarization approach is the one that has an accurate summary of cluster while the summary is as small as possible. To compare the CSGMM algorithm with other algorithms in terms of memory usage, we consider the number of samples that each of the methods preserves in the memory to represent the summary of the cluster. The sample preserved in the memory for the CSGMM is the number of core points for all clusters. The memory usage for the ABACUS algorithm is also the number of core pints found by this algorithm. In the naïve method all samples of each cluster are preserved that is not efficient. The Negative Selection algorithm preserves all samples which are representative of abnormal samples. It is obvious that negative selection and naïve approaches preserve many samples, and therefore, in terms of memory usage they are not efficient. Among two other approaches, which are CSGMM and ABACUS, the result in Table 5 shows that the CSGMM algorithm uses less memory in comparison with the ABACUS algorithm. This confirms that the CSGMM is efficient in terms of memory usage, while at the same time it does not compromise the accuracy of the summarization as already shown by the experimental results on the accuracy and goodness of clustering.

Table 5. Number of core points for ABACUS and SGMM methods.

Dataset/Algorithm	CSGMM	ABACUS
MagicGamma	131	261
Wave	20	158
Shuttle	215	359
Segment	25	80
P2P	98	210
KDD	46	272
Diabet	15	32
CMC	90	221
Synthetic	18	54

Another important factor in summarization approach is the processing time of algorithm. In this experiment, we measure the processing time spent by the CSGMM and ABACUS methods to find core points. The result of Naïve and negative selection algorithms are not included in the result because, as we mentioned, they preserve all samples in the clusters, and therefore, are not memory efficient. The results in Table 6 show that the CSGMM algorithm spends less time to calculate the core points in comparison to that of the ABACUS method. A significant amount of processing time in the ABACUS algorithm is spent on repeating the algorithm to relocate the core points to find the best ones. However, the result shows that CSGMM algorithm can find better core point in only one iteration over the data.

Table 6. The time complexity for ABACUS and CGMM methods.

Dataset/Algorithm	CSGMM	ABACUS
MagicGamma	654 s	1534 s
Wave	214 s	523 s
Shuttle	1152 s	2646 s
Segment	26 s	94 s
P2P	703 s	1834 s
KDD	222 s	836 s
Diabet	7 s	38 s
CMC	15 s	54 s
Synthetic	109 s	348 s

7 Conclusions

Calculating a proper summary of a cluster that preserves the general structure and characteristics of the cluster is a challenging task. This paper introduces a new approach to summarize clusters. Unlike the classical simple representation of a cluster using a radius and a center, our approach keeps more information on the shape and sample distribution of the cluster. The approach presented in this paper summarizes clusters as a Gaussian Mixture Model. This representation preserves data distribution in the cluster. However, finding proper GMMs, and specifically the number of GMM components, is challenging task. Our proposed algorithm finds the number of GMM components automatically, then represents each cluster as a GMM to summarize clusters generated by any clustering algorithms. This allows us to regenerate the original data with high accuracy. Moreover, the computed GMM of each cluster is used to calculate a membership value for new coming samples. Using this membership values, a new sample is assigned to the right cluster with low memory usage and processing time. We applied the proposed algorithm on various datasets, both synthetic and real datasets. We measured DUNN, DB, SD, and SSQ indexes to compare the goodness of the clusters regenerated from CSGMM and ABACUS summarized data. We also measured false alarm and detection rates when CSGMM is applied in anomaly detection, and compared its performance against well-known anomaly detection algorithms such as Negative Selection and Naïve models. Finally, we compared the memory usage and processing time of our proposed algorithms against others. The experimental results demonstrated superior performance of the CSGMM algorithm in terms of accuracy, detection rates, as well as memory usage and processing time.

Acknowledgement. This research was supported by NSERC Canada, Grant Nbr RGPIN/ 341811-2012.

References

1. Ester, M., Kriegel, H., Sander, J., Xu, X.: A density-based algorithm for discovering clusters in large spatial databases with noise. In: KDD
2. Wang, W., Yang, J., Muntz, R.R.: Sting: a statistical information grid approach to spatial data mining. San Francisco (1997)
3. Han, J., Kamber, M., Pei, J.: Data Mining: Concepts and Techniques: The Morgan Kaufmann Series in Data Management Systems, 3rd edn. Morgan Kaufmann Publishers, Burlington (2006)
4. MacQueen, B.J.: Some Methods for classification and Analysis of Multivariate Observations (1967)
5. Jain, A.K., Dubes, R.C.: Algorithms for Clustering Data. Prentice-Hall, Upper Saddle River (1988)
6. Kaufman, L., Rousseeuw, J.P.: Clustering by means of Medoids, in Statistical Data Analysis Based on the L_1–Norm and Related Methods. Y. Dodge, North-Holland (1987)
7. Karypis, G., Han, H.E., Kumar, V.: CHAMELEON: a hierarchical clustering algorithm using dynamic modeling. IEEE Comput. 32(8), 68–75 (1999)
8. Kaufman, L., Rousseeuw, P.J.: Finding Groups in Data: An Introduction to Cluster Analysis. John Wiley, New York (1990)
9. Agrawal, J., Gunopulos, D., Raghavan, P.: Automatic sub-space clustering of high dimensional data for data mining applications (1998)
10. Hinneburg, A., Keim, D.A.: An efficient approach to clustering in large multimedia databases with noise (1998)
11. Guha, S., Meyerson, A., Mishra, N., Motwani, R.: Clustering data streams: theory and practice. IEEE Trans. Knowl. Data Eng. 15(3), 505–528 (2003)
12. Bifet, A., Holmes, G., Pfahringer, B.: New ensemble methods for evolving data streams (2009)
13. Aggarwal, C.C., Han, J., Wang, J., Yu, P.S.: A framework for clustering evolving data streams (2003)
14. Yang, D., Elke, A., Matthew, O.W.: Summarization and matching of density-based clusters in streaming environments. Proc. VLDB Endowment 5(2), 121–132 (2011)
15. Cao, F., Ester, M., Qian, W., Zhou, A.: Density-based clustering over an evolving data stream with noise. In: SIAM Conference on Data Mining (2006)
16. Chaoji, V., Li, W., Yildirim, H., Zaki, M.: ABACUS: mining arbitrary shaped clusters from large datasets based on backbone identification. In: SIAM/Omnipress (2011)
17. He, Z., Xu, X., Deng, S.: Discovering cluster-based local outliers. Pattern Recogn. Lett. 24, 1641–1650 (2003)
18. Gaddam, S., Phoha, V., Balagani, K.: K-means+ID3: a novel method for supervised anomaly detection by cascading k-means clustering and ID3 decision tree learning methods. IEEE Trans. Knowl. Data Eng. 19(3), 345–354 (2007)

19. Mohammadi, M., Akbari, A., Raahemi, B., Nasersharif, B., Asgharian, H.: A fast anomaly detection system using probabilistic artificial immune algorithm capable of learning new attacks. Evol. Intel. **6**(3), 135–156 (2014)
20. Kersting, K., Wahabzada, M., Thurau, C., Bauckhage, C.: Hierarchical convex NMF for clustering massive data (2010)
21. Hershberger, J., Shrivastava, N., Suri, S.: Summarizing spatial data streams using ClusterHulls. J. Exp. Algorithmics (JEA) **13** (2009). doi:10.1145/1412228.1412238
22. Mohammadi, M., Akbari, A., Raahemi, B., Nasersharif, B., Asgharian, H.: A fast anomaly detection system using probabilistic artificial immune algorithm capable of learning new attacks. Evol. Intel. **6**(5), 135–156 (2014)
23. Gaddam, S., Phoha, V., Balagani, K.: K-means+ID3: a novel method for supervised anomaly detection by cascading k-means clustering and ID3 decision tree learning methods. IEEE Trans. Knowl. Data Eng. **19**(3), 345–354 (2007)
24. Dunn, J.C.: Well separated clusters and optimal fuzzy partitions. Cybernetics **4**, 95–104 (1997)
25. Davies, L.D., Bouldin, W.D.: A cluster separation measure. IEEE Trans. Pattern Anal. Mach. Intell. **1**(4), 224–227 (1979)
26. Halkidi, M., Vazirgiannis, M., Batistakis, Y.: Quality scheme assessment in the clustering process. In: Zighed, D.A., Komorowski, J., Żytkow, J.M. (eds.) PKDD 2000. LNCS (LNAI), vol. 1910, pp. 265–276. Springer, Heidelberg (2000)
27. Sande, P.C., Monroe, J.G.: Negative selection of immature b cells by receptor editing or deletion is determined by site of antigen encounter. Immunity **10**(3), 289–299 (1999)

Content-Based News Recommendation: Comparison of Time-Based System and Keyphrase-Based System

Servet Tasci and Ilyas Cicekli[✉]

Department of Computer Engineering, Hacettepe University, Ankara, Turkey
{servettasci,ilyas}@cs.hacettepe.edu.tr

Abstract. As internet resources are increasing at an unprecedented speed, users are tired of searching important ones among them. So, what users need and where they can find them are getting more important. Users require a personalized support in sifting through large amounts of available information according to their interests and recommendation systems try to answer this need. In this context, it is crucial to offer user friendly tools that facilitate faster and more accurate access to articles in digital newspapers. In this paper, a content-based news recommendation system for news domain is presented and contents of news articles are represented by words appearing in news articles or their keyphrases. News articles are recommended according to user dynamic and static profiles. User dynamic profiles reflect user past interests and recent interests play much bigger roles in the selection of recommendations. Our recommendation system is a complete content-based recommendation system together with categorization, summarization and news collection modules.

Keywords: Recommendation systems · Content-based filtering · Keyphrase-based recommendation · Text summarization · Text classification

1 Introduction

The abundance of information with their dynamic contents is available on the web [2]. As there are many documents and resources, it is difficult to find what we need and where we can find them. Besides, websites and web pages are doubled every year. People are overwhelmed by the large amount of information on the web, therefore information overload problem is worsening at an unprecedented speed. As a consequence, user modelling and personalized information access are becoming crucial. Users require a personalized support in sifting through large amounts of available information, according to their interests.

Many search engines have emerged to alleviate this problem. These search engines use very large databases to index websites, and reduce information overload problem by allowing users a centralized search. But still users are tired of looking which documents are useful. Research results have shown that users often give up their search in the first try, examining no more than ten documents. This shows us that users want from a web site recommendations rather than search results.

© Springer International Publishing Switzerland 2015
A. Fred et al. (Eds.): IC3K 2014, CCIS 553, pp. 84–98, 2015.
DOI: 10.1007/978-3-319-25840-9_6

When visiting a news website, users are looking for new information that they have not known before and that information will be an interest for them. The kind of information that will be an interest for a user depends on that user's past activities. User profiles are created in order to recommend related items to users. Since user profiles are inferred from past user activities, it is important to know how effective they would be to predict user behavior.

The problem of recommending items has been studied extensively, and two main paradigms have emerged. Content-based recommendation [15, 19, 20] try to recommend items similar to those that a given user has liked in the past, whereas collaborative recommendation paradigm identify users preferences are similar to those of the given user and recommend items that they have liked [4]. These two approaches have their own strengths and weaknesses. Hybrid recommender systems [7] combine these approaches. Li and Li [21] use hypergraph learning for news recommendation.

In this paper, we describe a content-based recommender system for news domain. A content-based recommender system consists of mainly three steps: *Content analyzer, profile learner* and *filtering component*. *Content-analyzer* extracts features from unstructured text to produce a structured item representation. *Profile learner* is the heart of a recommendation system and it uses user feedbacks which are usually inferred as explicit or implicit feedbacks. Explicit feedbacks are results of explicit evaluations of items by users and they indicate whether items are interesting or not to users. Implicit feedbacks try to extract user interests from user click behaviors. In our recommendation system, we use explicit feedbacks and five-level ratings for user feedbacks in evaluation. *Filtering component* implements some strategies to rank interesting items with respect to user profiles and generates recommendations.

News domains have some special characteristics [30] and recommendation systems have to deal with these characteristics. Since a news domain contains a large volume of information and documents are in unstructured format, recommendation systems require more computation power. News items typically have short self-lives and recommendations should be up-to-date. Since most news articles describe specific events, recommended articles should be related with interested events. The user interest can easily shift from one event to another event, and selection and ranking of news articles to be recommended should be done according to this fact.

In this paper, we present a content-based recommendation system for news domain named as *HaberAnalizi*. The presented recommender system uses dynamic time-based user profiles that are automatically updated. Recently read news articles play more important roles in the creation of user profiles. The recommender system described here is a complete web site that collects news articles from various online Turkish news websites and recommends articles to its users. Everyday approximately 1500 news articles are collected. Then, these news articles are classified into eight categories: *Magazine, Economics, Politics, Culture-Art, Techno-Science, Health, Sports and Education*. After that, news articles are recommended to users according to their dynamic and static profiles. An accurate profile of a user is critical for the success of news recommendation system. So how we construct user profiles plays an important role in the success of the system.

The rest of the paper is organized as follows. Section 2 presents the related work in news recommendation domain. Section 3 explains the details of our news recommendation system. Section 4 presents the performance evaluation of the system and Sect. 5 concludes the paper and gives some possible future work.

2 Related Work

News recommendation has been studied for years and some methods have been proposed and evaluated. News-dude [6] is a news recommendation system uses TF-IDF values combining K-Nearest Neighbor (KNN) algorithm supporting a series of feedback options such as "interesting", "not interesting", etc. Most of the collaborative filtering systems use KNN for recommendation [8]. Likewise in our news recommendation system, we use five categories to assess recommendations.

Liang and Lai [22] propose a time-based approach to build user profiles from browsing. They calculated the user's elapsed time while reading an article and also they took into account the article's length and positions of the words in the document. In contrast to our recommendation system, they do not use news classification. They also calculate a time range for document reading and this is not proper because reader's reading speed can vary. They can have the *first-rater problem* since no static profiling is used.

Tan and Teo [31] present a personalized news recommendation system named PIN. In this system, users choose a list of keywords to be used in the creation of their profiles and the system recommends news according to these profiles. Since user profiles are not updated, it is hard to make it dynamic, and user interest changes cannot be reflected in profiles.

Liu et al. [24] present that user current interests are critical for the success of a recommendation system. Their system is also a content-based recommendation system and they also do not use collaborative filtering. In their system, they construct user profiles automatically on any interaction that user makes with the system and this puts extra burden on their system. We update user profiles daily not on each user interaction. Their system infers user interests based on user click behavior on the website. No rating or negative votes are used for privacy reasons. In addition to the click distribution of individual users, they calculated the click distributions of the general public. When we compare to this method with our approach, we give more importance to ratings of last seven day clicks of individual user in the creation of user dynamic profiles and these profiles are updated daily. Public trend is not important for our system. Likewise their system and our system do not take into account of time spent on the page.

Gong [14] uses collaborative filtering based on user clustering and item clustering. First of all, they divide users into clusters using a clustering algorithm based on some similarity threshold and a user-item matrix is created. Then item clusters are created based on similarity measures. In order to solve sparsity problem, they use expensive Pearson similarity measure [25] for new items that user has not rated yet. News items are recommended using KNN algorithm and the created user-item matrix. In our approach we categorize news and for each category we use news items that user has rated last week or user static profiles if the user has not rated any news article.

Saranya and Sadhasivam [30] collect categorized news from news agencies and dump them into a database. They use only one news agency articles for categorization of articles and this means that an article can be categorized differently depending on which corpus is used. In our system we use all news articles for categorization. Likewise, they also use both static and dynamic profiles. Dynamic user profiles are constructed during every interactive search session. They use user clicks to update dynamic profiles every time user reads a news article. This can be costly to implement and users do not want to wait for new news items on online systems.

Li et al. [23] especially try to address issues as news selection, news representation, news processing, and user profiling. Their system consists of three major components which are news article clustering, user profile construction and news item recommendation. News items are clustered using hierarchical clustering. In order to build user profiles, they use three dimensions: news content, similar access patterns and preferred news entities. User profiles and news articles are represented using triplets in the form of <Topic vector, patterns, entities>. For recommendation, similarity between news item and user profile is computed. They assessed their system in 15 days. First 5 days 50 users read news and then successive 10 days they recommended news items to the volunteers. In our system we collected the data approximately for about 5 months.

3   Content-Based News Recommendation System

In this section we explain the details of our time-based news recommendation system. Our recommendation system is a complete system which is designed as a website which collects news articles from online Turkish newspapers and recommends to its users. Although our recommendation system domain is online Turkish news domain, it can be adaptable to other news domains because the only specific resource for Turkish language is a Turkish stemmer. Other parts of the system are not language dependent.

The major parts of our content-based news recommendation system are as follows:

- Getting News from News Sites
- News Classification
- Summarization of News
- Profile Construction
- Time-Based News Recommendation System
- Keyphrase-Based News Recommendation System

First of all, obtaining news articles from substantial news sites is explained, and then how categories of news items are found are explained. Our recommendation system can also produce summaries of news articles using a text summarization system [26]. Later we explain the creation of static and dynamic user profiles. Finally, we explain our news recommendation approach at the end of this section.

3.1   Getting News from News Sites

Obtaining news contents from online news sites is not a trivial job. Because news sites publish news articles as parts of HTML pages and we have to obtain contents of news articles. An HTML page contains advertorial part, content part, links part and other parts, etc. We need just news content of the page. In order get the content of a news article, we use XML Path Language *(XPath)* expressions. Each news site has its own characteristics and contents of news articles are extracted using news sites characteristics and XPath expressions. After the content part of a news article is extracted, it is feed into the rest of the recommendation system.

Approximately from twenty online news sites we obtain approximately 1500 news articles every day. Lengths of news articles vary from a couple of sentences to longer documents. Contents of news articles are extracted and they are prepared for the rest of the system in order to be recommended to users.

3.2 News Classification

Most of news sites do not classify news articles. Even if they classify, classifications of news articles are so different at each news site. For example while a news item's category is politics in a web site, the other web site could categorize it as economics. So we have to classify them for standardization and obtaining better recommendations to users. After news sites are analyzed, we have determined eight categories for classification: Magazine, Economy, Politics, Culture-Art, Techno-Science, Health, Sports and Education.

Most of classification algorithms are based on the single term analyses of the text, e.g. vector space model [29]. On this context every document can be thought as a vector. Each word in the document can be thought as an attribute of the document.

Some main online news sites categorize daily news, and we used 40,000 articles with known categories from these sites as our corpus for categorization. Using these categorized news articles, we determined subtle words computing their frequencies. A news article and each category can be shown as a vector of pairs of words and their normalized frequencies. The normalized frequency of a word is computed by dividing its frequency in document with the total number of words in that document. After that, we have tested different similarity measures for classification with our corpus of 40,000 news articles with known categories. The details of our news article corpus are given in Fig. 1. The classification methods that are tested with our corpus are Support Vector Machines

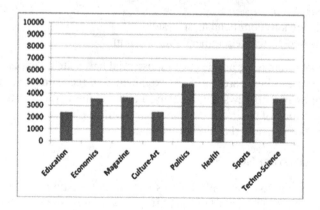

Fig. 1. News article corpus for classification.

(SVM), Euclidean Similarity, Cosine Similarity and Jaccard Similarity. Precision results of these methods on our corpus are given in Fig. 2 and it can be seen that Jaccard Method gives better results than other methods for our corpus. So we have decided to use Jaccard method for the classification of news articles in our system. We tried to eliminate the over-fitting problem using as much as many words as feature values.

	Jaccard	Cosine	Euclidean	SVM
Education	0,65	0,52	0,40	0,60
Economy	0,75	0,47	0,35	0,70
Magazine	0,65	0,50	0,42	0,60
Culture-Art	0,70	0,45	0,40	0,70
Politics	0,67	0,40	0,30	0,72
Health	0,82	0,45	0,60	0,75
Sports	0,80	0,70	0,65	0,70
Techno-Science	0,75	0,55	0,35	0,70
TOTAL	0,75	0,53	0,48	0,70

Fig. 2. Accuracy results of classification algorithms.

For classification we use stop word elimination and stemming. We remove approximately 150 stop words from our news articles. We also stem words using a Turkish stemmer named Zemberek [1].

3.3 Summarization of News Articles

News articles generally consist of few topics. These topics form main contents of news articles and some of these topics are explained deeply. A good summary of an article should cover major topics in that article. Summaries can have different forms [16] and they are commonly obtained by sentence extraction or abstraction methods [17]. Although abstraction methods can provide sophisticated summaries and they are hard to implement. On the other hand, sentence extraction approaches are easy to adapt and they can still produce useful summaries.

Since users want to read more news in short time, we use a summarizer in our recommender system to accomplish this need. One of the goals of our system is to provide users opportunity to reduce the information overload likewise multi-document summarization system [28]. We adapted a sentence extraction based summarization system [26] in order to generate summaries of news articles. We generate summaries whose lengths 30 % of original news articles using this summarization system.

3.4 Profile Construction

A personalized recommendation system needs user characteristics to recommend items. For that reason, the construction of user profiles is a crucial part of a recommendation system. In order to construct individual user profiles, the required information can be collected *explicitly* through direct user intervention, or *implicitly* through agents that monitor user activities. Profiles that can be modified or augmented are considered as *dynamic*, in contrast to *static* profiles that maintain the same information over time. Most recommender systems use both static and dynamic profiles. Static profiles are used to solve cold-start problem in recommender systems. If keywords are given for categories in a static profile, recommendations are based on these keywords and this solves some parts of cold-start problem. If no keywords are given in a static profile, we recommend most recommended news to that user in order to solve cold-start problem. As users read more news articles, their dynamic profiles improve. In our system, we use both dynamic and static profiles for recommendation. A comprehensive survey of various user profile construction techniques is provided in [13].

A static or dynamic profile consists of a set of probabilities for categories and a list of interest words for each category. Summation of all probability values in a profile is equal to 1 and the probability value of a category indicates the percentage of user's interest in that category. The probability value of a category determines the amount of news articles that will be recommended to that user. The interest words of a category are a set of pairs of words together with their normalized weights. The weight of an interest word is the normalized frequency of that word and reflects the amount of user interest to articles containing that word. The interest words of a category are used to select news articles in that category for recommendation to the user. A sample profile will be as follows:

```
Profile = {
  Magazine       : 0.20, MagazineWords,
  Economy          : 0.15, EconomyWords,
  Politics       : 0.25, PoliticsWords,
  Culture-Art     : 0,{},
  Techno-Science : 0.12, ScienceWords,
  Health         : 0.08, HealthWords,
  Sports         : 0.16, SportsWords,
  Education        : 0.04, EducationWords
}
```

In our recommender system, users fill profile forms and give interest words for categories. This information is used in the creation of a static profile. From given information, a static profile is created. The probability value of each category in a static profile is the same and this means that same amount of articles is recommended for each category if a static profile is used for recommendation. The interest words of a category are the words given for that category when the user fills a profile form. If the user does not give any interest words for a category, no articles in that category are recommended to that user when that static profile is used for recommendation.

Dynamic profiles are constructed according to news articles read by users. For each user, two dynamic profiles are created if that user reads news articles regularly. The first dynamic profile is the last week dynamic profile and it is constructed if the user has read news article during last seven days. The second dynamic profile is the last month dynamic profile and it is created according to news articles read by that user during last 30 days.

If a user has rated news items in last seven days the last week dynamic profile is created using the news items that are rated as *interesting* by the user in the last seven days, and news article recommendations are done according to this last week dynamic profile. If a last week dynamic profile is not constructed for a user, the last month dynamic profile of that user is used for recommendation if that profile is available. If no news articles are read by a user in the last month, no dynamic profiles will be available for that user and the static profile for that user is used for recommendation.

Dynamic profiles of a user are updated once every day. Recently read news articles are more likely to indicate user's current interests. For that reason, news articles read in the last days have more weights in profile constructions. Thus, the effect values of words appearing in recent day articles are multiplied with bigger weights than words in older day articles. Weights used in the construction of dynamic profiles decrease gradually towards past. These weight values are determined using news reading trends of users.

Category probability values are updated according to the articles read in that category. Before category interest words are updated with an *interested* article, stop words are eliminated from article words and article words are stemmed. Interest words in the profile are updated with these stemmed words. Frequency values of article words are used when updating weights of interest words in dynamic profiles.

3.5   Content-Based News Recommendation System

A typical recommendation system consists of an object set and a user set [33]. The aim of a recommendation system is to meet the requirements of users using this object set. In our recommendation systems, the object set is the set of daily news articles. Furthermore, a recommendation system should be able to suggest objects to users, which users would not discover for themselves [34] because of huge amount of objects. Our recommendation system tries to suggest news articles to users from a huge set of daily news articles.

Our recommendation system collects daily news articles from online newspaper sites. Everyday approximately 1500 news articles are collected from these newspaper sites. Our recommendation system recommends N daily articles to a user depending on that user's profiles where N is set to 50 in our evaluation sessions. Categories of daily collected news articles are computed by the categorization module of our system.

In order to recommend news articles to a user, first a profile of the user is determined for recommendation according to the user's history. If the user read and rated news articles in last seven days, the last week dynamic profile is used for recommendation. If the user has not read any article in the last week, the last month dynamic profile is used. If no dynamic profile available for the user, the static profile is used.

After a profile of a user is selected, news recommendations are done with respect to that profile. For each category Ci, the number (N_{Ci}) of news articles that will be recommended for that category is determined by multiplying the probability value of that category with the total number (N) of articles to be recommended. Then articles in category Ci of daily news articles are compared with the category interest words in the profile in order to measure similarities of articles with the category interest words. The most N_{Ci} similar articles are recommended to the user for the category.

The similarity of a news article with interest words is measured using the cosine similarity measure. Stop words are eliminated from article words and article words are stemmed before they are compared with interest words. Title words of an article have more weights in the computation of the similarity measure than regular article words.

3.6 News Recommendation System with Keyphrases

Users tend to read news articles that have similar contents. News article suggestions can be done using this idea. The usage of dynamic profiles and words appearing in previously read news articles in our content-based recommendation system tries to capture this fact indirectly.

Keyphrases of an article describes the semantic content of that article and what that article talks about. Since keyphrases of articles describe contents of those articles, content similarity of articles can be measured by comparing their keyphrases. Using this idea, we created a version of our recommendation system which employs keyphrases in recommendation.

A keyphrase can be a single word or a sequence of words which semantically describes the content of a text. In fact, a set of keyphrases for a text can be seen as a summary of that text. In addition to summarization, keyphrases can be used for indexing documents in order to be used in precise searches of those documents [3]. Unfortunately, most of digital documents are not associated with keyphrases and most of news articles do not have keyphrases. So, keyphrases of texts should be extracted using keyphrase extraction algorithms [10, 11]. Many keyphrase algorithms [18, 27] for Turkish texts are given in literature, and we use a Turkish keyphrase extraction algorithm [27] which is based on ideas in KEA keyphrase extraction algorithm [32] in order to extract keyphrases of Turkish news articles.

In order to create a good recommendation system based on keyphrases, some major steps should be done. A version of our recommendation system which is based on keyphrases employs following major steps.

- Extracting keyphrases of news articles.
- Finding similar news articles using keyphrases and similarity metrics.
- Recommending news articles to users using similarity metrics between daily news articles and news articles read before by users.

Since our recommendation system is designed for Turkish news domain, we use a Turkish keyphrase extraction algorithm [27] to extract keyphrases of Turkish news articles. This extraction algorithm returns a list of keyphrases which are sorted according to their importance for a given Turkish text. The top N keyphrases of a news article are used to represent that article.

We use keyphrases of news articles to measure similarities among them. In order to decide how many keyphrases should be used to represent news articles and which similarity metric gives the best result to measure similarity among news articles, we created a small test set. The test set contains two sets of articles and similarity algorithms find similar news articles in the second set for each article in the first set. Of course, articles in the second set are marked whether similar to any article in the first set. A set of different similarity metrics like *Cosine, Jaccard, Euclidean and Pearson Coefficient* are tried to determine similar news articles using their keyphrases. Figure 3 shows accuracy results of similarity algorithms when different numbers of keyphrases are used for articles. Cosine similarity has better results than other similarity metrics according to Fig. 3. According to the results in Fig. 3, when ten keyphrases are used for news articles, the best scores are obtained experimentally. As a result of experimental evaluation, cosine similarity metric and top ten keyphrases for news articles are decided to be used in our system

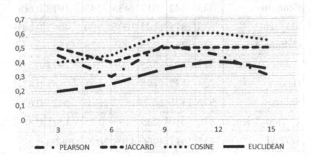

Fig. 3. Accuracy results of similarity algorithms for news article sets using different number of keyphrases.

In order to recommend daily news articles to users, similarities of daily articles with articles recently read by users are determined by the selected similarity algorithm using article keyphrases. Top similar daily articles are recommended to users. Comparison of our main recommendation system with our keyphrase based recommendation system is given in the evaluation section.

4 Evaluation

Evaluation of a recommendation system is important for determining the efficiency of the system There are many evaluation methods in order to evaluate a recommendation system such as precision and recall values, MAE (mean absolute error) value, RMSE (root mean squared error) value and ROC (receiver operating characteristic) analysis. For example, Cinematch system in the Netflix prize competition evaluates itself using RMSE [5]. For our recommendation system, we use precision, recall, f-measure and accuracy values for evaluation, and we also give the ROC analysis of our results.

In order to verify the efficiency of our proposed recommendation system, we tested it for five months on 30 users with approximately 1500 news items per day. Daily 30–50 news articles are recommended to each user. If a user has a dynamic profile, always 50 articles are recommended. After reading recommended news, users are asked to evaluate recommended news by clicking one of five rankings: *not interested, less interested, no comment, interested, mostly interested*. Users are assumed that they are interested with an article if they select one of three choices (*no comment, interested, mostly interested*) since *no-comment* choice in our evaluations reflects some interest in articles.

	Not Interested	Less Interested	No Comment	Interested	Mostly Interested	Number of Recommended Articles	Accuracy Results
Education	210	267	380	193	65	1115	0.57
Economy	343	442	613	435	145	1978	0.60
Magazine	638	815	1103	1006	500	4062	0.64
Culture-Art	345	267	225	295	125	1257	0.51
Politics	575	703	585	897	215	2975	0.57
Health	223	137	145	115	65	685	0.47
Sports	975	1370	1670	1572	820	6407	0.63
Techno-Science	336	610	245	195	135	1521	0.38
TOTAL	3645	4611	4966	4708	2070	20000	0.59

Fig. 4. Evaluation results for rated articles.

Then we randomly selected 20 days and 20 users who rated all of 50 recommended articles daily. Thus we collected 20,000 rated articles by users. All recommendations are made according their dynamic profiles. For these 20,000 rated articles, ratings and distributions over categories are given in Fig. 4. The average accuracy result for all categories is 0.59 and Magazine category has the highest accuracy result. According to Fig. 4, some interesting results can be inferred for each category. Users mostly interested news items in Sports, Magazine and Politics categories.

For any recommendation system it is hard to determine the number of items to be recommended to users. If the number of recommended items is high, the recommended items may contain too many unrelated items and users will be unhappy with this result. On the other hand, if fewer items are recommended, users may not be satisfied from this

Number of Recommended Articles	Precision	Recall	F-Measure	Accuracy
10	0.60	0.21	0.31	0.46
20	0.70	0.48	0.57	0.58
30	0.73	0.76	0.75	0.70
40	0.63	0.86	0.72	0.62
50	0.58	1.00	0.73	0.58

Fig. 5. Evaluation of news recommendation system and the computation of optimum news count.

result either. For this reason, we decided to compute optimum news count for recommendation in our system.

In order to compute optimum news count, we made a detailed analysis on the results given in Fig. 5. According to the accuracy result (0.59) in Fig. 5, users approximately found 29 articles to be *interesting* out of 50 recommended articles on the average. Recommended 50 articles were sorted according their similarity values and they were presented users in the decreasing order. We checked the number of articles found to be *interesting* in top 10, 20, 30, 40 and 50 recommended articles, and *Precision, Recall and F-Measure* results are presented in Fig. 5. According the results in Fig. 5, best precision and f-measure results are obtained in top 30 recommended articles and this indicates that optimum news count for our system is around 30. This means that users approximately find 22 articles interesting out of 30 recommended articles. Since the precision results for top 10 and 20 are not high enough, our similarity measure can be improved in order to move *interesting* articles into higher ranks.

As seen in the ROC curve graph representing *recall* against *fallout* in Fig. 6, recall value increasing rapidly until news count value reaches 30 then it increases slowly until value reaches 50. This means that 30 is the optimum news count for our system. ROC curves reflect precision/recall optimization for systems [9, 12].

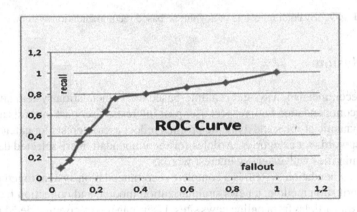

Fig. 6. ROC curve for optimum news count.

4.1   Evaluation of Recommendation System with Keyphrases

In order to compare the results of our main recommendation system with the results of the recommendation system based on keyphrases, 50 articles that are recommended to users during the evaluation of main recommendation system and evaluated by users are re-sorted using the recommendation system with keyphrases. With this evaluation process, our main objective is to see whether keyphrase based recommendation can move interested articles upward direction in recommendation lists or not. The results of this evaluation process are given in Fig. 7. When the results in Figs. 5 and 7 are compared, it can be seen that more interested articles are moved into top 10 and 20 recommended article list. According precision results in Figs. 5 and 7, 6 of top 10 recommended articles are found as interesting articles by users, 8 of top 10 recommended articles could have been found as interesting if keyphrase based system is used. According to the results, top 30 and 40 article lists are not improved by keyphrase based recommendation system. Although this evaluation process for keyphrase based recommendation system cannot be seen as a complete evaluation process, it reflects that the usage of keyphrases can improve the performance of the recommendation system.

Number of Recommended Articles	Precision	Recall	F-Measure	Accuracy
10	0.80	0.28	0.41	0.54
20	0.75	0.52	0.61	0.62
30	0.73	0.76	0.75	0.70
40	0.63	0.86	0.72	0.62
50	0.58	1.00	0.73	0.58

Fig. 7. Evaluation results of keyphrase based recommendation system.

5   Conclusion

Our news recommendation system is a time-based recommendation system and selects items for recommendation with respect user past interests. User past interests are represented by dynamic profiles and dynamic profiles reflect user interests for categories and user interest words for categories. Articles for recommendation are selected depending on their similarities with category interest words.

Our recommendation system is a complete recommendation system together with a news categorization module, a news summarization module and collection module for collecting news articles from online news sites. Users can read recommended daily news articles or they can browse all daily news in each category.

Users can rate news articles by explicitly giving a score between 1 and 5. User interest on article can be determined by an explicit evaluation by user same as in our system or user clicks together with amount of time spent to read that article. We selected the first approach because of its simplicity and clear rating mechanism. In bigger systems, user may not want to explicitly evaluate recommended articles, so the second approach may more appropriate for collection user interests.

The performance of a recommendation system can be affected by the performance of similarity measurement between articles and user profiles. Different similarity methods can be used in order to improve performance. Keyphrases are important for news recommendation systems and keyphrases can be used to measure similarity between documents. Our keyphrase based recommendation system improved performance of our main recommendation system.

Our news recommendation system focuses on selecting most interesting news articles that are similar news articles read by users before. However, readers can be burdened by more or less identical ones among returned news articles. Detection of news articles with identical contents can be useful to assist users in further reading and helping readers skip similar articles.

Acknowledgements. This paper is a revised and extended version of our KDIR-2014 paper whose title is "A Media Tracking and News Recommendation System".

References

1. Akın, A.A., Akın, M.D.: Zemberek, an open source NLP framework for Turkic languages (2007). http://zemberek.googlecode.com
2. Asanov, D.: Algorithms and Methods in Recommender Systems. Berlin Institute of Technology (2011). http://www.snet.tuberlin
3. Aygul, I., Cicekli, N., Ilyas Cicekli, I.: Searching documents with semantically related keyphrases. In: The Sixth International Conference on Advances in Semantic Processing (SEMAPRO 2012), Barcelona, Spain (2012)
4. Balabanovic, M., Shoham, Y.: Fab: content-based collaborative recommendation. Commun. ACM **40**(3), 66–72 (1997)
5. Bennett, J., Lanning, S.: The netflix prize. In: KKDD Cup and Workshop (2007)
6. Billsus, D., Pazzani, M.: A hybrid user model for news story classification. In: The Seventh International Conference on User Modeling, Banff, Canada, pp. 99–108 (1999)
7. Cantador, I., Bellogín, A., Castells, P.: Ontology-based personalised and context-aware recommendations of news items. In: The 2008 IEEE/WIC/ACM International Conference on Web Intelligence and Intelligent Agent Technology (2008)
8. O'Conner, M., Herlocker, J.: Clustering items for collaborative filtering. In: The ACM SIGIR Workshop on Recommender Systems, Berkeley, CA (1999)
9. Davis, J., Goadrich, M.: The relationship between precision recall and ROC curves. In: The 23rd International Conference on Machine Learning (ICML) (2006)
10. Ercan, G., Cicekli, I.: Keyphrase extraction through query performance prediction. J. Inf. Sci. **38**(5), 476–488 (2012)
11. Ercan, G., Cicekli, I.: Using lexical chains for keyword extraction. Inf. Process. Manag. **43**(6), 1705–1714 (2007)
12. Fisher, M.J., Fieldsend, J.E., Everson R.M.: Precision and recall optimisation for information access tasks. In: First Workshop on ROC Analysis in AL, European Conference on Artificial Intelligence (ECAI 2004), Valencia, Spain (2004)
13. Gauch, S., Speretta, M., Chandramouli, A., Micarelli, A.: User profiles for personalized information access. In: Brusilovsky, P., Kobsa, A., Nejdl, W. (eds.) Adaptive Web 2007. LNCS, vol. 4321, pp. 54–89. Springer, Heidelberg (2007)
14. Gong, S.: A collaborative filtering recommendation algorithm based on user clustering and item clustering. J. Softw. **5**(7), 745–752 (2010)
15. Goossen, F., IJntema, W., Frasincar, F., Hogenboom, F., Kaymak. U.: News personalization using the CF-IDF semantic recommender. In: The International Conference on Web Intelligence, Mining and Semantics (WIMS) (2011)
16. Hahn, U., Mani, I.: The challenges of automatic summarization. Computer **33**, 29–36 (2000)

17. Hovy, E., Lin, C.-Y.: Automated text summarization in SUMMARIST. In: Mani, I., Maybury, M.T. (eds.) Advances in Automatic Text Summarization, pp. 81–94. The MIT Press, Cambridge (1999)

18. Kalaycılar, F., Cicekli, I.: TurKeyX: Turkish keyphrase extractor. In: The 23rd International Symposium on Computer and Information Sciences (ISCIS 2008), Istanbul, Turkey (2008)

19. Kompan, M., Bieliková, M.: Content-based news recommendation. In: Buccafurri, F., Semeraro, G. (eds.) EC-Web 2010. LNBIP, vol. 61, pp. 61–72. Springer, Heidelberg (2010)

20. Li, L., Chu, W., Langford, J., Schapire, R.E.: Contextual-bandit approach to personalized news article recommendation. In: The 19th International Conference on World Wide Web, pp. 661–670 (2010)

21. Li, L., Li, T.: News recommendation via hypergraph learning: encapsulation of user behavior and news content. In: The Sixth ACM International Conference on Web Search and Data Mining, pp. 305–314 (2013)

22. Liang, T.-P., Lai, H-J.: Discovering user interests from web browser behavior: an application to internet news services. In: The 35th Annual Hawai'i International Conference on Systems Sciences. IEEE Computer Society Press (2002)

23. Li, L., Wang, D., Li, T., Knox, D., Padmanabhan, B.: SCENE: a scalable two-stage personalized news recommendation system. In: The 34th Annual International ACM SIGIR Conference on Research and Development in Information Retrieval, Beijing, China, pp. 124–134 (2011)

24. Liu, J., Dolan, P., Pedersen, E.R.: Personalized news recommendation based on click behavior. In: The 14th International Conference on Intelligent User Interfaces, Hong Kong, China (2010)

25. McLaughlin, M.R., Herlocker, J.L.: A collaborative filtering algorithm and evaluation metric that accurately model the user experience. In: The 27th Annual International ACM SIGIR Conference on Research and Development in Information Retrieval, Sheffield, United Kingdom (2004)

26. Ozsoy, M.G., Cicekli, I., Alpaslan, F.N.: Text summarization using latent semantic analysis. J. Inf. Sci. **37**(4), 405–417 (2011)

27. Pala, N., Cicekli, I.: Turkish keyphrase extraction using KEA. In: The 22nd International Symposium on Computer and Information Sciences (ISCIS 2007), Ankara, Turkey (2007)

28. Radev, D.R., Fan, W., Zhang, Z.: Web in essence: a personalized web-based multi-document summarization and recommendation system. In: The NAACL-01, pp. 79–88 (2001)

29. Salton, G., McGill, M.J.: Introduction to Modern Information Retrieval. McGraw-Hill Inc., New York (1986)

30. Saranya, K.G., Sadhasivam, G.S.: A personalized online news recommendation system. Int. J. Comput. Appl. **57**, 6–14 (2012)

31. Tan, A.-H., Toe, C.: Learning user profiles for personalized information dissemination. In: The 1998 IEEE International Joint Conference on Neural Networks, Alaska, pp. 183–188 (1998)

32. Witten, I.H., Pynter, G.W., Frank, E., Gutwin C., Nevill-Manning, C.G.: KEA: practical automatic keyphrase extraction. In: The Fourth ACM Conference on Digital Libraries, pp. 254–256 (1999)

33. Zhou, T., Ren, J., Medo, M., Zhang, Y.: Bipartite network projection and personal recommendation. Phys. Rev. E **76**, 046115 (2007)

34. Zhou, T., Zoltan, K., Liu, J., Medo, M., Wakeling, J.R., Zhang, Y.: Solving the apparent diversity-accuracy dilemma of recommender systems. Proc. Natl. Acad. Sci. U.S.A. **107**(10), 4511–4515 (2010). The National Academy of Sciences

Techniques for Processing LSI Queries Incorporating Phrases

Roger Bradford[✉]

Agilex Technologies Inc., Chantilly, VA, USA
r.bradford@agilex.com

Abstract. Latent semantic indexing (LSI) is a well-established technique that provides broad capabilities for information search, categorization, clustering, and discovery. However, there are some limitations that are encountered in using the technique. One such limitation is that the classical implementation of LSI does not provide a flexible mechanism for dealing with phrases. In the standard implementation of LSI, the only way that a phrase can be used as a whole in a query is if that phrase has been identified a priori and treated as a unit during the process of creating the LSI index. This requirement has greatly hindered the use of phrases in LSI applications. This paper presents a method for dealing with phrases in LSI-based information systems on an ad hoc basis – at query time, without requiring any prior knowledge of the phrases of interest. The approach is fast enough to be used during real-time query execution.

Keywords: Latent semantic indexing · LSI · Phrase-based retrieval · Phrase indexing · Phrase retrieval

1 Introduction

In 1988, researchers from Bellcore introduced the technique of latent semantic indexing (LSI) for text retrieval [1]. The technique relies on the notion of distributional semantics – specifically, that the meaning of a term in text is directly correlated with the contexts in which it appears. LSI accepts as input a collection of documents and produces as output a high-dimensional vector space. All of the documents in the collection are represented by vectors in this vector space. Similarly, all of the terms that comprise those documents are represented by vectors in this vector space (except for very frequently occurring terms that typically are treated as stopwords).

LSI employs the technique of singular value decomposition (SVD) to carry out large-scale dimensionality reduction (as described in the following section). This dimensionality reduction has two key effects:

- Terms that are semantically related are assigned representation vectors that lie close together in the LSI vector space.
- Documents that have similar conceptual content are assigned representation vectors that lie close together in the space.

© Springer International Publishing Switzerland 2015
A. Fred et al. (Eds.): IC3K 2014, CCIS 553, pp. 99–117, 2015.
DOI: 10.1007/978-3-319-25840-9_7

These characteristics of LSI spaces form the basis for a wide range of applications, ranging from automated essay scoring to literature-based discovery [2]. Typically, in these applications, little or no use is made of phrases. This is because, in the classical formulation of LSI, in order for a phrase to be used as such in a query, that phrase must be identified and treated as a unit in the initial stage of creating the LSI representation space. However, experience has shown that it is quite difficult to predetermine collections of phrases that will enhance performance in applications. This paper presents a technique for using phrases in LSI queries on an ad hoc basis – at query time, without requiring any changes in existing LSI spaces.

Although LSI primarily has been used with text, it is a completely general technique and can be applied to any collection of items composed of features. LSI has, for example, been used with great success in categorizing, clustering, and retrieving audio, image, and video data. Although this paper focuses on textual phrases, the approach described here is equally applicable to non-textual linked features in other types of media.

2   LSI Processing

The LSI technique applied to a collection of documents consists of the following primary steps [3]:

1. A matrix A is formed, wherein each row corresponds to a term that appears in the documents, and each column corresponds to a document. Each element $a_{m,n}$ in the matrix corresponds to the number of times that the term m occurs in document n.
2. Local and global term weighting is applied to the entries in the term-document matrix. This weighting may be applied in order to achieve multiple objectives, including compensating for differing lengths of documents and improving the ability to distinguish among documents. Some very common words such as *and*, *the*, etc. typically are deleted entirely (i.e., treated as stopwords).
3. Singular value decomposition (SVD) is used to reduce this matrix to a product of three matrices:

$$A = U \Sigma V^{T} \tag{1}$$

Let A be composed of t rows corresponding to terms and d columns corresponding to documents. U is then a t*t orthogonal matrix having the left singular vectors of A as columns. V is a d*d orthogonal matrix having the right singular vectors of A as columns. Σ is a t*d diagonal matrix whose elements are the singular values of A (the non-negative square roots of the eigenvalues of AA^{T}).

4. Dimensionality is reduced by deleting all but the k largest values of Σ, together with the corresponding columns in U and V, yielding an approximation of A:

$$A_k = U_k \Sigma_k V_k^{T} \tag{2}$$

which is the best rank-k approximation to A in a least-squares sense.

5. This truncation process provides the basis for generating a k-dimensional vector space. Both terms and documents are represented by k-dimensional vectors in this vector space.

6. New documents (e.g., queries) and new terms are represented in the space by a process known as folding-in [1]. To add a new document, for example, that document is first subjected to the same pre-processing steps (e.g., stopword removal) as those applied to the original documents used in creating the space. The document then is assigned a representation vector that is the weighted average of the representation vectors for the terms of which it is composed. A similar process is employed to fold in new terms (see Sect. 5.2 for more detail).

7. The similarity of any two objects represented in the space is reflected by the proximity of their representation vectors, generally using a cosine measure. Results of queries are sorted by cosine: the higher the cosine, the more similar the returned object (term or document) is to the query.

Extensive experimentation has shown that proximity of objects in such a space is an effective surrogate for conceptual similarity in many applications [4].

3   Word n-Grams

In creating an LSI representation space, the objects treated as terms do not have to be individual words. In natural language, there are many combinations of two or more words that act as a unit. Named entities are a particularly important example of this, and are discussed in the next section. All other word n-grams of interest will be referred to here using the generic term *phrase*.

3.1   Named Entities

Named entities are "atomic units" in text that belong to specific categories, such as persons, locations, or organizations. For nearly a decade now, treating multi-word named entities as individual features has been a well-established procedure in the creation of LSI spaces [5]. Typically, PERSON, LOCATION, and ORGANIZATION names are identified using entity extraction software. In some cases, other entities also are extracted, such as telephone numbers or equipment designators. The identified named entities then are marked up so that each is treated as a unit in the creation of the term-document matrix used to produce the LSI index. For example, the person name *John Brown* might be marked up as *p_John_Brown_p*. In the creation of a term-document matrix from such marked-up text, the LSI software will treat this contiguous character string as a single term.

Preprocessing of named entities has two salutary effects for LSI applications. The first, and most pervasive effect, is that processing multi-word named entities as single terms significantly improves the representational fidelity of the resulting LSI space. Without such extraction, LSI will, for example, conflate all occurrences of the term *John* in a collection of documents. In even a moderately large text collection, this could imply incorrect contextual linkages in thousands of instances. Typical text collections in large LSI applications will contain references to millions of named entities. The great majority of these entity names will consist of multi-word expressions. If these names are not treated as logical units, the LSI processing will, in aggregate, incorporate millions of incorrectly imputed associations.

The LSI technique has been demonstrated to have a remarkable degree of resiliency to many different types of noise [6–8]. However, it is not immune to the effects of the millions of errors that can occur if tokens of named entities are not treated appropriately. The effect is strongest when relations of importance in the LSI space depend heavily on contextual associations among entities (e.g., people, locations, and organizations). Such relations are often of key importance in applications, and this fact has led to routine use of entity extraction as a pre-processing step in creating LSI spaces for analytic applications.

The second beneficial effect of entity preprocessing is that it allows users to formulate more effective LSI queries. For example, when entity preprocessing has been employed, a user query containing the name *John Brown* will retrieve documents related to that specific individual. In a classical LSI implementation, such a query will retrieve documents that may be related to anyone named *John*, and to any objects that happen to be *brown*. Similarly, in many relatively recent document collections, an LSI search for *George Cheney* (the internationally known speaker and writer on organizational communication) would yield poor associations, because the many references to *George Bush* and *Dick Cheney* would create representation vectors for *George* and *Cheney* that were closely associated with politics as opposed to the educational context of *George Cheney*. This can be a major limitation in discovery applications where names of people provide key context.

In addition, entity preprocessing allows users to ask questions in a manner more closely aligned with their worldview. Marking up different types of entities in a distinctive manner allows filtering of results in ways that have proven to be highly useful. For example, an LSI space created in the classical manner only allows users to employ queries of the form: *What **terms** are most closely related to a given set of **terms**?* With entity preprocessing and corresponding results filtering, it is possible for a user to employ queries of the form: *What **people** are most closely associated with a given **organization**?*

The performance of contemporary entity extraction software leaves much to be desired. Many contemporary LSI applications encompass millions to tens of millions of documents. Processing such collections using current entity extraction software can yield millions of entity extraction errors. Nonetheless, experience has shown that, in most cases, it is better to extract the entities than to not process entities at all.

The remainder of this paper will focus on phrases, where the situation is much more complicated. However, it should be noted that the ad hoc phrase processing technique described in Sect. 5 is completely applicable to named entities. There are several ways it can be used to compensate for entity extraction errors in LSI applications.

3.2 Phrases

People directly perceive phrases as having utility in representing the information content of documents. For example, Fig. 1 shows the distribution of the number of tokens constituting indexing elements chosen by professional indexers for a set of technical journal articles in INSPEC [9]. Less than 14 % of the indexing elements chosen consisted of single tokens.

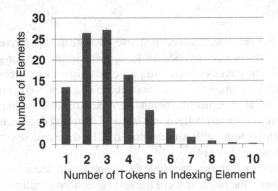

Fig. 1. Human choice of phrases as indexing elements.

Phrase search is widely used in information retrieval. In [10] it is reported that as many as 10 % of web queries are explicit phrase queries (i.e., entered within quotes) and many more are implicit phrase queries entered without double quotes. Use of phrases in queries is a standard capability in most contemporary text retrieval systems (the great majority of which are based on Boolean keyword retrieval).

With the classical formulation of LSI, simply entering the words of a phrase in a query typically will yield poor results. If, for example, *life cycle analysis* is used as a query, the system will retrieve the LSI representation vectors for each of these individual words and then form a query vector that is the weighted average of these three vectors. Since the words *life*, *cycle*, and *analysis* are used in many contexts other than that indicated by the phrase *life cycle analysis*, the query vector thus formed will constitute a poor approximation of an appropriate vector for representing the specific concept of *life cycle analysis*.

It has long been recognized that treating phrases as units could have beneficial effects in LSI applications. In one of the earliest papers on LSI, the inventors of the technique noted the potential value of use of phrases:

*"We think we can greatly improve performance by **incorporating short phrases**..."* [3] (emphasis added).

Subsequent papers noted the specific potential value of incorporation of phrases into LSI in order to improve precision:

*"Although we typically use only single terms to describe documents, **phrases could also be included** in the matrix...there are many variations in weighting, number of dimensions, **use of phrases**, etc. that we need to examine...A lack of specificity appears to be the main reason for false alarms (highly ranked but irrelevant documents). This is not surprising because LSI was designed as a recall-enhancing method, and we have not added precision-enhancing tools"* [11] (emphasis added).

4   Related Work

There is a long history of investigation of the utility of phrases in information retrieval. Broadly treating phrases in a document collection as indexing elements has yielded generally disappointing results.

4.1 Phrases in IR

As early as 1975, Salton et al. showed that use of statistically-derived phrases could improve retrieval performance for SMART, a then-current vector-space-based text retrieval system [12]. The authors obtained improvements in average precision of 17–39 %, but were dealing with very small text collections (each less than 500 documents). In 1989, Fagan carried out similar experiments, obtaining mixed results, varying from −11 % to +20 %, with somewhat larger test collections (up to 12 thousand documents) [13]. In 1997, Mitra conducted similar tests with a collection of 211 thousand documents, obtaining precision improvements of 1 % [14].

There are two primary factors responsible for the observed decline over time of phrase impact in these studies – the larger sizes of the test collections employed and the fact that the comparison baseline (retrieval performance using individual terms) improved significantly over that 20-year time frame. In discussing results from the first five TREC conferences, Mitra et al. noted that: *"In the past five years of TREC, overall retrieval effectiveness for SMART has more than doubled, but the added effectiveness due to statistical phrases has gone down from 7 % to less than 1 %"* [14].

In 1999, Turpin and Moffat provided additional evidence that, as performance of vector-space-based information retrieval systems employing individual terms improved, the utility of adding phrases diminished [15].

Summarizing the first eight years of TREC testing, Harman noted: *"the investigation of the use of phrases in addition to single terms ... has long been a topic for research in the information retrieval community, with generally unsuccessful results. ... almost all* (TREC) *groups have experimented with phrases. In general, these experiments have been equally unsuccessful"* [16].

In 2006, Metzler, Strohman, and Croft reported results of their work over three years on the TREC Terabyte Tracks. During that period they performed a variety of tests using statistically-derived phrases. None of the techniques tried yielded improvement over a bag-of-words baseline [17].

Multiple approaches have been tried for automatically identifying phrases of interest within a document collection. In general, these produce far more candidate phrases than is desirable. A number of approaches have been tried for pruning long lists of candidate phrases. Somewhat surprisingly, applying filtering criteria such as mutual information and entropy generally has not resulted in lists of phrases that yield significant retrieval enhancement.

There has been some success with selecting specific types of phrases. Ogawa et al. employed two-term noun phrases in tests using TREC-8 queries applied to the WT2 g document collection (247 thousand documents) and TREC-9 queries applied to the WT10 g document collection (1.7 million documents). They achieved improvements in average precision of 3–12 % using phrases derived from the TREC queries. For phrases derived from query expansion, they achieved improvements for the smaller test set, but not for the larger one [18]. Zhai et al. achieved similar results for noun phrases with TREC-5 queries [19]. Kraaij and Pohlmann found modest improvements using proper names as indexing units [20]. Jiang et al. had good success with classifier-thing bigrams (word pairs in which the first word effectively selects for a subclass of the type referred to by the second word) [21].

Some techniques for creating lists of phrases derive the candidates from existing data collections. In particular, WordNet and Wikipedia have been popular sources of candidate phrases. However, these approaches typically have not shown a significant advantage over other techniques.

Counterintuitively, human selection of phrases also generally has not produced significant improvements in retrieval performance. In 2010, Broschart et al. conducted experiments using the TREC GOV2 collection (25 million documents), three sets of TREC queries, and five users. For all five users, incorporation of their chosen phrases as indexing units yielded overall *lower* average precision than the baseline comparison (BM25F). They found that users frequently disagreed on phrases in the TREC queries. On average, two users highlighted the same phrase only 47 % of the time [22]. Such inconsistencies also were noted by Kim and Chan. They had ten human subjects read articles averaging 1300 words in length and choose the ten most meaningful phrases from each one. On average, only 1.3 phrases matched those chosen by others [23].

4.2   LSI Testing Using Phrases

There has been limited reported testing of phrase pre-processing for LSI. In 2000, Lizza and Sartoretto incorporated two-word phrases into LSI indexes for small document collections. They reported a 9 % improvement in average precision for the TIME collection (423 documents), but no useful increase for the MEDLINE (1033 documents) and CISI (1460 documents) collections [24].

Weimer-Hastings investigated the effect of incorporating noun and verb phrases when using LSI to grade student answers to questions. He employed manually identified phrases in comparing student answers with expected answers. He found a decrease in correlation with human evaluations when the phrases were used [25].

Wu and Gunopulos tested LSI with two-word phrases, selected based on a threshold for either document frequency or information gain. In their testing, they employed the R118 subset of the Reuters 21578 test set. They only were able to increase F1 for that test set from .8417 to .8449, which was not statistically significant [26].

Table 1. LSI phrase indexing results – synonym test.

Indexed terms	% Correct
Individual words	59.55%
Individual words + two-word phrases	61.59%
Individual words + two-word phrases + three-word phrases	62.73%

Nakov et al. examined the impact of different text pre-processing steps (lemmatization, stopword removal, etc.) in using LSI for document categorization. They employed a test set of 702 Bulgarian-language documents assigned to 15 categories. Altogether they tested 120 combinations of local weight, global weight, and LSI dimensions. In 87 of these cases, incorporating phrases into the LSI index yielded no change in microaveraged categorization accuracy for the collection. In 27 cases there was an improvement; for 9 cases the results were worse. Most improvements were 1 % or less. In all cases where the best-performing local and global weights were used, phrase processing had no effect [27].

Grönqvist examined LSI performance on a synonym test of 440 Swedish queries, with 10.9 % of the queries and 35.5 % of the answers consisting of phrases. Results from creating the LSI index with and without phrases are shown in Table 1.

He noted that the difference between the first case and the second was not statistically significant, but that the difference between the first and the third was [28].

Overall, incorporation of two- and three-word phrases [29]:

- Corrected 30 queries that had been incorrectly evaluated by classical LSI.
- Incorrectly judged 15 queries that had been correctly evaluated by classical LSI.

In his thesis, Grönqvist also performed document retrieval tests with 101 CLEF topics for Swedish and 133 TREC topics for English. For Swedish, there was a small decrease in performance when incorporating phrases. For English, the results were essentially unchanged from the individual word case [30].

Olney used LSI for determining the semantic equivalence of sentence pairs. He employed two test sets - the Microsoft Research Paraphrase Corpus and a corpus based on the novel 20,000 Leagues under the Sea. The training set consisted of the TASA corpus (~70 MB of text) combined with the Wall Street Journal section of the Penn Treebank (~50,000 words). In creating the term-document matrix he included all word bigrams that occurred more than once. This increased the number of rows in the matrix by an order of magnitude (from 75,640 to 756,741). He found a negligible difference in task performance for both test sets. A variant in which word trigrams were used as context, rather than documents, also yielded no significant difference [31].

The results of these tests with LSI are summarized in Table 2.

Table 2. Summary of LSI phrase testing.

Year	Authors	# of Docs	Δ in Performance
2000	Lizza & Sartoretto	423	+ 9%
2000	Lizza & Sartoretto	1033	≈ 0
		1460	≈ 0
2000	Weimer-Hastings	192	-18%
2002	*Wu and Gunopulos*	9603	+ .4%
2003	*Nakov, Valchanova, and Angelova*	702	≤ 1%
2005	Grönqvist	440	+ 5%
2006	Grönqvist	29K	≈ 0
2009	Olney	≈ 45K	≈ 0

Collectively, these studies suggest that pre-selection of phrases for LSI indexing has limited utility, at least using current phrase selection techniques. This result is consistent with the previously-cited studies of use of phrases with other vector-space-based information retrieval techniques.

4.3 Updated LSI Phrase Testing

In order to provide more up-to-date information regarding LSI and phrases, a test was carried out that applied contemporary resources – NLTK and WordNet. Figure 2 shows the results of applying two approaches to phrase identification and two approaches to markup in LSI preprocessing, using these resources. The document collection in this testing consisted of 50,000 news articles.

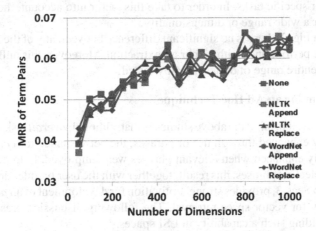

Fig. 2. Phrase preprocessing technique comparison.

The test procedure employed a simple measure of the representational fidelity of an LSI space. It is based on the fact that one of the fundamental operations in an LSI space is to compare a term with another term. The degree of relatedness of the two terms is based on LSI's integrated analysis of the contexts in which those terms occur. For an LSI space to be useful, it must indicate a high degree of association between pairs of terms that have a close association in the real world. For example, there should be a high degree of indicated association between the terms *Paris* and *France*. The test used here employs a set of 250 term pairs, each pair having a close real-world association. The chosen pairs are examples of multiple types of associations (country-capital pairs, common synonyms, person-country pairs, person-occupation pairs, words and their plural forms, nouns and their adjectival forms, noun-verb pairs, verb declensions, words and their contractions, and contextually related terms (e.g., astronomy-telescope, volcano-lava).

In the testing, for each term pair of interest, one term was treated as a query in the LSI space. Vectors for other terms were ranked in relationship to this query, based on the cosine measure between those vectors and the query vector. The mean reciprocal rank (MRR) of the paired terms in these result lists was used as the measure of relatedness between the terms. MRR was used to allow comparisons across different collections and across LSI spaces with differing values of dimensionality. (Cosine values, per se, could not be used as a measure for comparing term pairs, because cosine values are not directly comparable across collections).

Previous experience has shown that this type of comparison provides a useful quick indication of the fidelity with which an LSI space reflects real-world associations [32].

A large list of phrases was generated using the Natural Language Tookit (NLTK) [33] to process the news articles. A smaller list was created using WordNet as a source of candidate phrases. In each case, two preprocessing approaches were tried – replacing the individual terms with the phrase and leaving the original terms, simply appending the phrase. The dimensionality chosen for an LSI space can have a major effect on performance for specific tasks. In order to take this factor into account, the testing was carried out over a wide range of dimensionality.

As shown in Fig. 2, there is no significant difference between any of the results when compared to the performance with no phrase extraction. Moreover, this null result holds true across the entire range of dimensionality tested.

4.4   Motivation for an Ad Hoc Technique

One consistent result in all of the above studies is that, although *overall* task performance was not notably enhanced through use of phrases, the performance for *certain queries* was significantly enhanced when relevant phrases were employed. In light of the difficulties in pre-selecting phrases, this result, together with the user predilection to employ phrases noted in Sect. 3, provides strong motivation for development of an *ad hoc* phrase query capability for vector space retrieval. The following discussion presents a novel method of providing such a capability in LSI spaces.

5   Ad Hoc Phrase Processing Method

The fundamental characteristics of an LSI space enable a query-time phrase processing capability for LSI. The key relevant characteristic of the LSI technique is the duality condition described below.

5.1   Duality Condition

In any LSI space, the following duality condition holds:

- The LSI representation vector for any *document* corresponds to the weighted average of the LSI vectors for the *terms* contained in that document.
- The LSI representation vector for any *term* corresponds to the weighted average of the LSI vectors for the *documents* that contain that term.

5.2   Phrase Vector Creation Method

For a given LSI space, the proposed procedure for generating a representation vector for an arbitrary given phrase is as follows:

1. Identify the documents from the indexed collection that contain the given phrase.
2. Retrieve the LSI vectors corresponding to those documents.

3. Calculate the weighted sum of those representation vectors. (Using the notation of Sect. 2, let P be the 1*d vector whose components indicate frequency of occurrence of the phrase in the documents of the collection, weighted by the same global and local weighting algorithms as used in creating the original LSI index. The new phrase vector is then given by $PV_k\Sigma_k^{-1}$).

The resulting vector will be a good approximation of the vector that *would have been created* if that phrase had been treated as a unit during the LSI pre-processing stage. It is not *exactly* the same vector, because of the complex balance that is created between term and document vectors when the SVD algorithm is applied to the collection as a whole during the creation of the LSI space. Changing what are considered to be terms within a set of documents has transitive impacts that affect the entire LSI space created from those documents. These changes will create differences in retrieval results. However, as shown in the following examples, the approximation vector produced using this method is quite accurate, at least for an LSI space of reasonable size.

In order for this technique to be applied at query time, the process of creating the approximation vector for a phrase must be quite fast (on the order of seconds). Thus all three steps described above must be carried out rapidly:

- For a given phrase, it is necessary to rapidly determine which documents contain that phrase. This necessitates use of an inverted index of term occurrences. This easily can be implemented using a variety of open source or commercial software packages. With modern hardware and software implementations, providing response times on the order of seconds for this function is straightforward, even for large document collections. Moreover, modern LSI applications frequently incorporate a Boolean retrieval capability for added flexibility. In such cases, the needed inverted index will already be available. For example, the LSI engine employed in the testing described here comes bundled with the DT Search commercial Boolean text retrieval system. Many extant LSI applications employ the Lucene open source text retrieval engine for Boolean retrieval. Such software can retrieve postings lists for terms (and combine them into occurrence lists for phrases) with sub-second response, even for large document collections.
- Once the set of documents containing the phrase has been identified, retrieving the corresponding LSI vectors (and their term weights) is a straightforward database lookup process, using the same database used for other LSI retrieval functions.
- Combining the vectors can be carried out rapidly in local memory.

In contemporary hardware environments it is quite feasible for this series of operations to be carried out in near-real time (seconds) while processing a user query.

6   Test Results

The most straightforward method of demonstrating the efficacy of an ad hoc technique such as this is to directly compare the results of example LSI queries involving phrases for three cases:

1. Where there is no attempt at specific phrase processing (classical LSI processing, where the terms of the phrase are treated independently).
2. Where the phrase is treated as a unit during pre-processing of the text of the document collection (so that the phrase is treated as a single feature in creating the term-document matrix).
3. Where the phrase is processed at query time using the procedure described above.

The following examples were generated using LSI spaces constructed from a collection of 1.6 million news articles from the time frame 2012–2013. This document set was a convenient size for experimentation, although at the lower end of collection size for contemporary enterprise LSI applications, most of which tend to range from millions to tens of millions of documents [34]. The LSI indexes were created using the following processing parameters: logarithmic weighting locally, entropy weighting globally, 300 dimensions, and pruning of terms that only occurred once. (These are typical parameters for contemporary LSI applications).

A decision was made to carry out entity extraction for PERSON, LOCATION, and ORGANIZATION entities as a pre-processing step for the baseline LSI case. The ad hoc query technique described here is eminently suitable for dealing with named entities. If these entities had not been included in the baseline case, quite striking examples could have been presented here for ad hoc retrieval of person, organization, and location names. However, it has become routine practice in recent years for entity extraction to be applied as a pre-processing step in creating LSI indexes for applications. (Named entities are one exception to the problem of pre-determining multiword units. Even with the error rates of contemporary entity extractors, incorporating entity extraction into LSI pre-processing has been shown to have significant beneficial effect on LSI performance in most applications). Since entity extraction is so widely used, it was felt that the most meaningful baseline for comparisons for this paper would be an LSI index that included entity pre-processing.

The entity extraction was carried out using a commercial entity extraction product (Rosoka version 3.0). The LSI indexing was carried out using a commercial LSI engine (Content Analyst version 3.10). Index creation time was approximately 45 min employing 16 m1.xlarge Amazon AWS EC2 instances for creation of the term-document matrix and one cr1.8xlarge instance for the SVD calculation.

In each of the following examples, three sets of results are presented for each phrase query:

- The first set of results was generated using a standard LSI query against a classical LSI index, with no specific phrase processing.
- The second set of results was produced by taking the given phrase and building a new LSI space, identical to that of the previous case, with the single exception of treating that one phrase as a unit during the LSI processing.
- The third set of results was obtained from the first LSI space by employing the ad hoc phrase retrieval technique described in this paper.

In each of the following three examples, the tables show the ten terms in the LSI space that were ranked closest to the query vector corresponding to the given phrase (i.e., those terms having the highest cosine value between their representation vectors and that of the query). For the news articles used in the testing, there are many similar terms. Requiring exact matches within the top ten terms out of the 1.5 million total terms in the collection is a rather stringent measure of performance.

Only the top ten terms are shown here, due to space limitations. However, examination of the top 100 terms demonstrated comparable degrees of similarity in each case. In general, the higher the indicated overlap for the top terms in each example, the greater the similarity would be for the results of any query related to or containing the given phrase.

6.1 Example 1

For this example, the query phrase was *gross national product*. Table 3 shows the closest ten term results for this phrase used as a query, for each of the three phrase processing variants.

Table 3. Top ten term results for query = *gross national product*.

	PHRASE PROCESSING		
	NONE	PRE-PROCESSED	AD HOC
1.	product	lithuanian_ministry_of_finance	gdp
2.	national	expenditures	gross
3.	gross	estonian_statistics__office	expenditures
4.	projected	icelandic_treasury	lithuanian_ministry_of__finance
5.	flajs	unipolarization	dependent
6.	andrej_flajs	spkef	dependency
7.	forecasts	economy	gdps
8.	projections	economic	estonian_statistics_office
9.	grow	pajula	icelandic_treasury
10.	slower	asfinag	expenditure

For the case where there was no phrase processing, the top three (closest) terms correspond to the individual terms in the phrase. This is classical LSI - treating the phrase as a straightforward combination of the meanings of its individual terms. Terms five and six refer to Andrej Flajs, who was the lead author on a study of gross national income which is frequently referenced in the articles of the collection. Terms four and seven through ten are all generic terms that occur frequently in articles in the context of discussion of gross national product (GNP).

For the case where the phrase was treated as a unit in pre-processing, the top ten results are quite different. Three of the top four terms are names of organizations. During the time frame of the articles, there was extensive news coverage of the European financial crisis. These three organizations were frequently mentioned in news articles related to GNP changes in that time frame. SpKef was the largest savings bank in Iceland, which failed. Hardo Pajula is an economic analyst who wrote extensively on the Euro

Zone crisis in this time period. ASFiNAG is a corporation that was a significant contributor to high Austrian debt in this time frame.

It is interesting that there is *no* overlap among the top ten terms for these two cases. The classical implementation of the LSI query yields terms that are generically related to a discussion of gross national product. However, as indicated, identifying those three words as constituting an important phrase, and treating it as a unit in the LSI pre-processing, yields much more specific results. The results focus on key entities central to discussions of GNP in these specific documents.

For the ad hoc phrase processing approach described in this paper, four of the top ten terms overlap with the top ten terms retrieved in the full phrase pre-processing instance. In fact, these four overlap terms are the top four terms retrieved in the pre-processed case. The acronym GDP, standing for gross domestic product, is a closely related term.

6.2   Example 2

For this example, the query phrase was *rare earth element*. (Rare earth elements are members of a subgroup of the periodic table of elements, plus scandium and yttrium). Table 4 shows the closest ten terms for this phrase for each of the three phrase processing variants.

Table 4. Top ten term results for query = *rare earth element*.

	PHRASE PROCESSING		
	NONE	PRE-PROCESSED	AD HOC
1.	earth	praseodymium	dysprosium
2.	earths	dysprosium	rare
3.	rare	rhodia	bastnasite
4.	planetary	association_of_china_rare_earth	praseodymium
5.	planets	molycorp	molycorp
6.	comets	scandium	superfund
7.	asteroids	nechalacho	molycorp_inc
8.	jpl	molycorp_inc	association_of_china__rare_earth
9.	hi_tech_co	jia_yinsong	nechalacho
10.	asteroid	bastnasite	su_bo

For the case of no phrase processing, it is clear that the results are driven by the term *earth*. The terms *rare* and *element* are used in multiple contexts, but, in this collection, *earth* is discussed almost always in the context of being a planet. Accordingly, seven of the top ten terms in this case have the context of *celestial bodies*. JPL is an acronym for Jet Propulsion Laboratory, a leading US organization for space exploration.

As in the first example, there is *no* overlap between the top ten results for the case of no phrase processing and the results for the pre-processed case. In the pre-processed case, the top two terms and the sixth term are names of rare earth elements. *Rhodia* is the name of a leading rare earth production company. The fourth term in the listing is the *Association of China Rare Earth Industry*, from which the entity extractor has dropped the word *industry*. In the time frame of these news articles, there was extensive discussion of Chinese mining and use of rare earth elements. *Molycorp* is a mining company that is a major supplier of rare earth ores. *Nechalacho* is the site of a major mining activity for rare earth ores. *Jia Yinsong* is chief of the Rare Earth Office of the Ministry of Industry and Information Technology in China. Bastnasite is the most abundant rare earth element mineral.

In this example, the contexts of the two result lists for the no-phrase-processing and the phrase-pre-processing instances are completely different. Whereas in Example 1, a user might have been willing to work with the more general results given by classical LSI for *gross national product*; in this case, the results with no phrase processing are completely unrelated to the intent of the query *rare earth element*.

In this example, the ad hoc approach has performed very well. Seven of the top ten terms from the phrase-pre-processing instance occur among the top ten terms for the ad hoc case. This includes the top two terms from the pre-processed case. Su Bo is Vice Minister of Industry in China. The news articles contain excerpts from a number of speeches that he made about rare earth elements in this time frame.

In this case, the phrase is highly non-compositional; i.e., its meaning is not a simple combination of the meanings of its constituent words. (Actually it is not a simple combination of the dominant senses of the constituent terms in this collection). The technique described here has its greatest impact for such phrases. For this type of query, use of the technique described here would have a significant beneficial impact on user satisfaction.

6.3   Example 3

For this example, the query phrase was *highly enriched uranium*. Table 5 shows the closest ten terms for this phrase for each of the three phrase processing variants.

Table 5. Top ten term results for query = *highly enriched uranium*.

		PHRASE PROCESSING	
	NONE	PRE-PROCESSED	AD HOC
1.	uranium	heu	plutonium
2.	enriched	plutonium	heu
3.	enrichment	fissile	fissile
4.	weapons-grade	nuclear_threat_initiative	weapons-grade
5.	fordo	weapons-grade	nonproliferation
6.	international_atomic energy_agency	robert_gallucci	enriched
7.	iaea	nonproliferation	nuclear
8.	centrifuges	cppnm	gary_samore
9.	centrifuge	miles_pomper	nuclear_threat_initiative
10.	enriching	hippel	siegfried_hecker

For the case with no phrase processing, the terms are generally related to uranium enrichment. Fordo is the site of a uranium enrichment plant in Iran.

For the case where the phrase was treated as a unit in pre-processing, the top ten results are quite different – there is only one term that also occurred in the top ten for the no-phrase-processing case.

With phrase pre-processing, the acronym for *highly enriched uranium* (HEU) is the top term, as might be expected. Nuclear Threat Initiative is a nonprofit organization dedicated to reducing the spread of weapons of mass destruction. CPPNM is an acronym for Convention on the Physical Protection of Nuclear Material. Robert Gallucci is president of the MacArthur foundation. Miles Pomper is a Senior Research Associate at the Center for Nonproliferation Studies. Frank N. von Hippel is Co-Director of the Program on Science and Global Security at Princeton University. Multiple news articles from the collection dealt with papers that these individuals had written and speeches that they had given related to nuclear proliferation.

The ad hoc approach performed well for this phrase. Six of the top ten result terms from full phrase pre-processing occur in the ad hoc result list, including the top five terms from the pre-processed case.

The other terms presented also have the appropriate context. Gary Samore is the Executive Director for Research at the Belfer Center for Science and International Affairs. Siegfried Hecker was co-director of the Center for International Security and Cooperation from 2007–2012. Multiple articles from the collection discussed speeches given and papers written by these two individuals that dealt with nuclear proliferation.

6.4 Summary Term Overlap Results

A summary of the term overlap results from these examples is shown in Table 6. For the case of no phrase preprocessing, there is little or no overlap in the top 10 terms as compared to the case of full phrase preprocessing. For the ad hoc technique, in contrast, the overlap of the top ten terms compared to full phrase preprocessing is much higher.

Table 6. Comparison of result sets.

Phrase	Overlap of Result Sets with those for full phrase pre-processing			
	No Phrase Pre-processing		Ad Hoc Technique	
	Top 10 Terms	Top 100 Docs	Top 10 Terms	Top 100 Docs
gross national product	0	12%	40 %	52%
rare earth element	0	15%	70%	79%
highly enriched uranium	10%	3%	60%	47%

An interesting aspect of the term results is their specificity. Considering the three examples taken together, only 23 % of the terms produced by classical LSI corresponded to named entities. More than three-fourths were general terms. The pre-processed results were much more specific. More than half (57 %) corresponded to named entities. The ad hoc technique described here produced results intermediate between these – 40 % of those results were named entities.

6.5 Document Result Comparisons

Examination of the results for retrieved documents showed patterns similar to those for terms. In these tests, each of the three example phrases was used as a query and the 100 highest-ranked documents were retrieved. These results also are shown in Table 6. They show differences similar to those for terms. For the news articles used in the testing, there are many similar documents. Requiring exact matches within the top 100 documents out of 1.6 million is a relatively strict measure of performance.

Although space limitations preclude showing other examples, all phrases tested have shown results similar to those in Table 6. In general, the results are most striking when the query phrase is highly non-compositional. Overall, for term retrieval, the average overlap between the no-phrase-processing results and those for full phrase pre-processing is only a few percent. For document retrieval, the average overlap is only about 10 %. For the technique described in this paper, the average overlap with full phrase pre-processing for both types of retrieval is close to 60 %.

The test data demonstrate that the retrieval results generated using the proposed technique constitute a useful approximation of those that would have been obtained if the given phrases had been treated as indexing units during the creation of the LSI spaces.

7 Conclusions

The method described here provides an approach for using arbitrary phrases as queries in an LSI space after the LSI index has been created. Tests demonstrate that the results of such queries are a useful approximation of the results that would have been obtained if those phrases had been treated as indexing units during the creation of the LSI space. This is true both when retrieving closest terms and closest documents. In contemporary hardware environments, the technique is fast enough that it can be used for near-real-time processing at query time.

The implementation of this technique in LSI-based information systems can be anticipated to yield significant improvements in user satisfaction. Users are accustomed to being able to use phrase searches in Boolean retrieval systems and find their absence in LSI systems a limitation. There also are clear improvements in performance for many queries. In general, the precision of results obtained using the technique described here will be much greater than that obtained by simply including the individual terms of the phrase in a standard LSI query. This is particularly true for highly non-compositional phrases.

The technique described here is equally applicable for English- and for foreign-language text. It also has direct analogues for dealing with coupled features in LSI spaces used to represent non-textual data.

Some advanced knowledge discovery applications employ workflows in which LSI queries are automatically generated. Such applications typically emphasize generation of high-precision result sets. The technique described here is particularly applicable to such environments.

Acknowledgements. The author would like to thank the members of the Semantic Engineering staff at Agilex Technologies who participated in the reported testing.

References

1. Furnas, G., et al.: Information retrieval using a singular value decomposition model of latent semantic structure. In: Proceedings 11th SIGIR, pp. 465–480 (1988)
2. Dumais, S.: Latent semantic analysis. In: ARIST Review of Information Science and Technology, vol. 38, Chap. 4 (2004)
3. Dumais, S., et al.: Using latent semantic analysis to improve access to textual information. In: Proceedings of CHI 1988, 15–19 June 1988, Washington, DC, pp.281–285 (1988)
4. Bradford, R.: Comparability of LSI and human judgment in text analysis tasks. In: Proceedings of Applied Computing Conference, Athens, Greece, pp. 359–366 (2006)
5. Bradford, R.: Relationship discovery in large text collections using latent semantic indexing. In: Proceedings of the Fourth Workshop on Link Analysis, Counterterrorism and Security, SIAM Data Mining Conference, Bethesda, MD, 20–22 April 2006
6. Price, R., Zukas, A.E.: Application of latent semantic indexing to processing of noisy text. In: Kantor, P., Muresan, G., Roberts, F., Zeng, D.D., Wang, F.-Y., Chen, H., Merkle, R.C. (eds.) ISI 2005. LNCS, vol. 3495, pp. 602–603. Springer, Heidelberg (2005)
7. Xu, L., et al.: Functional cohesion of gene sets determined by latent semantic indexing of PubMed abstracts. PLoS ONE **6**(4), e18851 (2011)
8. Huang, J., Kumar, S., Zabih, R.: An Automatic hierarchical image classification scheme. In: Proceedings of the Sixth ACM International Conference on Multimedia, pp.219–228 (1998)
9. Hulth, A.: Combining machine learning and natural language processing for automatic keyword extraction. Ph.D. thesis, Stockholm University, April (2004)
10. Manning, C., Raghavan, P., Schütze, H.: Introduction to Information Retrieval, p. 36. Cambridge University Press, New York (2008)
11. Caid, W., Dumais, S., Gallant, S.: Learned vector-space models for document retrieval. Inf. Process. Manage. **31**(3), 419–429 (1995)
12. Salton, G., Yang, C., Yu, T.: A theory of term importance in automatic text analysis. JASIS **26**(1), 33–44 (1975)
13. Fagan, J.: The effectiveness of a nonsyntactic approach to automatic phrase indexing for document retrieval. JASIS **40**(2), 115–132 (1989)
14. Mitra, M., et al.: An analysis of statistical and syntactic phrases. In: Proceedings of RIAO 1997, Montreal, Canada, pp. 200–214 (1997)
15. Turpin, A., Moffat, A.: Statistical phrases for vector-space information retrieval. In: Proceedings of SIGIR 1999, Berkley, CA, August 1999, pp. 309–310 (1999)
16. Harmon, D.: The TREC ad hoc experiments. In: Voorhees, E.M., Harmon, D.K. (eds.) TREC Experiment and Evaluation in Information Retrieval. MIT Press, Cambridge (2005)
17. Metzler, D., Strohman, T., Croft, W.: Indri at TREC 2006: lessons learned from three terabyte tracks. In: Proceedings of Fifteenth Text REtrieval Conference. NIST Special Publication SP 500-272 (2006)
18. Ogawa, Y., et al.: Structuring and expanding queries in the probabilistic model. In: Proceedings of Ninth Text REtrieval Conference (TREC-9), pp. 427–435. NIST Special Publication 500-249 (2000)

19. Zhai, C., et al.: Evaluation of syntactic phrase indexing – CLARIT NLP track report. In: Proceedings of Fifth TExt Retrieval Conference, pp. 347–358. NIST Special Publication 500-238 (1996)
20. Kraaij, W., Pohlmann, R.: Comparing the effect of syntactic vs. statistical phrase indexing strategies for Dutch. In: Nikolaou, C., Stephanidis, C. (eds.) ECDL 1998. LNCS, vol. 1513, pp. 605–617. Springer, Heidelberg (1998)
21. Jiang, M., et al.: Choosing the right bigrams for information retrieval. In: Proceeding of the Meeting of the International Federation of Classification Societies, pp.531–540 (2004)
22. Broschart, A., Berberich, K., Schenkel, R.: Evaluating the potential of explicit phrases for retrieval quality. In: Gurrin, C., He, Y., Kazai, G., K, Udo, Little, S., Roelleke, T., Rüger, S., van Rijsbergen, K. (eds.) ECIR 2010. LNCS, vol. 5993, pp. 623–626. Springer, Heidelberg (2010)
23. Kim, H.-R., Chan, P.: Identifying variable-length meaningful phrases with correlation functions. In: Proceedings of 16th IEEE International Conference on Tools with Artificial Intelligence, pp.30–38 (2004)
24. Lizza, M., Sartoretto, F.: A comparative analysis of LSI strategies. In: Berry, M. (ed.) Computational Information Retrieval, SIAM, pp. 171–181 (2001)
25. Weimer-Hastings, P.: Adding syntactic information to LSA. In: Proceedings of the 22nd Annual Meeting of the Cognitive Science Society (2000)
26. Wu, H., Gunopulos, D.: Evaluating the utility of statistical phrases and latent semantic indexing for text classification. In: Proceedings ICDM, pp. 713–716 (2002)
27. Nakov, P., Valchanova, E., Angelova, G.: Towards deeper understanding of the LSA performance. In: Proceedings of Recent Advances in Natural Language Processing 2003, pp. 311–318 (2003)
28. Grönqvist, L.: An evaluation of bi- and tri-gram enriched latent semantic vector models. In: Proceedings of ELECTRA Workshop, Methodologies and Evaluation of Lexical Cohesion Techniques in Real-World Applications, Salvador, Brazil, 19 August 2005, pp. 57–62 (2005)
29. Grönqvist, L.: Evaluating latent semantic vector models with synonym tests and document retrieval. In: Proceedings of ELECTRA Workshop, Methodologies and Evaluation of Lexical Cohesion Techniques in Real-World Applications, Salvador, Brazil, 19 August 2005, pp. 86–88 (2005)
30. Grönqvist, L.: Exploring latent semantic vector models enriched with n-grams. Ph.D. thesis, Växjö University, Sweden (2006)
31. Olney, A.: Generalizing latent semantic analysis. In: Proceedings, 2009 IEEE International Conference on Semantic Computing, pp. 40–46 (2009)
32. Bradford, R.: An empirical study of required dimensionality for large-scale latent semantic indexing applications. In: Proceedings of ACM 17th Conference on Information and Knowledge Management (CIKM) Napa, California, 26–30 October 2008
33. Bird, S., Klein, E., Loper, E.: Natural Language Processing with Python - Analyzing Text with the Natural Language Toolkit. O'Reilly Media, San Francisco (2009)
34. Bradford, R.: Implementation techniques for large-scale latent semantic indexing applications. In: Proceedings of ACM Conference on Information and Knowledge Management, Glasgow, Scotland, October 2011

Determining the Relative Importance of Webpages Based on Social Signals Using the Social Score and the Potential Role of the Social Score in an Asynchronous Social Search Engine

Marco Buijs[✉] and Marco Spruit

Business Informatics, Utrecht University,
Domplein 29, 3512 JE Utrecht, Netherlands
marcobuijs@gmail.com, m.r.spruit@uu.nl
http://www.uu.nl

Abstract. There are many ways to determine the relative importance of webpages. Specifically, a method that has proven to be very successful in practice is to value a webpage based on its position in the hyperlinked graph of the web. However, there is no generally applicable algorithm to determine the value of webpages based on an arbitrary number social signals such as likes, tweets and shares. By taking such social signals into account a more democratic method arises to determine the value of webpages. In this article we propose an algorithm named the Social Score that takes into account an arbitrary number of social signals to determine the relative importance of a webpage. Also, we present a worldwide top fifty of webpages based on the Social Score. Last, the potential role of the Social Score in an asynchronous Social Search engine is evaluated.

Keywords: Social Score · Asynchronous Social Search · PageRank · Web search · Top-K ranking · Quality assessment · Data analytics · Information extraction

1 Introduction

It has been proven that top-K ranking in Web Search cannot be based only on matching queries to documents [13]. Although many techniques for ordering Webpages based on query-document similarity are present, it is not sufficient to get good results in Web Search. There are too many resources that look too much alike to which documents are more relevant than others. Search Engines turn to other factors to determine the general importance of Webpages and take that into account especially when many documents contain approximately the same content. What appears to work best is to use query-document matching ranking techniques in combination with query-independent ranking techniques. This way, Search Engines are able to determine what search results should be shown on

© Springer International Publishing Switzerland 2015
A. Fred et al. (Eds.): IC3K 2014, CCIS 553, pp. 118–131, 2015.
DOI: 10.1007/978-3-319-25840-9_8

top, even when a query matches many documents more or less equally well. Important concepts in query-document matching are Term Frequency-Inverse Document Frequency (TF-IDF) ranking and metadata extraction [10,18]. Those are then combined with query-independent ranking techniques to gain better results. One such factor to determine the general importance of Webpages in which the match between a document and the query is taken out of the equation is PageRank [3]. With PageRank, the value of pages is determined by the Link structure on the Web. When a page is linked to from many important Webpages, the page itself is also considered to be important. Using a recursive algorithm, the relative value of every Webpage is calculated. This way is based on how the value of scientific papers can also be estimated: based on the number of times your paper is cited by other papers that all also have their own relative value in the community. PageRank has been proven very successful and it is often referred to as the algorithm that made Google gain its huge market share in Web Search. Such a system in which Links determine the value of Webpages is also referred to as the Link Economy [6]. In the Link Economy, the power to influence the ranking of search results is with the authors on the Web. Before the Link Economy, there was the Hit Economy, in which the value of Webpages was mainly based on the visit counters that were available on Webpages. In the Hit Economy, the number of visits determined the value of Webpages and the power to influence the rankings was directly in the hands of all internet users. We propose a method to give back the power to influence Web Search rankings directly to all internet users instead of only web authors. The system is based on the Like Economy, in which the value of Webpages is based on online Social Signals such as likes, tweets, mentions, shares, bookmarks and pins. For every Webpage we calculate a Social Score, which is based on Social Signals from multiple Social Media platforms.

Based on work from Evans and Chi [5] and Golovchinsky, Pickens and Back [8], we define asynchronous Social Search as

"Information seeking supported by a network of people where collaboration takes place in a nonconcurrent way."

Important concepts in asynchronous Social Search are user-generated content and user feedback. Therefore, in this paper we will specifically evaluate the potential of the Social Score to contribute to an asynchronous Social Search engine. To get a better understanding of what is important for an asynchronous Social Search engine to perform well, we also devoted a section to the critical success factors of asynchronous Social Search.

2   Related Work

To our knowledge, not much research has been performed to improve web search using explicit Social Signals such as likes and tags. However, related work has been performed in the field of the Semantic Web. The Semantic Web aims at adding logic to the World Wide Web. The idea behind this is that the Web becomes better readable for machines. This way, machines would be able to get a better understanding of how pages are related to each other and where they have to look for certain information. Furthermore, the Semantic Web enables machines to aggregate data from different pages and present this aggregated data to users in a clear overview [2].

A truly personal approach to Information Retrieval on the WWW has been taken by Delicious. On Delicious people can create an account, add bookmarks to it and retrieve those bookmarks later on based on tags that can be assigned to bookmarks. They can also befriend people and search in the bookmarks of their friends. Several studies were performed on whether such an approach could improve web search and the results differed [9,15,19].

Bookmarking on Delicious is a form of collaborative tagging. Golder and Huberman performed research in this field of study and they define collaborative tagging as

"the process by which many users add metadata in the form of keywords to shared content" [7].

During their research they observed that people use a great variety of tags, but also consensus is reached in such a way that stable patterns emerge in tag proportions with respect to tagged resources. They also identify the main reason behind tagging, which is personal use. They conclude that the stable patterns in tagging can be used to organise and describe how web resources relate to each other. Tags can be seen as a form of Social Signals that could be taken into account in determining the relative importance of Webpages. Not all Social Signals assign words to a resource. Social Signals can be less complex, such as a like. A like only indicates positivity with respect to, for example, a web resource.

In 2007, Bao, Wu, Fei, Xue, Su and Yu saw the potential of social annotations to determine the value of Webpages [1]. Although they took a different approach with their ranking method that they call SocialPageRank, the idea is rather similar to the Social Score method as proposed in this paper. One difference between the approaches is that SocialPageRank makes use of more complex mathematical calculations whereas the Social Score makes use of simpler math and is easier to understand. Furthermore, the computational complexity of the Social Score method to calculate the Social Score of one Webpage is $O(1)$ whereas in the SocialPageRank method it is not possible to calculate any individual Score for a Webpage without calculating the other scores for the other Webpages. This is because SocialPageRank makes use of recursive Matrix multiplications just like PageRank does to converge to a stable scoring model. In each iteration the computational complexity is $O(|U||W| + |s||W| + |U||s|)$ where $|U|$ is the number of users U of the Social Media platform, $|W|$ is the number of Webpages W in a Corpus C and $|s|$ is the number of social annotations or Social Signals. The number of iterations determines the accuracy of the resulting scores for the Webpages. The last and most important difference is that SocialPageRank only makes use of data from one Social Media platform what leaves more open space for bias. The Social Score method is more generic and can take into account as many Social Signals from as many Social Media Platforms as desired.

3   Critical Success Factors of Asynchronous Social Search

The Critical Success Factor (CSF) that holds specifically for an asynchronous Social Search method is user involvement. Without involvement of users asynchronous Social Search methods will never work because the method relies on input from users. Therefore, involvement should be easy, fun and make the user more productive. Examples of successful systems that rely on user involvement are Wikipedia and Stack Overflow. Although Wikipedia is considered to be one of be biggest websites in the world it had only 40 000 active contributors in 2006 [16]. This indicates that only a fraction of people in the world have to become active contributors to serve the entire population with decent quality. One way of getting users involved is by keeping the effort required as low as possible [7]. Therefore, efforts to tag and like web resources should be minimal. Also, the system should heavily rely on implicit feedback instead of explicit feedback to reduce efforts of users. To be able to rank resources based on both implicit and explicit user feedback, the Hit economy and the Like economy as described by Gerlitz and Helmond need to be supported by the system to determine the value of webpages [6]. A good option to support the Hit economy is that users should be able to enable automatic browser tracking behaviour so that every website that is visited by the user automatically gets indexed. Benefits of active participation should be maximised by providing active participants with search results they indicated were the most relevant to them regarding certain keywords. Results should be provided on an instant basis, changing with every key pressed on the keyboard by the user. Ninety per cent of the queries should return the most relevant results for that user within the first three key-strokes.

A CSF that holds for virtually all search methods is spam and malicious use prevention. People want their website to come up at the highest rank in the search engine and try to achieve this in all sorts of ways. In the case of PageRank for example, one might try to create heaps of webpages referring to one page that you want to come up on top in search engines. Another CSF is performance. Queries of users must be handled in milliseconds to be able to provide search results to users instantly. Because the web is very large and still expanding every moment, scalability is also very important. Although the asynchronous Social Search method can be applied in small communities, its real strength lies in massive collaboration. The amount of resources to index will grow with the number of active participators, but also over time. Just like other search methods, it is also critical to have your search method always available to everyone on the internet. Therefore, 100 % uptime should be strived for. Just like any other system, the user interface of the search method has a critical impact on whether people will use the search method or not. The focus should be on ease of use and simplicity. Although people do not seem to like the idea most of the time, one CSF of a search method appears to be personalised search results. The search method should enable personalised search in a transparent way such that users can understand and manipulate their personal results to queries. Furthermore, users should be given the explicit choice to enable or disable personalised search. People should also be able to create and join groups that maintain their

Table 1. The key success factors of asynchronous Social Search.

Factor	Description
Involvement	Without involvement of users asynchronous Social Search methods will never work because the method relies on input from users [17]. A cycle of credit, much freedom for users and tagging support could support user involvement [7,12,16].
Implicit feedback	Implicit feedback such as clicks and visits has been proven to be very useful for both discovery and ranking of search results [14]. This all comes down to supporting the Hit economy [6].
Explicit feedback	Explicit feedback such as likes, shares and pins gives people the possibility to directly influence the rankings according to their point of view. This enables true democratic ranking [6].
General search success factors	More general success factors such as precision, recall, personalisation and good discovery capabilities also hold for asynchronous Social Search. [13]
General IT system success factors	Even more general success factors that hold for most Information Technology systems that also hold for asynchronous Social Search are scalability, availability, performance and ease of use [4].

own corpus of resources. Joining such groups should influence the personalised ranking results. It should also be possible to search in the corpus that a group maintains. Table 1 provides a summary of the critical success factors of asynchronous Social Search. Figure 1 gives an overview of the main CSFs of asynchronous Social Search in the form of a model.

4   Social Score

Just like PageRank, the Social Score is used next to existing techniques like TF-IDF. That is what makes the algorithm so similar to Pagerank: it calculates the value of a Webpage, completely independent of any query. Additional algorithms are required for both PageRank and the Social Score to actually use this information in search engines, because also the query has to be taken into account. Only if algorithms such as TF-IDF result in many hits, which is often the case on the Web, the Social Score can be used to determine which resources should be returned first. In opposite to PageRank, the Social Score can be calculated for every Webpage individually, without having to recompute all the other scores for all other resources. An arbitrary number of Social Signals from different Social Media Platforms can be taken into account. To prevent bias towards a certain group of internet users or a certain domain, it is good practice to take as many

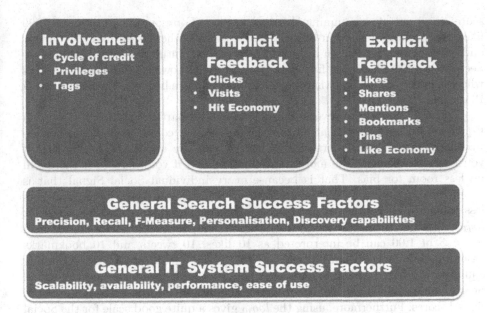

Fig. 1. The key success factors of asynchronous Social Search.

signals from as many Social Media Platforms as possible into account. To calculate the Social Score S for a Webpage W, we take into account n Social Signals related to Webpage W. The Social Score takes into account a list L of n Social Signal Scores s, where s is the number of Social Signals from one Social Media Platform. For example, s could be the number of shares of a Webpage W on Facebook or the number of tweets about W on Twitter. Now the Social Score S is calculated as defined in Eq. 1.

$$S = \frac{\sum_{i=1}^{n} \log_{10}(1 + L_i)}{n} \tag{1}$$

An example of calculating the Social Score S for a Webpage W only taking into account two Social Media platforms is as follows: let's say that the website "example.com" has 99 likes on Facebook and has been mentioned in 9 tweets on Twitter. Then, $L_1 = 99$, $L_2 = 9$ and $n = 2$. We calculate the sub scores per Social Signal and divide by n. The sub score for likes on Facebook is $\log_{10}(1 + 99) = 2$ and the sub score for tweets on Twitter is $\log_{10}(1 + 9) = 1$. To calculate Social Score S for Webpage W we take the average resulting in $S = 1.5$. As you can see in this example, the Social Score increases with the number of likes and shares. Table 2 gives some more examples in which three Social Signals are taken into account: the number of likes on Facebook, the number of tweets on Twitter and the number of bookmarks on Delicious. From this table we can infer that S is higher when Social Signal Scores are in balance. When they are out of balance, the same total number of likes, tweets and bookmarks result in a lower Social Score S. This happens because the log_{10} is taken of every individual Social Signal score s and not after summing them all up first. For example, only having 999 likes on Facebook results in a Social Score of 1.00 whereas 99 likes, 99, tweets and 99 bookmarks result in a Social Score of 2.00. Intuitively this makes sense because it looks like in the first case there is a bias towards Facebook and likes whereas in the second case there seems to be a balance between the Social Media Platforms.

Although there were several reasons to choose particularly for the log_{10} in Eq. 1, we did not experiment with other $logs$ and it could be that another log would perform better in practice. What we can say about the log, is that if you lower it, there will be more room for bias and if you increase it there will be less room for bias. That is because every individual Social Signal that is taken into account gets more influence on S if a lower log is taken and gets less influence on S when a higher log is taken. The first reason we chose for log_{10} is that it is relatively easy to interpret for people. For example, a Social Score S of 1.00 can be interpreted as 10 likes, 10 tweets and 10 bookmarks assuming a balanced distribution over the Social Media platforms. Assuming equally distributed Social Signals over the Social Media platforms, the Social Score can be explained as the order of magnitude for the underlying Social Signal scores. Furthermore, using the log_{10} gives a quite good scale for the Social Score. When a Webpage would have one billion Social Signals per Social Media platform that is taken into account, that page would have a Social Score S of 9.00. Currently there are no such Webpages present in the world that have that many Social Signals related to them on any Social Media platform. Therefore, we can safely assume that the Social Score will always produce values between 0 and 9 disregarded which Social Media platforms are used to calculate the Social Score.

Table 2. Examples of calculating Social Score S given three Social Signal Scores s.

Likes	Tweets	Bookmarks	Social Score S
0	0	0	0.00
999	0	0	1.00
99	99	99	2.00
333	333	333	2.52
10^8	10^7	10^6	7.00

The Social Score S of one specific web resource can be calculated in $O(1)$ time by making use of the Application Programming Interfaces (APIs) of the Social Media Platforms. Now let's consider a Corpus C consisting of indexed Webpages W. To determine all the Social Scores of all Webpages W in C, this would take $O(|C|)$ time. Therefore, we can say that computational complexity increases linearly with the size of Corpus C. Notice that the Social Score S of a Webpage W will generally change over time because people keep interacting with the Webpage W via Social Media Platforms.

5 Experiment

To be able to validate whether the Social Score S can accurately determine the importance of Webpages, an experiment was performed in which a Corpus C of over 120 000 Webpages was gathered. The experiment was part of a larger experiment about asynchronous Social Search engine quality in which a prototype was built and tested.

There were three ways in which results could be added to the prototype. The first one was manually, by filling in a URL, title, description and keywords.

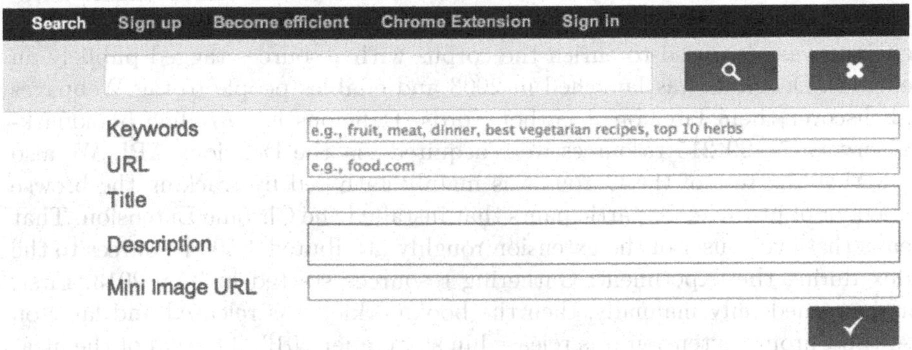

Fig. 2. Screenshot of how a link could be added manually.

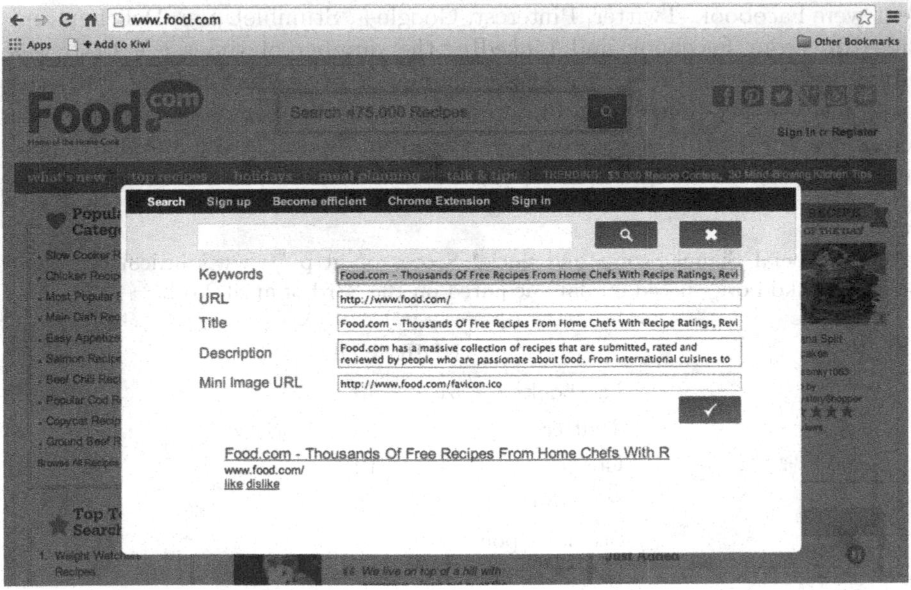

Fig. 3. Example of how a link could be added using the bookmarklet.

Figure 2 provides a screenshot of what this way looked like in practice. The second was by adding a bookmarklet to your favourites in your web browser. When a user had the bookmarklet in his favourites list in his web browser and he visited a website, he could click on the bookmarklet. This resulted in a popup of the search engine with a form shown to add a result to the search engine. In this form, the URL, title, description and keywords are already filled in based on the page that the user is currently visiting. This second way of adding Webpages to the search engine is less time consuming than the first. An example is shown in Fig. 3. The third way to add search results to the search engine was by installing an extension for the Chrome web browser. By installing this extension, all the websites that were visited by the user were added to the search engine automatically. To guarantee a decent corpus size, the API of bookmarking website Delicious was also used to enrich the corpus with resources tagged publicly on Delicious. Delicious was launched in 2003 and enables people to tag Webpages and discover them later on. In other words, Delicious is an online bookmarking service [7]. 29 215 resources were acquired via the Delicious API. We also know that the rest of the Corpus was mainly gathered by tracking the browse behaviour of just over 20 participants that installed the Chrome Extension. That means that every user of the extension roughly attributed 4 500 resources to the index during the experiment. Gathering resources started in june 2013. First, this happened only manually, then the bookmarklet was released and later on also the Chrome extension was released in september 2013. The end of the measurement period for the experiment was the seventh of january 2014.

For every page a Social Score S was calculated based on Signal Scores s from seven Social Media Platforms. The Social Media platforms used in the experiment were Facebook, Twitter, Pinterest, Google+, StumbleUpon, Delicious and LinkedIn. From Facebook and LinkedIn, the number of times the URL was shared was acquired. From Twitter the number of tweets in which the URL was mentioned was acquired. From Pinterest, the total number of times that items were pinned on the Webpage was acquired. From Google+, the number of people that +1'd a URL was acquired. From StumbleUpon, the number of times a

Table 3. Social Signal Scores and Social Scores of http://www.kdir.ic3k.org/ and http://www.kdd.org/ based on data acquired on the 23rd of april 2014.

	KDIR	KDD
Facebook	51	45
Twitter	1	11
Google+	1	12
Pinterest	0	0
StumbleUpon	2	1
Delicious	8	39
LinkedIn	0	1
Social Score	0.54	0.87

URL was stumbled upon was acquired. Last, from Delicious, the total number of times a URL had been bookmarked was retrieved. Two example calculations are shown in Table 3. Here, two Webpages from different conferences are rated and it appears that http://www.kdd.org/ is more important than http://www.kdir. ic3k.org/ according to their Social Scores S. What is important is of course a subjective matter. The Social Score takes the viewpoint that websites are more important than others when they experience more social interaction, preferably equally distributed over the Social Media platforms that are taken into account. When the Social Score would actually be used in a search engine, first algorithms like TF-IDF are used to find matching resources and only when there are many results that match a query approximately equally well, the Social Score should be used to identify the most important ones. Although the results of this experiment indicated a statistically significant improvement in ranking quality compared to the baseline method, this particular experiment could unfortunately not be used to prove that this improvement was caused by the Social Score. That is because the baseline method also had other differences in the ranking algorithm than only the Social Score. This does not mean that the results of that experiment were useless, the experiment just had another purpose: comparing a baseline method with an asynchronous Social Search method. From the Corpus a top 50 was assembled based on Social Score S. Figure 5 provides an overview of the most important websites worldwide according to the Social Score S. Notice that there were ten duplicates in the list, such as https://twitter.com and http:// twitter.com which both refer to the same content. Such duplicates were removed from the list. In large extent the list feels intuitively right. The most disturbing about the list is that Wikipedia has been ranked only 31st. In a PageRank algorithm Wikipedia would probably score top five, but apparently Wikipedia is not a source that many people frequently share or like via Social Media compared to the number of back Links created to Wikipedia by authors on the Web. Another interesting fact is that there are two Youtube videos in the top 50. Both are very popular songs that went viral via Social Media and therefore mainly scored high

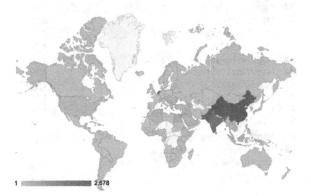

Fig. 4. Distribution of sessions over the world measured from the first of june 2013 till the seventh of january 2014. Acquired using Google Analytics.

#	URL	S
1	http://www.google.com/	5,90
2	http://www.facebook.com/	5,73
3	https://twitter.com/	5,50
4	http://www.youtube.com/	5,33
5	http://www.flickr.com/	5,14
6	http://www.amazon.com/	5,10
7	http://espn.go.com/	5,07
8	http://www.ted.com/	4,97
9	http://grooveshark.com/	4,94
10	http://www.pandora.com/	4,88
11	http://www.nytimes.com/	4,85
12	http://www.yahoo.com/	4,85
13	http://9gag.com/	4,79
14	http://www.ebay.com/	4,75
15	http://www.etsy.com/	4,74
16	http://www.apple.com/	4,70
17	http://www.imdb.com/	4,68
18	http://www.youtube.com/watch?v=9bZkp7q19f0	4,58
19	http://maps.google.com/	4,57
20	http://www.pinterest.com/	4,50
21	http://mashable.com/	4,49
22	http://www.nationalgeographic.com/	4,49
23	http://www.time.com/time/	4,48
24	http://www.linkedin.com/	4,40
25	http://www.rollingstone.com/	4,40
26	http://www.speedtest.net/	4,40
27	http://www.mtv.com/	4,39
28	http://www.codecademy.com/	4,35
29	http://www.kickstarter.com/	4,29
30	http://www.wix.com/	4,28
31	http://www.wikipedia.org/	4,26
32	http://www.fcbarcelona.com/	4,26
33	http://www.youtube.com/watch?v=jofNR_WkoCE	4,25
34	http://dictionary.reference.com/	4,25
35	http://translate.google.com/	4,25
36	http://www.indeed.com/	4,25
37	http://www.ign.com/	4,25
38	http://instagram.com/	4,21
39	http://www.asos.com/	4,21
40	http://digg.com/	4,19
41	http://www.last.fm/	4,18
42	http://www.stereomood.com/	4,17
43	http://imgur.com/	4,17
44	http://thenextweb.com/	4,16
45	http://www.picmonkey.com/	4,15
46	http://edition.cnn.com/	4,15
47	http://www.apple.com/iphone/	4,14
48	https://mail.google.com/mail/	4,14
49	http://weavesilk.com/	4,13
50	http://www.weather.com/	4,12

Fig. 5. Top 50 URLs worldwide according to Social Score S on march 27th 2014.

on Facebook and Twitter. Theoretically there could be sources missing that have never been indexed by the search engine. That would be rather unlikely though, because indexing is mainly based on visits and you would expect that the most popular Webpages on the Web would have been visited at least once during this study by one or more users. It could be the case that there is a website in a large country or continent of which no users participated in the experiment. Although we do not have exact data on where people came from that installed the extension that supported automatic browse behaviour tracking, we do have data about where the visitors of the search engine prototype came from. A map showing the distribution of sessions over the world is shown in Fig. 4. The second biggest source of resources was Delicious. Delicious is used by a way broader audience which also decreases the chance that we are missing an important URL in our Top 50 according to the Social Score. There could very well be a bias in the ranking however, towards webpages that are popular in the Western world. That is because the Social Media platforms that were taken into account are most popular in the Western world. Obviously, the ranking presented in Fig. 5 changes over time. With every like, share or other form of Social Media interaction with respect to a Webpage, the Social Score of the resource changes. It was outside the scope of this research to determine how often the Social Score should be updated.

6   Conclusions

The proposed concept of a Social Score for every web resource based on online Social Signals from Social Media platforms such as likes and shares is a promising alternative to existing methods to determine the query-independent importance of Webpages. We conclude that the proposed Social Score S is a good alternative for the more computational expensive PageRank and SocialPageRank methods in which iterative matrix multiplications are required. The Social Score of one Webpage W can be calculated in constant time and calculating the Social Scores for all Webpages W in a Corpus C would therefore be calculated in $O(|W|)$ time, so linear in the number of Webpages. Furthermore, contrary to PageRank and SocialPageRank, updating one score of one Webpage is actually possible using the Social Score method without any overhead. The Social Score could be a complementary property to take into account in existing search methods. Specifically, it would be a very suitable component for an asynchronous Social Search method because using the Social Score enables a search method to make use of a broad set of explicit feedback on Webpages without any additional effort required from users. By making use of the Social Score a shift can be made towards the Like Economy, away from the Link Economy. This way, all internet users will gain more direct influence on the ranking of results. It would indicate a shift from an Aristocracy in which the power is in the hands of the technically skilled web authors and developers to a direct democracy for Web Search.

More research should be performed to determine the quantity of likes compared to the quantity of links available on the Web. Also the quality of Social

Signals with respect to links could be compared in an experiment with a double blind test in which two identical search engines are used except one is using PageRank and the other is using the Social Score. When a user poses a query, both search methods are queried and the results are mixed as described by Joachim [11]. Based on which search engine receives the most number of clicks it can be determined which search method is performing better. This way, PageRank could be used as a baseline method to be able to measure the performance of the Social Score. The same kind of experiment could be performed to compare the Social Search method with SocialPageRank. It would also be interesting to study the effects of taking into account different numbers of Social Signals from different numbers of Social Media platforms. It could be investigated whether Facebook Signals are from higher quality than Twitter Signals and whether bias can actually be removed by taking into account more Social Signals and Social Media platforms. More traditional experiments could also be performed to measure the impact on precision and recall of the proposed Social Score.

A potential problem that might have impact on the performance of search engines making use of the Social Score could be a bias towards Social Media websites, in particular the ones that are used to calculate the Social Score. One could imagine that a Facebook page on average receives significantly more Facebook likes than a traditional Webpage. One solution could be to take into account platforms such as Delicious, which are all about traditional Webpages. Also, on most Social Media platforms it is possible to share traditional Webpages. A few examples are Twitter, LinkedIn and Facebook. The effects on ranking could be investigated in future research.

Also, the proneness of the Social Score to malicious use and spam should be evaluated. Would it be easier or harder to influence the ranking of results when Likes are used instead of Links to determine the value of Webpages? Research could be performed to investigate how such malicious use of the Social Score could be prevented effectively. One direction for a solution might be to use many Social Media Platforms to force spammers to spam many different systems. One advantage of the Social Score is that spam and malicious use is not only the problem of the Search Engine, but also a direct problem for the Social Media platforms at hand.

Last, the moments in time on which likes were assigned to resources could be taken into account. Future research could be performed to measure the benefits of such a change to the system. However, calculations would become more complex and most APIs of Social Media platforms will not provide this information. Therefore, in practice such a change to the proposed algorithm would probably cost much effort to achieve in practice and result in a less flexible and less generally applicable concept.

References

1. Bao, S., Xue, G., Wu, X., Yu, Y., Fei, B., Su, Z.: Optimizing web search using social annotations. In: Proceedings of the 16th International Conference on World Wide Web, pp 501–510. ACM (2007)

2. Berners-Lee, T., Hendler, J., Lassila, O., et al.: The semantic web. Sci. Am. **284**(5), 28–37 (2001)
3. Brin, S., Page, L.: The anatomy of a large-scale hypertextual web search engine. Comput. Netw. ISDN Syst. **30**(1), 107–117 (1998)
4. Chung, L., Nixon, B.A., Yu, E., Mylopoulos, J.: Non-functional requirements in software engineering, vol. 5. Springer Science & Business Media (2012)
5. Evans, B.M., Chi, E.H.: Towards a model of understanding social search. In: Proceedings of the 2008 ACM Conference on Computer Supported Cooperative Work, pp 485–494. ACM (2008)
6. Gerlitz, C., Helmond, A.: The like economy: social buttons and the data-intensive web. New Media Soc. **15**(8), 1348–1365 (2013)
7. Golder, S.A., Huberman, B.A.: Usage patterns of collaborative tagging systems. J. Inf. Sci. **32**(2), 198–208 (2006)
8. Golovchinsky, G., Pickens, J., Back, M.: A taxonomy of collaboration in online information seeking. arXiv preprint arXiv:0908.0704
9. Heymann, P., Koutrika, G., Garcia-Molina, H.: Can social bookmarking improve web search? In: Proceedings of the International Conference on Web Search and Web Data Mining, pp 195–206. ACM (2008)
10. Hu, Y., Xin, G., Song, R., Hu, G., Shi, S., Cao, Y., Li, H.: Extraction from bodies of html documents and its application to web page retrieval. In: Proceedings of the 28th Annual International ACM SIGIR Conference on Research and Development in Information Retrieval, pp 250–257. ACM (2005)
11. Joachims, T.: nbiased evaluation of retrieval quality using clickthrough data. In: SIGIR Workshop on Mathematical/Formal Methods in Information Retrieval, vol. 354. Citeseer (2002)
12. Latour, B., Woolgar, S.: Laboratory Life: The Social Construction of Scientific Facts. Princeton University Press, Princeton (1979)
13. Manning, C.D., Raghavan, P., Schütze, H.: Introduction to Information Retrieval, vol. 1. Cambridge University Press, Cambridge (2008)
14. Matthijs, N., Radlinski, F.: Personalizing web search using long term browsing history. In: Proceedings of the Fourth ACM International Conference on Web Search and Data Mining, pp 25–34. ACM (2011)
15. Noll, M.G., Meinel, C.: Web search personalization via social bookmarking and tagging. In: Aberer, K., Choi, K.-S., Noy, N., Allemang, D., Lee, K.-I., Nixon, L.J.B., Golbeck, J., Mika, P., Maynard, D., Mizoguchi, R., Schreiber, G., Cudré-Mauroux, P. (eds.) ASWC 2007 and ISWC 2007. LNCS, vol. 4825, pp. 367–380. Springer, Heidelberg (2007)
16. Ortega, F., Gonzalez-Barahona, J.M., Robles, G.: On the inequality of contributions to wikipedia. In: Proceedings of the 41st Annual Hawaii International Conference on System Sciences, pp 304–304. IEEE (2008)
17. Robey, D., Farrow, D.: User involvement in information system development: a conflict model and empirical test. Manage. Sci. **28**(1), 73–85 (1982)
18. Salton, G., McGill, M.J.: Introduction to Modern Information Retrieval. McGraw-Hill, New York (1983)
19. Yanbe, Y., Jatowt, A., Nakamura, S., Tanaka, K.: Can social bookmarking enhance search in the web? In: Proceedings of the 7th ACM/IEEE-CS Joint Conference on Digital Libraries, pp 107–116. ACM (2007)

A Novel Knowledge-Based Architecture
for Concept Mining on Italian and English Texts

Dante Degl'Innocenti[(✉)], Dario De Nart, and Carlo Tasso

Department of Mathematics and Computer Science, Artificial Intelligence Lab,
University of Udine, Udine, Italy
deglinnocenti.dante@spes.uniud.it, {dario.denart,carlo.tasso}@uniud.it
http://ailab.uniud.it

Abstract. Manually annotating unstructured texts for finding significant concepts is a knowledge intensive process and, given the amount
of data available on the Web and on digital libraries nowadays, it is
not cost effective. Therefore automatic annotators capable to perform
like human experts are extremely desirable. State of the art systems
already offer good performance but they are often limited to one language, one domain of application, and can not entail concepts that do
not appear but are logically/semantically implied in the text. In order to
overcome this shortcomings, we propose here a novel knowledge-based,
language independent, unsupervised approach towards keyphrase generation. We developed DIKpE-G, an experimental prototype system which
integrates different kinds of knowledge, from linguistic to statistical,
meta/structural, social, and ontological knowledge. DIKpE-G is capable to extract, evaluate, and infer meaningful concepts from a natural
language text. The prototype performs well over both Italian and English texts.

Keywords: Concept extraction · Keyphrase extraction · Information
extraction · Italian language · Natural language processing · Text analysis · Text classification · Text summarization

1 Introduction

Due to the growth of the amount of unstructured text data available on the Web
and in digital libraries, the demand for automatic summarization and real-time
information filtering has rapidly increased. However, such systems need metadata that can precisely and compactly represent the content of a document.
Even though a huge number of different metadata formats has been proposed
and Semantic Web technologies have grown bigger and bigger over the last few
years, the most common way to represent these metadata is still constituted by
KeyPhrases. A KeyPhrase (herein KP) is a short phrase, typically made of one to
four words which identifies a *concept*. Such representation bears several advantages: it is simple to understand, yet expressive and less exposed to polisemy
issues than a single-term-keyword representation; moreover it has an high cognitive plausibility, since it is known [1] that KPs are more informative features

© Springer International Publishing Switzerland 2015
A. Fred et al. (Eds.): IC3K 2014, CCIS 553, pp. 132–142, 2015.
DOI: 10.1007/978-3-319-25840-9_9

than single words for representing the content of a text. Associating meaningful KPs to a text is a trivial task for humans, however, even by exploiting social Web collaborative technologies, one cannot expect the whole Web to be manually annotated, therefore automatic KP generation techniques are highly desirable. As shown in Sect. 2, several authors have already addressed the problem of KP generation in English texts, but little work has been done with other languages. Italian, in particular, though being the ninth most used language on the Web [2] has never received much attention. In this work, we present DIKpE-G an experimental system specifically built for performing KP Extraction and Inference from Italian and English documents. The proposed system exploits a knowledge-based approach combining various classes of knowledge, in part language-dependent, in part independent and it is designed to emulate some of the cognitive processes that are exploited when a human expert is asked to summarize or classify a text.

The paper is organized as follows: in Sect. 2 we briefly illustrate some related work; in Sect. 3 we present our keyphrase generation approach; in Sect. 4 we give a brief description of the DIKpE-G prototype, in Sect. 5 we expose some experimental results and, finally, in Sect. 6 we conclude the paper.

2 Related Work

Several authors in the literature have already addressed the problem of extracting keyphrases from natural language documents and a wide range of approaches have been proposed. The authors of [3] identify four types of keyphrase extraction strategies:

- *Simple Statistical Approaches*: these techniques assume that statistical information is enough to identify keywords and KPs, thus they are generally simple and unsupervised; the most widespread statistical approaches consider word frequency, TF-IDF or word co-occurrency [4]. It is important to note how TF-IDF based methods require a closed document corpora in order to evaluate inverse frequencies, therefore they are not suitable to an open world scenario, where new items can be included in the corpora at any time.
- *Linguistic Approaches*: these techniques rely on linguistic knowledge to identify KPs. Proposed methods include lexical analysis [5], syntactic analysis [6], and discourse analysis [7].
- *Machine Learning Approaches*: since KP extraction can be seen as a classification task, machine learning techniques can be used as well [8–10]. The usage of Naive Bayes, SVM and other supervised learning strategies has been widely discussed and applied in systems such as KEA [11], LAKE [12], and GenEx [9].
- *Other Approaches*: other strategies exist which do not fit into one of the above categories and most of the times they are hybrid approaches combining two or more of the above techniques. Among others, heuristic approaches based on knowledge-based criteria [13], and meta-knowledge over the domain [14] have been proposed.

Also the problem of defining multi-language approaches has been discussed by several authors. In [15] it is presented a multilingual approach towards sentence extraction for summarization purposes based on a machine learning approach. The authors of [16] introduce a multilingual KP extraction system exploiting a statistical approach based on word frequency and a reference corpus in 11 different European languages, including Italian. The performance of such system, however, relies on the quality of the reference corpus since phrases not included in the corpus will never be extracted from the text. Moreover, its accuracy proved to be highly variable over the 11 considered languages and overall poor. The authors of [17] propose a more sophisticated approach based on a set of heuristic rules for identifying a set of potentially good candidate KPs; candidate KPs are then selected according to a TF-IDF based score metric. The system exploits two language dependant resources: a stopwords list and a stemmer. Upon a suitable substitution of such language dependant resources, the system proved to perform well in different languages.

Keyphrase extraction from Italian texts has received little attention. The authors of [18] propose TAGME, a system whose purpose is to annotate documents with hyperlinks to Wikipedia pages by identifying *anchors* in the text. The task of identifying text anchors can be seen as a naive KP extraction technique and is capable to identify and propose KPs only if they are also in Wikipedia. The system by [16], previously mentioned, is also capable of extracting KPs from Italian text, however it features a very limited accuracy.

3   A Knowledge-Based Approach to KeyPhrase Generation

In order to accomplish our goals and to take into consideration our previous work on keyphrase extraction for English texts [19], we propose here a *Knowledge-Based* KP extraction technique based upon (i) exploitation of several kinds of knowledge, (ii) consideration of the specific languages addressed, and (iii) typical/common writing styles. An initial design work of knowledge engineering allowed us to identify four classes of knowledge which can be exploited to recognize meaningful phrases in a text:

1. *Statistical Knowledge:* this knowledge deals exclusively with the quantitative aspects of natural language, such as the frequency of a given word in a text or its inverse document frequency in a corpus; though lacking of a clear semantic meaning, it can be useful to identify terms and phrases that characterize a text.

2. *Linguistic Knowledge:* this knowledge comes from the specific language considered and deals with morphological and grammatical aspects of the text; examples of linguistic knowledge are Part-Of-Speech (POS) tags, the information on whether a given word is a stopword or not, or whether a given sequence of words is constituted by an acceptable pattern of POS tags for a KP (such as, for instance: "noun-noun" or "adjective-noun").

3. *Meta/Structural Knowledge:* this knowledge consists of heuristics over the general structure of the text and typically deals with the position of a phrase in the considered document; an example of meta-knowledge is knowing that phrases appearing in the abstract of an article may be more representative than the ones included in its body. This knowledge corresponds to various writing styles exploited by the author of the text. Another example of exploitable meta-knowledge is constituted by some specific metadata inserted in a document by the author (such as the "topic" meta-tag in Web pages and the "subject" meta-tag in a *PDF* file).

4. *Semantic/Social Knowledge:* this knowledge comes from sources external to the considered text. Semantic knowledge deals with the meaning of the terms present in the candidate KPs and with the typical conceptual context where they are used. An ideal source of semantic knowledge is constituted by ontologies, which describe concepts, their properties, and their mutual relationships, together with the natural language terminology usually exploited for linguistically referring to them. Other common sources of such kind of knowledge are dictionaries, thesauri, classification schema, etc. This knowledge is useful for recognizing terms belonging to a specific jargon and for resolving polysemic words. Other relevant examples of sources of semantic knowledge, which are becoming more and more popular in the participative Web (Web 2.0), are fast growing collaborative dictionaries, thesauri and knowledge bases, such as DBpedia. They feature a very wide conceptual coverage and they provide a way to socially validate candidate KP: for a candidate KP being an entry of one of these sources, means that other humans have already identified it as a meaningful way to linguistically refer to the underlined concept. This is the reason why we consider appropriate to attach to this kind of knowledge also the term "social".

It is important to point out how such classes of knowledge differ from each other in terms of domain and language dependency: as shown in Fig. 1 statistical knowledge is both domain and language independent, linguistic knowledge is domain independent, but language dependent, meta/structural knowledge is domain dependent, and, finally semantic/social knowledge may be both domain and language dependent. Domain and language dependency are very different. Domain dependency can be sensibly reduced by considering only general assumptions, such as assuming that most of the interesting concepts of a document will be introduced in its first section. It can also be turned down by taking into account information gathered from dictionaries or ontologies with a very broad scope (such as Wikipedia). Language dependency, on the other hand, cannot be relaxed: language dependent knowledge, indeed, needs dedicated modules and/or knowledge bases.

When reading a text with the purpose of extracting relevant concepts a human expert typically performs various kinds of evaluations and we believe that, in order to match the performance of a human, an automatic system should try to follow the same process. To this purpose, the overall KP extraction process is organized into three stages: in the first phase, the text is analysed in order to

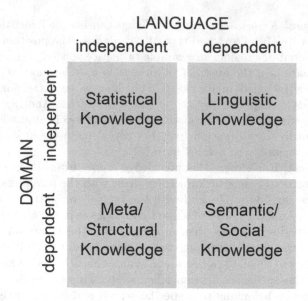

Fig. 1. Dependencies of the various kinds of knowledge considered.

identify all the possible candidate KPs to be possibly extracted from the text. Later, in a second phase, each candidate KP is scored by associating it to a set of features which are the result of applying the various kinds of knowledge described above to the specific candidate KP. More specifically, each class of knowledge is mapped into one or more features and the final selection criterion of candidate KPs takes into account all the features. The chosen features are then combined to produce a final decision associated to the candidate KP: this can be performed, for instance, by means of a unique score or of a multi-dimensional classification technic. This knowledge based approach can be used both in a supervised and an unsupervised scenario. In a supervised scenario the feature combination function could be the result of a training activity of a machine learning algorithm (e.g.: Bayesian classifier, Support Vector Machine, Artificial Neural Network, etc.), while in an unsupervised approach it is explicitly known and may be the result of a knowledge engineering activity. Finally, in the third phase other relevant KPs are generated once the major concepts included in the text have been extracted. In this stage, a domain-dependent inference process takes place, able to identify other (usually more general or related) concepts that are derived starting from the concepts (KPs) extracted in the first two stages and by exploiting external semantic/social knowledge.

4   System Overview

In order to support our claims we have developed *DIKpE-G*, a revised extended version of the system presented in [19,20]. DIKpE-G stands for *Domain Independent Keyphrase Extractor - Generator*. Figure 2 shows the overall organization of the system.

The data workflow mimics the 3-phase cognitive process described in the previous section. First of all the text is read and the *KP Extraction Module* (*KPEM*) discovers and ranks concepts (KPs) that appear in the text, then the *KP Inference Module* (*KPIM*) augments the set of extracted KPs with new linked, related or implied concepts. Operation of DIKpE-G is also supported by *External Knowledge Sources* (*EKS*): in the current implementation we exploit *Wikipedia*[1] and *Wordnik*[2]. The generated KPs represent tacit and explicit knowledge because part of them is explicitly contained in the text and the rest of them are inferred starting from the ones already present in the text.

In order to identify the KPs, the KPEM relies on a series of *Language Specific Resources* (*LSR*). They consist of a *POS-Tagger* module, a *Stemmer* module and two repositories: one for stopwords and one for POS-Patterns that typically characterize KPs. Decoupling the language dependent part from the rest of the architecture allows us to easily port the system to other languages. All the necessary language dependent modules are in fact widely available for all major languages: for example, the *Snowball stemmer* library[3] provides functionality for over twenty languages and the *TreeTagger*[4] provides POS tagging for over fifteen languages.

The extraction task is organized in two steps: the candidate KPs selection and the ranking phase. In the first step all possible sequences of one, two, three, and four words are considered, but only the ones matching a valid POS pattern are chosen as candidate KPs. Identification of valid POS patterns is a knowledge engineering task and can be carried out by considering widely used patterns (indicated as "valid") in a large enough set of human generated KPs (human

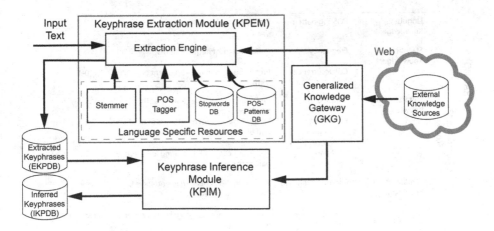

Fig. 2. Architecture of the DIKpE-G System.

[1] www.wikipedia.org.

[2] www.wordnik.com.

[3] snowball.tartarus.org.

[4] www.cis.uni-muenchen.de/~schmid/tools/TreeTagger.

generated such as the author KPs included in scientific papers). The number of POS patterns depends on the considered tag set. Currently we have a dozen POS patterns for the Italian language and about 40 for the English language. The difference is due to the different granularity of the employed TAG set.

In the following second step, each candidate KP is assessed by means of a set of features, which are computed by exploiting the various classes of knowledge previously described in Sect. 3. In the current implementation of DIKpE-G, we are experimenting the set of features introduced in [20]. More specifically, in Fig. 3, we show, for the various steps of the extraction, the different classes of knowledge taken into account, the relative features considered and, for each of them, their purposes and value range.

As it can be noticed in Fig. 3, each feature has a value varying in various ranges. Once for each KP a specific set of values have been computed for its features, a final ranking step is performed, which is aimed at producing a final global rank for each KP. The result is a ranked list of KPs: the highest ranked are proposed as relevant keyphrases for the input text. In our vision, the ranking step can be performed in various ways, ranging from (i) a strictly numerical approach to (ii) a more sophisticated and general knowledge-based assessment based on both qualitative and quantitative reasoning. The highly modular archi-

		Knowledge Class	Feature	Purpose	Value Range
KP EXTRACTION	Candidate KP Identification	Linguistic Knowledge	POS Tag patterns	Excluding certain patterns	
			Stop-word list	Excluding certain words	
			Stemming	Working on common stems	
	Candidate KP Scoring	Linguistic Knowledge	POS Tag patterns	Preferring typical patterns	0-1
		Statistical Knowledge	Frequency	Preferring most frequent terms	0-1
			Co-Occurrence	Preferring common co-occurrent patterns	0-1
		Meta/Structural Knowledge	Phrase depth	Preferring concepts appearing at the beginning of the text	0-1
			Phrase last Occurrance	Preferring concepts mentioned at the end of the text	0-1
			Life Span	Preferring concepts appearing in a large part of the text	0-1
		Semantic/Social Knowledge	Flag of presence in EKS	Preferring KPs appearing in ontologies, dictionaries, thesauri, ...	boolean
			Flag of presence in Web 2.0 EKS	Preferring concepts recognized by other human actors	boolean
KP INFERENCE		Semantic/Social Knowledge	Navigation paths in EKS	Inferring new KPs related to many extracted KPs	
			Navigation paths in EKS	Disambiguating polysemic inferred KPs	

Fig. 3. Usage of the various classes of knowledge proposed in DIKpE-G.

tecture of DIKpE-G, allows a seamless substitution of the modules and sub-modules devoted to ranking, permitting in such a way the experimentation of alternative approaches. The current DIKpE-G prototype follows the approach proposed in [19], which adheres to a numerical approach: each feature is given a numerical value and all the features are then combined in order to compute a unique index called *keyphraseness*, which represents how much a candidate KP is considered suitable and significant for representing the content of the input text. The keyphraseness index is computed in the current DIKpE-G prototype as a weighted linear combination of the features values. The features weights are currently experimentally obtained. However we are exploring new approaches, namely (i) rule based reasoning for mapping the various features in an n-dimensional space, where different regions of space are associated to different levels of the keyphraseness index and (ii) machine learning techniques for associating (by means of training based on ad-hoc annotated data sets) the set of the features' values of the single KPs to the corresponding level of keyphraseness.

The final phase is devoted to inferring new KPs (i.e. KPs which are not already present in the input text) starting from the topmost ranked extracted KPs. The KPIM considers each extracted KP in order to match it against the entries of the available EKSs: if a match is found (i.e. the considered KP is also an entry of a specific EKS), all the concepts (terms) present in the EKS and linked to the matching entry are considered as candidate *inferred* KPs. All the candidate inferred KPs collected from all the extracted KPs are then ranked according to the sum of the keyphraseness values of the extracted KPs from which they have been derived. Note that inferred KPs can be obtained both from hi-ranked or low-ranked extracted KPs. For instance the system can infer a KP that is linked to a large number of low-ranked KPs rather than a KP that is linked to a little number of hi-ranked ones. The top-n inferred KPs are finally returned as output together with the extracted KPs identified by the KPEM.

5 Evaluation

In order to support and validate our approach several experiments have been performed. To evaluate the performance when considering English texts, the original version [19] was benchmarked against the KEA algorithm on a set of 215 English documents labelled with keyphrases generated by the authors and by additional experts. The comparison was performed only on the KP extraction capabilities and not on the inference ones. For each document, the KP sets returned by the two compared systems were matched against the set of human generated KPs. Each time a machine-generated KP matched a human-generated KP, it was considered a correct KP; the number of correct KPs generated for each document was then averaged over the whole data set. Various machine-generated KP set sizes were tested. As shown in Table 1, the DIKpE system significantly outperformed the KEA baseline and the improvement increases as the KPs set size increases.

When the DIKpE prototype has been extended into the current DIKpE-G prototype, we have added knowledge bases in order to cover also the Italian

Table 1. Performance of DIKpE compared to KEA.

Extracted Keyphrases	Average number of correct KPs	
	KEA	DIKpE
7	2.05	3.86
15	2.95	5.29
20	3.08	5.92

language. The initial experimental evaluation activity has concerned the Italian language and it has shown very encouraging results. Due to the lack of extensive labelled corpora and available baseline systems, the evaluation of DIKpE-G on the Italian language has followed so far a qualitative approach. A set of 50 papers was gathered, and 11 to 16 KPs were automatically extracted from each paper. A dozen of human experts of various ages and gender were then asked to read all the texts and to assess the quality of extracted KPs. The main goal of the experiment was to identify common pitfalls of the KP extraction process and to classify unsatisfactory KPs extracted. Table 2 shows the seven classes identified and their relative frequency. A significant number of KPs were perceived as "too generic" by our experts; in particular these KPs are generally made of a single word with a very generic meaning such as "catene" (chains) or "funzione" (function) and often were included in other KPs made of multiple words (such as "catene montuose", that means "mountain ranges"). Another frequent flaw in the extracted KPs by DIKpE-G was the presence of incomplete phrases such as "spaziale Orion". However also these KPs were often part of a longer phrase that was returned as well ("navicella spaziale Orion"). These observations led us to introduce a simple heuristic consisting in not returning short phrases which are included in longer ones already in the extracted set. This simple mechanism allowed us to significantly increase to 75 % the fraction of good KPs as they were presented again to the expert pool.

Results gathered so far are promising, however development is still in progress and further more systematic evaluation activities are planned: we want to evaluate the KP inference capabilities for both the English and the Italian language.

Table 2. Results of user evaluation.

Evaluation	Frequency
Good	56.28 %
Too Generic	14.72 %
Too Specific	2.27 %
Incomplete	9.85 %
Not Relevant	9.85 %
Meaningless	7.03 %

6   Conclusions and Future Work

In this paper, we present a novel knowledge-based multilingual approach for concept mining that can be easily extended to any given Western language due to the actually large availability of linguistic resources such as POS taggers and stemming algorithms. The preliminary evaluation of the experimental results suggests that once a satisfactory set of language-specific resources is available, the overall quality of the generated KPs is not affected by the language switch. The four different classes of knowledge considered provide a conceptual framework with a higher level of abstraction than other state-of-the-art systems, featuring a clear separation between language dependent and independent KP selection criteria. Such framework allows us to overcome several shortcomings of the current systems which often consider only one or two classes of knowledge. Moreover, the unsupervised nature of our approach allows our system to accomplish its task with no need of training data, which is a major advantage for non-English languages because of the tremendous lack of annotated data corpora that we are experiencing nowadays.

Results gathered so far show a promising outlook and the system can be effectively employed in several application domains, such as digital libraries and recommender systems.

Our future work will therefore address all the major issues highlighted by the expert evaluation, such as a still high number of KPs perceived as too generic. We also aim at improving the overall underlined conceptual model of human KP generation, by further analysing the four knowledge classes identified and by refining the reasoning process exploited in the system. We plan to observe how experts identify KPs, for instance, by thinking-aloud interviews. The user interaction should be improved as well, since the system actually acts as a black box giving little or no hints to the final user of the process that selected a particular KP, and this encourages distrust in the system. In order to address this issue, the development of an interactive explanation and result tracking interface is ongoing. Finally, specific attention will be devoted to the evaluation issues, both (i) for improving and completing the evaluation of our approach and (ii) for contributing to the development of a methodological standard for evaluating KP extraction and KP inference capabilities systems.

References

1. Velardi, P., Navigli, R., Cucchiarelli, A., D'Antonio, F.: A new content-based model for social network analysis. In: ICSC, pp. 18–25. IEEE Computer Society (2008)
2. W3Techs: Usage of content languages for websites (2014). http://w3techs.com/technologies
3. Zhang, C.: Automatic keyword extraction from documents using conditional random fields. J. Comput. Inf. Syst. **4**, 1169–1180 (2008)
4. Matsuo, Y., Ishizuka, M.: Keyword extraction from a single document using word co-occurrence statistical information. Int. J. Artif. Intell. Tools **13**, 157–169 (2004)

5. Barker, K., Cornacchia, N.: Using noun phrase heads to extract document keyphrases. In: Hamilton, H.J. (ed.) Canadian AI 2000. LNCS (LNAI), vol. 1822, pp. 40–52. Springer, Heidelberg (2000)
6. Fagan, J.: Automatic phrase indexing for document retrieval. In: Proceedings of the 10th Annual International ACM SIGIR Conference on Research and Development in Information Retrieval, SIGIR 1987, pp. 91–101. ACM, New York (1987)
7. Krapivin, M., Marchese, M., Yadrantsau, A., Liang, Y.: Unsupervised key-phrases extraction from scientific papers using domain and linguistic knowledge. In: Third International Conference on Digital Information Management, ICDIM 2008, pp. 105–112 (2008)
8. Frank, E., Paynter, G.W., Witten, I.H., Gutwin, C., et al.: Domain-specific keyphrase extraction. In: Proceedings of the Sixteenth International Joint Conference on Artificial Intelligence, pp. 668–673. Morgan Kaufmann Publishers (1999)
9. Turney, P.D.: Learning algorithms for keyphrase extraction. Inf. Retrieval 2, 303–336 (2000)
10. Hulth, A.: Improved automatic keyword extraction given more linguistic knowledge. In: Proceedings of the 2003 Conference on Empirical Methods in Natural Language Processing, EMNLP 2003, Stroudsburg, PA, USA, pp. 216–223. Association for Computational Linguistics (2003)
11. Witten, I. H., Paynter, G. W., Frank, E., Gutwin, C., Nevill-Manning, C.G.: Kea: practical automatic keyphrase extraction. In: Proceedings of the fourth ACM conference on Digital libraries, pp. 254–255. ACM (1999)
12. DAvanzo, E., Magnini, B., Vallin, A.: Keyphrase extraction for summarization purposes: the lake system at duc-2004. In: Proceedings of the 2004 Document Understanding Conference (2004)
13. Liu, Z., Li, P., Zheng, Y., Sun, M.: Clustering to find exemplar terms for keyphrase extraction. In: Proceedings of the 2009 Conference on Empirical Methods in Natural Language Processing: Volume 1 - Volume 1. EMNLP 2009, Stroudsburg, PA, USA, pp. 257–266. Association for Computational Linguistics (2009)
14. Danilevsky, M., Wang, C., Desai, N., Guo, J., Han, J.: Kert: Automatic extraction and ranking of topical keyphrases from content-representative document titles (2013). arXiv preprint arXiv:1306.0271
15. Litvak, M., Last, M., Friedman, M.: A new approach to improving multilingual summarization using a genetic algorithm. In: Proceedings of the 48th Annual Meeting of the Association for Computational Linguistics, pp. 927–936. Association for Computational Linguistics (2010)
16. Paukkeri, M.S., Nieminen, I.T., Pöllä, M., Honkela, T.: A language-independent approach to keyphrase extraction and evaluation. In: COLING (Posters), pp. 83–86 (2008)
17. El-Beltagy, S.R., Rafea, A.: Kp-miner: a keyphrase extraction system for english and arabic documents. Inf. Syst. 34, 132–144 (2009)
18. Ferragina, P., Scaiella, U.: Tagme: on-the-fly annotation of short text fragments (by wikipedia entities). In: Proceedings of the 19th ACM International Conference on Information and Knowledge Management, CIKM 2010, pp. 1625–1628. ACM, New York (2010)
19. Pudota, N., Dattolo, A., Baruzzo, A., Ferrara, F., Tasso, C.: Automatic keyphrase extraction and ontology mining for content-based tag recommendation. Int. J. Intell. Syst. 25, 1158–1186 (2010)
20. De Nart, D., Tasso, C.: A domain independent double layered approach to keyphrase generation. In: WEBIST 2014 - Proceedings of the 10th International Conference on Web Information Systems and Technologies, pp. 305–312. SCITEPRESS Science and Technology Publications (2014)

A Generic and Declarative Method
for Symmetry Breaking in Itemset Mining

Belaïd Benhamou[1]([✉]), Saïd Jabbour[2], Lakhdar Sais[2], and Yacoub Salhi[2]

[1] Aix-Marseille Université, Laboratoire des Sciences de l'information et des Systèmes
(LSIS), Domaine Universitaire de Saint Jérôme, Avenue Escadrille Normandie
Niemen, 13397 Marseille Cedex 20, France
`belaid.benhamou@univ-amu.fr`
[2] Centre de Recherche en Informatique de Lens (CRIL), Université d'Artois,
Rue Jean Souvenir, SP 18, 62307 Lens Cedex, France
`{jabbour,sais,salhi}@cril.fr`

Abstract. The search of frequent patterns in a transaction database is
a well-known problem in the field of data mining. Several specific methods for solving this problem and its variants have been developed in the
data mining community. A generic and declarative alternative approach
to these targeted and specific methods was recently introduced. It consists in representing the data mining problems as constraint networks,
then use an appropriate solver as a black box to solve the encoded problem. For instance, several works expressed the frequent itemset mining
problem and its variants as constraint networks or Boolean satisfiability,
and thus offer the possibility of using the associated efficient solvers to
solve the problem. On the other hand, the symmetry notion was very
invested and shown to be efficient in the fields of constraint programming and propositional satisfiability. The principle of symmetry could
be exported to other areas where some structures can be exploited effectively. Especially, in the field of data mining where several tasks can be
expressed as constraint networks. In this work, we propose a generic and
declarative method to eliminate symmetries in data mining problems
expressed as Boolean constraints. We show how the symmetries between
items of a transaction database can be detected and eliminated by adding
symmetries breaking predicates (SBP) to the Boolean encoding of the
considered data mining problem.

Keywords: Data mining · Itemset mining · Symmetry · Satisfiability ·
Constraint programming

1 Introduction

In this paper, we investigate the notion of symmetry elimination in Frequent
Itemset Mining (FIM) [1]. The itemset mining problem has several applications
and remains central in the Data mining research field. The most known example
is the one considered by large retail organizations called *basket data*. A record of

© Springer International Publishing Switzerland 2015
A. Fred et al. (Eds.): IC3K 2014, CCIS 553, pp. 143–160, 2015.
DOI: 10.1007/978-3-319-25840-9_10

such data contains essentially the customer identification, the transaction date and the items bought by the customer. Advances in bar-codes technology, the use of credit cards of frequent-customer card make it now possible to collect and store a great amounts of sale data. It is then important for the retail firms to know the set of items that are frequently bought by customers. This is the frequent itemset mining problem. Since its introduction in 1993 [1], several highly scalable algorithms are introduced ([2,15,21,25,35,43,44,46]) to enumerate the sets of frequent itemsets. The two challenging questions investigated in such algorithms are: in one hand how to compute all the frequent itemsets in a reasonable CPU time and in the other hand how to compact the output and reduce its size when there is a huge number of frequent itemsets. Many other data mining tasks exist, such as the association rule mining, the frequent pattern, clustering and episode mining, but almost all of them are closely in relationship to itemset mining which looks to be the canonical problem. A lot of efficient and scalable algorithms are developed for target and specific mining tasks. As stated in [41], different methods for the itemset mining are provided. Mainly they differ from each other in the way they explore the search space, the data structure they use, the exploitation of the anti-monotonicity property. The other important point is the size of the output of such algorithms. Some solutions are found, for instance one can enumerate only the closed, the maximal, the condensed, the preferred, or discriminative itemsets instead of all the frequent itemsets.

Data mining community introduced the *constraint-based mining* framework in order to specify in terms of constraints the properties of the patterns to be mined ([12–14,37]). A wide variety of constraints are successfully integrated and implemented in different specific data mining algorithms.

Recently De Raedt et al. ([23,39]) introduced the alternative of using constraint programming in data mining. They showed that a such alternative can be efficiently applied for a wide range of pattern mining problems. Most of the pattern mining constraint (e.g. frequency, closeness, maximality, and anti-monotonicity constraints) had been expressed in a declarative constraint programming language. The data mining problem is modeled as a constraint satisfaction problem (CSP) and a solver (e.g. Gecode) is then used to enumerate solutions corresponding to the set of interesting patterns. A strength point here is that different constraints can be combined without the need to modify the solver, unlike in the existing specific data mining algorithms. Since the introduction of this declarative approach, there is a growing interest in finding generic and declarative approaches to model and solve data mining tasks. For instance, several works expressed data mining problems as propositional satisfiability ([26,28,29,31,33,40]) and used efficient modern SAT solvers as black-box to solve them. More recently, a constraint declarative framework for solving Data mining tasks called MiningZinc [22], had been introduced.

On the other hand, symmetry is by definition a multidisciplinary concept. It appears in many fields ranging from mathematics to Artificial Intelligence, chemistry and physics. It reveals different forms and uses, even inside the same field. In general, it returns to a transformation, which leaves invariant (does not modify its fundamental structure and/or its properties) an object (a figure, a molecule, a physical system, a formula or a constraints network...). For instance, rotating a chessboard up to 180 degrees gives a board that is indistinguishable from the original one. Symmetry is a fundamental property that can be used to study these various objects, to finely analyze these complex systems or to reduce the computational complexity when dealing with combinatorial problems.

As far as we know, the principle of symmetry has been first introduced by Krishnamurthy [32] to improve resolution in propositional logic. Symmetries for Boolean constraints are studied in depth in [8,9]. The authors showed how to detect them and proved that their exploitation is a real improvement for several automated deduction algorithms efficiency. Since that, many research works on symmetry appeared. For instance, the static approach used by James Crawford et al. in [16] for propositional logic theories consists in adding constraints expressing global symmetry of the problem. This technique has been improved in [6] and extended to 0–1 Integer Logic Programming in [3]. The notion of interchangeability in Constraint Satisfaction Problems (CSPs) is introduced in [19] and symmetry for CSPs is studied earlier in [7,38].

In the context of constraint programming, Guns et al. [24] used symmetry breaking constraints to impose a strict ordering on the patterns in k-pattern set mining. More recently, symmetry detection and elimination are integrated in itemset mining problems [27,30]. Two different approaches are proposed. In the first one, symmetries are eliminated by rewriting the transaction database (eliminating items), while in the second approach the authors integrate symmetry elimination in Apriori-like algorithms. For other previous studies on symmetries in data mining, we refer the reader to the related work section.

The work that we investigate in this paper, goes in this direction. It consists in detecting and eliminating symmetries in the itemset mining problem expressed as a Boolean satisfiability. We will show how *global symmetries*[1] of the given transaction database are detected and expressed in terms of symmetry breaking predicates. Such predicates are added to the boolean encoding of the itemset mining problem in a preprocessing step and a SAT solver is used as a black box to enumerate the non-symmetrical solutions (the non-symmetrical frequent itemsets). In most of the data mining tasks, we usually need to enumerate interesting patterns and this usually lead to an output of huge size. Eliminating symmetries might reduce the size of the output and lead to discover the non-symmetrical patterns which are the most important and representative of the knowledge.

The rest of the paper is organized as follows. In Sect. 2, we give some necessary background on the satisfiability problem, permutations and the necessary notion on itemset mining problem. We study the notion of symmetry in itemset mining represented as boolean constraints in Sect. 3. In Sect. 4 we show how symmetries can be detected by means of graph automorphism. We show in Sect. 5 how this symmetry can be eliminated by adding symmetry breaking predicates to the Boolean encoding. Section 6 gives experiments on different data-sets to show the advantage of using symmetries in itemset mining. Section 7 investigates the related works and Sect. 8 concludes the work.

---

[1] Symmetries that are present in the initial formulation of the problem.

2   Background

We summarize in this section some background on the satisfiability problem, permutations, and itemset mining problem.

2.1   Propositional Satisfiability (SAT)

We shall assume that the reader is familiar with propositional logic. We give here, a short description. Let V be the set of propositional variables called only variables. Variables will be distinguished from literals, which are variables with an assigned parity 1 or 0 that means $True$ or $False$, respectively. This distinction will be ignored whenever it is convenient, but not confusing. For a propositional variable p, there are two literals: p the positive literal and $\neg p$ the negative one.

A clause is a disjunction of literals such that no literal appears more than once, nor a literal and its negation at the same time. This clause is denoted by $p_1 \vee p_2 \vee \ldots \vee p_n$. A formula $\mathcal{F}$ in conjunctive normal form (CNF) is a conjunction of clauses.

A truth assignment to a CNF $\mathcal{F}$ is a mapping ρ defined from the set of variables of $\mathcal{F}$ into the set $\{True, False\}$. If $\rho[p]$ is the value for the positive literal p then $\rho[\neg p] = \neg \rho[p]$. The value of a clause $p_1 \vee p_2 \vee \ldots \vee p_n$ in ρ is $True$, if the value $True$ is assigned to at least one of its literals in ρ, $False$ otherwise. By convention, we define the value of the empty clause ($n = 0$) to be $False$. The value $\rho[\mathcal{F}]$ is $True$ if the value of each clause of $\mathcal{F}$ is $True$, $False$, otherwise. We say that a CNF formula $\mathcal{F}$ is satisfiable if there exists some truth assignments ρ that assign the value $True$ to $\mathcal{F}$, it is unsatisfiable otherwise. In the first case I is called a model of $\mathcal{F}$. Let us remark that a CNF formula which contains the empty clause is unsatisfiable.

It is well-known [42] that for every propositional formula $\mathcal{F}$ there exists a formula $\mathcal{F}'$ in conjunctive normal form (CNF) such that $\mathcal{F}'$ is satisfiable iff $\mathcal{F}$ is satisfiable. In the following we will assume that the formulas are given in a CNF.

2.2   Permutations

Let $\Omega = \{1, 2, \ldots, N\}$ for some integer N, where each integer might represent a propositional variable. A permutation of Ω is a bijective mapping σ from Ω to Ω that is usually represented as a product of cycles of permutations. We denote by $Perm(\Omega)$ the set of all permutations of Ω and $\circ$ the composition of the permutation of $Perm(\Omega)$. The pair $(Perm(\Omega), \circ)$ forms the permutation group of Ω. That is, $\circ$ is closed and associative. The inverse of a permutation is a permutation and the identity permutation is a neutral element. A pair $(T, \circ)$ forms a sub-group of $(S, \circ)$ iff T is a subset of S and forms a group under the operation $\circ$. The orbit $\omega^{Perm(\Omega)}$ of an element ω of Ω on which the group $Perm(\Omega)$ acts is $\omega^{Perm(\Omega)} = \{\omega^\sigma | \omega^\sigma = \sigma(\omega), \sigma \in Perm(\Omega)\}$. A generating set of the group $Perm(\Omega)$ is a subset Gen of $Perm(\Omega)$ such that each element of $Perm(\Omega)$ can be written as a composition of elements of Gen. We write $Perm(\Omega) = < Gen >$. An element of Gen is called a generator. The orbit of $\omega \in \Omega$ can be computed by using only the set of generators Gen.

2.3 Frequent, Closed and Maximal Itemset Mining Problems

Let $\mathcal{I} = \{0,\ldots,m-1\}$ be a set of m items and $\mathcal{T} = \{0,\ldots,n-1\}$ a set of n transactions (transaction identifiers). A subset $I \subseteq \mathcal{I}$ is called an itemset and a transaction $t \in \mathcal{T}$ over $\mathcal{I}$ is in fact, a pair (t_{id}, I) where t_{id} is the transaction identifier and I the corresponding itemset. In the *basket data* example, t_{id} represents the customer identification and I the set of items he put in his basket (he bought). Usually, when there is no confusing, a transaction is just expressed by its identifier. A transaction database $\mathcal{D}$ over $\mathcal{I}$ is a finite set of transactions such that no different transactions have the same identifier. Such a data set expresses in the *basket data* the different transactions made by the customers. A transaction database can be seen as a binary matrix $n \times m$, where $n = |\mathcal{T}|$ and $m = |\mathcal{I}|$, with $\mathcal{D}_{t,i} \in \{0,1\}$ forall $t \in \mathcal{T}$ and forall $i \in \mathcal{I}$. More precisely, a transaction database is expressed by the set $D = \{(t,I) \mid t \in \mathcal{T}, I \subseteq \mathcal{I}, \forall i \in I : \mathcal{D}_{t,i} = 1\}$. The *coverage* $C_{\mathcal{D}}(I)$ of an itemset I in a transaction database $\mathcal{D}$ is the set of all transactions in which I occurs. That is, $C_{\mathcal{D}}(I) = \{t \in \mathcal{T} \mid \forall i \in I, \mathcal{D}_{t,i} = 1\}$. The *support* $S_{\mathcal{D}}(I)$ of an itemset I in $\mathcal{D}$ is the number $|C_{\mathcal{D}}(I)|$ of transactions supporting I. It is just the cardinality of its coverage set. Moreover, the *frequency* $F_{\mathcal{D}}(I)$ of I in $\mathcal{D}$ is defined by $\frac{|C_{\mathcal{D}}(I)|}{|\mathcal{D}|}$.

Table 1. An instance of a transaction database.

t_id	itemset
001	Beer, Wine, Whisky, Vodka, Cognac, Water
002	Beer, Wine, Whisky, Vodka, Gin, Water
003	Beer, Wine, Whisky, Water
004	Beer, Wine, Vodka, Water
005	Ricard, Coke, Pepsi, Water
006	Pastis, Pepsi, Coke, Water
007	Shweps, Orangina, Pepsi
008	Shweps, Orangina, Coke
009	Juice, Orangina, Pepsi
010	Juice, Orangina, Coke

Example 1. Consider the transaction database $\mathcal{D}$ of Table 1 made over the set of drink items $\mathcal{I} = \{Beer, Wine, Whisky, Cognac, Vodka, Pastis, Ricard, Gin, Co-ke, Pepsi, Shweps, Juice, Water, Orangina\}$. For example, we can see in Table 1 that the itemset $I = \{Beer, Wine\}$ has $C_{\mathcal{D}}(I) = \{001, 002, 003, 004\}$, $S_{\mathcal{D}}(I) = |C_{\mathcal{D}}(I)| = 4$, and $F_{\mathcal{D}}(I) = 0,4$.

Given a transaction database $\mathcal{D}$ over $\mathcal{I}$, and θ a minimal support threshold, an itemset I is said to be frequent if $S_{\mathcal{D}}(I) \geq \theta$. I is a closed frequent itemset if in addition to the frequency constraint it satisfies the following constraint: for all itemset J such that $I \subset J$, $S_{\mathcal{D}}(I) > S_{\mathcal{D}}(J)$. I is said to be a maximal frequent itemset if in addition to the frequency constraint it satisfies the following constraint: for all itemset J such that $I \subset J$, $S_{\mathcal{D}}(J) < \theta$. Both closed and maximal itemsets are two known condensed representation for frequent itemsets. The data mining tasks we are dealing with in this work are defined as follows:

Definition 1.   1. *The frequent itemset mining task consists in computing the following set* $\mathcal{FIM}_{\mathcal{D}}(\theta) = \{I \subseteq \mathcal{I} | S_{\mathcal{D}}(I) \geq \theta\}$.
2. *The closed frequent itemset mining task consists in computing the following set* $\mathcal{CLO}_{\mathcal{D}}(\theta) = \{I \in \mathcal{FIM}_{\mathcal{D}}(\theta) | \forall J \subseteq \mathcal{I}, I \subset J, S_{\mathcal{D}}(I) > S_{\mathcal{D}}(J)\}$.
3. *The maximal frequent itemset mining task consists in computing the following set* $\mathcal{MAX}_{\mathcal{D}}(\theta) = \{I \in \mathcal{FIM}_{\mathcal{D}}(\theta) | \forall J \subseteq \mathcal{I}, I \subset J, S_{\mathcal{D}}(J) < \theta\}$.

The anti-monotonicity property in itemset mining expresses the fact that all the subsets of a frequent itemset are also frequent itemsets. More precisely:

Proposition 1 (Anti-monotonicity). *Let θ be a minimal support threshold, if the itemset I is such that $S_{\mathcal{D}}(I) \geq \theta$, then $\forall J \subseteq I$, $S_{\mathcal{D}}(J) \geq \theta$.*

3   Symmetry in Itemset Mining

Both constraint programming and Satisfiability are two known declarative programming frameworks where the user has just to specify the problem he want to solve rather than specifying how to solve it. The frequent itemset mining tasks and some of its variants (closed, maximal, etc.) had been encoded for the first time in [23,39] as constraint programming tasks where a constraint solver could be used as a black box to solve them. Since that, other works ([26,28,29,31,33,40]) expressed the data mining tasks as a satisfiability problem where the mining tasks are represented by propositional formulas that are translated into their conjunctive normal forms (CNF) which will be given as inputs to a SAT solver. In this work we use the encoding proposed in [29] which we augment by the symmetry breaking predicates that are used to avoid enumerating the symmetrical models or the symmetrical no-goods of the resulting CNF encoding.

The general idea behind the CNF encoding of an itemset mining task defined on a transaction database $\mathcal{D}$ is to encode each of its interpretations as a pair (I', T') where the set of item variables I' represent an itemset $I \subseteq \mathcal{I}$ and the set of transaction variables T' express its corresponding covering $T \subseteq \mathcal{T}$ in the transaction database $\mathcal{D}$. To do that, a boolean variable I'_i is associated with each item $i \in \mathcal{I}$ and a variable T'_t is associated with each transaction $t \in \mathcal{T}$. The itemset I is then defined by all the variables I'_i that are true. That is $I'_i = 1$, if $i \in I$, and $I'_i = 0$ if $i \notin I$. The set of transaction T covered by I is then defined by the set of variable T'_t that are true. That is, $T'_t = 1$ if $t \in C_{\mathcal{D}}(I)$ and $T'_t = 0$ if $t \notin C_{\mathcal{D}}(I)$.

For instance, the $\mathcal{FIM}_{\mathcal{D}}(\theta)$ task can be seen as the search of the set of models $M = \{(I', T') \mid I \subseteq \mathcal{I}, T \subseteq \mathcal{T}, T = C_{\mathcal{D}}(I), |T| \geq \theta\}$ of the corresponding

CNF encoding. We have to encode both the covering constraint $T = C_{\mathcal{D}}(I)$ and the frequency constraint $|T| \geq \theta$. These constraints are expressed by the following boolean and pseudo-boolean constraints:

$$\bigwedge_{t \in T} (\neg T'_t \leftarrow \bigvee_{i \in \mathcal{I}, \mathcal{D}_{t,i}=0} I'_i)$$

$$\sum_{t \in T} T'_t \geq \theta$$

The frequent closed itemset task is specified by adding to the two previous constraints the following constraints:

$$\bigwedge_{t \in T} (\neg T'_t \rightarrow \bigvee_{i \in \mathcal{I}, \mathcal{D}_{t,i}=0} I'_i)$$

$$\bigwedge_{i \in \mathcal{I}} ((\bigwedge_{t \in T} (T'_t \rightarrow \mathcal{D}_{t,i} = 1)) \rightarrow I'_i)$$

The maximal frequent itemset mining is specified by adding the following constraint:

$$\bigwedge_{i \in \mathcal{I}} ((\sum_{t \in T} T'_t \times \mathcal{D}_{t,i} \geq \theta) \rightarrow I'_i)$$

We denote by $CNF(k, \mathcal{D})$, the CNF formula encoding the data mining task k over the transaction database $\mathcal{D}$, where k refers to $\mathcal{FIM}_{\mathcal{D}}(\theta)$, $\mathcal{CLO}_{\mathcal{D}}(\theta)$ or $\mathcal{MAX}_{\mathcal{D}}(\theta)$. We also use a predicate $P^k_{\mathcal{D}}$ to represent the task k in $\mathcal{D}$. That is, an itemset $I \subseteq \mathcal{I}$ having $T \subseteq \mathcal{T}$ as a cover verifies $P^k_{\mathcal{D}}$ ($P^k_{\mathcal{D}}(I, T) = true$) if I is an itemset which is an answer to the data mining task k and T is its cover.

Remark 1. We recall that a model J of $CNF(k, \mathcal{D})$ is a pair (I', T') where the item variable part I' expresses the itemset $I \subseteq \mathcal{I}$ which is an answer to the considered task k and the transaction variable part T' encodes its cover $T \subseteq \mathcal{T}$. More precisely each literal I'_i which is true in I' represents the item i in the itemset I which is an answer to the task k and each literal T'_t which is true in T' represents the transaction t in T which is the corresponding cover of I. In the sequel we denote by the pair (I, T) the itemset and its cover that are extracted from an interpretation $J = (I', T')$ of $CNF(k, \mathcal{D})$.

Symmetry is well studied in constraint programming and propositional satisfiability. Since Krishnamurthy's [32] symmetry definition and the one given by Benhamou et al. in [10,11] in propositional logic, several other definitions are given by the CP community.

Symmetry has already been defined in itemset mining [27,30]. We give in the following a similar definition and show how to eliminate such symmetry by means of symmetry breaking predicates that we add to the Boolean encoding to solve efficiently some data mining tasks like frequent, closed or maximal itemset mining.

Definition 2. *Let $\mathcal{D}$ be a transaction database over a set of items $\mathcal{I}$. A symmetry of $\mathcal{D}$ is a permutation σ defined on $\mathcal{I}$ such that $\sigma(\mathcal{D}) = \mathcal{D}$.*

Remark 2. It is obvious to see that a permutation on the set of items $\mathcal{I}$, induces a permutation $\sigma_{\mathcal{T}}$ on the set of transactions $\mathcal{T}$ and a permutation $\sigma_{\mathcal{D}}$ on the data-set $\mathcal{D}$ itself. We denote such permutations only by σ when there is no confusion.

A symmetry of $\mathcal{D}$ is an item permutation that leaves $\mathcal{D}$ invariant. If we denote by $Perm(\mathcal{I})$ the group $(Perm(\mathcal{I}), \circ)$ of permutations of $\mathcal{I}$ and by $Sym(\mathcal{I}) \subset Perm(\mathcal{I})$ the subset of permutations of $\mathcal{I}$ that are the symmetries of $\mathcal{D}$, then $(Sym(\mathcal{I}), \circ)$ is trivially a sub-group of $Perm(\mathcal{I})$ denoted only by $Sym(\mathcal{I})$.

Theorem 1. *Let σ be a symmetry of a transaction database $\mathcal{D}$, $I \subseteq \mathcal{I}$ an itemset having a cover $T \subseteq \mathcal{T}$, and $P_{\mathcal{D}}^k$ the predicate expressing the data mining task k in $\mathcal{D}$, then $P_{\mathcal{D}}^k(I, T) = true$ iff $P_{\mathcal{D}}^k(\sigma(I), \sigma(T)) = true$.*

Proof. It is trivial to see that a symmetry of $\mathcal{D}$ verifies such property. Indeed, if σ is a symmetry of $\mathcal{D}$, then $\sigma(\mathcal{D}) = \mathcal{D}$, thus it results that $\mathcal{D}$ and $\sigma(\mathcal{D})$ have the same itemsets and covers satisfying the predicate $P_{\mathcal{D}}^k$. Thus σ must transform each itemset I with a cover T verifying the predicate $P_{\mathcal{D}}^k$ to an itemset $\sigma(I)$ with a cover $\sigma(T)$ verifying the predicate $P_{\mathcal{D}}^k$.

In other words the symmetry σ of $\mathcal{D}$ transforms each itemset I having a cover T which is a solution to the data mining task k into a symmetrical itemset $\sigma(I)$ having a cover $\sigma(T)$ which is also a solution for the task k. It also transforms each itemset which is not a solution to the task k into a symmetrical itemset which will not be a solution to the task k. For instance if the task k concerns the frequent itemset mining problem, then by applying σ to a frequent itemset I we obtain a symmetrical frequent itemset $\sigma(I)$. If I is not frequent, then $\sigma(I)$ will not be frequent too.

Example 2. Consider the transaction database defined in Table 1 of Example 1 and the permutation $\sigma = (Whisky, Vodka)\,(Cognac, Gin)\,(Ricard, Pastis)$ $(Wine, Beer)\,(Shweps, Juice)\,(Pepsi, Coke)$ which is defined on the set of items $\mathcal{I}$ of $\mathcal{D}$. We can see that $\sigma(\mathcal{D}) = \mathcal{D}$, then σ is a symmetry of $\mathcal{D}$.

Now, we give an important property which establishes a relationship between the symmetries of a transaction database $\mathcal{D}$ and the Boolean encoding $CNF(k, \mathcal{D})$ of the data mining task k defined over $\mathcal{D}$.

Proposition 2. *Let $\mathcal{D}$ be a transaction database, $CNF(k, \mathcal{D})$ the Boolean encoding of the data mining task k, σ a symmetry of $\mathcal{D}$ and $J = (I', T')$ an interpretation of $CNF(k, \mathcal{D})$, then J is a model of $CNF(k, \mathcal{D})$ iff $\sigma(J)$ is a model of $CNF(k, \mathcal{D})$.*

Proof. Let σ be a symmetry of the transaction database $\mathcal{D}$ and $J = (I', T')$ a model of the Boolean encoding $CNF(k, \mathcal{D})$. It results that the corresponding pair itemset and cover (I, T) verify the predicate $P_{\mathcal{D}}^k$ of the data mining task k, that is $P_{\mathcal{D}}^k(I, T) = true$. We have to prove that $\sigma(J) = (\sigma(I'), \sigma(T'))$ is also a model of $CNF(k, \mathcal{D})$. The permutation σ is a symmetry of $\mathcal{D}$, thus by Theorem 1, it results that the pair $(\sigma(I), \sigma(T))$ verifies the predicate $P_{\mathcal{D}}^k$, that is $P_{\mathcal{D}}^k(\sigma(I), \sigma(T)) = true$. Therefore $\sigma(J)$ is also a model of $CNF(k, \mathcal{D})$, since the pair $(\sigma(I), \sigma(T))$ verifying the predicate $P_{\mathcal{D}}^k$ is extracted from the model $\sigma(J) = (\sigma(I'), \sigma(T'))$ of $CNF(k, \mathcal{D})$.

Remark 3. The previous proposition allows us to use the symmetries of a transaction database $\mathcal{D}$ in its corresponding Boolean encoding $CNF(k, \mathcal{D})$ in order to detect symmetrical models and consider only one element in each symmetrical equivalent class. This gives an important alternative for symmetry exploitation in constraint-based data mining methods. Indeed, we can just compute the symmetries of $\mathcal{D}$ instead of computing those of its Boolean $CNF(k, \mathcal{D})$ which could be time consuming. This could accelerate the symmetry detection as the size of the transaction database $\mathcal{D}$ is generally substantially smaller than the size of its corresponding boolean encoding $CNF(k, \mathcal{D})$.

In Example 1, if we consider $\theta = 2$ and the symmetry σ of Example 2, then there will be symmetrical frequent itemsets in $\mathcal{D}$. For instance, both $I_1 = \{Beer, Wine, Wisky, Water\}$ and $I_2 = \{Shweps, Pepsi\}$ are frequent itemsets in $\mathcal{D}$. By the symmetry σ we can deduce that $\sigma(I_1) = \{Beer, Wine, Vodka, Water\}$ and $\sigma(I_2) = \{Juice, Coke\}$ are also frequent itemsets. These are what we call symmetrical frequent itemsets of $\mathcal{D}$ which correspond to symmetrical models[2] in $CNF(k, \mathcal{D})$. A symmetry σ transforms each frequent itemset (a model of the CNF encoding) into a frequent itemset and each non frequent itemset (a no-good of the CNF encoding) into a non frequent itemset. Symmetry elimination offers the advantage to enumerate only non-symmetrical patterns (like I_1 and I_2 here) which are considered as the most pertinent to the user for understanding the data.

4   Symmetry Detection

The most known technique to detect syntactic symmetries for CNF formulas in satisfiability is the one consisting in reducing the considered formula into a graph [3,5,6,16] whose automorphism group is identical to the symmetry group of the original formula. We adapt the same approach here to detect the syntactic symmetries of a transaction database $\mathcal{D}$. As it is done in [30], we represent the database $\mathcal{D}$ by a graph $G_{\mathcal{D}}$ that we use to compute the symmetry group of $\mathcal{D}$ by means of its automorphism group. When this graph is built, we use a graph automorphism tool like Saucy [5] to compute its automorphism group which gives the symmetry group of $\mathcal{D}$. We summarize bellow the construction of the graph which represent the transaction database $\mathcal{D}$. Given a transaction database '$\mathcal{D}$, the associated colored graph $G_{\mathcal{D}}(V, E)$ is defined as follows:

---

[2] Here, we omitted the part T of the model representing the cover of I.

– The set of colored vertices $V = \mathcal{I} \cup \mathcal{T}$ is build as follows:
   1. Each item $i \in \mathcal{I}$ is represented by a vertex $i \in V$ of the color 1 in $G_{\mathcal{D}}(V, E)$.
   2. Each item $t \in \mathcal{T}$ is represented by a vertex $t \in V$ of the color 2 in $G_{\mathcal{D}}(V, E)$.
– The set of edges E is defined by $E = \{(t, i) \mid \mathcal{D}_{t,i} = 1\}$. That is, an edge connects each transaction vertex $t \in \mathcal{T}$ to each vertex representing an item supported by t.

Example 3. Consider the transaction database $\mathcal{D}$ of Table 1 given in Example 1. Its corresponding graph $G_{\mathcal{D}}(V, E)$ is shown in Fig. 1. We can see for instance that the vertex permutation $\gamma = (Wisky, Vodka)\ (Cognac, Gin)$ $(Ricard, Pastis)\ (Pepsi, Coke)\ (Shweps, Juice)\ (001, 002)\ (003, 004)\ (005, 006)$ $(007, 010)\ (008, 009)$ is one among the automorphisms of $G_{\mathcal{D}}(V, E)$. The restriction of the automorphism γ to $\mathcal{I}$ represents the symmetry $\sigma = (Wisky,$ $Vodka)\ (Cognac, Gin)\ (Ricard, Pas - tis)\ (Pepsi, Coke)\ (Shweps, Juice)$ that we used in Example 2.

An important property of the graph $G_{\mathcal{D}}(V, E)$ is that it preserves the group of symmetries of $\mathcal{D}$. That is, the symmetry group of $\mathcal{D}$ is identical to the automorphism group of its graph representation $G_{\mathcal{D}}(V, E)$. Thus, we could use a graph automorphism system like Saucy on $G_{\mathcal{D}}(V, E)$ to detect the symmetry group of $\mathcal{D}$. The graph automorphism system returns a set of generators Gen of the symmetry group from which we can deduce each symmetry of $\mathcal{D}$.

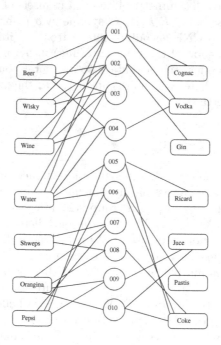

Fig. 1. The graph of the transaction Database of Table 1.

5   Symmetry Elimination

Here we deal with the global symmetry which is present in the formulation of the given problem that is represented by the transaction database $\mathcal{D}$. Global symmetry can be eliminated in a static way in a pre-processing phase by just adding the symmetry breaking predicates to the Boolean encoding $CNF(k, \mathcal{D})$ and use a SAT solver as a black box on the resulting CNF formula.

We shall compute the Lex-Leader Symmetry Breaking Predicate (LL-SBP) induced by the automorphisms of $G_{\mathcal{D}}$. More precisely, the group of automorphisms $Aut(G_{\mathcal{D}})$ of the graph $G_{\mathcal{D}}$ (or the symmetry group $Sym(\mathcal{D})$ of $\mathcal{D}$) induces an equivalence relation on the set of interpretations of $CNF(k, \mathcal{D})$. That is, an interpretation I_1' is equivalent to another interpretation I_2' of $CNF(k, \mathcal{D})$ if there exists a symmetry σ of $\mathcal{D}$ such that $I_2' = \sigma(I_1')$. The symmetry breaking predicates are chosen such that they are true for exactly one interpretation in each equivalent class (the least interpretation in the lex ordering). In general, we introduce an ordering on the the variables I_i' corresponding to the items of $\mathcal{I}$ and use it to construct a lexicographical order on the set of interpretations.

The construction of the symmetry-breaking predicate is based on the lex-leader method introduced by Crawford et al. [16]. Given a symmetry group $Sym(\mathcal{D}) = \{\sigma_1, \sigma_2, \ldots, \sigma_k\}$ of $\mathcal{D}$ and a total ordering $I_1' < I_2' < \cdots < I_n'$ on the variables of $CNF(k, \mathcal{D})$ corresponding to the items of $\mathcal{I}$. The partial lex-leader symmetry-breaking predicate (PLL-SBP) [4] that we have to add to $CNF(k, \mathcal{D})$ is expressed as follows:

$$PP(\sigma_l) = \bigwedge_{1 \leq i \leq n} [\bigwedge_{1 \leq j \leq i-1} (I_j' = I_j'^{\sigma_l}) \rightarrow (I_i' \leq I_i'^{\sigma_l})]$$

$$PLL - SBP(Sym(\mathcal{D})) = \bigwedge_{\sigma_l \in GEN(Sym(\mathcal{D}))} PP(\sigma_l)$$

$PP(\sigma_l)$ is the permutation predicate corresponding to the symmetry generator σ_l and the expression $(I_i' \leq I_i'^{\sigma_l})$ denotes the clause $(I_i' \rightarrow I_i'^{\sigma_l})$.

The $PLL - SBP$ is translated to a linear size CNF formula by introducing auxiliary variables e_j to represent the expressions $(I_j' = I_j'^{\sigma_l})$. For example, $e_j \leftrightarrow (I_j' = I_j'^{\sigma_l})$ gives rise to the following clauses:

$$(\neg I_j' \vee \neg I_j'^{\sigma_l} \vee e_i), (I_j' \vee I_j'^{\sigma_l} \vee e_i)$$

$$(\neg I_j' \vee \neg e_i \vee I_j'^{\sigma_l}), (I_j' \vee \neg e_i \vee \neg I_j'^{\sigma_l})$$

Some optimizations such that ones studied in Aloul [4] could be done to get a more compact CNF $PLL - SBP$.

6   Experiments

In this section, we present an experimental analysis of our symmetry breaking approach for SAT based itemset mining.

6.1   Input Data-Sets

We choose for our experiments two classes of data-sets:

- **Simulated Data-sets:** In this class, we use the simulated data-sets, generated specifically to involve interesting symmetries. The data is available at http://www.cril.fr/ decMining.
- **Public Datasets:** The datasets used in this class are well known in the data mining community and are available at https://dtai.cs.kuleuven.be/CP4IM/ datasets/.

6.2   The Experimented Methods

As we aim to enumerate all the frequent/closed itemsets on the SAT based encoding, our experiments are conducted using MiniSAT-Enum dedicated to the enumeration of all models of a given CNF formula. MiniSAT-Enum is obtained from MiniSAT 2.2[3] as follows: each time a model is found a no-good (clause) is generated and added to the formula in order to avoid enumerating the same models. MiniSAT-Enum takes as input a CNF formula and a set of items variables and returns the set of frequent/closed itemsets.

The methods that we experimented and compared are the following:

1. **MiniSAT-Enum:** search without symmetry breaking on the CNF encoding of the data mining task $CNF(k, \mathcal{D})$
2. **MiniSAT-Enum-SBP:** search with symmetry breaking. This method generates in a pre-processing phase the symmetry-breaking predicates, then apply MiniSAT-Enum to the resulting CNF instance $CNF(k, \mathcal{D}) + PLL - SBP$. The CPU time of *MiniSAT-Enum-SBP* includes the time spent to generate the $PLL - SBP$.
3. **MiniSAT-Enum-ISB:** this method [30], called ItemPair symmetry breaking (ISB), eliminates symmetries in a preprocessing step, by rewriting the transaction database $\mathcal{D}$ as a $\mathcal{D}'$ by eliminating symmetric items. MiniSAT-Enum is then applied on the CNF formula $CNF(k, \mathcal{D}')$ encoding the new transaction database.

In our experiments, we exploit Saucy[4], a new implementation of the Nauty system. It is originally proposed in [5] and significantly improved in [17]. The latest version of Saucy outperforms all the existing tools by many orders of magnitude, in some cases improving run time from several days to a fraction of a second.

We are interested on the CPU time and on the number of models or closed/frequent itemsets found with and without symmetry breaking. All our programs are run on Intel Xeon quad-core machines with 32 GB of RAM running at 2.66 Ghz.

---

[3] MiniSAT: http://minisat.se/.

[4] Saucy2: Fast symmetry discovery - http://vlsicad.eecs.umich.edu/BK/SAUCY/.

6.3   The Obtained Results

In Fig. 2, we present the results obtained on a simulated data *dataset-gen-jss-5*. The experiment show the comparison of MiniSAT-Enum (CFIM), MiniSAT-Enum-SBP (CFIM-SBP) and MiniSAT-Enum-ISB (CFIM-ISB) w.r.t. CPU time in seconds (the curves on the left in Fig. 2) and the number of patterns (the curves on the left in Fig. 2). As we can see, by breaking symmetries, we significantly reduce both the number of closed frequent itemsets (output) and CPU-time. Such reduction of the size of the output induces a significant reduction of the search time. Interestingly, breaking symmetries using by adding SBP on the CNF encoding of the itemset mining task (CFIM-SBP) is clearly better than eliminating symmetric items on the original transaction database (CFIM-ISB). This experiment show that our approach break more symmetries than the one proposed in [30].

The second experiment is conducted on well-know academic datasets. In this experiment, we are interested on the frequent itemsets mining problem. In Fig. 3, we present the comparative results w.r.t. the computation time. No reduction is observed on the number of frequent itemsets. On these datasets, most of found symmetries involves items in the same transactions. This explains

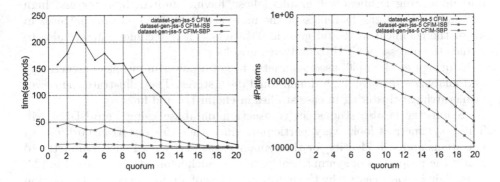

Fig. 2. Simulated data -(closed freq itemsets): CPU time/number of patterns.

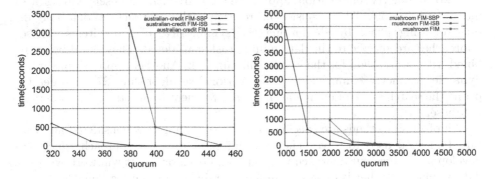

Fig. 3. Public data - *Australian (left) and Muchroom (right)*-(freq itemsets): CPU time.

why these particular symmetries does not reduce the number of closed/frequent itemsets. However, even when the size of the output is not reduced, breaking symmetries using our approach significantly reduce the search space, since all the symmetrical no-good interpretations are not explored. In general symmetry breaking reduces dramatically the search space and the corresponding CPU time for this declarative approach, but did not reach the performances of optimized dedicated algorithms like FPgrowth for example.

7 Related Works

The purpose of eliminating symmetry in data mining tasks is in general either to obtain a more compact output or to decrease the necessary CPU time for its generation or to handle new mining properties to find interesting frequent patterns. Some symmetry works are introduced in the field of Data mining following this direction.

Symmetries in graph mining are studied in Desrosiers et al. [18], and in Vanetik [45]. The area of graph mining has a great importance in many applications. In Desrosiers et al. [18] symmetry is exploited to prune the search space of sub-graph mining algorithms. However, in Vanetik [45], symmetry is used to find interesting frequent sub-graphs (those having limited diameter and high symmetry). Such graphs represent the more structurally important patterns in all of the chemical, text and genetic data-sets. Their technique allows also to reduce the necessary CPU to find such graphs.

Murtagh et al. in [36] used symmetry to get a powerful means of structuring and analyzing massive, high dimensional data stores. They illustrate the power of hierarchical clustering in case studies in chemistry and finance.

Symmetry is also studied in transaction database using Zero-BDDs [34]. These symmetries looks very particular, since they are just transpositions of two items and still identity for the remain items. They used such symmetry to study the properties of symmetrical patterns. Such symmetries are used in [20] to explain in some cases why the number of rules of a minimal cover of a relation is exponential in number of items.

Two symmetry elimination approaches for frequent itemset mining are introduced in [30]. They consist in rewriting the transaction database in pre-processing phase by eliminating the symmetrical of some items. These approaches are specific to the data mining task considered. They could be combined with our method for the itemset mining task. Another approach integrate dynamic symmetry elimination in the Apriori-like algorithm [27] in order to prune the search space of enumerating all the frequent item sets of a transaction database.

All of these methods are specific to the data mining task considered and the target method used to solve. They are different from the approach which we develop here, since our approach is generic and declarative. It will work with all data mining task that is expressed in a constraint programming language.

8 Conclusion

We studied in this work the notion of symmetry for data mining tasks expressed as declarative constraints. We showed how the symmetries of the given transaction database can be detected and eliminated by adding symmetry-breaking predicate to the constraint encoding of the considered data mining task. We showed that even though such symmetries could not be syntactically the symmetries of the CNF encoding of the data mining problem, they conserve the set of its models (the set of interesting patterns). Detecting symmetry on the given transaction database rather than the CNF encoding of the considered data mining task could result in a great save of efforts in the symmetry detection. Indeed, the size of the transaction database is in general smaller then its corresponding CNF encoding. The transaction database is represented by a colored graph that is used to compute its symmetries. The symmetry group of the transaction database is identical to the automorphism group of the corresponding graph. The graph automorphism tools SAUCY is naturally used on the obtained graph to detect the group of symmetries of the transaction database. This symmetry is eliminated statically by adding in a pre-processing phase the well known lex order symmetry breaking predicates to the CNF encoding of the considered data mining task. We then applied as a black box a SAT model enumeration algorithm on this resulting encoding to solve the corresponding data mining problem.

The proposed symmetry breaking method is implemented and experimented on a variety of transaction data-sets. The first experimental results confirmed that eliminating symmetry is profitable for the considered data mining tasks.

As a future work, we are looking to eliminate symmetry in other data mining problems and try to extend symmetry exploitation to the local symmetry that could exists at some nodes of the search tree. Both kind of exploitation could be complementary, then one can naturally think on the advantage of combining them.

Acknowledgements. Many thanks for the KDIR-2014 reviewers for their insightful remarks which increases the level and the quality of this manuscript.

References

1. Agrawal, R., Imieliński, T., Swami, A.: Mining association rules between sets of items in large databases. In: Proceedings of the 1993 ACM SIGMOD International Conference on Management of Data, SIGMOD 1993, pp. 207–216. ACM, New York (1993)
2. Agrawal, R., Srikant, R.: Fast algorithms for mining association rules in large databases. In: Proceedings of the 20th International Conference on Very Large Data Bases, VLDB 1994, pp. 487–499. Morgan Kaufmann Publishers Inc., San Francisco (1994)
3. Aloul, F.A., Ramani, A., Markov, I.L., Sakallak, K.A.: Symmetry breaking for pseudo-boolean satisfiabilty. In: ASPDAC 2004, pp. 884–887 (2004)

4. Aloul, F.A., Markov, I.L., Sakallah, K.A.: Shatter: efficient symmetry-breaking for boolean satisfiability. In: DAC, pp. 836–839. ACM (2003)
5. Aloul, F.A., Ramani, A., Markov, I.L., Sakallah, K.A.: Solving difficult SAT instances in the presence of symmetry. In: Proceedings of the 39th Design Automation Conference (DAC 2002), pp. 731–736. ACM Press (2002)
6. Aloul, F.A., Ramani, A., Markov, I.L., Sakallah, K.A.: Solving difficult instances of boolean satisfiability in the presence of symmetry. IEEE Trans. CAD Integr. Circuits Syst. **22**(9), 1117–1137 (2003)
7. Benhamou, B.: Study of symmetry in constraint satisfaction problems. In: PPCP 1994, pp. 246–254 (1994)
8. Benhamou, B., Sais, L.: Theoretical study of symmetries in propositional calculus and application. In: CADE 2011, pp. 281–294 (1992)
9. Benhamou, B., Sais, L.: Tractability through symmetries in propositional calculus. JAR **12**, 89–102 (1994)
10. Benhamou, B., Sais, L.: Theoretical study of symmetries in propositional calculus and applications. In: CADE, pp. 281–294 (1992)
11. Benhamou, B., Sais, L.: Tractability through symmetries in propositional calculus. J. Autom. Reasoning **12**(1), 89–102 (1994)
12. Besson, J., Boulicaut, J.F., Guns, T., Nijssen, S.: Generalizing itemset mining in a constraint programming setting. In: Džeroski, S., Goethals, B., Panov, P. (eds.) Inductive Databases and Constraint-Based Data Mining, pp. 107–126. Springer, New York (2010)
13. Bonchi, F., Lucchese, C.: Extending the state-of-the-art of constraint-based pattern discovery. Data Knowl. Eng. **60**(2), 377–399 (2007)
14. Bucilă, C., Gehrke, J., Kifer, D., White, W.: Dualminer: A dual-pruning algorithm for itemsets with constraints. Data Mining and Knowledge Discovery **7**(3), 241–272 (2003)
15. Burdick, D., Calimlim, M., Gehrke, J.: Mafia: a maximal frequent itemset algorithm for transactional databases. In: ICDE, pp. 443–452 (2001)
16. Crawford, J., Ginsberg, M., Luks, E., Roy, A.: Symmetry-breaking predicates for search problems. In: Knowledge Representation (KR), pp. 148–159. Morgan Kaufmann (1996)
17. Darga, P.T., Sakallah, K.A., Markov, I.L.: Faster symmetry discovery using sparsity of symmetries. In: Proceedings of the 45th Annual Design Automation Conference, DAC 2008, pp. 149–154. ACM, New York (2008)
18. Desrosiers, C., Galinier, P., Hansen, P., Hertz, A.: Improving frequent subgraph mining in the presence of symmetry. In: MLG (2007)
19. Freuder, E.: Eliminating interchangeable values in constraints satisfaction problems. AAAI 1991, pp. 227–233 (1991)
20. Gély, A., Medina, R., Nourine, L., Renaud, Y.: Uncovering and reducing hidden combinatorics in guigues-duquenne bases. In: Ganter, B., Godin, R. (eds.) ICFCA 2005. LNCS (LNAI), vol. 3403, pp. 235–248. Springer, Heidelberg (2005)
21. Grahne, G., Zhu, J.: Fast algorithms for frequent itemset mining using FP-trees. IEEE Trans. Knowl. Data Eng. **17**(10), 1347–1362 (2005)
22. Guns, T., Dries, A., Tack, G., Nijssen, S., Raedt, L.D.: Miningzinc: a modeling language for constraint-based mining. In: International Joint Conference on Artificial Intelligence. Beijing, China, August 2013
23. Guns, T., Nijssen, S., De Raedt, L.: Itemset mining: a constraint programming perspective. Artif. Intell. **175**(12–13), 1951–1983 (2011)
24. Guns, T., Nijssen, S., de Raedt, L.: k-pattern set mining under constraints. IEEE TKDE **99**(PrePrints) (2011)

25. Han, J., Pei, J., Yin, Y.: Mining frequent patterns without candidate generation. In: Proceedings of the 2000 ACM SIGMOD International Conference on Management of Data, SIGMOD 2000, pp. 1–12. ACM, New York (2000)
26. Henriques, R., Lynce, I., Manquinho, V.M.: On when and how to use sat to mine frequent itemsets. CoRR abs/1207.6253 (2012)
27. Jabbour, S., Khiari, M., Sais, L., Salhi, Y., Tabia, K.: Symmetry-based pruning in itemset mining. In: 25th International Conference on Tools with Artificial Intelligence (ICTAI 2013). IEEE Computer Society, Washington November 2013
28. Jabbour, S., Sais, L., Salhi, Y.: Boolean satisfiability for sequence mining. In: CIKM, pp. 649–658 (2013)
29. Jabbour, Said, Sais, Lakhdar, Salhi, Yakoub: The Top-k frequent closed itemset mining using Top-k SAT problem. In: Železný, Filip, Blockeel, Hendrik, Kersting, Kristian, Nijssen, Siegfried (eds.) ECML PKDD 2013, Part III. LNCS, vol. 8190, pp. 403–418. Springer, Heidelberg (2013)
30. Jabbour, S., Sais, L., Salhi, Y., Tabia, K.: Symmetries in itemset mining. In: 20th European Conference on Artificial Intelligence (ECAI 2012). pp. 432–437. IOS Press, August 2012
31. Khiari, M., Boizumault, P., Crémilleux, B.: Constraint programming for mining n-ary patterns. In: Cohen, D. (ed.) CP 2010. LNCS, vol. 6308, pp. 552–567. Springer, Heidelberg (2010)
32. Krishnamurthy, B.: Short proofs for tricky formulas. Acta Inf. **22**(3), 253–275 (1985)
33. Métivier, J.P., Boizumault, P., Crémilleux, B., Khiari, M., Loudni, S.: A constraint language for declarative pattern discovery. In: Proceedings of the 27th Annual ACM Symposium on Applied Computing, SAC 2012. pp. 119–125. ACM, New York (2012)
34. Minato, S.: Symmetric item set mining based on zero-suppressed BDDs. In: Todorovski, L., Lavrač, N., Jantke, K.P. (eds.) DS 2006. LNCS (LNAI), vol. 4265, pp. 321–326. Springer, Heidelberg (2006)
35. Minato, S.I., Uno, T., Arimura, H.: Fast generation of very large-scale frequent itemsets using a compact graph-based representation (2007)
36. Murtagh, F., Contreras, P.: Hierarchical clustering for finding symmetries and other patterns in massive, high dimensional datasets (2010). CoRR abs/1005.2638
37. Pei, J., Han, J., Lakshmanan, L.V.S.: Pushing convertible constraints in frequent itemset mining. Data Min. Knowl. Discov. **8**(3), 227–252 (2004)
38. Puget, J.F.: On the satisfiability of symmetrical constrained satisfaction problems. In: Komorowski, J., Raś, Z.W. (eds.) ISMIS 1993. LNCS, vol. 689, pp. 350–361. Springer, Heidelberg (1993)
39. Raedt, L.D., Guns, T., Nijssen, S.: Constraint programming for itemset mining. In: KDD, pp. 204–212 (2008)
40. Raedt, L.D., Guns, T., Nijssen, S.: Constraint programming for data mining and machine learning. In: AAAI (2010)
41. Tiwari, A., Gupta, R., Agrawal, D.: A survey on frequent pattern mining: current status and challenging issues. Inform. Technol. J. **9**, 1278–1293 (2010)
42. Tseitin, G.S.: On the complexity of derivation in propositional calculus. In: Structures in the constructive Mathematics and Mathematical logic, pp. 115–125. H.A.O Shsenko (1968)
43. Uno, T., Asai, T., Uchida, Y., Arimura, H.: Lcm: An efficient algorithm for enumerating frequent closed item sets. In: Proceedings of Workshop on Frequent itemset Mining Implementations (FIMI03) (2003)

44. Uno, T., Kiyomi, M., Arimura, H.: Lcm ver. 2: Efficient mining algorithms for frequent/closed/maximal itemsets. In: FIMI (2004)
45. Vanetik, N.: Mining graphs with constraints on symmetry and diameter. In: Shen, H.T., Pei, J., Özsu, M.T., Zou, L., Lu, J., Ling, T.-W., Yu, G., Zhuang, Y., Shao, J. (eds.) WAIM 2010. LNCS, vol. 6185, pp. 1–12. Springer, Heidelberg (2010)
46. Zaki, M.J., Hsiao, C.J.: Efficient algorithms for mining closed itemsets and their lattice structure. IEEE Trans. Knowl. Data Eng. **17**(4), 462–478 (2005)

Enhancing Online Discussion Forums with Topic-Driven Content Search and Assisted Posting

Damiano Distante[1]([✉]), Alejandro Fernandez[2], Luigi Cerulo[3],
and Aaron Visaggio[3]

[1] Unitelma Sapienza University, Rome, Italy
damiano.distante@unitelma.it
[2] LIFIA, CIC/F.I., National University of La Plata, La Plata, Argentina
alejandro.fernandez@lifia.info.unlp.edu.ar
[3] University of Sannio, Benevento, Italy
{lcerulo,visaggio}@unisannio.it

Abstract. Online forums represent nowadays one of the most popular and rich repository of user generated information over the Internet. Searching information of interest in an online forum may be substantially improved by a proper organization of the forum content. With this aim, in this paper we propose an approach that enhances an existing forum by introducing a navigation structure that enables searching and navigating the forum content by topics of discussion. Topics and hierarchical relations between them are semi-automatically extracted from the forum content by applying Information Retrieval techniques, specifically Topic Models and Formal Concept Analysis. Then, forum posts and discussion threads are associated to discussion topics on a similarity score basis. Moreover, to support automatic moderation in websites that host several forums, we propose a strategy to assist a user writing a new post in choosing the most appropriate forum into which it should be added. An implementation of the topic-driven content search and navigation and assisted posting forum enhancement approaches for the Moodle learning management system is also presented in the paper, opening to the application of these approaches to several real distance learning contexts. Finally, we also report on two case studies that we have conducted to validate the two approaches and evaluate their benefits.

Keywords: Online discussion forums · Information search · Information extraction · Text mining · Topic modeling · Navigability · Searchability · Assisted posting · E-learning · Learning management systems · Moodle

1 Introduction

Online discussion forums represent one of the main sources of user generated content (i.e., social media) and asynchronous communication means in the form of message posts over the Internet. Most visited websites, including blogs and

A. Fred et al. (Eds.): IC3K 2014, CCIS 553, pp. 161–180, 2015.
DOI: 10.1007/978-3-319-25840-9_11

social networks, use forums to support user interaction and knowledge sharing. In several domains ranging from e-commerce [12,22], to news [18], and health-care [27], discussion forums constitute rich and widely accessed repositories of information for Internet users.

As an example, software developers forums are an effective source of information where programmers search for and describe solutions to specific problems[1]. In e-learning contexts, discussion forums enable asynchronous communication student-to-student, and teacher-to-student, e.g., to support collaborative learning and group work [15,26]. Whatever the forum domain, discussions held in a certain period of time become a source of information for any user accessing the forum afterwards.

In general online forums organize messages into a chronological order. A user starts a new discussion by posting an initial message, other users post their replies or comments to it, and the list of messages forms a *discussion thread*. If users are allowed to reply to other users' replies in additional to the original message, discussions take the form of trees, with discussion branches.

The effectiveness of a discussion forum as information source mainly depends on the forum richness in information, but also on the forum organization and on the searching paradigm users can adopt to find contents of their interest.

Search features usually provided with online discussion forums are limited to full-text search which returns a list of forum messages that include (and/or do not include) one or more of the query keywords in their body and/or their title. Such a search feature may return too many or too few results (depending on the forum size and the query keywords) and may miss messages which are semantically related to the query keywords but do not actually include them [1].

Hierarchical graphs constitute an effective paradigm to represent users' knowledge [34]. In a previous work [7] we have introduced an approach to improve information retrieval and content navigation in online discussion forums by introducing in them a complementary hierarchical topic-driven navigation structure. Information Retrieval (IR) techniques, specifically Topic Models [4] and formal concept analysis (FCA) [10], are used to discover discussion topics and hierarchical relations between them in the forum content. Then, forum messages and discussion threads are associated to discussion topics based on a similarity score, thus to enable searching and navigating them on a topic-driven basis, additional to conventional chronological order and full-text search approaches.

In this paper we present an implementation of this approach as a plugin for the Moodle learning management system which makes the topic-driven navigation approach accessible and evaluable in several e-learning contexts. We also present a case study that provides a first qualitative assessment of the benefits of topic-driven navigation and search of forum content, with respect to traditional full-text search.

There are scenarios, such as large communities of interest or learning spaces, that call for the organization of interactions into multiple forums. Creating

---

[1] An example of such forum is the Microsoft MSDN Developer Network forum. http://social.msdn.microsoft.com/Forums/en/categories/.

multiple forums aims at making each of them more focused, and manageable in terms of frequency of updates. The Moodle English Community[2], for example, organises discussions in 57 forums. Even if the conversation space is split in multiple forums, some topics can be cross-cutting. For example, "Scorm", the Sharable Content Object Reference Model, is the main focus of the "Scorm" forum[3] of the Moodle English Community. However, the topic is also discussed in the "Comparisons and advocacy", "General Help", and "General Developers" forums, among others. Topic-driven navigation helps users discover and navigate such connections.

In these scenarios, the lack or absence of moderation normally results in a poor organization of the forums content, with duplication of content and difficulties in finding relevant information among them. The organization of content in forums is the outcome of the choices users make when posting new messages. They choose the forum to post to, and decide whether to start a new discussion or contribute to an existing one. As the result of these choices, forums turn more or less cohesive, and topics become more or less scattered. To mitigate this problem, in this paper we also propose a strategy to assist a user while writing a new message (i.e., creating a new discussion) in deciding in which forum to post it. We evaluate our strategy against forums of the Moodle public community[4] and against on-line discussions in Stack Exchange[5], a network of question and answer websites on diverse topics that counts with a community focused curation process.

This paper is a revised and extended version of our earlier work presented in [9]. Particularly, new contributions of this paper are: *i.* an extended version of the topic-driven forum enhancement approach that now allows analyzing several forums at once to build one single topic-driven navigation structure that spans all their content; *ii.* an approach to assist users in choosing the most appropriate forum/thread to which to post a new message, thus to support the automatic moderation of a forum; *iii.* the implementation of both approaches via our TDForum plugin for the Moodle learning management system.

The rest of the paper is organized as follows. Section 2 describes our approach, earlier introduced in [7], to enhance an existing forum with topic-driven content search and navigation capabilities. Section 3 discusses the approach to assist users in selecting the most appropriate forum in which to add a new post. Section 4 presents the implementation of both these approaches for the Moodle[6] learning management system. Section 5 reports on a case study conducted to qualitatively assess the benefits of the topic-driven forum enhancement approach in searching forums for information of interest for the user. It also reports on a experiment that assesses the performance of our assisted

---

[2] https://moodle.org/course/view.php?id=5.
[3] https://moodle.org/mod/forum/view.php?f=365.
[4] https://moodle.org/course/.
[5] http://stackexchange.com.
[6] www.moodle.org.

posting approach. Section 6 overviews related work, while Sect. 7 draws conclusions and introduces future works.

2   The Topic-Driven Forum Navigation Enhancement Process

The topic-driven forum navigation enhancement process, introduced by Cerulo and Distante in [7], is shown in Fig. 1. It consists of four main steps represented in the figure as rectangles and described briefly in the following subsections.

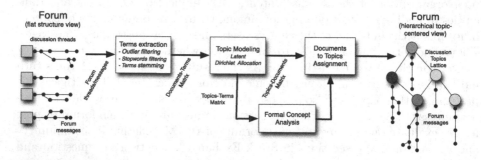

Fig. 1. The topic-driven forum navigation enhancement process [7].

2.1   Terms Extraction

We represent a forum message as a vector of indexing terms, $\{t_1, \ldots, t_m\}$, extracted, from the corpus of n messages, $\{d_1, \ldots, d_n\}$, through a standard text analysis pipeline usually adopted in Information Retrieval that comprises: outlier filtering, stopwords filtering, and stemming [1].

The outcome of this step is a document-term matrix $\mathbb{DT}$, where each element $\{\mathbb{DT}\}_{jp}$ is the *tf-idf* of the term t_p in the forum message d_j [1].

2.2   Topic Modeling

Topic modeling, in particular Latent Dirichlet Allocation (LDA), is a statistical technique that is able to extract frequently co-occurring terms, known as *topics*, from a corpus of documents [4]. The input is the document-term matrix, $\mathbb{DT}$, obtained from the previous task, while the output is a topic-document matrix, $\mathbb{TD}$, and a topic-term matrix, $\mathbb{TT}$.

The number of topic k is a parameter that controls the granularity of the topics and must be fixed a priori.

Intuitively, the top terms of a topic are semantically related and represent some real-world concepts. For example the concept related to problems e-mails setup is represented by the terms "mail", "problem", "setup". The topic membership of a document describes which concepts are present in that document. Table 1 shows and example of topic-document and topic-term matrices.

Table 1. Examples of topic-term and topic-document matrices.

Topic	Topic-term (top terms)	Topic-document			
		d_1	d_2	d_3	d_4
z_1	problem, email, setup	0.6	0.7	0.1	
z_2	problem, email, connection, setup	0.3	0.1	0.1	0.5
z_3	problem	0.1			
z_4	problem, video, decoder, setup	0.1	0.2	0.8	0.5
z_5	problem, video	0.1	0.2		

2.3 Formal Concept Analysis

Using the topic membership of a term, we prune a topic lattice by means of Formal Concept Analysis (FCA). FCA is a computational way to derive a concept hierarchy or formal ontology from a collection of objects and their properties [3,10].

We model the topics as the objects of a formal context and the terms as their attributes. The relation R of the formal context is computed from the topic-term matrix $\mathbb{TT}$ by means of a decision threshold h_T, i.e., a term (attribute) t_p belongs to a topic (object) z_i, $(z_i, t_p) \in R$ iif $\{\mathbb{TT}\}_{ip} \geq h_T$.

As a clarification example consider the formal context shown in Table 3 and the topic lattice obtained from such a formal context shown in Fig. 2. Topics are mapped on circles and hierarchical relationships are represented by arcs. Large circles are mapped on topics extracted with the topic modeling approach, while small circles are intermediate topics extracted with the formal concept analysis. The lattice shows the hierarchical relationships between topics. In the lattice the top most topic (z_3) is the most general topic. A path starting from the top most topic is a more specific topic. For example z_2 is reachable by the path from z_3 (problem), setup, z_1 (email), and z_2 (connection), and represent the more specific topic of problems related to the email connection setup.

Table 2. The formal context obtained from the topic-term matrix shown in Table 1.

	problem	email	connection	video	decoder	setup
z_1	×	×				×
z_2	×	×	×			×
z_3	×					
z_4	×			×	×	×
z_5	×		×	×		

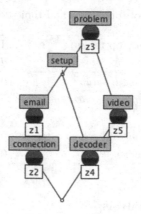

Fig. 2. The topic-lattice pruned from the formal context shown in Table 2.

2.4   Documents to Topics Assignment

During this step each document (*i.e.* forum message/thread) is mapped onto the topics to which it is more likely to belong by estimating the probabilities of each topic for that message (topic-document matrix). For this purpose we adopted the topic-document matrix $\mathbb{TD}$ and a decision threshold h_D, *i.e.*, a document d_j belongs to a topic z_i iif $\{\mathbb{TD}\}_{ij} \geq h_D$.

2.5   Parameter Setting and Accuracy Evaluation

Selecting the number of topic k is one of the most problematic modeling choice in topic models [31]. We adopt a metric, introduced by Meilă [20] for clustering comparison, that measures the *Variation of Information* as the entropies the mutual information associated with cluster assignments. Intuitively, the entropy measures the uncertainty of allotting an item to a cluster, while the mutual information measures the reduction of such uncertainty when the allocation in the other cluster is known. Following the approach adopted by Wallach *et al.* [31] the assignment of documents (forum messages or threads in our context) to topics can be assimilated to a sort of cluster assignment. In our previous work [7] we showed that above a certain value of k no significant increment of Variation of Information can be observed in a specific context. We consider such a value the optimal number of topic in that context.

We evaluated the document assignment task to check whether the documents assigned to topics by the Latent Dirichlet Allocation were congruent with the semantics of their content [2]. In our previous work [7] we addressed this question with a controlled experiment obtaining in average a precision ranging between 52 % and 74 %.

3   Assisted Posting

To assist users in deciding where to submit their posts, we change the accustomed posting workflow. Instead of first deciding where to post and then writing the message, users first write and then let an algorithm suggest the most adequate forum. The algorithm ranks available forums according to some measure of relevance. The user can decide to submit the post to the highest ranked forum, or to a different one. To calculate the relevance of a post to a given forum we have explored two alternatives. The "One-like-this" approach assumes that the post is most relevant to the forum that contains the most similar post. The "Centroids" approach assumes that the post is most relevant to the forum that, as a whole, is most similar to the content of the post. To operationalise these alternatives, we looked into the theory of Vector Space Models [24].

3.1   Vector Space Models: One-like-this and Centroids

As explained in Sect. 2.1, the extraction of terms from posts results in a matrix $\mathbb{DT}$, where each element $\{\mathbb{DT}\}_{jp}$ is the *tf-idf* of the term t_p in the forum message d_j. Thus, row $\{\mathbb{DT}\}_j$ in the matrix, represents the *tf-idf* vector for forum message d_j. The messages in a forum constitute as a space of vectors where similar posts have similar vectors (according to some similarity function such as Cosine Similarity). Both strategies we propose to assess the relevance of a post to a forum build upon this model. They therefore require, as the first step, the construction of the *tf-idf* vector $\{\mathbb{DT}\}_n$ for the new post d_n.

   To find the forum that contains the post most similar to the one the user attempts to submit (One-like-this), we compute the Cosine Similarity between vector $\{\mathbb{DT}\}_n$ and the *tf-idf* vectors of all messages in available forums. For each forum we take the similarity coefficient of the most similar post. Forums are ranked according to these coefficients, suggesting the user to post to the one with the highest value. Although this strategy is intuitive and straightforward to implement, it incurs in high computation cost for large forums.

   We define similarity between a post and a forum in terms of the Cosine Similarity between the post's *tf-idf* vector and the centroid of the latter. During the indexing process, we compute the centroid of each forum by taking that average between the *tf-idf* vector of all its messages. To find the most similar forum to the new post d_n we compute the similarity coefficients between the $\{\mathbb{DT}\}_n$ vector and the centroids of all available forums. Forums are ranked according to these coefficients, suggesting the user to post to the one with the highest value. The number of comparisons required at the time of posting linearly depends on the number of forums available and not on the number of posts. This approach incurs in high computation cost, for large forums, to calculate centroids. However, the larger the number of messages in a forum, the less impact a few new messages have in its centroid's position. Therefore, centroid calculation can be performed periodically as an off-line batch process.

3.2 Assisted Posting Workflow

Figure 3 summarizes the modified posting workflow and the key elements of the assistance algorithm. First, the user writes the post in a form similar to the one used by regular Moodle Forums. When the user submits the post, its content is analysed and it's *tf-idf* vector created. Cosine Similarity is used to compare the vector to those of the centroids of each forum in the TDForum activity. Centroids are updated periodically, independently of the posting workflow. The similarity coefficients for each forum are used to rank the suggestions of the most adequate forum according to the new post's content. Although the system's recommendation is to post to the highest ranked forum, the user makes the final decision.

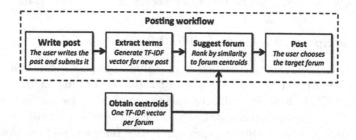

Fig. 3. Posting workflow modified to include assisted selection of the most appropriate forum.

4 TDForum: A Plugin for the Moodle Learning Management System

Topic-Driven Forum (TDForum) is a Moodle plugin (particularly, an *activity module*) that implements the topic-driven forum navigation enhancement approach described in Sect. 2 for the Moodle open-source learning management system.

In Moodle, *activity* is a general name for a group of features in a course. Usually an activity is something that a student will do that interacts with other students and/or the teacher. Assignments, quizzes, surveys, workshops, chats, and forums are examples of activities that can be created in a course and that are provided in Moodle by default. Each activity is implemented by a software module (plugin) located in the *mod* sub-folder of the Moodle instance. Additional activities can be included by installing the corresponding Moodle plugin[7].

From a source code point of view, each Moodle activity module consists of a series of mandatory files (e.g., install.xml, lib.php, and view.php) used to install the module and integrate it within the Moodle system, and other files specific to the plugin.

---

[7] A rich and up-to-date list of Moodle plugins can be found in the Moodle Plugins Directory at http://www.moodle.org/plugins.

Figure 4 shows the architecture of the TDForum Moodle plugin that we developed. In the figure, we can distinguish the components representing the plugin *front-end* (the graphical interfaces that Moodle users interact with), and those that are part of the plugin *back-end*.

The plugin front-end comprises the components *Main View* and *Discussion Topics View* corresponding to the two possible views on the forum content: (*i*) standard chronological list of discussions augmented with discussion topics and scores, and (*ii*) navigable hierarchical discussion topics graph. The last view is built using the *JavaScript InfoVis Toolkit*[8]. It also includes the *Admin User Interface* component which lets administrators manage the forum data processing and customize the visualization plugin parameters.

The plugin back-end contains the components implementing the forum analysis and indexing process described in Sect. 2 to build the additional topic-driven navigation structure. In particular, the *Process Controller* controls the process by executing the commands provided through the plugin admin user interface. It also exports forum content from the Moodle database into a local temporary csv text file and imports the data on the new navigation structure from the local filesystem into the Moodle database.

The *Data Processing* component includes the following sub-components:

- *Data Preprocessing:* a Perl script which extracts threads and messages from the csv file into separated text files and performs terms extraction and text filtering such as stopwords and stemming (cf. Sect. 2.1).
- *Topics Identification and Documents to Topics Assignment:* a R[9] script which uses the Topic Model library[10] to perform discussion topics identification and documents to topics assignment. The matrices Topics-Terms and Topics-Documents of the detected forum discussion topics and scores associated to them are generated in this step (cf. Sects. 2.2 and 2.4).
- *Formal Concept Analysis and Topics Graph Export:* this component uses the FcaStone[11] Formal Concept Analysis command-line utility to generate the lattice representative of the hierarchy of topics and to export the topics graph used in the graph view of the plugin (cf. Sect. 2.3).

The TDForum activity implemented by our plugin offers the same features provided by a standard Moodle forum (particularly, a *main view* which lists forum discussions and messages organized in a chronological order, the functionality of posting new messages or replying to existing ones, full-text search of messages, etc.) and adds to them a *discussion topics view* which acts as a topic-driven navigation index to the forum content.

The *main view* (Fig. 5) presents the list of discussion threads of the forum in a chronological order and adds to each of them the list of discussion topics in it identified, and the calculated similarity score (column 'Discussion topics' in the

---

[8] http://philogb.github.io/jit/.
[9] http://cran.r-project.org/.
[10] http://cran.r-project.org/web/packages/topicmodels/.
[11] http://fcastone.sourceforge.net/.

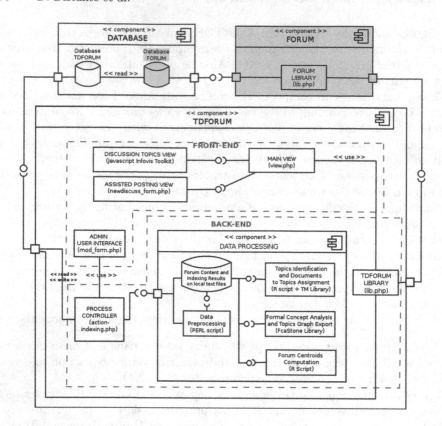

Fig. 4. Architecture of the TDForum Moodle plugin (with a gray background color, standard Moodle components).

figure). Score values range between 0 and 1 (with 1 representing the maximum similarity value) and the list of topics associated to a discussion is ordered by score. By right-clicking one of the topics of the list, the user can search for discussions or messages which are related to the selected topic. The results of this search is presented sorted by decreasing values of score.

The *discussion topics view* (Fig. 6) shows the list of discussion topics found by the analysis process for the considered forum (scrollable list on the left side of the figure) and a graph that the user can pan and zoom which highlights the hierarchical relations between the identified topics. The user can navigate the discussion topics graph or the topics list and once she finds a topic of her interest she can retrieve the list of discussions/messages associated to it with a click.

The plugin has been designed to extend a standard Moodle forum and, at the same time, to be independent from it. As such, if it is installed, applied on a forum, and then deactivated, none of the content of the original forum are lost, nor the additional messages/discussions that will have been added in it after the plugin instantiation. Moreover, the current version of the plugin now allows to

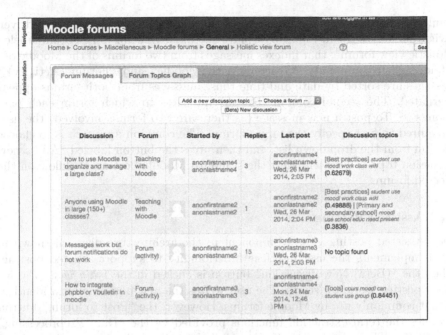

Fig. 5. The main view of TDForum showing the list of forum discussion threads enhanced with discussion topics and scores associated to them.

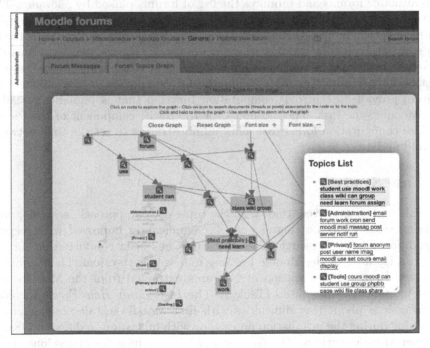

Fig. 6. The discussion topics view of TDForum showing the topics list and the hierarchical topics graph.

simultaneously index posts from various forums, thus supporting topic-driven navigation across them. Figure 5 displays the main view for a TDForum labeled "Holistic view forum", that indexes messages from two forums of the Moodle.org online english community, "Teaching with Moodle" and "Forum (Activity)". Messages are sorted by date and time thus, messages from both forums appear alternated. The second column (Forum) indicates to which forum each post belongs to. To post a new message (as there are two forums involved) the user is required to first specify the target forum. The common approach is to choose a forum from the drop down list, and then press the button labeled "Add a new discussion topic". This will start the standard Moodle posting workflow on the selected forum.

4.1   Assisted Posting

The "Assisted posting view" component in the architecture diagram shown in Fig. 4 implements the user interface functionality to support assisted posting. When the "(Beta) New discussion" button is clicked in the *main view* (Fig. 5), a new posting workflow starts. The user writes the new post in a form identical to that commonly use by Moodle forums. However, the "Post to forum" button invoques the term extraction functions provided by the "Data Preprocessing" component. The resulting *tf-idf* vector is fed to the similarity function, in the "Forum Centroids Computation" component, that compares it to the centroids of all available forums and produces the list of forums ranked by relevance. The function additionally filters out all those forums the user does not have permission to post to. The user then selects the destination forum and proceeds to the next step in the posting workflow where forum specific checks, such as permission to include attachments, are made. Then, a final step confirms the user's submission. The "Forum Centroids Computation" component in the architecture diagram provides a centroids computation function that is periodically called by a Moodle sync task. The "Admin User Interface" component of the Plugin provides configuration forms that allow administrators to set the time between centroid updates.

5   Case Study

We evaluated qualitatively that, with the topic-driven approach, searching and browsing tasks of forum contents can be significantly improved with respect to traditional full-text search. The case study has been conducted on forums inside an instance of the Moodle learning management system. The context is composed by a reduced version of 2 forums extracted from the Moodle user and development communities (Table 3). The *Installation Help Forum* includes all discussions about user difficulties with first Moodle installations or errors happening during the installation process, or with migration to different OSs, or to newer Moodle versions. The *General Help Forum* includes discussions about problems not included into other Moodle community forums, such as problems with database access, file upload, block modules and student enrollment.

Table 3. Case study context.

Forum	# threads	# posts	# users	Time period
Installation help	253	777	78	May 1, 2013–May 24, 2013
General help	115	714	107	Jul 15, 2013–Aug 8, 2013

We evaluated effectiveness in 11 searching tasks in terms of (i) the number of items (forum posts) the user had to inspect in order to satisfy the information need, and (ii) the time spent to accomplish the task (Table 4). The nature of the 11 searching tasks has been defined by the first two authors of this paper. For each task the search goal, *i.e.,* the expected posts that should be retrieved, is known beforehand. The other two authors performed the searching tasks with two complementary approaches: (i) full-text search, and (ii) topic driven navigation. The first approach is accomplished with the default full-text search engine implemented in the Moodle platform which performs a full-text search from a set of user defined keywords. By default, the keywords are linked by an AND operator and the system retrieves the list of all posts containing all searched terms. The second approach is executed with the TDForum Moodle plugin introduced in Sect. 4. While performing the tasks, we collected the number of items inspected and the time needed to achieve the search goal. With the Moodle full-text search the number of inspected items is computed by counting the number of posts examined before finding the correct expected posts. With the topic-driven navigation approach the number of inspected items is the sum of two quantities: the number of links followed to reach the closest topic in the Discussion Topics View of the TDForum plugin and the number of posts examined before finding the correct expected posts.

Table 5 reports the results obtained by executing the evaluation protocol on the 11 tasks. The table reports for each search goal the number of inspected items and the time spent to find the correct posts. For Moodle full-text search we also reported the number of search attempts (queries) performed with different search keywords necessary to reach the goal. In general the number of items inspected with full-text search is in average higher than the number of items inspected with TDForum (14 vs. 9). The time necessary to obtain the correct answer is in average less in TDForum (137 s vs. 170 s) because with full-text search more time is spent to choose the correct search keywords. The difference is not statistically significant due to the limited number of samples, thus further experiment are necessary to draw more general conclusions.

5.1   Assisted Posting Performance

In order to validate the approach and assess how well each of the alternatives performed, we evaluated and compared the "Centroids" and the "One-like-this" algorithms on existing forum data. To conduct the evaluation we obtained posts

Table 4. Search tasks definition.

ID	Search goal	Adopted keywords
1	Retrieve the 10 posts related to css problems in the General help forum	*css, problem*
2	Retrieve the 5 posts related to login issues in the General help forum	*login, problem*
3	Retrieve the 3 posts related to uploading files problems in the General help forum	*file, not, upload*
4	Retrieve the 4 posts related to changing Moodle fonts in the General help forum	*change, font*
5	Retrieve the 5 posts related to not sent enrollment email in the General help forum	*enrollment*
6	Retrieve the 3 posts related to editing Moodle theme in the General help forum	*change, theme*
7	Retrieve the 5 posts related to web hosting in the General help forum	*moodle, web, hosting*
8	Retrieve the 10 posts related to Moodle upgrading problems in Installation help forum	*problem, moodle, upgrade*
9	Retrieve the 5 posts related to editing admin password in the Installation help forum	*admin, password*
10	Retrieve the unique post related to missing files after Moodle migration in the Installation help forum	*missing, files, after, migration*
11	Retrieve the 2 post related to slower system after upgrade in the Installation help forum	*css, problem*

from two forums in the Moodle Community, namely "Teaching with Moodle"[12], and "Forum"[13], (a forum about the *Forum activity*). In total they had 422 posts. In addition, we conducted an evaluation on forums obtained from Stack Exchange. We built two datasets from it. One of them consists of two forums with closely related topics, namely "Webapps"[14] and "Webmasters"[15]. It contains 34,248 posts. The other dataset combines four forums with different topics, namely "Beer"[16], "Aviation"[17], "Politics"[18], and "Space Exploration"[19]. It contains 4,642 posts.

---

[12] https://moodle.org/mod/forum/view.php?id=41.
[13] https://moodle.org/mod/forum/view.php?id=732.
[14] http://webapps.stackexchange.com.
[15] http://webmasters.stackexchange.com.
[16] http://beer.stackexchange.com.
[17] http://aviation.stackexchange.com.
[18] http://politics.stackexchange.com.
[19] http://space.stackexchange.com.

Table 5. Case study results (time in seconds).

Task ID	Moodle full-text search			TDForum search	
	# queries	# items	Time	# items	Time
1	2	15	201	20	254
2	1	5	109	5	92
3	1	17	131	8	168
4	3	12	230	7	135
5	3	13	225	11	187
6	5	40	275	9	113
7	1	5	134	4	62
8	2	42	287	10	131
9	1	4	122	6	90
10	1	1	86	16	192
11	1	0	70	6	85
Average	2	14	170	9	137

Table 6. Evaluation results for assisted posting - error percentages.

Data-set	Split (train - test)	Centroids	One-Like-This
Moodle community	30 % - 70 %	9.03 %	54.49 %
Moodle community	50 % - 50 %	7.02 %	47.66 %
Moodle community	70 % - 30 %	6.68 %	48.62 %
StackExchange similar	30 % - 70 %	16.03 %	23.19 %
StackExchange similar	50 % - 50 %	15.35 %	22.65 %
StackExchange similar	70 % - 30 %	15.49 %	22.44 %
StackExchange different	30 % - 70 %	4.79 %	43.14 %
StackExchange different	50 % - 50 %	3.89 %	37.12 %
StackExchange different	70 % - 30 %	4.07 %	33.80 %

We evaluated both algorithms using cross validation. A percentage of the messages in each forum, selected randomly, was set apart as the test set, the rest used for training. We used both algorithms to obtain the most appropriate target for each of the posts in the test set and we recorded the number of times the algorithm suggested a forum different from the one the post was in Table 6 depicts the results we observed. To cope with variability in the results, we performed 6 runs for each percentage split and reported the average results. The "Centroids" approach performed consistently better, by a large margin, than the simpler "One-like-this". Moreover, the "Centroids" approach performed substantially better when the forums had different topics.

6   Related Work

Recently, on-line education systems are becoming widespread tools adopted by both historical and newly founded educational institutions. E-learning and e-teaching are new contexts for education where large amounts of information are generated and ubiquitously available. Most of generated information has the form of free text without a structure crucial for automating knowledge retrieval.

Data Mining has been historically used to extract knowledge from free text [1]. Knowledge extraction from e-learning systems, in particular from user generate data, has been introduced in [6,13]. Patterns of system usage by teachers and learning behavior by students has been investigated in [29]. Data clustering was suggested to promote group-based collaborative learning and to diagnose students incrementally [5].

Web Mining techniques to meet some of the current challenges in distance education was presented in [28] where a clustering of forum messages are in fact grouped into similar discussion topic classes. Association Rules mining has been widely adopted in e-learning, in particular recommendation systems [32,33], learning material organization [30], student learning assessments [23], course adaptation to the students behavior [14], and evaluation of educational websites [25]. In educational research the development of cooperative learning and knowledge sharing inside student groups constitute recent research trends [16]. To this aim, Web technologies should grasp the opportunities raised by mixing the Social and the Semantic Web [11] and on adopting Semantic and Artificial Intelligence techniques for discovering information objects and restructure large digital collections [19]. Concept maps and their use for navigation in educational contexts has been investigated in the recent past by different authors. As a representative of this research effort we cite the work of Dicheva and Aroyo [8]. In this work the authors propose a framework and a set of tools for the development of ontology-aware repositories of learning materials. While the idea and use of *concept maps* is similar to our topic-driven navigation structure, in our approach topics are extracted from free text in a semi-automatic way, by leveraging information retrieval techniques and then validated by the user, while concepts have to be manually defined by the authors of the learning materials in the work of Dicheva and Aroyo.

Information retrieval and topic modeling have been used in the context of on-line conversations, for example, to help users in on-line communities obtain faster and better answers. This has been achieved by routing new messages to those users that are potential experts in the topic of the request and more likely to answer [35]. In this work, the authors compute expertise by using both the content of messages and the structure of the network. Moreover, posting assistance has been a research topic in the context of Issue Tracking Systems. Jalbert *et al.* [17] propose a mechanism to automatically classify duplicate bug reports upon creation, thus saving developer time. They use surface features, textual semantics, and graph clustering to predict duplicate status. Similarly, Nguyen *et al.* [21] use a combination of traditional information retrieval strategies and topic modeling to identify duplicate bug reports even when different terms are used. Although these works share owr goal of keeping conversation focused and cohesive, our approach is not concerned with duplicates neither with obtaining prompt replies. Still, future work could incorporate such concerns.

7 Conclusions and Future Work

Online discussion forums are one of the main asynchronous communication means and repositories of user generated content over the Internet. Learning management systems (LMSs), such as Moodle, use forums to support interaction and collaboration between students and students-to-teachers. Discussions taken place in a forum at some time represent a source of information for users accessing the forum afterwards. However, the effectiveness of a forum as a source of information for its users, additionally to be closely related to its richness in content, is also influenced by the way its contents are organized made searchable.

In this paper we presented an approach and a plugin for the Moodle LMS that enhances content navigation and information search in online discussion forums with a topic-driven navigational paradigm. The approach enables the automatic recovery of a lattice of discussion topics from the forum content, and the introduction of an additional navigation structure and graphical user interface which enable navigating and searching forum contents by topics of discussion. The plugin also supports the indexing of multiple forums into a single topic-driven enhanced forum, which is useful when conversation is split in various independent forums, with crosscutting topics.

While the approach has proven correctness for both the identified topics and the document-to-topics assignment [7], in this paper we have also shown with a case study that the additional navigation structure significantly improves the search of information stored in forum discussions.

In order to keep forums focused and cohesive, the plugin integrates a functionality that assists users in choosing the most appropriate forum to post to. Evaluation of the underlying assisted posting algorithms showed high accuracy, specially when dealing with forums discussing different topics.

In the future we aim to apply our approach in the context of social networks, in order to explore how it could improve social organization and user interaction. As a matter of fact, social networks are increasingly used in e-learning as side means for connecting students and teachers. Additionally, we plan to explore the applicability of "assisted posting" at a thread granularity level, to identify specific threads to post to within a given forum or a set of forums. This would increase cohesion of discussions and, potentially, reduce content duplication.

References

1. Baeza-Yates, R.A., Ribeiro-Neto, B.: Modern Information Retrieval. Addison-Wesley Longman Publishing Co., Inc., Boston (1999)
2. Bakalov, A., McCallum, A., Wallach, H.M., Mimno, D.M.: Topic models for taxonomies. In: Proceedings of the 12th ACM/IEEE-CS Joint Conference on Digital Libraries, JCDL 2012, Washington, D.C., USA, 10–14 June 2012, pp. 237–240 (2012)

3. Birkhoff, G. (ed.): Lattice Theory, vol. 25, 3rd edn. American Mathematical Society Colloquium Publications, Providence (1967)
4. Blei, D.M.: Introduction to probabilistic topic models. Commun. ACM **55**, 77–84 (2011). http://www.cs.princeton.edu/blei/papers/Blei2011.pdf
5. Castro, F., Nebot, A., Mugica, F.: Extraction of logical rules to describe students' learning behavior. In: Proceedings of the Sixth Conference on IASTED International Conference Web-Based Education, WBED 2007, vol. 2, pp. 164–169. ACTA Press, Anaheim (2007). http://dl.acm.org/citation.cfm?id=1323159.1323189
6. Castro, F., Vellido, A., Nebot, A., Mugica, F.: Applying data mining techniques to e-learning problems. In: Jain, L., Tedman, R., Tedman, D. (eds.) Evolution of Teaching and Learning Paradigms in Intelligent Environment. Studies in Computational Intelligence, vol. 62, pp. 183–221. Springer, Heidelberg (2007)
7. Cerulo, L., Distante, D.: Topic-driven semi-automatic reorganization of online discussion forums: a case study in an e-learning context. In: Proceedings of IEEE Global Engineering Education Conference (EDUCON 2013), pp. 303–310, March 2013
8. Dicheva, D., Dichev, C.: Tm4l: Creating and browsing educational topic maps. Br. J. Educ. Technol. **37**(3), 391–404 (2006). http://dx.doi.org/10.1111/j.1467-8535.2006.00612.x
9. Distante, D., Cerulo, L., Visaggio, C.A., Leone, M.: Enhancing online discussion forums with a topic-driven navigational paradigm: a plugin for the moodle learning management system. In: Proceedings of the 6th International Conference on Knowledge Discovery and Information Retrieval, KDIR 2014, pp. 97–106. Scitepress (2014)
10. Ganter, B., Wille, R.: Formal Concept Analysis: Mathematical Foundations. Springer, Heidelberg (1999)
11. Ghenname, M., Ajhoun, R., Gravier, C., Subercaze, J.: Combining the semantic and the social web for intelligent learning systems. In: Proceedings of IEEE Global Engineering Education Conference (EDUCON 2012), pp. 1–6, April 2012
12. Gruen, T.W., Osmonbekov, T., Czaplewski, A.J.: eWOM: the impact of customer-to-customer online know-how exchange on customer value and loyalty. J. Bus. Res. **59**, 449–456 (2006)
13. Hanna, M.: Data mining in the e-learning domain. Campus-Wide Inf. Syst. **21**(1), 29–34 (2004)
14. Hogo, M.A.: Evaluation of e-learning systems based on fuzzy clustering models and statistical tools. Expert Syst. Appl. **37**(10), 6891–6903 (2010). http://dx.doi.org/10.1016/j.eswa.2010.03.032
15. Hrastinski, S.: What is online learner participation? A literature review. Comput. Educ. **51**(4), 1755–1765 (2008)
16. Jakobsone, A., Kulmane, V., Cakula, S.: Structurization of information for group work in an online environment. In: Proceedings of IEEE Global Engineering Education Conference (EDUCON 2012), pp. 1–7, April 2012
17. Jalbert, N., Weimer, W.: Automated duplicate detection for bug tracking systems. In: IEEE International Conference on Dependable Systems and Networks with FTCS and DCC, 2008, DSN 2008, pp. 52–61, June 2008
18. Li, Q., Wang, J., Chen, Y.P., Lin, Z.: User comments for news recommendation in forum-based social media. Inf. Sci. **180**, 4929–4939 (2010)
19. Martin, A., Leon, C.: An intelligent e-learning scenario for knowledge retrieval. In: Proceedings of IEEE Global Engineering Education Conference (EDUCON 2012), pp. 1–6, April 2012

20. Meilă, M.: Comparing clusterings by the variation of information. In: Schölkopf, B., Warmuth, M.K. (eds.) COLT/Kernel 2003. LNCS (LNAI), vol. 2777, pp. 173–187. Springer, Heidelberg (2003)
21. Nguyen, A.T., Nguyen, T.T., Nguyen, T.N., Lo, D., Sun, C.: Duplicate bug report detection with a combination of information retrieval and topic modeling. In: Proceedings of the 27th IEEE/ACM International Conference on Automated Software Engineering, ASE 2012, pp. 70–79. ACM, New York (2012). http://doi.acm.org/10.1145/2351676.2351687
22. Otterbacher, J.: Searching for product experience attributes in online information sources. In: Proceedings of the International Conference on Information Systems (ICIS 2008). Association for Information Systems, December 2008
23. Romero, C., Ventura, S., Bra, P.D.: Knowledge discovery with genetic programming for providing feedback to courseware authors. User Model. User-Adap. Inter. **14**(5), 425–464 (2005). http://dx.doi.org/10.1007/s11257-004-7961-2
24. Salton, G., Wong, A., Yang, C.S.: A vector space model for automatic indexing. Commun. ACM **18**(11), 613–620 (1975). http://doi.acm.org/10.1145/361219.361220
25. Machado, L.D.S., Becker, K.: Distance education: a web usage mining case study for the evaluation of learning sites. In: 2003 IEEE International Conference on Advanced Learning Technologies (ICALT 2003), Athens, Greece, 9–11 July 2003, pp. 360–361. IEEE Computer Society (2003)
26. Stefan, H.: A theory of online learning as online participation. Comput. Educ. **52**(1), 78–82 (2009)
27. Sudau, F., Friede, T., Grabowski, J., Koschack, J., Makedonski, P., Himmel, W.: Sources of information and behavioral patterns in online health forums: qualitative study. J. Med. Internet Res. **16**, e10 (2014)
28. Sung, H.H., Sung, M,B., Sang, C.P.: Web mining for distance education. In: Proceedings of the 2000 IEEE International Conference on Management of Innovation and Technology (ICMIT 2000), vol. 2, pp. 715–719. IEEE (2000). http://dx.doi.org/10.1109/ICMIT.2000.916789
29. Tang, T., McCalla, G.: Smart recommendation for an evolving e-learning system: architecture and experiment. Int. J. e-Learning **4**(1), 105–129 (2005)
30. Tsai, C.-J., Tseng, S.S., Lin, C.-Y.: A two-phase fuzzy mining and learning algorithm for adaptive learning environment. In: Alexandrov, V.N., Dongarra, J., Juliano, B.A., Renner, R.S., Tan, C.J.K. (eds.) ICCS-ComputSci 2001. LNCS, vol. 2074, pp. 429–438. Springer, Heidelberg (2001)
31. Wallach, H.M., Mimno, D.M., McCallum, A.: Rethinking LDA: why priors matter. In: 23rd Annual Conference on Neural Information Processing Systems 2009. Advances in Neural Information Processing Systems, vol. 22, pp. 1973–1981 (2009)
32. Yang, Q., Sun, J., Wang, J., Jin, Z.: Semantic web-based personalized recommendation system of courses knowledge research. In: Proceedings of the 2010 International Conference on Intelligent Computing and Cognitive Informatics, ICICCI 2010, pp. 214–217. IEEE Computer Society, Washington, D.C. (2010). http://dx.doi.org/10.1109/ICICCI.2010.54
33. Zaíane, O.R.: Building a recommender agent for e-learning systems. In: Proceedings of the International Conference on Computers in Education, ICCE 2002, pp. 55–59. IEEE Computer Society, Washington, D.C. (2002). http://dl.acm.org/citation.cfm?id=838238.839230

34. Zhang, K., Peck, K.: The effects of peer-controlled or moderated online collaboration on group problem solving and related attitudes. Can. J. Learn. Technol./La revue canadienne de l'apprentissage et de la technologie **29**(3), 93–112 (2003)
35. Zhou, Y., Cong, G., Cui, B., Jensen, C.S., Yao, J.: Routing questions to the right users in online communities. In: Proceedings of the 2009 IEEE International Conference on Data Engineering, ICDE 2009, pp. 700–711. , IEEE Computer Society, Washington, D.C. (2009). http://dx.doi.org/10.1109/ICDE.2009.44

Random Perturbations of Term Weighted Gene Ontology Annotations for Discovering Gene Unknown Functionalities

Giacomo Domeniconi[1(✉)], Marco Masseroli[2], Gianluca Moro[1],
and Pietro Pinoli[2]

[1] DISI, Università degli Studi di Bologna, Via Venezia 52, 47521 Cesena, Italy
{gianluca.moro,gianluca.moro}@unibo.it
[2] DEIB, Politecnico di Milano, Piazza L. Da Vinci 32, 20133 Milan, Italy
{marco.masseroli,pietro.pinoli}@polimi.it

Abstract. Computational analyses for biomedical knowledge discovery greatly benefit from the availability of the description of gene and protein functional features expressed through controlled terminologies and ontologies, i.e. of their controlled annotations. In the last years, several databases of such annotations have become available; yet, these annotations are incomplete and only some of them represent highly reliable human curated information. To predict and discover unknown or missing annotations existing approaches use unsupervised learning algorithms. We propose a new learning method that allows applying supervised algorithms to unsupervised problems, achieving much better annotation predictions. This method, which we also extend from our preceding work with data weighting techniques, is based on the generation of artificial labeled training sets through random perturbations of original data. We tested it on nine Gene Ontology annotation datasets; obtained results demonstrate that our approach achieves good effectiveness in novel annotation prediction, outperforming state of the art unsupervised methods.

Keywords: Gene ontology · Biomolecular annotation prediction · Bioinformatics · Knowledge discovery · Supervised learning · Term weighting

1 Introduction

A common machine learning task often performed in several application domains, including bioinformatics, is the prediction of associations between items and features characterizing them. It well supports knowledge discovery, particularly when the considered features are described by means of controlled term, especially if such terms are related within ontologies. Indeed, several terminologies and ontologies exist and are used to describe structural and functional features of biomolecular entities, mainly genes and proteins. Among them, the Gene Ontology (GO) [1] is the most developed and considerable one; it is widely used to annotate, i.e. characterize, genes and proteins by associating them to its terms.

This research is part of the "GenData 2020" project funded by the Italian MIUR.

A. Fred et al. (Eds.): IC3K 2014, CCIS 553, pp. 181–197, 2015.
DOI: 10.1007/978-3-319-25840-9_12

The GO consists of three sub-ontologies that overall include more than 40,000 controlled terms, which characterize species-independent Biological Processes (BP), Molecular Functions (MF) and Cellular Components (CC). These terms are hierarchically related, mainly through *"is a"* or *"part of"* relationships, within a Directed Acyclic Graph (DAG) and are designed to capture orthogonal features of genes and proteins. In the GO DAG, each node represents a GO term and each directed edge from a node a to a node b represents a relationship existing from a child term a to its parent term b.

Controlled annotations are very valuable for high-throughput and computationally intensive bioinformatics analyses. Yet, some of them are less reliable, or may even be incorrect, since automatically inferred without any human curation, which is highly time consuming. Furthermore, available biomolecular annotations are incomplete, since several gene and protein features of many organisms are still to be discovered and annotated. In this scenario, computational methods able to predict new annotations and estimate incorrectness of available ones are paramount, specially the methods that provide ranked lists of inferred annotations; they can, for instance, quicken the curation process by focusing it on the prioritized novel annotations [2].

Here, we first apply different supervised algorithms to discover new GO term annotations of different organism genes based on available GO annotations; then, we benchmark them with an unsupervised method previously used to this purpose. Since in this context it is not available a labeled set of instances to train a supervised algorithm, we propose to assign labels to the originally unlabeled GO annotations based on a random perturbation of the annotation matrix that switches off some known annotations. In this way, we create a novel matrix with missing annotations; thus, we can train the model to recognize from this artificial matrix the real annotations. This allows applying supervised methods to available gene annotations and predicting new gene function annotations with better performance than the previously used unsupervised methods. We introduced this general approach in [3]; in this paper we propose its extension with the use of real values, instead of binary ones, to represent the biomolecular annotations. The proposed data representation is exactly the same as that of a classification problem represented in a Vector Space Model; thus we can apply a weighting scheme to increase the effectiveness of the predictive model. We conducted experiments with several weighting measures to analyze the behavior of the proposed method with real-valued matrices. Despite multiple heterogeneous data could be leveraged to predict gene functions through sophisticated techniques previously proposed, with the methods here presented we confirm that simpler analytical frameworks, which use faster methods based only on available annotations, are as much effective and useful.

The rest of the paper is organized as follows. Section 2 describes the annotation datasets used in our experiments. Section 3 exposes the methods used to predict new annotations. Section 4 illustrates the performed experiments and reports their results, benchmarking them with those of a previous work. Section 5 reports an overview of other works about genomic functions prediction. Finally, in Sect. 6 we discuss our contribution and foresee possible future developments.

2   Genomic Datasets

In order to have easy access to subsequent versions of gene annotations to be used as input to the considered algorithms or to evaluate the results that they provide, we took advantage of the Genomic and Proteomic Data Warehouse (GPDW) [4]. In GPDW several controlled terminologies and ontologies, which describe genes and gene products related features, functionalities and phenotypes, are stored together with their numerous annotations to genes and proteins of many organisms. These data are retrieved from several well known biomolecular databases. In the context of developing and testing machine learning methods on genomic annotations, GPDW is a valuable source since it is quarterly updated and old versions are kept stored. We leveraged this feature in our method evaluation by considering differed versions of the GO annotations of the genes of three organisms. In GPDW they are available with additional information, including an *evidence code* that describes how reliable the annotation is. We leveraged it by filtering out the less reliable annotations, i.e. those with *Inferred from Electronic Annotation (IEA)* evidence, from the datasets used for our evaluation. Table 1 gives a quantitative description of the considered annotations.

In GPDW, as in any other biomolecular database, only the most specific controlled annotations of each gene are stored. This is because, when the controlled terms used for the annotation are organized into an ontology, as for the GO, biologists are asked to annotate each gene only to the most specific ontology terms representing each of the gene features. In this way, when a gene is annotated to a term, it is implicitly indirectly annotated also to all the more generic terms, i.e. all the ancestors of the feature terms involved in its direct annotations. This is called *annotation unfolding*.

All direct and indirect annotations of a set of genes can be represented by using binary matrices. Let $\mathcal{G}$ be the set of genes of a certain organism and $\mathcal{T}$ a set of feature terms. We define the annotation matrix $\mathbf{A} \in \{0,1\}^{|\mathcal{G}| \times |\mathcal{T}|}$ as the matrix whose columns correspond to terms and rows to genes. For each gene

Table 1. Quantitative characteristics of the nine considered annotation datasets. Figures refer to the sum of direct and indirect annotations not inferred from electronic annotation, i.e. without IEA evidence code.

	Gallus gallus			Bos taurus			Danio rerio		
	CC	MF	BP	CC	MF	BP	CC	MF	BP
# considered genes	260	309	275	497	540	512	430	699	1,528
# considered terms	123	134	610	207	226	1,023	131	261	1,176
# annotations (July 2009)	3,442	1,927	8,709	7,658	3,559	18,146	4,813	4,826	38,399
# annotations (May 2013)	3,968	2,507	10,827	9,878	5,723	24,735	5,496	6,735	58,040
Δ annotations between GPDW versions									
#Δ annotations	526	580	2,118	2,220	2,164	6,589	683	1,909	19,641
%Δ annotations	15.28	30.10	24.32	29.00	60.80	36.31	14.19	39.56	51.15

$g \in \mathcal{G}$ and for each term $t \in \mathcal{T}$, the value of the $\mathbf{A}(g,t)$ entry of the annotation matrix is set according to the following rule:

$$\mathbf{A}(g,t) = \begin{cases} 1, & \text{if } g \text{ is annotated either to } t \\ & \text{or to any of } t \text{ descendants} \\ 0, & \text{otherwise} \end{cases} \qquad (1)$$

Examples of two versions of these matrices are shown in Fig. 1a and b, where $\mathbf{A}_1$ is an updated version of $\mathbf{A}_0$. Each GPDW update contains some number of new discovered annotations, namely new 1 in the matrix.

3   Annotation Discovery Methods

3.1   Data and Problem Modelling

The discovery of new genomic annotations can be modeled as a supervised problem in which, given a feature term t, you want to predict if a gene g is likely to be, or not to be, annotated to that term t, i.e. if the element $\mathbf{A}(g,t)$ of the annotation matrix is likely to be 1, or 0, basing on known annotations to other terms of the gene g, as in Fig. 1c.

All the terms $t \in \mathcal{T}$ must be predicted, i.e. all the columns of the matrix, thus the problem can be modeled as a supervised multi-label classification, with the difference that we do not have a distinct set of features and labels, but we have a set of terms that are both classes and features. To address this problem, we use the most common approach in the literature, i.e. transform it into a set of binary classification problems, which can then be handled using single-class classifiers. Henceforth, for simplicity of exposition, we will refer to a single supervised task concerning the discovery of a new annotation of the gene g to the term t (for instance the term *GO:005737* in Fig. 1), which is then repeated iteratively for all other genes and terms.

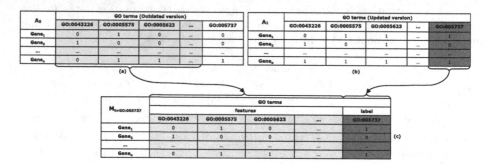

Fig. 1. Illustrative diagram of the data representation. The data set (c) is created with an older annotation version $\mathbf{A}_0$ (a) for the features and an uptdated version $\mathbf{A}_1$ (b) for the labels.

Let's now see how to assign a label to each instance of the data model. Given an annotation matrix, our proposal is to use as input a version of the matrix with less annotations (referred as outdated matrix, since it may resemble an outdated annotation dataset version); then, to derive from such input matrix the features of the data model, and consider as label of each record the presence or absence of an annotation in a more complete matrix (referred as updated matrix, since it may resemble a newer annotation dataset version). This representation is sketched in Fig. 1. Given the feature term t considered for the prediction, called *class-term*, the representation of the data is created by taking as features, for each gene, all the annotations to all the other terms in an outdated version of the matrix $\mathbf{A}_0$, while the label is given by the value of the class-term in the updated version of the matrix $\mathbf{A}_1$. Henceforth, we refer to this representation matrix as $\mathbf{M}_t$, where t is the class-term of the model.

This data representation is exactly the same as that of a supervised classification problem represented in a Vector Space Model. Thus, a classic supervised task could be envisaged by subdividing this new matrix $\mathbf{M}_t$ horizontally and using a part of the genes to train the model and the remaining part to test it. In this domain, however, this approach is not applicable because it implies the availability of at least the part of the updated matrix to train the model, but new datasets are only released as a whole and not partially. Thus, the purpose is to predict which annotations are missing in the entire matrix, rather than on some part of it. The data representation matrix $\mathbf{M}_t$ requires information from two different annotation dataset versions. Thus, since the aim is to make predictions over the entire dataset, to train the model we use a matrix $\mathbf{M}_t^{\text{train}}$ that is created by using the information from both the latest version currently available at training time, i.e. $\mathbf{A}_1$, and an older version of the matrix with missing annotations, i.e. $\mathbf{A}_0$. With this two different versions of the matrices, the training set is created by using the features derived from the outdated version $\mathbf{A}_0$ and the labels from the updated one $\mathbf{A}_1$. Then, the validation of the classification model has to be made by discovering new annotations, missing in the current state of the matrix. Therefore, the features regarding the current version $\mathbf{A}_1$ and labeled with the values of a future updated matrix $\mathbf{A}_2$ are used to create the validation matrix $\mathbf{M}_t^{\text{valid}}$. The training and validation data representation process is sketched in Fig. 2.

3.2   Random Perturbation

The supervised problem modelling described in the previous subsection requires, at training time, two versions of the annotation matrix to create the supervised model, i.e. $\mathbf{A}_0$ and $\mathbf{A}_1$. However, biologists typically have available only the most updated version of the annotation matrix, not keeping stored the outdated versions for space reasons, given the large amount of data. Thus, with reference to Fig. 1, there is available only one version of the matrix, i.e. only the current version $\mathbf{A}_1$, with which the training data representation $\mathbf{M}_t^{\text{train}}$ is created.

To overcome the problem just mentioned, we start from the observation that also the input matrix $\mathbf{A}_1$ contains missing annotations. Therefore, we could use

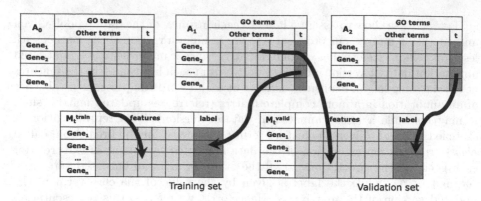

Fig. 2. Illustrative diagram of the dataset representation for the prediction model of the annotations to a term t. The training set ($\mathbf{M}_t^{\text{train}}$) is created with an older annotation version $\mathbf{A}_0$ for the features and the current annotation version $\mathbf{A}_1$ for the labels. Similarly, the validation set ($\mathbf{M}_t^{\text{valid}}$) is created using $\mathbf{A}_1$ and a future updated annotation matrix $\mathbf{A}_2$.

only this matrix to obtain the representation $\mathbf{M}_t$, assuming $\mathbf{A}_0 = \mathbf{A}_1$. However, the classification model will have to discover new gene-term annotations starting from an outdated matrix; thus, it will be more effective if it is trained with a training set in which the features are taken from an outdated matrix, with a greater number of missing annotations than the matrix version from which the labels of the instances are obtained. If we consider that the annotations of genes to features are discovered by teams of biologists that work independently from each other, a reasonable hypothesis is that the new annotations discovered by the entire scientific community, on the whole, do not have any kind of bond or rule. This should be equivalent to a random process of discovery of new annotations. Such considerations led to our thesis that new gene annotations can be better discovered by artificially increasing the number of missing annotations in the input matrix $\mathbf{A}_0$. Since, as mentioned, usually only the input matrix $\mathbf{A}_1$ is available, this can be achieved by randomly deleting known annotations in the matrix $\mathbf{A}_1$ to obtain a new matrix $\mathbf{A}_0$ artificially perturbed.

Thus, to get the data to train the classification model, we propose to randomly perturb the matrix $\mathbf{A}_1$ to create a new matrix $\mathbf{A}_0$, in which some annotations are eliminated with a probability p. In this way we obtain the matrix $\mathbf{A}_0 = random\_perturbation(\mathbf{A}_1, p)$. Formally, for each gene g and term t, the perturbation is done as follows:

$$\mathbf{A}_0(g,t) = \begin{cases} 0 & \text{if } \mathbf{A}_1(g,t) = 1 \wedge random \leq p \\ 1 & \text{if } \mathbf{A}_1(g,t) = 1 \wedge random > p \\ 0 & \text{if } \mathbf{A}_1(g,t) = 0 \end{cases} \qquad (2)$$

Once the perturbed matrix $\mathbf{A}_0$ is generated, to ensure its correctness with respect to the unfolding of the annotations, the matrix $\mathbf{A}_0$ is corrected by switch-

ing to 0 also all the annotations to the same gene of all the descendants of the
ontological terms with modified gene annotation; we call this process *perturbation unfolding*. It is important to note that, depending on this correction,
the percentage of the actual modified annotations of the matrix $\mathbf{A}_0$ will hence
be greater than the percentage derived from p. The overall data representation
process is the same as that shown in Fig. 2, with the difference that the matrix
$\mathbf{A}_0$ is created by perturbing randomly $\mathbf{A}_1$.

Considering the annotation unfolding in the GO, in order to avoid trivial
predictions (i.e. 1 if a child is 1), in the set of features of the dataset $\mathbf{M}_t$ all the
descendants or ancestors of the term t are not taken into consideration. Once
created the training matrix $\mathbf{M}_t^{train}$, we can use any supervised algorithm, capable
of returning a probability distribution, to train the prediction model and then
validate it with $\mathbf{M}_t^{valid}$. The prediction model provides a probability distribution
$pd(g, t)$, called *likelihood*, concerning the presence of an annotation of the gene
g to the term t. To provide predictions of only new annotations, only those
annotations that were missing in the outdated version of the matrix are taken
into account. The supervised process described above is repeated for all the terms
$t \in \mathcal{T}$, giving as final output a list of predictions of new gene annotations ordered
according to their likelihood; the illustrated annotation discovery workflow is
sketched in Fig. 3.

3.3   Likelihood Correction

As shown above, the output of the supervised model is a list of predicted annotations, each one with a likelihood degree. According to the hierarchical structure
of GO, when a gene is annotated to an ontological term, it must be also annotated to all the ancestors of that term; this constraint is also known as *True Path
Rule* [5]. The supervised classifier, however, provides a likelihood for each gene
annotation regardless of the predictions of the annotation of other GO terms to
the same gene. This can result in possible cases of anomalies in which a gene
shall be annotated to a term, but not to one or more of its ancestor terms, thus
violating the True Path Rule. To obtain a likelihood that takes into account the
hierarchy of the terms, once obtained the likelihood of each gene-term association, we proceed as follows:

1. For each novel gene-term annotation, to the probability given by the model
   we add the average of all the probabilities of the novel annotations of the
   gene to all the ancestors of the term. Note that, since the classification model
   provides in output a probability distribution ranging between 0 and 1, the
   hierarchical likelihood of each gene-term annotation shall be between 0 and
   2, as follows:

$$pd^H(g, t) = \frac{\sum\limits_{t_a \in ancestors(t)} pd(g, t_a)}{|ancestors(t)|} + pd(g, t) \tag{3}$$

2. Once the likelihood is made hierarchical, the correction of the possible anomalies regarding the True Path Rule is taken into account. An iterative process is carried on from the leaf terms to the root term of the hierarchy, upgrading each likelihood with the maximum likelihood value of the descendant terms, as follows:

$$l(g,t) = \max\{pd^H(g,t), \max_{t_c \in children(t)} \{pd^H(g,t_c)\}\} \qquad (4)$$

In such a way, for each ontology term, the likelihood of a gene to be annotated to that term is always greater than or equal to the likelihood of the gene to be annotated to the term descendants.

3.4   Term Weighting

To increase the associative power of the gene-term matrix, we can give a weight to each existing association between genes and terms. This approach is similar to a classic term weighting in information retrieval [6,7]. Some weighting schemes have already been appliated to the prediction of genomic annotations [8]; here we tested these and other different schemes from information retrieval and data classification realms, in order to weigh each known annotation in the representation matrix $\mathbf{M_t}$.

Fixed the *class-term* t_c of the representation matrix $\mathbf{M_t}$, for each feature term $t \in \mathcal{T} : t \neq t_c$ four elements (A, B, C and D) shall be defined in order to describe the term weighting schemes: A denotes the number of genes associated both with t_c and t; B denotes the number of genes associated with t_c and not with t; C denotes the number of genes associated with t and not with t_c and D denotes the number of genes associated neither with t_c or t. The sum of all the genes is denoted with $N = A + B + C + D = |\mathcal{G}|$.

Generally, a term weighting scheme is based on three factors: (i) *term frequency factor* or local weight; (ii) *collection frequency factor* or global weight; (iii) *normalization factor*. The *term frequency factor* measures how important a feature, namely an ontology term, is to a certain gene. For each gene g and feature term t, it can be expressed as $tf(g,t) = 1 + M$, where M is the number of descendant terms of t which are associated with the gene g, both directly or indirectly (i.e. derived from the unfolding procedure). Considering that this tf is measured in a different way than in the standard information retrieval methods, we also consider the case in which the local factor consists in a simple binary value, regarding the presence or the absence of the association between t and g.

The *collection frequency factor* may be taken from virtually all proposed weighting schemes in information retrieval or data mining. An interpretation of the common *idf* (inverse document frequency, [9]) is the *igf* (inverse gene frequency, [8]); for each term t its value is:

$$igf(t) = \ln \frac{|\mathcal{G}|}{|genes\ annotated\ to\ t|} \equiv \log\left(\frac{N}{A+C}\right) \qquad (5)$$

The combination of these two measures, *term* and *collection frequency factors*, with the *normalization factor* generates several possible weighting schemes [10, 11]. The contribution of seven of these generated schemes to the gene annotation prediction using an unsupervised method is studied in [8], where a substantial improvement is shown by using some of them. Out of all the schemes analyzed in that work, we focus on the combination regarding no-transformation in the tf factor and the maximum, cosine and none normalization (i.e. the schemes named NTM, NTC and NTN). In this work, we refer to these three schemes as $tf.igf^M$, $tf.igf^C$ and $tf.igf^N$, respectively. In our experiments, we also tested the igf alone, using only a binary value (bin) as term frequency factor; we refer to these schemes as igf^M, igf^C and igf^N. In addition, we tested also the two term frequency measures, i.e. tf and bin, alone, without a collection frequency factor.

Furthermore, we use some weighting schemes derived directly from the information retrieval [12], such as χ^2 or ig (information gain). These two schemes are calculated as follows:

$$\chi^2 = N \cdot \frac{(A \cdot D - B \cdot C)^2}{(A+C) \cdot (B+D) \cdot (A+B) \cdot (C+D)} \tag{6}$$

$$ig = -\frac{A+B}{N} \cdot \log \frac{A+B}{N} + \frac{A}{N} \cdot \log \frac{A}{A+C} + \frac{B}{N} \cdot \log \frac{B}{B+D} \tag{7}$$

Finally, we also tested the *relevance frequency* (rf) scheme proposed by [13] for the text classification task. The rf of a term t is based on the idea that the higher the concentration of genes associated both with t and the *class-term*, the greater the contribution of t in the prediction model.

$$rf = \log \left(2 + \frac{A}{\max(1, C)} \right) \tag{8}$$

3.5 Evaluation

In our experiments we tested the effectiveness of supervised models in discovering new functional gene annotations from the available annotations. Since the proposed method is applicable to any supervised algorithm that returns a probability distribution, we tested different types of existing algorithms in order to measure their effectiveness, in particular: *Support Vector Machines*, *nearest neighbors*, *decision trees*, *logistic regressions* and *naive bayes*, using the implementations provided by Weka[1] in its 3.7.9 version. In the experiments we tested the Weka classifiers: *IBk* (with $k = 3$), *J48*, *Logistic*, *Naive Bayes* (*NB*), *Random Forest* (*RF*) and *SMO*. For each algorithm we used the default parameter settings provided by Weka; no tuning of parameters has been done for time reasons.

We measured the effectiveness of the predictions in the same way it was done in [8], in order to be able to directly compare our results with those in that work; the overall procedure was as follows.

---

[1] http://www.cs.waikato.ac.nz/ml/weka/.

1. We extracted the input annotations from an outdated version of the GPDW (July 2009), excluding from those annotations the ones less reliable, i.e. with IEA *evidence* code.
2. We randomly perturbed the unfolded annotation matrix to get a modified version of it, with some missing annotations.
3. We applied a weighting scheme on the representation matrix.
4. By running the prediction algorithm, we got a list of predicted annotations ordered by their confidence value (i.e. their corresponding likelihood $l(g,t)$).
5. We selected the top P predictions (we use $P = 250$) and we counted how many of these P predictions were found confirmed in the updated version of the GPDW (May 2013 version), regardless their evidence code.
6. For each experiment, steps 2, 3, 4 and 5 were repeated 10 times by varying the random seed. The effectiveness of each experiment was determined by averaging the counts obtained in all the experiment repetitions.

We depict the training and validation procedure workflows in Fig. 3.

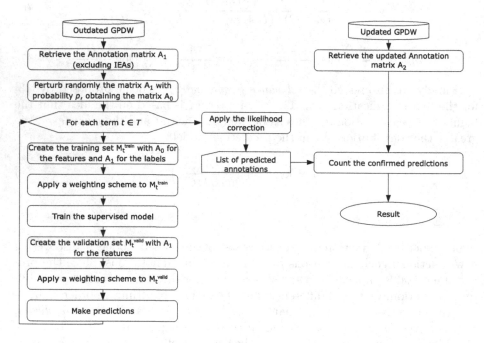

Fig. 3. Workflow of the training and validation processes.

4   Results

Table 2 shows the results obtained by varying the supervised algorithm used to train the prediction model, always using a fixed random perturbation probability $p = 0.05$ and without weighting the term associations, i.e. using the *binary* scheme. State of art methods [8] reach a total of 447 correct predictions; Table 2 shows that, with the proposed method, 3 out of 6 of the tested algorithms outperform them. Obtained results are excellent if we consider that they are obtained without any tuning of the algorithm parameters; therefore there is margin to improve them with an appropriate tuning. According to the results in Table 2, we can infer that using the standard parameterization provided by Weka, the algorithm that obtains the best results is IBk, with an improvement of 44.8 % compared with the results of [8]. IBk results also 6.2 % better than $Random$ $Forest$ and 6.3 % better than $J48$, the other two supervised algorithms that result better than the state of art.

Table 3 shows the results obtained by using the IBk algorithm with different term weighting schemes in the representation matrix. Differently from the work of Pinoli and colleagues [8], where weighting schemes improved their method, from our results we can infer that using weighting schemes does not lead to an improvement of the prediction effectiveness. The comparisons between the weighting schemes considered provided fluctuating results, but, in general, the best scheme appears to be the $binary$ one. We can note that the schemes with the tf factor do not achieve good results, as well as, although with better performance, the information retrieval classical supervised measures, namely χ^2 and ig; the best weighting scheme results the rf, which however is, in total, slightly worst than the $binary$ one.

Table 2. Validation results of the predictions obtained by varying the supervised algorithm used to build the prediction model. The results show, for each of the nine considered datasets, the amount of the top 250 predicted gene annotations that have been found confirmed in the updated GPDW version. The setup of these experiments was done with random perturbation of the training matrix with probability $p = 0.05$ and the binary term weighting scheme. The first column (SIM) reports the results obtained in [8] with the SIM best configuration. Each result is reported as the average and corresponding standard deviation of 10 experiments repeated by changing the random perturbation seed. In bold the best result for each dataset.

Dataset	SIM	IBk	J48	Logistic	NB	RF	SMO
Gallus g. - BP	**86**	58.6 ± 20.2	47.2 ± 4.7	32.7 ± 6.8	25.4 ± 4.4	52.7 ± 12.1	28.7 ± 9.3
Gallus g. -MF	24	58.0 ± 5.6	**79.7 ± 12.7**	40.0 ± 10.4	14.2 ± 1.6	54.4 ± 9.6	50.7 ± 14.3
Gallus g. -CC	50	**81.5 ± 8.2**	73.4 ± 8.5	31.9 ± 6.4	23.5 ± 3.7	55.2 ± 11.3	29.6 ± 4.0
Bos t. -BP	55	48.9 ± 6.8	49.7 ± 5.1	37.0 ± 6.5	28.4 ± 4.2	**62.4 ± 7.6**	31.2 ± 4.6
Bos t. -MF	28	58.2 ± 4.4	**58.8 ± 10.5**	27.5 ± 4.3	15.7 ± 2.9	57.5 ± 11.2	36.9 ± 4.4
Bos t. -CC	91	**112 ± 9.7**	94.3 ± 9.8	38.2 ± 5.3	8.2 ± 2.0	93.7 ± 10.4	48.4 ± 6.8
Danio r. -BP	35	**70.9 ± 15.9**	59.8 ± 6.1	31.0 ± 4.8	25.2 ± 3.3	58.1 ± 5.1	16.6 ± 2.3
Danio r. -MF	35	77.5 ± 10.3	75.8 ± 7.1	54.4 ± 11.0	41.2 ± 2.7	**83.1 ± 9.6**	79.7 ± 8.7
Danio r. -CC	44	81.5 ± 8.5	69.3 ± 8.7	27.6 ± 7.6	26.2 ± 6.6	**92.3 ± 11.0**	30.2 ± 6.6
Total	447	**647.1**	608.8	320.3	207.9	609.4	352.0

Table 3. Validation results of the predictions obtained using IBk as supervised algorithm, $p = 0.05$ as probability of perturbation and varying the term weighting scheme.

Dataset	BIN	tf	$tf.igf^N$	$tf.igf^C$	$tf.igf^M$	igf^N	igf^C	igf^M	χ^2	ig	rf
$G.g.$-BP	58.6	52.5	48.0	46.4	51.2	58.6	62.0	**66.6**	47.0	50.2	47.2
$G.g.$-MF	58	53.0	57.6	58.4	53.2	51.6	**64.4**	60.8	62.6	64.0	61.4
$G.g.$-CC	**81.5**	55.6	48.2	42.2	58.4	61.0	45.4	50.8	68.0	69.6	67.2
$B.t.$-BP	48.9	61.8	52.8	34.4	48.0	45.8	**72.6**	65.0	39.0	40.2	56.6
$B.t.$-MF	58.2	39.0	38.4	48.0	38.6	73.0	60.6	64.0	72.8	**75.4**	73.6
$B.t.$-CC	112.0	90.2	93.8	72.6	80.4	93.0	103.4	90.0	101.8	103.4	**121.4**
$D.r.$-BP	**70.9**	68.6	63.4	59.9	61.2	70.6	67.2	69.2	69.3	68.3	67.6
$D.r.$-MF	**77.5**	45.2	57.2	46.0	23.6	60.8	55.4	53.0	57.2	52.4	62.2
$D.r.$-CC	81.5	74.2	30.6	49.2	40.4	61.4	77.6	60.8	75.2	73.6	**82.6**
Total	**647.1**	540.1	490.0	457.1	455.0	575.8	608.6	580.2	592.9	597.1	639.8

Table 4. Validation results of the predictions obtained using the IBk supervised algorithm, binary term weighting scheme and varying the probability p of random perturbation of the training matrix.

Dataset	$p = 0$	$p = 0.05$	$p = 0.10$	$p = 0.15$	$p = 0.20$	$p = 0.25$	$p = 0.30$
Gallus g.- BP	42	**58.6 ± 20.2**	54.8 ± 16.2	51.3 ± 12.5	55.9 ± 10.4	50.2 ± 10.2	47.4 ± 9.7
Gallus g.- MF	50	58 ± 5.6	61.8 ± 11	59.5 ± 13	58.3 ± 10.2	**63.6 ± 13.5**	64.2 ± 8.4
Gallus g.- CC	75	81.5 ± 8.2	77.5 ± 9.7	**82.2 ± 8.1**	78.1 ± 7.5	73.3e ± 13.2	78.8e ± 12
Bos t.- BP	43	48.9 ± 6.8	51.7 ± 10.1	50.4 ± 8.4	**53.1 ± 9.6**	52 ± 12.5	52.2 ± 15.4
Bos t.- MF	58	58.2 ± 4.4	62.7 ± 7.7	71.4 ± 10.9	73 ± 12.6	74.7 ± 11.6	**77 ± 13**
Bos t.- CC	108	112 ± 9.7	114.3 ± 11	118.6 ± 13	118.1 ± 13	**119 ± 13.1**	116.7 ± 22
Danio r.- BP	55	70.9 ± 15.9	70.6 ± 16.5	74.8 ± 13.9	85.7 ± 25.6	83.1 ± 16.3	**90.6 ± 19.4**
Danio r.- MF	76	**77.5 ± 10.3**	72.5 ± 7.1	67.7 ± 10.1	62 ± 7.6	58.4 ± 8.7	51.4 ± 15.1
Danio r.- CC	79	81.5 ± 8.5	84.7 ± 8.7	**90.7 ± 10**	85.6 ± 13.5	83.3 ± 14.5	75.8 ± 19.9
Total	586	647.1	650.6	666.6	**669.8**	661.6	654.1

The proposed method introduces a new parameter: the probability p of the random perturbation of the training matrix. Table 4 shows the results obtained by varying this probability p and using the best supervised algorithm from Table 2, namely IBk, and the *binary* weighting scheme. Table 4 results show that the best predictions are obtained with $p = 0.2$. Considering the *perturbation unfolding*, this p value leads to a perturbed matrix $\mathbf{A}_0$ with more than 20% of annotations less than in $\mathbf{A}_1$ (empirically they are about 30% less). Such percentage is very close to the average value of the variation of number of annotations between $\mathbf{A}_2$ and $\mathbf{A}_1$, i.e. 33.4%, notable in Table 1. Moreover, the probability p that gets the best results for each dataset seems to have a relationship with the dataset annotation variation between $\mathbf{A}_2$ and $\mathbf{A}_1$. This result leads to the conjectures that (i) representing new annotations randomly leads to train a classifier able to predict the actual new annotations between two different annotation versions; (ii) the more the amount of artificial missing annotations introduced in the training set is comparable to the actual missing annotations in the validation set, the more the predictions are accurate. Another result deducible from Table 4 is that using $p = 0$, namely the annotation matrix is not perturbed ($\mathbf{A}_0 = \mathbf{A}_1$), we get anyway good results, higher than those in [8]. This is important since it allows to avoid the parameter p and the tuning of the system for any considered dataset when not top performance is required. For a graphical view, the results discussed are also shown in Fig. 4, grouped by organism. Our approach outperforms the best accuracy achieved in [8] of 49.66 %, in particular we obtain the highest improvement in large datasets, i.e. in the *Danio rerio* dataset there is an improvement of 104.56 % of the correct annotations predicted.

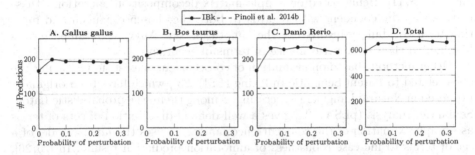

Fig. 4. Validation results of the predictions obtained by varying the probability of perturbation p, compared with those obtained in [8]. The results show, for each organism in the A, B and C charts, the sum of the predicted annotations that have been found confirmed in the updated GPDW version of the three GO ontologies. The chart D shows the total values for all the organism.

5   Related Works

Different methods have been proposed to predict biomolecular annotations. In [14], decision trees and Bayesian networks were suggested to learn patterns from available annotation profiles and predict new ones. Tao and colleagues [15] improved the results by using a k-nearest neighbour (k-NN) classifier to make a gene inherit the annotations that are common among its nearest neighbour genes in a gene network, where distance between genes is based on the semantic similarity of the GO terms used to annotate them.

Novel gene annotations can also be inferred based on multiple data sources. In [16], gene expression levels from microarray experiments are used to train a Support Vector Machine (SVM) classifier for each gene annotation to a GO term; consistency among predicted annotation terms is then enforced through a Bayesian network mapped onto the GO structure. Conversely, in [17,18], the authors took advantage of textual information by mining the literature and extracting keywords that are then mapped to GO concepts. This approach has the disadvantage to require a preparatory data integration step in order to be performed; this both adds complexity to the framework and reduces its flexibility.

In [19,20], Khatri and colleagues suggested a prediction algorithm based on the Singular Value Decomposition (SVD) method of the gene-to-term annotation matrix, which is implicitly derived from the count of co-occurrences between pairs of terms in the available annotation dataset. This prediction method based on basic linear algebra was then extended in [21], by incorporating gene clustering based on gene functional similarity computed on Gene Ontology annotations. It was further enhanced by automatically choosing its main parameters, including the SVD truncation level, based on the evaluated data [22]. The SVD has also been used with annotation co-occurrence weights based on gene-term frequencies [8,11]. Being based on simple matrix decomposition operations, these methods are independent with regards to both the chosen organism and function term vocabulary involved in the annotation set. Anyway, obtained results highlighted their poor performance in terms of accuracy.

Other methods based on evaluation of co-occurrences exist; in particular the ones related to Latent Semantic Indexing (LSI) [23], which have been originally proposed in Natural Language Processing. Among them, the probabilistic Latent Semantic Analysis (pLSA) [24] gives a well defined distribution of sets of terms as an approximation of the co-occurrence matrix. It uses the *latent* model of a set of terms to increase robustness of annotation prediction results. In [25,26], pLSA proved to provide general improvements with respect to the truncated SVD method of Khatri and colleagues ([19]).

In bioinformatics, *topic modeling* has been leveraged also by using the Latent Dirichlet Allocation (LDA) algorithm [27]. In [28,29], LDA was used to subdivide expression microarray data into clusters. Very recently, Pinoli et al. [30] took advantage of the LDA algorithm, together with the Gibbs sampling [31–33], to predict gene annotations to GO terms. These methods strongly overcome the ones based on linear algebra, but the complexity of the underlying model and the slowness of the training algorithms make these approaches ill-suited when the size of the dataset grows.

In summary, previously proposed methods for biomolecular annotation prediction either are general and flexible, but provide only limited accuracy mainly due to the simple model used, or improve prediction performance by either leveraging a complex integrative analytical framework, which often is difficult and time consuming to be properly set up, or adopting a more complex model, which in turn significantly slows the prediction process in particular in the usual case of many data to be evaluated.

6   Conclusions

In this paper we propose a method to discover new GO term annotations for genes of different organisms, based on available GO annotations of these genes.

Our idea is to train a model to recognize the presence of novel gene annotations using the obsolete annotation profile of the gene, labeling each term of an outdated annotation profile of a gene with a label taken from an updated version of it. This approach requires two different versions of the annotation matrix to build the training data representation. However, biologists typically have available only the most updated version of the gene annotation matrix. Given this constrain, we have proposed a method to overcome this lack; creating a different annotation matrix, representing an older version of the input one, by perturbing the known annotation matrix in order to randomly remove some of its annotations. This allows the use of supervised algorithms even in datasets without labels and the comparison with results obtained by unsupervised methods on the same originally unlabeled datasets.

Obtained results are very encouraging, since they show a great improvement compared with unsupervised techniques. Furthermore, these results could be even better with an appropriate tuning of the parameters of the supervised algorithms used; our purpose is to thoroughly investigate this aspect in the future. The extension, using weighted real values to represent gene-term associations, did not yield better results with respect to the binary value representation. Thus, we found that the computationally simpler scheme, namely the binary scheme, achieves results generally better than other more complex schemes.

From the obtained results we can see that, by increasing the number of perturbed (removed) annotations, the results improve, reaching a peak when the number of artificially missing annotations in the training set is comparable to the number of those in the validation set, i.e. when the variety of missing annotations has been fully mapped in the training set. Furthermore, it is noteworthy also the case where we do not perturb the training matrix, avoiding the tuning of the parameter p, which gets anyway good results.

We plan to further verify the effectiveness of the proposed approach, also considering the possibility to train the prediction model using genes and annotations relating to a different organism with respect the target one.

References

1. GO Consortium, et al.: Creating the gene ontology resource: design and implementation. Genome Res. **11**, 1425–1433 (2001)
2. Pandey, G., Kumar, V., Steinbach, M.: Computational approaches for protein function prediction: A survey. Technical report, Minneapolis, MN, USA (2006)
3. Domeniconi, G., Masseroli, M., Moro, G., Pinoli, P.: Discovering new gene functionalities from random perturbations of known gene ontological annotations. In: International Conference on Knowledge Discovery and Information Retrieval (KDIR 2014) (2014)
4. Canakoglu, A., Ghisalberti, G., Masseroli, M.: Integration of biomolecular interaction data in a genomic and proteomic data warehouse to support biomedical knowledge discovery. In: Biganzoli, E., Vellido, A., Ambrogi, F., Tagliaferri, R. (eds.) CIBB 2011. LNCS, vol. 7548, pp. 112–126. Springer, Heidelberg (2012)
5. Tanoue, J., Yoshikawa, M., Uemura, S.: The genearound go viewer. Bioinformatics **18**, 1705–1706 (2002)
6. Manning, C.D., Raghavan, P., Schütze, H.: Introduction to Information Retrieval. Cambridge University Press, Cambridge (2008)

7. Domeniconi, G., Moro, G., Pasolini, R., Sartori, C.: Cross-domain text classification through iterative refining of target categories representations. In: Proceedings of the 6th International Conference on Knowledge Discovery and Information Retrieval (2014)

8. Pinoli, P., Chicco, D., Masseroli, M.: Weighting Scheme Methods for Enhanced Genomic Annotation Prediction. In: Formenti, E., Tagliaferri, R., Wit, E. (eds.) Computational Intelligence Methods for Bioinformatics and Biostatistics. LNCS (LNBI), vol. 8452, pp. 76–89. Springer, Heidelberg (2014)

9. Sparck Jones, K.: Document Retrieval Systems, pp. 132–142. Taylor Graham Publishing, London (1988)

10. Domeniconi, G., Moro, G., Pasolini, R., Sartori, C.: Iterative refining of category profiles for nearest centroid cross-domain text classification. In: Fred, A., et al. (eds.) IC3K 2014. CCIS, vol. 553, pp. 50–67. Springer, Heidelberg (2015)

11. Done, B., Khatri, P., Done, A., Draghici, S.: Semantic analysis of genome annotations using weighting schemes. In: IEEE Symposium on Computational Intelligence and Bioinformatics and Computational Biology, CIBCB 2007, pp. 212–218. IET (2007)

12. Debole, F., Sebastiani, F.: Supervised term weighting for automated text categorization. In: Proceedings of SAC-03, 18th ACM Symposium on Applied Computing, pp. 784–788. ACM Press (2003)

13. Lan, M., Tan, C.L., Su, J., Lu, Y.: Supervised and traditional term weighting methods for automatic text categorization. IEEE Trans. Pattern Anal. Mach. Intell. **31**, 721–735 (2009)

14. King, O.D., Foulger, R.E., Dwight, S.S., White, J.V., Roth, F.P.: Predicting gene function from patterns of annotation. Genome Res. **13**, 896–904 (2003)

15. Tao, Y., Sam, L., Li, J., Friedman, C., Lussier, Y.A.: Information theory applied to the sparse gene ontology annotation network to predict novel gene function. Bioinformatics **23**, i529–i538 (2007)

16. Barutcuoglu, Z., Schapire, R.E., Troyanskaya, O.G.: Hierarchical multi-label prediction of gene function. Bioinformatics **22**, 830–836 (2006)

17. Raychaudhuri, S., Chang, J.T., Sutphin, P.D., Altman, R.B.: Associating genes with gene ontology codes using a maximum entropy analysis of biomedical literature. Genome Res. **12**, 203–214 (2002)

18. Pérez, A.J., Perez-Iratxeta, C., Bork, P., Thode, G., Andrade, M.A.: Gene annotation from scientific literature using mappings between keyword systems. Bioinformatics **20**, 2084–2091 (2004)

19. Khatri, P., Done, B., Rao, A., Done, A., Draghici, S.: A semantic analysis of the annotations of the human genome. Bioinformatics **21**, 3416–3421 (2005)

20. Done, B., Khatri, P., Done, A., Draghici, S.: Predicting novel human gene ontology annotations using semantic analysis. IEEE/ACM Trans. Comput. Biol. Bioinf. (TCBB) **7**, 91–99 (2010)

21. Chicco, D., Tagliasacchi, M., Masseroli, M.: Genomic annotation prediction based on integrated information. In: Biganzoli, E., Vellido, A., Ambrogi, F., Tagliaferri, R. (eds.) CIBB 2011. LNCS, vol. 7548, pp. 238–252. Springer, Heidelberg (2012)

22. Chicco, D., Masseroli, M.: A discrete optimization approach for SVD best truncation choice based on ROC curves. In: 2013 IEEE 13th International Conference on Bioinformatics and Bioengineering (BIBE), pp. 1–4. IEEE (2013)

23. Dumais, S.T., Furnas, G.W., Landauer, T.K., Deerwester, S., Harshman, R.: Using latent semantic analysis to improve access to textual information. In: Proceedings of the SIGCHI Conference on Human factors in Computing Systems, pp. 281–285. ACM (1988)

24. Hofmann, T.: Probabilistic latent semantic indexing. In: Proceedings of the 22nd Annual International ACM SIGIR Conference on Research and Development in Information Retrieval, pp. 50–57. ACM (1999)

25. Masseroli, M., Chicco, D., Pinoli, P.: Probabilistic latent semantic analysis for prediction of gene ontology annotations. In: The 2012 International Joint Conference on Neural Networks (IJCNN), pp. 1–8. IEEE (2012)

26. Pinoli, P., Chicco, D., Masseroli, M.: Enhanced probabilistic latent semantic analysis with weighting schemes to predict genomic annotations. In: 2013 IEEE 13th International Conference on Bioinformatics and Bioengineering (BIBE), pp. 1–4. IEEE (2013)

27. Blei, D.M., Ng, A.Y., Jordan, M.I.: Latent dirichlet allocation. J. Mach. Learn. Res. **3**, 993–1022 (2003)

28. Bicego, M., Lovato, P., Oliboni, B., Perina, A.: Expression microarray classification using topic models. In: Proceedings of the 2010 ACM Symposium on Applied Computing, pp. 1516–1520. ACM (2010)

29. Perina, A., Lovato, P., Murino, V., Bicego, M.: Biologically-aware latent Dirichlet allocation (BaLDA) for the classification of expression microarray. In: Dijkstra, T.M.H., Tsivtsivadze, E., Marchiori, E., Heskes, T. (eds.) PRIB 2010. LNCS, vol. 6282, pp. 230–241. Springer, Heidelberg (2010)

30. Pinoli, P., Chicco, D., Masseroli, M.: Latent Dirichlet allocation based on gibbs sampling for gene function prediction. In: Proceedings of the International Conference on Computational Intelligence in Bioinformatics and Computational Biology, pp. 1–7. IEEE Computer Society (2014)

31. Griffiths, T.: Gibbs Sampling in the Generative Model of Latent Dirichlet Allocation, Technical report, Stanford University (2002)

32. Casella, G., George, E.I.: Explaining the gibbs sampler. Am. Stat. **46**, 167–174 (1992)

33. Porteous, I., Newman, D., Ihler, A., Asuncion, A., Smyth, P., Welling, M.: Fast collapsed gibbs sampling for latent Dirichlet allocation. In: Proceedings of the 14th ACM SIGKDD International Conference on Knowledge Discovery and Data Mining, pp. 569–577. ACM (2008)

Building MD Analytical Stars Using a Visual Linked Data Query Builder

Victoria Nebot[✉] and Rafael Berlanga

Computer Languages and Systems, Universitat Jaume I,
Campus Riu Sec s/n, 12071 Castellón, Spain
{romerom,berlanga}@uji.es

Abstract. In this paper we present the foundations and a prototype tool for Linked Data (LD) analysis and visualization. The goal is to provide users with a tool enabling them to easily analyze Linked Data without knowledge of the SPARQL query language. The success of the multidimensional (MD) model for data analysis has been mainly due to its simplicity. The MD model views data in terms of dimensions and measures. Therefore, we propose the notion of MD analytical stars as the foundation to query LD. These stars are MD conceptual patterns that summarize LD (i.e., dimensions and measures). We have developed a tool that enables the user to graphically build MD analytical stars as queries by suggesting possible dimensions and measures from a LD set. The query is automatically translated to SPARQL and executed. Visualization of results as charts, tables, etc. is performed using web-based technologies. We show the prototype tool and a running example.

Keywords: Linked Data · RDF · Multidimensional models · Linked data querying

1 Introduction

During the last years, communities from different areas have published data in the cloud of LD following the publication guidelines, providing the basis for creating and populating the Web of Data. Currently, there are approximately 74 billion triples over 1000 datasets[1].

The increasing availability of these semi-structured and semantically enriched datasets has prompted the need for new tools able to explore, query, analyze and visualize these semi-structured data [5]. While several different tools such as graph-based query builders, semantic browsers and exploration tools [2–4,8] have emerged to aid the user in querying, browsing and exploring LD, these approaches have a limited ability to summarize, aggregate and display data in the form that a scientific or business user expects, such as tables and graphs. Moreover, they fall short when it comes to provide the user an overview of the data that may be of interest from an analytical viewpoint.

---

[1] http://stats.lod2.eu.

© Springer International Publishing Switzerland 2015
A. Fred et al. (Eds.): IC3K 2014, CCIS 553, pp. 198–211, 2015.
DOI: 10.1007/978-3-319-25840-9_13

LD constitutes a valuable source of knowledge worth exploiting using analytical tools. Business Intelligence (BI) uses the MD model to view and analyze data in terms of dimensions and measures, which seems the most natural way to arrange data. BI has traditionally been applied to internal, corporate and structured data, which is extracted, transformed and loaded (ETL) into a pre-defined and static MD model. The relational implementation of the MD data model is typically a star schema. The dynamic and semi-structured nature of LD poses several challenges to both potential analysts and current BI tools. On one hand, exploring the datasets using the available browsers and tools to find MD patterns is cumbersome due to the semi-structured nature of the data and the lack of support for obtaining summaries of the data. Moreover, as the datasets are dynamic their structure may change or evolve, making the one-time MD design approach unfeasible.

In this paper, we propose MD analytical stars as the foundation to query LD. A MD analytical star is a MD star-shaped pattern at the concept level that encapsulates an interesting MD analysis [14]. The star is focused on a subject of analysis and is composed by one or several measures (i.e., measurable attributes on which calculations can be made) and dimensions (i.e., the different analytical perspectives). These stars reflect relevant patterns in the dataset, as the measures and dimensions that compose them are calculated following a statistical approach [15]. To ease the composition of MD analytical stars we have developed a web-based prototype tool that allows the user to easily compose them by selecting suggested dimensions and measures for a given subject of analysis. With this user-friendly query builder we are freeing the user from the cumbersome task of browsing, exploring and building queries in specific languages to find interesting analytical patterns in large LD sets. Moreover, the tool is able to translate the user graphical query to SPARQL, execute it and display the results using different charts and diagrams.

We summarize our contribution as follows:

- We define the concept of MD analytical star as a mapping of the MD model to LD. That is, we identify the subject of analysis, dimensions and measures that compose a MD analytical star in LD.
- We introduce the notion of aggregation power for the dimensions and measures and make an estimation to filter dimensions and measures according to this score.
- We have developed a user-friendly web-based tool that allows users to query and visualize results from LD sets by dynamically building MD analytical stars.

The structure of the paper is as follows. In Sect. 2 we review the literature related to the problem of analyzing LD. Section 3 presents the main foundations that underlie our approach. In Sect. 4 we present a model for MD analytical stars over LD sources. Section 5 summarizes how dimensions and measures are calculated from LD. Section 6 explains the prototype implemented and shows a running example. Finally, Sect. 7 gives some conclusions and future work.

2   Related Work

We have performed a thorough review on the related literature to find out that the majority of approaches use querying, exploration and only light-weight analytics over LD.

For querying LD, SPARQL has become the de-facto standard. However, directly querying a dataset using SPARQL interface cannot be considered an end-user task as it requires familiarity with its syntax and the structure of the underlying data. Graph-based query builders such as [3] can help users build triple patterns by using auto-completion to express queries. However, users do not always have explicit queries upfront, but need to explore the available data first in order to find out what information might be interesting to them. Sgvizler[2] allows to render results of SPARQL queries as charts, maps, etc. However, it requires SPARQL knowledge and focuses only on the visualization part.

The review in [5] about visualization and exploration of LD concludes that most of the tools are designed for technical users and do not provide an overview or summary of the data.

LD browsers such as [2,4,19] are designed to display one entity at a time and do not support the user in aggregation tasks. Most of them use faceted filtering to better guide the user in exploration tasks. However, the user gets overview of only a small part of the dataset. On the other hand, browsers such as [17,18] provide a more powerful browsing environment, but are tailored to a specific application.

Graph-based tools such as RDF-Gravity[3], IsaViz[4] or Relfinder [8] provide node-link visualizations of the datasets and the relationships between them. Although this approach can obtain a better understanding of the data structure, graph visualization does not scale well to large datasets.

The CODE Query Wizard and Vis Wizard developed under the CODE project[5] are a web-based visual analytics platform that enables non-expert users to easily perform exploration and lightweight analytic tasks on LD. Still, the user has to browse the data to find interesting analytical queries. Payola [12] is a framework that allows any expert user to access a SPARQL endpoint, perform analysis using SPARQL queries and visualize the results using a library of visualizers.

We claim that existing tools for exploration and analysis of LD provide little or no support for summaries so that the user can have an idea of the structure of the dataset and the parts that seem more interesting for analysis. In that line, we have also looked into approaches that provide graph summaries over LD using different techniques such as bisimulation and clustering [1,10]. However, these graph summaries are produced without an analytical focus, therefore, the resulting summaries may not be useful for analysis purposes.

---

[2] http://dev.data2000.no/sgvizler/.
[3] http://semweb.salzburgresearch.at/apps/rdf-gravity/.
[4] http://www.w3.org/2001/11/IsaViz/.
[5] http://code-research.eu/.

Recently, there have been some attempts to analyze LD that go beyond querying and browsing. [14] proposes MD analysis over LD under the OWL formalism. However, most LD sets lack this semantic layer. Other approaches [6,9] have proposed MD analysis over LD relying on the previous manual annotation of the MD elements (dimensions and measures) and using previously defined MD vocabularies.

3   Background

In this section, we review background concepts on which our approach is based on.

3.1   Linked Data

LD is a set of common practices and general rules to contribute to the Web of Data [7]. The basic principles are that each entity should be assigned a unique URL identifier, the identifiers should be dereferenceable by HTTP and the entity representations should be interlinked together to form a global LD cloud.

The most adopted standard to implement the Web of Data is RDF [13], which allows us to make statements about entities. It assumes data modeled as triples with three components: subject, predicate and object. We consider only valid RDF triples using URIs (U), blank nodes (B) and literals (L). These triples can also be viewed as graphs, where vertices correspond to subjects and objects, while labeled edges represent the triples themselves. SPARQL [16] has become the standard for querying RDF data and it is based on the specification of triple patterns.

In RDF there is no technical distinction between the schema and the instance data, even though it provides terminology to express class membership or categorization (`rdf:type`). The RDFS extension allows to create taxonomies of classes and properties. It also extends definitions for some of the elements of RDF, for example it sets the domain and range of properties and relates the RDF classes and properties into taxonomies using the RDFS vocabulary. OWL extends RDFS and allows for expressing further schema definitions in RDF. The formal semantics of RDFS and OWL enrich RDF with implicit information that can be reasoned over. Throughout the paper, we refer both to the explicit and implicit triples, which have been derived using some reasoning mechanism. We use the naming convention of OWL referring to classes, properties and individuals to homogenize terminology.

3.2   Multidimensional Models

The MD model is the conceptual abstraction mostly used in BI. The observations or facts are analyzed in terms of dimensions and measures [11]. They focus on a subject of analysis (e.g., sales) and define a series of dimensions or different analysis perspectives (e.g., location, time, product), which provide contextual information. Facts are aggregated in terms of a series of measures (e.g., average sales). As a result, analysts are able to explore and query the resulting data cube applying OLAP operations. A typical query would be to display the evolution of the sales during the current year of personal care products by city.

BI has traditionally been applied to internal, corporate and structured data, which is extracted, transformed and loaded into a pre-defined and static MD model. The relational implementation of the MD data model is typically a star schema, where the fact table containing the summarized data is in the center and is connected to the different dimension tables by means of functional relations.

4   MD Analytical Stars

In this section we explain how we model MD analytical stars from LD.

We formalize the representation of an RDF graph using graph notation.

Definition 1. *(RDF graph) An RDF graph G is a labeled directed graph $G = \langle V, E, \lambda \rangle$ where:*

- *V is the set of nodes, let V^0 denote the nodes in V having no outgoing edge, and let $V^{>0} = V \backslash V^0$;*
- *$E \subseteq V \times V$ is the set of directed edges;*
- *$\lambda : V \cup E \to U \cup B \cup L$ is a labeling function such that $\lambda_{|V}$ is injective, with $\lambda_{|V^0} : V^0 \to U \cup B \cup L$ and $\lambda_{|V^{>0}} : V^{>0} \to U \cup B$, and $\lambda_{|E} : E \to U$.*

Typical analysis usually involves investigating a set of particular facts according to relevant criteria (dimensions) and measurable attributes (measures). Here, we use the notion of basic graph pattern (BGP) queries, which is a well-known a subset of SPARQL. A BGP is a set of triple patterns, where each triple has a subject, predicate and object, some of which can be variables. We are specially interested in rooted BGP queries, as they resemble the star-shaped pattern typical of MD analysis.

Definition 2. *(Rooted query) Let q be a BGP query, $G = \langle V, E, \lambda \rangle$ its graph and $v \in V$ a node that is a variable in q. The query q is rooted in v iff G is a connected graph and any other node $v' \in V$ is reachable from v following the directed edges in E.*

Example 1. (Rooted query) The query q is a rooted BGP query, with x_1 as root node.

$$q(x_1, x_2, x_3, x_5) :- x_1 \text{ Annual\_Carbonemissions\_kg } x_3,$$
$$x_1 \text{ Country } x_2,$$
$$x_1 \text{ Fuel\_type } x_4$$

The query's graph representation below shows that every node is reachable from the root x_1.

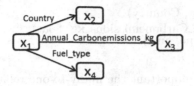

Even though rooted queries express data patterns by means of the predicate chains, these are still vague as the variable nodes can match any element in $U \cup B \cup L$. To narrow down the scope of the patterns we define the notion of typified rooted queries as follows:

Definition 3. *(Typified rooted query) A typified rooted query q' is a rooted query with graph $G = \langle V, E, \lambda \rangle$ where each variable node $v_x \in V$ has an associated class or datatype. That is, each variable v_x has an outgoing edge (v_x, v_y) such that $\lambda((v_x, v_y)) = rdf\colon type$ and $\lambda(v_y) \in U$ and v_y has an outgoing edge (v_y, v_z) such that $\lambda((v_y, v_z)) = rdf\colon type$ and $\lambda(v_z) \in \{rdfs\colon Class, rdfs\colon Datatype\}$.*

Example 2. (Typified rooted query) The previous query q can be typified as follows:

$q(x_1, x_2, x_3, x_5)$:- x_1 rdf:type Powerplant, Powerplant rdf:type rdfs:Class,

$\quad x_1$ Country x_2, x_2 rdf:type Country, Country rdf:type rdfs:Class,

$\quad x_1$ Annual_Carbonemissions_kg x_3,

$\quad x_3$ rdf:type xsd:float, xsd:float rdf:type rdfs:Datatype,

$\quad x_1$ Fuel_type x_4, x_4 rdf:type Fuel, Fuel rdf:type rdfs:Class

From now on, we omit the type edges and represent typified rooted queries with the (data)type's name in the variable node.

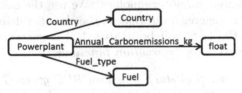

It is immediate to see that a typified rooted query is composed by a set of typified paths that go from the root to a sink node (node with no outgoing edges). The root node represents a class and the sink node represents either a class or a datatype. We formalize this notion next:

Definition 4. *(Typified path) Given a typified rooted query q with x_1 as root node and graph $G = \langle V, E, \lambda \rangle$, a typified path is a sequence $p = c_1 - r_1 - c_2 - r_2 - \ldots - r_{n-1} - c_f$ where $\lambda(x_1) = c_1$ is the root class, c_f is a sink class or datatype, every r_i is a property and every c_i has an associated class.*

Example 3. (Typified path) In the previous query q we can identify the following typified paths:

(Powerplant, Country, Country)
(Powerplant, Annual_Carbonemissions_kg, float)
(Powerplant, Fuel_type, Fuel)

In MD modeling it is important the many-to-one relation between facts and dimensions to ensure aggregation power. That is, one fact must be associated with one dimension value, whereas a dimension value can and should be associated to multiple facts. In our LD scenario, we define the aggregation power of a typified path as follows:

Definition 5. *(Aggregation power) Given a typified path $p = c_1 - r_1 - c_2 - r_2 - \ldots - r_{n-1} - c_f$, the aggregation power is calculated as the ratio between the number of different individuals (or literals) of the sink class (or datatype) c_f that satisfy the path and the number of individuals of the root class c_1 that satisfy the path.*

Example 4. (Aggregation power) Given the following paths, we calculate the aggregation power as follows:

$$(\text{Powerplant}_{(74561)}, \text{Country}, \text{Country}_{(200)}) \rightarrow \frac{200}{74561} = 0.0027$$
$$(\text{Powerplant}_{(19796)}, \text{Fuel\_type}, \text{Fuel}_{(30)}) \rightarrow \frac{30}{19796} = 0.0015$$

Notice that an aggregation power closer to 0 exhibits a high aggregation capacity, meaning that the sink individuals act as categories for the root individuals of the path. Aggregation power closer to 1 means low aggregation power of the path.

We are now ready to introduce MD analytical stars. For this, we make use of traditional data warehousing terminology. We use the notion of *classifier* to denote the level of data aggregation, that is, the classifier defines the dimensions according to which the facts will be analyzed. The *measure* allows obtaining values to be aggregated using *aggregation functions*.

Definition 6. *(MD analytical star) Given an RDF graph $G = \langle V, E, \lambda \rangle$, a MD analytical star rooted in the node $x \in V$ is a triple: $S = \langle c(x, d_1, ..., d_n), m(x, v), \bigoplus \rangle$ where:*

- *$c(x, d_1, ..., d_n)$ is a typified query rooted in the node r_c of its graph G_c, with $\lambda(r_c) = x$ and each path $x - ... - d_i$ is a typified path. This is the classifier of x w.r.t. the n dimensions $d_1, ..., d_n$. The node x is the subject of analysis.*
- *$m(x, v)$ is a typified query rooted in the node r_m of its graph G_m, with $\lambda(r_m) = x$. This query is only composed by a typified path $x - ... - v$[6]. This is called the measure of x.*
- *$\bigoplus$ is an aggregation function over a set of values, that is, the aggregator for the measure of x w.r.t. its classifier.*

---

[6] For the sake of simplicity, we assume that an MD analytical star has only one measure.

– *Each of the typified paths of the classifier has an aggregation power below a threshold δ.*

Notice that typified rooted queries (and therefore, typified paths) are the building block to suggest MD analytical stars.

Example 5. (MD analytical star) The MD analytical star below asks for the average of annual carbon emission of powerplants, classified by country and fuel type.

$$\langle c(x, x_1, x_3), m(x, x_4), average \rangle$$

where the classifier and measure queries are:

$$c(x, x_1, x_3) \; : - \; x \text{ rdf:type Powerplant,}$$
$$x \text{ Country } x_1, \; x_1 \text{ rdf:type Country,}$$
$$x \text{ Fuel\_type } x_2, \; x_2 \text{ rdf:type Fuel}$$
$$m(x, x_4) \; : - \; x \text{ rdf:type Powerplant,}$$
$$x \text{ Annual\_Carbonemissions\_kg } x_4, \; x_4 \text{ rdf:type xsd:float}$$

The answer to an MD analytical star is a set of tuples of dimension values found in the answer of the classifier query, together with the aggregated result of the measure query. Therefore, it can be represented as a cube of n dimensions, where each cell contains the aggregated measure.

5   Calculating Dimensions and Measures

MD analytical stars are composed by the classifier and measure typified queries rooted in a potential class acting as subject of analysis. These typified queries are composed by typified paths with a certain aggregation power. For space restrictions, we omit the process of calculating the set of typified paths from a LD. The method is based on probabilistic graphical models and we make use of the statistics about instance data to generate the paths. We refer the reader to Sect. 5 of [15] for details and to Sect. 6 for checking the experimental evaluation. As a result, a set of typified paths with aggregation power below a threshold δ are obtained.

Paths are classified into dimensions and measures based on the aggregation power of the path and the type of the sink node. Paths ending in numeric datatypes (i.e., xsd:integer, xsd:float, xsd:double, etc.) and with low aggregation power are considered measures, whereas the rest of paths (i.e., paths ending in classes or datatypes with high aggregation power) are considered dimensions.

The set of calculated dimensions and measures from a LD dataset is used by the prototype tool to help the user build MD analytical stars. In the next section, we show the implemented prototype along with a running example.

6   Prototype Tool

The current prototype for building and executing MD analytical stars over LD sets has been implemented as a client/server web application. The architecture is shown in Fig. 1. We distinguish the client, the server, an external LD endpoint, and the catalogue of calculated dimensions and measures. For the client we use AJAX and HTML5/CSS3. The server is implemented as a RESTFUL API in PHP. It has three main tasks: (1) handle requests from the user about the dimensions and measures of the catalogue, (2) act as a proxy to send queries to the LD endpoint and (3) process the results of the endpoint. The external LD endpoint is a SPARQL endpoint that we access to execute the MD queries. The catalogue is the application data, that is, the dimensions and measures calculated for the LD set. We have implemented the catalogue as an independent RESTful web service to make it portable.

This prototype helps the user build MD analytical stars by suggesting possible dimensions and measures for a specific subject of analysis (steps 1–4). Moreover, it is completely functional, as the queries graphically built by the user are automatically translated to SPARQL queries over the dataset endpoint, which returns the results that can be either displayed or exported in different formats (steps 5–7).

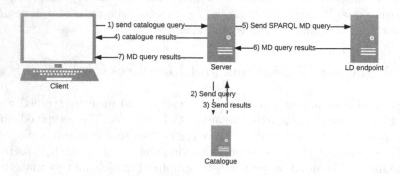

Fig. 1. Prototype architecture.

For demonstration purposes, we have selected the Enipedia[7] LD dataset, which is an initiative aimed at providing a collaborative environment through the use of wikis and the Semantic Web for energy industry issues. The dataset provides energy-related data from different open data sources structured and linked in RDF. The dataset contains around 5M triples.

Next, we show the different functionality of the prototype. The home page is shown in Fig. 2. The left navigation menu shows the different steps to perform a MD query, which are: (1) show most interesting subjects of analysis, (2) select

[7] http://enipedia.tudelft.nl/wiki/Main_Page.

one, (3) select measures and dimensions and (4) show results. We will go through all of them by means of a running example.

In the home page the user can ask to show the n most interesting subjects of analysis if (s)he has no knowledge of the dataset, or directly type the name of a class, which comes with the autocompletion feature. Interestingness is measured by the number of individuals of the class. In Fig. 2 we select *powerplant* as subject of analysis as we are interested in analyzing the annual carbon emission rates of powerplants from different perspectives.

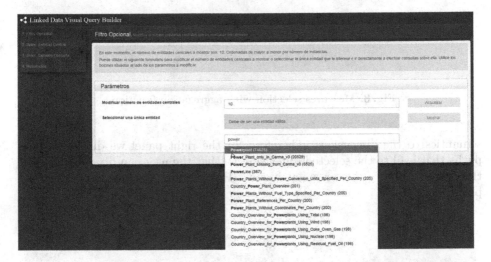

Fig. 2. Home page. The user selects to display the n most interesting subjects of analysis or types one.

The next step consists in selecting measures and dimensions for the subject of analysis. These are suggested by the prototype and the user is in charge only of selecting the ones required. Figure 3 shows the panel where all available measures for the subject *powerplant* are displayed and the user can select one or more by ticking the box and selecting the appropriate aggregation function. In this case we select *Annual_Carbonemissions2000_kg* and *average* as the aggregation function.

The process of selecting the dimensions is shown in Figs. 4 and 5. The left panel of Fig. 4 shows the possible dimensions. In this case, the user selects two dimensions, *Country* and *Fuel*, meaning that the analysis (s)he has in mind consists in displaying the average carbon emission levels of the powerplants by country and the type of fuel of each powerplant. The right panel serves the purpose of disambiguating a selected dimension. Notice that a dimension is defined by a typified path from the subject of analysis to a sink class. By only selecting the sink class as dimension there can be ambiguities as a sink class can be reached by different paths and the properties of the paths may give different

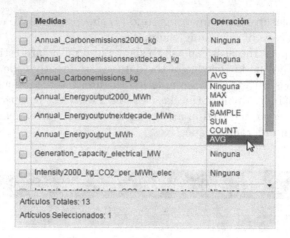

Fig. 3. Measures selection with aggregation function.

semantics to the dimensions. Therefore, in the right panel we display all the paths that lead to the selected dimension so that the user can disambiguate. In the case of *Country* there is no ambiguity. However, for *Fuel* the user select the path *Fuel_type*.

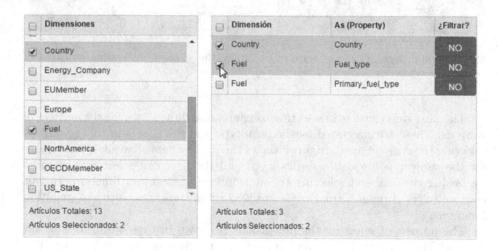

Fig. 4. Dimensions selection and optional filtering.

The prototype also offers the feature of dimension filtering. This is shown in Fig. 5 and can be enabled with the button next to each dimension. In case we activate this option, a new panel will appear (i.e., right panel) displaying the individuals that belong to the selected dimension. In this case, the user wants

Fig. 5. Filter dimensions by value.

to filter the values of the dimension *Country* and selects only specific countries for the analysis.

	France	Germany	Italy	Portugal	Spain
Biogas	336565.5	341477	1600275		10503150
Biomass	16545372	87120684	0	0	0
Coal	1433121363	2265279333	2654453333	7659395000	3973609333
Coke_Oven_Gas	92480333	493275000	3539700000		
Diesel_Oil	10596218		3352990000	74923900	
Fuel_Oil	503953345	115512289	2219679533		891870666
Gasoil	38607100			98718000	114193000
Geothermal	0	0	0	0	
Hard_Coal	1711578454	3031621259	2904350500		2780718333
Heavy_Fuel_Oil	194461001	1089806250	1409179777	473732700	616783740
Hydro	0	6712	7209062	0	0
Landfill_Gas	642243	587027	1142146	0	1997395
Municipal_Solid_Waste	2307644	302552634	955507	20651200	21148824
Natural_Gas	72921075	478840983	644805216	616774500	253712046
Nuclear	0	0	0	0	0
Residual_Fuel_Oil	335359466	1089806250	1285028214		616783740
Solar_Radiation	0	0	0	0	0
Tidal	0				
Wave	0			0	
Wind	0	0	0	0	325994

Fig. 6. MD query results.

Finally, Fig. 6 shows the results of the MD query. These can be displayed in different formats and using different visualizations or exported to a file.

7   Conclusions and Future Work

In this paper we have presented the foundations and a prototype tool for LD analysis and visualization following a MD approach. We have proposed MD analytical stars as the foundation to enable the user to easily query LD sets. The MD patterns (i.e., dimensions and measures) suggested by the web-based developed tool to create the stars are based on the semantics of the data (i.e., they provide a conceptual summary of the data), follow the MD model (i.e., information is modeled in terms of analysis dimensions and measures) and are extracted following a statistical approach. The tool has been implemented using web-based technologies.

As future work, it would be interesting to group semantically similar dimensions into categories to offer a cleaner view to the user. Another interesting functionality is to provide dimension hierarchies to be able to perform the classical roll-up/drill-down operations of OLAP over the results. In a near future we will also address the current issue of having a batch process for calculating the dimensions and measures off-line. We aim to apply query sampling methods directly over the LD endpoint to be able to dinamically calculate dimensions and measures. This process will be smoothly integrated in the web-based tool.

Acknowledgements. The work has been partially funded by the CICYT project TIN2011-24147 from the Spanish Ministry of Economy and Competitiveness (MINECO). We would like to thank Iván Posilio for his truly involvement in the development of the prototype.

References

1. Alzogbi, A., Lausen, G.: Similar structures inside RDF-graphs. In: LDOW. CEUR Workshop Proceedings, vol. 996 (2013)
2. Araújo, S., Schwabe, D.: Explorator: a tool for exploring RDF data through direct manipulation. In: LDOW. CEUR Workshop Proceedings, vol. 538 (2009)
3. Auer, S., Lehmann, J.: What have Innsbruck and Leipzig in common? Extracting semantics from Wiki content. In: Franconi, E., Kifer, M., May, W. (eds.) ESWC 2007. LNCS, vol. 4519, pp. 503–517. Springer, Heidelberg (2007)
4. Berners-Lee, T., Chen, Y., Chilton, L., Connolly, D., Dhanaraj, R., Hollenbach, J., Lerer, A., Sheets, D.: Tabulator: exploring and analyzing linked data on the semantic Web. In: Proceedings of the 3rd International Semantic Web User Interaction (2006)
5. Dadzie, A., Rowe, M.: Approaches to visualising linked data: a survey. Semant. Web **2**(2), 89–124 (2011)
6. Etcheverry, L., Vaisman, A.A.: Enhancing OLAP analysis with Web cubes. In: Simperl, E., Cimiano, P., Polleres, A., Corcho, O., Presutti, V. (eds.) ESWC 2012. LNCS, vol. 7295, pp. 469–483. Springer, Heidelberg (2012)
7. Heath, T., Bizer, C.: Linked Data: Evolving the Web into a Global Data Space. Synthesis Lectures on the Semantic Web. Morgan & Claypool Publishers, San Rafael (2011)
8. Heim, P., Lohmann, S., Stegemann, T.: Interactive relationship discovery via the semantic Web. In: Aroyo, L., Antoniou, G., Hyvönen, E., ten Teije, A., Stuckenschmidt, H., Cabral, L., Tudorache, T. (eds.) ESWC 2010, Part I. LNCS, vol. 6088, pp. 303–317. Springer, Heidelberg (2010)

9. Kämpgen, B., Harth, A.: Transforming statistical linked data for use in OLAP systems. In: I-SEMANTICS 2011, Graz, Austria, 7–9 September 2011. ACM International Conference on Proceedings Series, pp. 33–40 (2011)
10. Khatchadourian, S., Consens, M.P.: ExpLOD: summary-based exploration of interlinking and RDF usage in the linked open data cloud. In: Aroyo, L., Antoniou, G., Hyvönen, E., ten Teije, A., Stuckenschmidt, H., Cabral, L., Tudorache, T. (eds.) ESWC 2010, Part II. LNCS, vol. 6089, pp. 272–287. Springer, Heidelberg (2010)
11. Kimball, R., Ross, M.: The Data Warehouse Toolkit: The Complete Guide to Dimensional Modeling. Wiley, New York (2011)
12. Klímek, J., Helmich, J., Nečaský, M.: *Payola*: collaborative linked data analysis and visualization framework. In: Cimiano, P., Fernández, M., Lopez, V., Schlobach, S., Völker, J. (eds.) ESWC 2013. LNCS, vol. 7955, pp. 147–151. Springer, Heidelberg (2013)
13. Klyne, G., Carroll, J.J.: Resource description framework (RDF): concepts and abstract syntax (2004). http://www.w3.org/TR/rdf-concepts
14. Nebot, V., Berlanga, R.: Building data warehouses with semantic Web data. Decis. Support Syst. **52**(4), 853–868 (2012)
15. Nebot, V., Berlanga, R.: Towards analytical MD stars from linked data. In: Proceedings of the International Conference on Knowledge Discovery and Information Retrieval (KDIR 2014), KDIR 2014, pp. 117–125 (2014)
16. Prudhommeaux, E., Seaborne, A.: SPARQL query language for RDF (2008). http://www.w3.org/TR/rdf-sparql-query
17. Schraefel, M.C., Shadbolt, N.R., Gibbins, N., Harris, S., Glaser, H.: CS AKTive space: representing computer science in the semantic Web. In: WWW, pp. 384–392. ACM, New York (2004)
18. Stadler, C., Lehmann, J., Höffner, K., Auer, S.: LinkedGeoData: a core for a Web of spatial open data. Semant. Web J. **3**(4), 333–354 (2012)
19. Zviedris, M., Barzdins, G.: ViziQuer: a tool to explore and query SPARQL endpoints. In: Antoniou, G., Grobelnik, M., Simperl, E., Parsia, B., Plexousakis, D., De Leenheer, P., Pan, J. (eds.) ESWC 2011, Part II. LNCS, vol. 6644, pp. 441–445. Springer, Heidelberg (2011)

Modeling Sentiment Polarity
with Meta-features to Achieve
Domain-Independence

Octavian Lucian Hasna$^{(\boxtimes)}$, Florin Cristian Măcicăşan,
Mihaela Dînşoreanu, and Rodica Potolea

Computer Science Department, Technical University of Cluj-Napoca,
26-28 Gh. Bariţiu St., Cluj-Napoca, Romania
{Octavian.Hasna, Florin.Macicasan, Mihaela.Dinsoreanu,
Rodica.Potolea}@cs.utcluj.ro

Abstract. Opinion mining has become an important field of text mining with high impact in numerous real-world problems. The limitations most solutions have in case of supervised learning refer to domain dependence: a solution is specifically designed for a particular domain. Our method overcomes such limitations by considering the generic characteristics hidden in textual information. We identify the sentiment polarity of documents that are part of different domains by using a uniform, cross-domain representation. It relies on three classes of original meta-features that characterize the datasets belonging to various domains. We evaluate our approach using datasets extensively referred in the literature. The results for in-domain and cross-domain verification show that the proposed approach handles novel domains increasingly better as its training corpus grows, thus inducing domain-independence.

Keywords: Sentiment detection · Meta-features · Classification · Text mining · Evaluation · Supervised learning · Domain independence

1 Problem Statement

In recent years, the increased accessibility and freshness of electronic word-of-mouth made us consider it even more. We can easily access the opinions and impressions of others whenever we plan to make a customer decision. As any other experience, opinions are filled with subjectivity. They can reflect a positive experience, underline dissatisfaction or attempt to maintain a balance between the two. This relation is especially of interest for the field of opinion mining.

A similar evolutionary process happens at the other end of the product experience chain. Companies who refuse to listen to what customers have to say tend to lose in the long run. On the other hand, companies who see customer opinions as a source of insight and early feedback [1] build a stronger bond.

Both perspectives leverage electronic word-of-mouth as a decision support system. Whether it's what to buy next or what to sell in a month, customer experiences distilled in opinions are an important tool in the world we live in. In this context, an automated solution for tagging opinions with their sentiment orientation is beneficial.

© Springer International Publishing Switzerland 2015
A. Fred et al. (Eds.): IC3K 2014, CCIS 553, pp. 212–227, 2015.
DOI: 10.1007/978-3-319-25840-9_14

We propose an approach to domain independent document-level sentiment polarity identification that leverages a combination of three meta-feature classes. The novelty of the approach relies on the feature-vector characteristics: as the classification instances are characterized via meta-features, the model gains in generality being domain independent.

We utilize the following three classes of meta-features:

- *Part-of-speech* patterns represent syntactic constructs with increased sentiment promise;
- *Polarity histograms* group the words of a document in buckets based on their sentiment polarity;
- *Sentiment lexicons* represent a proven collection of words annotated with polarity information.

2 Research Goals

We aim to analyze and identify the sentiment orientation of documents that contain information specific to different domains. Movies, books, DVDs, kitchen appliances or electronics are just subsets of the entire collection of domains to be considered.

The cross-domain diversity of sentiment expressiveness is a problem that continuously peaks the interest of researchers in this field. Among existing approaches, a limitation induced by this diversity is domain dependence. A solution is specifically designed or at least specifically tuned for a given domain. Such solutions consider as distinctive features the intimate structural details of a specific domain.

We strive for an approach that manages to overcome such limitations by relying on generic characteristics hidden in textual information, with the goal of domain independence. Figure 1 describes this process as a domain-level sentiment orientation funnel. Unstructured documents from heterogeneous domains are to be reduced to a common representation thus inducing structure. Furthermore, the funneling process reveals document-level sentiment orientation by further analyzing this common representation. An interesting event is the apparition of a document from a new domain. In this case, we must be able to funnel its documents and properly identify their polarity.

Our objectives can be summarized in terms of providing answers for the next three questions:

- How to uniformly structure documents from different domains in a classification-friendly format?
- How to identify the sentiment orientation of classification-friendly representations?
- How to handle novel domains with an existing model?

In order to achieve a common domain-independent representation some intrinsic aspects of the sentiment orientation problem such as objectivity and negation are to be considered. Viewed as a binary problem, sentiment polarity identification assumes subjectivity in the input documents. The problem of negation detection represents an important sub-goal. Negation has the effect of inversing the polarity of constructs it determines. This generates local polarity shifts that bubble up at document level.

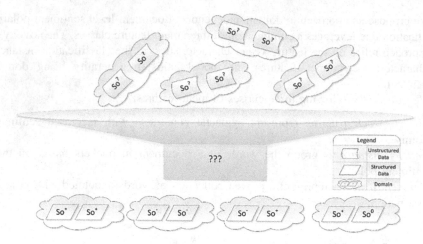

Fig. 1. The Domain-level sentiment orientation funnel.

The goal of a uniform representation throughout domains may be achieved by avoiding to incorporate features with increased domain-specific expressiveness. The aim is a domain independent representation that pads the road towards a domain independent classifier. This journey continues towards consistently handling new domains with increasingly low cross-domain verification penalties.

This work does not research sentence or aspect-level sentiment orientation. Moreover, it does not consider the identity of the opinion holder or any aspects related to one's opinion (subjectivity) patterns. We leave these perspectives on polarity for future research. We focus our approach on the opinionated documents themselves in order to be able to generalize across domains.

3   Related Work

In [2] three dimensions of sentiment analysis are underlined: document, sentence and aspect. A document-level analysis is interested in the whole expressed opinion. The implicit assumption is that the entire document expresses an opinion on a single entity. Sentence-level analysis is closely related to subjectivity classification. It requires the identification of subjective views and opinions. At aspect-level, sentiment is expressed with respect to various components of an entity. Entity features are selected from frequent nouns or noun phrases and their sentiment orientation is measured using lexicon-based approaches. The same conceptual feature can be represented with different textual representations. This is why synonyms become an important tool for aspect-based sentiment analysis. At the core of each approach is identifying the sentiment orientation of individual words.

In [3], the analysis of the sentiment orientation of short phrases, like Twitter comments is performed. The goal is to analyze the compositional effect of sentiment in a given short phrase. To this purpose the authors propose the Sentiment Treebank, a corpus of labelled parse trees. It leverages the Recursive Neural Tensor Network model that represents a phrase through a word vector and a parse tree. A tensor-based compositional function is used to associate sentiment polarity to individual tree nodes. The consolidated polarity of the root gives the orientation of the phrase. They leverage seven sentiment orientations (three degrees of negative, neutral and three degrees for positive).

In [4], a set of heuristics for extracting expressions with increased sentiment affinity based on dependency relations is proposed, with focus on range, trend and negation indicators. The range indicators are viewed as the members of the WordNet synset of "above" while trend indicators are modelled around "increase". Furthermore, they detect negation and cluster part-of-speech and grammatical relations to increase generality.

Ensemble techniques for sentiment classification are explored in [5]. Feature sets and classification algorithms are integrated with the aim of improving accuracy by defining three POS-based feature classes: adjectives (JJx), nouns (NNx) and verbs (VBx). Furthermore, the authors utilize word dependency parsing together with unigrams and bigrams as WR-based feature sets.

Domain independent sentiment lexicons are used in the work of [6] to tune a classifier on different domains. They propose both fix scoring schemas and sum-based predictors that boost results on the analyzed domains. Class-imbalanced recall is viewed as an issue for scenarios when misclassification costs vary with class. They propose a term frequency-based score adjustment metric as a possible solution.

The problem of domain adaptation is investigated in [7] where a correspondence technique for learning structural similarities between lexicons specific to different domains is proposed. In [8] the problem of domain adaptation is approached by using an annotated subset of the target domain to tune a single-domain classifiers.

4   Meta-Features for Sentiment Classification

Our approach strives for domain-independence by inferring sentiment polarity from high-level meta-features not directly tied to the contents of the analyzed document. The goal is to generalize the components of the feature vector so as to achieve domain independence. We propose three original classes of meta-features: *sentiment lexicons*, *part-of-speech patterns* and *polarity histograms*. In the following, we detail the theoretical foundation behind them.

To clarify our terminology, we start with an initial set of notions. The *sentiment polarity* of an entity (sp_e) describes an ordered distribution over the following sentiment orientations: *Positive* (ρ), *Negative* (η) and *Objective* (o) such that $sp_e = \langle \rho, \eta, o \rangle$ and $\rho + \eta + o = 1$. It measures the degree to which a given entity (a word or a document) exhibits one of the possible sentiment orientations. The *sentiment orientation* of an entity so_e is described as $so_e = \text{argmax}_{x \in sp_e}(x)$ where $so_e \in \{0, +, -\}$. A *word* is an element of a vocabulary V. A *word* can have a *sentiment polarity* sp_w. A *document* (d) is an entity described by a sequence of N words ($W_d = \{w_1, w_2, \ldots, w_N\} = W_{1:N}$).

4.1 Sentiment Lexicons

A sentiment lexicon (SL) is represented by a collection of words annotated with their sentiment polarity. It is essentially a 2-tuple that associates a positive or negative weight to each word from the lexicon's vocabulary, depending on the word's sentiment orientation. The weights are limited to the unit interval and respect the sign of the orientation. Formally it can be described as follows:

$$SL_V = \{\langle w, sp_w \rangle | w \in V \land sp_w \in [-1, 1]\} \tag{1}$$

The power of sentiment lexicons is within the knowledge associated to their vocabulary. Individual words reflect their associated polarity whenever they appear in documents. Combining sentiment lexicons is an important technique for building upon their individual power and expertise. Therefore we also define the basic set operations (union, intersection and difference) on sentiment lexicons. An operation applied on a SL is equivalent with applying that operation on their associated vocabulary. In case of vocabulary overlaps, collisions are resolved by selecting the sentiment polarity of the SL with the highest priority (an input). Thus, we define the following:

$$SL_{V_i} \cup SL_{V_j} = \left\{ \langle w, sp_{w,x} \rangle \middle| w \in V_i \cup V_j \land x = argmax\left(P_{SL_{V_i}}, P_{SL_{V_j}}\right) \right\} \tag{2}$$

$$SL_{V_i} \cap SL_{V_j} = \left\{ \langle w, sp_{w,x} \rangle \middle| w \in V_i \cap V_j \land x = argmax\left(P_{SL_{V_i}}, P_{SL_{V_j}}\right) \right\} \tag{3}$$

$$SL_{V_i} \backslash SL_{V_j} = \left\{ \left\langle w, sp_{w,SL_{V_i}} \right\rangle \middle| w \in V_i \backslash V_j \right\} \tag{4}$$

Our approach uses as lexicon a combination between two collections commonly used in literature. The first lexicon is proposed in [9] and represents a list of English words with a well-defined positive or negative orientation. Depending on orientation, we associated to each word in this lexicon one of the following polarities: $sp_w^+ = \langle 0.9, 0.05, 0.05 \rangle$ or $sp_w^- = \langle 0.05, 0.9, 0.05 \rangle$. We denote this sentiment lexicon as SL_{HuLiu}.

The second resource we leverage is SentiWordNet (SWN) [10]. SWN is the result of an automatic annotation of all WordNet synsets (ss). As a result, each synset receives a positive and a negative polarity. SWN uses the WordNet structure which groups similar meanings of different words in a synset. A word can be part of multiple synsets by exhibiting different senses. So a word can have different sentiment polarities based on the sense it plays in the analyzed document. Let $sense_w$ be the set of synsets associated to a word in SWN.

To build a sentiment lexicon associated to SWN (SL_{SWN}) we associate a single polarity to each word. As words might be associated with multiple synsets ($|sense_w| \geq 1$), we define the multi-synset fall-back schema (MSFB) which associates to each word a polarity. Their synset's, in case the word is part of a single synset or otherwise, the polarity of the synset that maximizes the absolute difference between their positive (ρ) and negative (η) polarity. This relation is defined as follows:

$$SL_{SWN} = \{\langle w, MSFB(w)\rangle | w \in V_{SWN}\} \tag{5}$$

$$\text{MSFB}(w) = argmax_{ss_i \in sense_w} \left| \rho_{sp_{ss_i}} - \eta_{sp_{ss_i}} \right| \tag{6}$$

We also explore an alternative to MSFB which relies on the definition of each SWN synset (def_{ss}), a short description of the concept modelled by the synset that provides meaning to the senses of words linked to ss. Thus we define the multi-synset window schema (MSW): it associates to each word of a document (W_d) the polarity of the synset that maximizes the overlap between the definition of the synset and a window of a given size (Δ_w) in the vicinity of that word within the document. We view def_{ss} as a mini-document so it has a sequence of words ($W_{def_{ss}}$) associated to it. For a word $w_i \in W_d$, MSW is the synset that maximizes the size of the intersection between $W_{def_{ss}}$ and the Δ_w-size window around w_i (7).

$$MSW(w_i, W_d) = argmax_{ss_i \in sense_w} \left| W_{def_{ss_i}} \cap W_{i-\lfloor \frac{\Delta_w}{2} \rfloor : i + \lceil \frac{\Delta_w}{2} \rceil} \right| \tag{7}$$

A challenge introduced by MSW is the fact that it is tied to the document currently analyzed which induces a strong dependency between the generative processes that build SL_{SWN} and the domain at hand. Although it might seem as a drawback in our quest for domain independence, MSW better captures the context in which a word is used and selects the polarity of the most appropriate sense for a given word. In this respect, MSFB selects the synset with the strongest absolute difference between the positive and negative scores, a process that completely disregards the context in which a word is used and the sense it exhibits. Both generative processes have inherent drawbacks. We aim to reduce their impact by proposing a way of reducing the number of considered senses for a given word.

From the polarity point of view, a word is distinguished if one of the positive or negative orientations is strictly greater than the other two orientations. Thus the sentiment lexicon defined on a distinguished vocabulary (dV) will only contain words with a well-defined orientation. Formula (8) describes this relation.

$$w \in dV \equiv w \in V \wedge sp_w = \langle \rho, \eta, o \rangle \wedge \rho \neq \eta \wedge (\rho > o \vee \eta > o) \tag{8}$$

Since many of the synsets in SWN are objective, we choose to define dSWN as the subset of SWN with synsets that respect the distinguishability property. This reduces the number of synsets and helps us underline the sentiment baring words. We call such sentiment lexicons strongly distinguishable. A sentiment lexicon SL_V is strongly distinguishable if, for any $w \in V$ the condition in relation (8) holds true. A lexicon that exhibits this property is SL_{HuLiu}. Applying one of the multi-synset schemas, we build SL_{dSWN} as the sentiment lexicon associated to $dSWN$. An interesting consequence is that the percentage of words with a single synset grows, thus drastically reducing the impact of the drawbacks associated with the two multi-synset schemas.

Furthermore, we are interested in analyzing the vocabulary overlap between SL_{dSWN} and SL_{HuLiu}. With the help of relations (2)–(4) we can further refine lexicon combinations and analyses their impact on the overall classification process. In the rest of the paper, we will refer to combinations between sentiment lexicons based on the applied set of operations (e.g. the union between SL_{HuLiu} and SL_{dSWN} becomes $SL_{HuLiu\_\cup\_dSWN}$).

We leverage sentiment lexicons as domain-independent meta-features. They represent a fixed set of words that are to be searched in the document instances that are part of each of the analyzed domains. We measure a Boolean meta-feature (i.e. whether or not an element of the lexicon appears in the document instance). The feature vector of a document associated to a sentiment lexicon ($fSL_V(d)$) is the tuple of Boolean features that mark the existence of a word from V within the analyzed document instance. It is described as follows:

$$fSL_V(d) = \left\langle i = \begin{cases} 1, w \in W_d \\ 0, w \notin W_d \end{cases} \middle| w \in V \right\rangle \tag{9}$$

Apart from being used as meta-features, sentiment lexicons are word-level polarity providers for the other two classes of meta-features. An interesting aspect to consider in this situation is negation. Any word might appear in a negated context which inverses its sentiment polarity:

$$inverse(sp_w) = \{\langle \rho', \eta', o' \rangle | \rho' = \eta \wedge \eta' = \rho \wedge o' = o\} \tag{10}$$

4.2   Part-of-Speech Patterns

Part-of-speech patterns ($POSp(w_1, w_2)$) represent a specialized combination of words tagged with POS information. In [11] five such patterns are used to extract specific bigrams with increased sentimental promise. Turney analyses trigrams that respect the part-of-speech patterns represented in Table 1 and selects bigrams based on the priority induced by the third word (the first POS pattern has a higher priority than the next three). The selected bigram instances represent the feature vector.

Table 1. POS patterns proposed by Turney.

	First word POS (w_1)	Second word POS (w_2)	Third word (not extracted)
1	Adjective	Noun	Anything
2	Adverb	Adjective	Not noun
3	Adjective	Adjective	Not noun
4	Noun	Adjective	Not noun
5	Adverb	Verb	Anything

Starting from the POS bigram instances concept, we generalize and propose as features the templates of the polarized five POS bigrams. We say that a bigram ($\langle w_j, w_{j+1} \rangle$) will match the ith part-of-speech pattern ($POSp_i$) if the following relation holds true:

$$POSp_i(w_j, w_{j+1}) = \langle pos_{w_j} pos_{w_{j+1}} \rangle \equiv POSp_i \qquad (11)$$

The polarity of a POS pattern is computed based on a linear combination between the sentiment polarities of the two words that are part of the pattern. Instances with a distinguished positive polarity count for $POSp_+$, while those with negative polarity count for $POSp_-$. The polarity of a word is given by the underlying sentiment lexicon. We describe the linear combination relation as follows:

$$sp_{w_j, w_{j+1}} = \omega * sp_{w_j} + (1 - \omega) * sp_{w_{j+1}} \qquad (12)$$

where ω is an experimentally computed coefficient and sp_{w_j} and $sp_{w_{j+1}}$ are retrieved from the underlying lexicon. We've determined that a good ω would be 0.5 for $POSp_2$ and $POSp_3$. For the other three, we associate 0.8 for the adjective or adverb. We treat negation for $POSp$ by considering $inverse(sp_{w_j, w_{j+1}})$ as the pattern's polarity.

We propose the usage of these patterns as meta-features in two instantiations: one for the positive orientation ($POSp_+$) and one for the negative ($POSp_-$) thus generating 10 new meta-features. For each of them, the actual value of the feature in the feature vector is the associate number of patterns of that type that appear in the document.

The value of an individual $POSp_i$ instance of a given orientation (o) in a document (d) is described by the cardinal of the set of instantiations of that pattern in the sequence of words describing d, as follows:

$$cnt_{POSp}(d, i, o) = |\{j|w_j \in W_d \wedge POSp_i(w_j, w_{j+1}) \wedge so_{w_j, w_{j+1}} = o\}| \qquad (13)$$

where w_j is a word from d and the ith pattern is instantiated with the proper sentiment orientation (o).

The part-of-speech patterns represent a 10-tuple with the combination of pattern instance count for all five patterns and each of the two orientations. This relation is described by the following relation:

$$fPOSp(d) = \langle cnt_{POSp}(d, i, o) \mid i = \overline{1,5}, o \in \{+, -\} \rangle \qquad (14)$$

4.3 Polarity Histograms

Polarity Histograms are a measure of the degree to which a document contains words of different sentiment polarities. We analyze words from the document that exhibit sentimental promise based on the underlying polarity lexicon. We group them in buckets of size Δ on a two-dimensional lattice. The actual bucket size depends on the polarity values reported by the sentiment lexicon.

The diameter of a disc in Figs. 2 and 3 represent the number of words that have a positive sentiment polarity within $[x, x + \Delta_x)$ and a negative polarity within $[y, y + \Delta_y)$. Figure 2 depicts the polarity histogram of a positive document. It uses $SL_{HuLiu\_\cup\_dSWN}$ as sentiment lexicon. As the figure shows, there are no words in buckets below 0.5

because the lexicon is strongly distinguishable. In Fig. 3 we represent the polarity histogram resulted from processing a negative document using the $SL_{HuLiu\_\cup\_SWN}$ lexicon (a lexicon that lacks the distinguishability property).

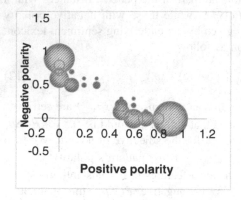

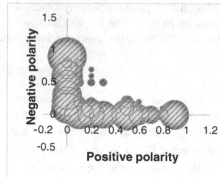

Fig. 2. Polarity histogram for a document with positive sentiment orientation using $SL_{HuLiu\_\cup\_dSWN}$.

Fig. 3. Polarity histogram for a document with negative sentiment orientation using $SL_{HuLiu\_\cup\_SWN}$.

We adopt polarity histograms as the third set of meta-features as they capture the overall polarity information of a document, normalized to a given lexicon. In the polarity context, negations have the same impact as for POS patterns. Thus, we inverse the sentiment polarity for a word if it occurs negated in the document. In our experiments, the total number of buckets is 66 (as describe in relation (15)).

$$66 = \left| \left\{ \langle [i, i+0.1), [j, j+0.1) \rangle \; \middle| \; \begin{matrix} i+j \leq 1 \wedge i = x * 0.1 \wedge j = y * 0.1 \wedge \\ x, y = \overline{0, 10} \end{matrix} \right\} \right| \quad (15)$$

The number of words that fall in a given bucket is computed following the formula in (16). It counts the number of words (w_j) from a given document (d) that have their polarity point within the area of *bucket*.

$$cnt_{PH}(bucket, d) = \left| \{ j | w_j \in W_d \wedge sp_{w_j} \in bucket \} \right| \quad (16)$$

The polarity histogram feature vector that corresponds to a document d is described in relation (17). It defines a tuple of dimensionality $|Buckets|$.

$$fPH(d) = \langle cnt_{PH}(b, d) \mid b \in Buckets \rangle \quad (17)$$

4.4   The Sentiment Polarity Identifier

The **Sentiment Polarity Identifier** (SPI) sketched in Fig. 4 is responsible for associating a sentiment polarity (sp_d) to a document (d).

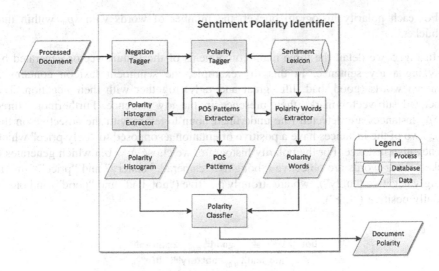

Fig. 4. Sentiment Polarity Identifier.

Documents are analyzed with respect to the three classes of meta-features. To better measure the sentiment orientation of a word, we start with *negation detection* for which a naïve approach is applied: it searches for words that are part of a *negation lexicon*. So far only explicit negations are considered (words negated by *not*, *don't* and similar). Each time a negation is detected its determined word is marked as *negated* and its polarity value reversed.

Next, the *polarity tagging* step attaches to each processed word its associated sentiment polarity using the configured sentiment lexicon. The polarity of the document is aggregated from the document's words which belong to the lexicon as well. The sentiment polarity of a word is inversed if it is preceded by a negation. Polarity enriched words are the input for all three meta-feature extractors.

At this point we start collecting instances of our three meta-feature classes. The POS patterns extractor collects instances of the 10 POS patterns. Then we apply the polarity histogram extractor which starts filling in the defined buckets based on the individual polarity of words in the analyzed document.

Finally, we apply the polarity words extractor that selects the words that are part of the sentiment lexicon. At this level, we treat negation by doubling the size of the lexicon's vocabulary (each word gets negated). The feature vector of a document (fv(d)) used by the polarity classifier is formally described as

$$\mathrm{fv}(d) = fPOSp(d) \cup fPH(d) \cup fSL(d) \tag{18}$$

It contains the following meta-features:

- For each word in the polarity lexicon, a Boolean marker describing its membership in W_d;
- 10 meta-features whose values represent the number of part-of-speech pattern instances of each type found in the document;

- For each polarity histogram bucket, the number of words with sp_w within that bucket.

In Fig. 5 we detail the three main components of the feature vector generated by analyzing a toy sentence. In this toy example, the sentiment lexicon contains 5 polarized words (good, bad, life, great and ugly) together with their negation. The associated sub-vector marks their presence in the toy sentence. Furthermore, three $POSp_1$ instances are detected (the underlined noun together with the adjective on the left). Two of the instances have a positive orientation as opposed to "ugly price" which is a negative instance. For the polarity histogram, we chose $\Delta = 0.5$ which generates 6 buckets. Four words are objective ("bought", "camera", "battery" and "price"), one is strongly negative ("ugly"), two are strongly positive ("not_bad" and "good") and one is partially positive ("life").

I^{PRP} $bought^{VBD}$ a^{DT} $good^{JJ}_{(1,0,0)}$ $camera^{NN}$

$with^{IN}$ a^{DT} $not\_bad^{JJ}_{(1,0,0)}$ $battery^{NN}$ $life^{NN}_{(.5,0,.5)}$

at^{IN} an^{DT} $ugly^{JJ}_{(0,1,0)}$ $price^{NN}$

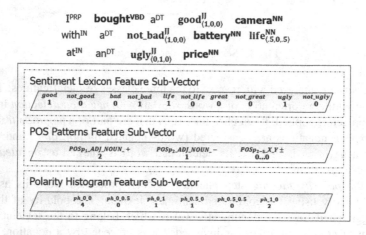

Fig. 5. Instances of meta-features in a toy sentence.

5   Results and Evaluation

We are first interested in the effect of applying the distinguishability property on SL_{SWN}. We analyse the dimensionality reduction induced by this property with a focus on the proportion of words that end up being associated with a single synset. In Table 2 we show the number of synsets (#syn) and words (#w) in both sentiment lexicons. SL_{dSWN} has 94.1 % less words and 95.1 % less synsets then SL_{SWN}. Furthermore, we measure the distribution of words (SynPerWord) that are associated to a single synset, 2 or 3 synsets or more. Table 2 shows that for 85.4 % of the words from SL_{dSWN} a unique sentiment polarity can be associated from their corresponding synset. This means that we rely on the multi-synset fallback schema 4 times less than for SL_{SWN}.

The second aspect that we analyze is the best configuration for the feature vector. Based on preliminary results for initial evaluations on different classifiers (NB, SVM and C4.5), we restricted the evaluations to the Naïve Bayes (implementation available in Weka) configured to use a kernel separator [12]. While evaluating the configuration

Table 2. Comparison between SWN and dSWN.

SL_V	#syn	#w	SynPerWord		
			1	2 or 3	More
SWN	117659	147306	.401	.124	.475
dSWN	5736	8548	.854	.134	.012

of the classifier we use the version 2.0 of the Movie Review Dataset (MR) first introduced by [13]. It consists of 1000 negative and 1000 positive movie reviews crawled from the IMDB movie archive. The average document length is 30 sentences.

For validation, we randomized the dataset, hide 10 % for evaluation and split the remaining 90 % into 10 folds and tested in a 10 fold cross-validation technique. We repeat this process with multiple random seeds.

The evaluations were performed on the data set with different features combinations. The first set of features is the elements of a sentiment lexicon. The candidate lexicons are SL_{dSWN}, SL_{HuLiu} and the lexicons obtained by applying the basic set operations on the two (denoted correspondingly in Table 3). By structurally analyzing the lexicons, we measured the number of words (#w) in each and the distribution of positive (#pw) and negative (#nw) word sentiment polarities. It's interesting to note for all lexicons the distribution of negative words is greater than the distribution of words with a positive sentiment orientation.

Table 3. Sentiment lexicons comparison.

SL_V	Structure			Evaluation			
	#w	#pw	#nw	p	σ_p	r	σ_r
HuLiu	6786	.295	.705	75.5	2.6	74.6	2.6
dSWN	8548	.359	.641	72.7	1.6	72.4	1.6
HuLiu_ ∪ _dSWN	13080	.336	.664	76.7	2.9	75.8	2.9
HuLiu_ ∩ _dSWN	2254	.302	.698	71.1	1.3	70.8	1.4
HuLiu_ \ _dSWN	4532	.291	.708	66.8	2.5	66.4	2.6
dSWN_ \ _HuLiu	6294	.380	.620	62.9	2.1	62.7	1.8

Using only sentiment lexicons as features while classifying instances from MR we measured the average weighted precision (p) and recall (r) for the positive and negative classes together with their standard deviation (σ_p and σ_r). The evaluation results from Table 3 suggest that the best sentiment lexicon choice is the union between the two lexicons. Furthermore, the vocabulary intersection SL_V is more valuable than any of the two difference SL_V.

We further aim to find the best feature set, using as candidates our three meta-feature categories: sentiment lexicon (SL), part-of-speech patterns (POSP) and polarity histograms (PH). Seven experiments were performed each considering a different combination of the three categories of feature vector candidates. For polarity histograms we set the bucket sizes Δ_x and Δ_y to 0.1. This allows for a lower granularity in the analysis. There are only 10 degrees for the positive orientation and 10 degrees for the negative orientation.

The results in Table 4 show that each of the three meta-feature classes brings an incremental improvement in precision and recall. The biggest impact is brought by the sentiment lexicon. The best feature vector contains the combination between part-of-speech patterns, polarity histograms and the sentiment lexicon.

Table 4. Feature vector composition from meta-features.

Configuration	p	σ_p	r	σ_r
POSP	64.2	2.7	63.7	3.1
PH	64	2.9	62.4	2.8
SL	76.7	2.9	75.8	2.9
SL + POSP	81.6	1.4	81.4	1.4
SL + PH	82.5	1.6	82	1.6
PH + POSP	67.1	2.7	6.4	3.3
SL + POSP + PH	84.1	1.5	82.9	1.6

To asses domain independence we have tested the feature vector configuration on other domains. Proposed by [7], the Multi-Domain Sentiment (MDS) dataset is a collection of Amazon reviews from multiple domains. It consists of 26 domains with labelled positive and negative reviews. We've considered in our experiment 14 domains that have more than 800 positive and 800 negative labelled reviews. In literature, the initial 4 domains are extensively used for evaluation. They cover the Book (B), DVD (D), Kitchen (K) and Electronics (E) and have 1000 positive and 1000 negative reviews. For this experiment we also measure classification accuracy (acc) because this is the metric used for comparison in other studies using the MDS-4. Table 5 reports the results for 10 random seeds with two outliers excluded (min & max precision) on both MR and the 15 domains of MDS. The relative balance of the measured precision and recall (0.8 % difference) suggest that our approach does not affect sensitivity and is able to consistently identify polarity in different domains.

Table 5. In-domain verification using multiple domains.

Dataset	MR	Book (b)	DVD (d)	Electro (e)	Kitchen (k)	Apparel (a)	Baby (ba)	Camera (c)	Health (h)	Magazine (m)	Music (mu)	Software (s)	Sports (sp)	Toys (t)	Video (v)	AVERAGE
p	84.1	74.0	80.7	80.1	84.9	85.3	83.7	85.5	81.0	85.7	79.2	82.5	81.9	82.9	76.2	81.8
σ_p	1.5	2.5	2.3	2.5	1.9	1.9	2.2	2.2	2.5	1.8	2.3	2.9	2.6	2.1	3.8	4.0
r	82.9	71.2	80	79.6	84.7	85.1	83.6	85.1	80.8	85.2	78.9	81.8	81.6	82.5	74.1	81.1
σ_r	1.6	2.9	2.6	2.4	1.8	1.9	2.1	2.3	2.4	1.9	2.4	3.2	2.7	2.3	4.7	4.6
acc	82.8	71.8	80	79.5	84.6	85.1	83.6	85.1	80.8	85.2	78.9	81.8	81.6	82.5	74.1	81.2

We compare our approach with other studies that leveraged the same datasets for in-domain classification (training and validation on the same domain). In Table 6 we compare against the results of in-domain testing of [5, 14].

Table 6. Comparative analysis of accuracy.

	MR	B	D	E	K
Lin2012	76.6	70.8	72.5	75	72.1
Xia2011PoS	82.7	76.7	78.85	81.75	82.4
Xia2011WR	85.80	81.2	81.7	84.15	87.5
Our solution	82.87	71.18	80.03	79.59	84.69

As we were interested in experimenting with cross-domain verification, we split all the domains (n) into 10 % for validation and 90 % for training using different random seeds. This generated an in-domain test (used as "golden standard") and $n - 1$ cross-domain tests (test on other domains). Table 7 covers our results. The datasets are referred to by their initials from Table 5. We measure the relative loss (ζ) in classification accuracy due to cross-domain verification. Let a_{train}^{test} be the accuracy of training on the domain train and testing on the domain test. Thus the relative loss is $\zeta_a^b = a_a^b - a_a^a$, the difference between the cross-domain accuracy for training on a and testing on b and the "golden standard" on a. A line in Table 7 contains the values for in domain testing (a_a^a – on the diagonal) and the relative loss for testing in other domains (ζ_a^b) where $a \neq b$. We also consolidate the average relative loss (Δ_ζ) for training on a given domain and its standard deviation (σ_ζ). We look for domains that minimize the average relative loss (Δ_ζ). Excluding the MR outlier dataset, the average loss throughout domains from the in-domain average of 81.2 % is −7.66 % with a standard deviation of 2.51 %.

Furthermore, we attempt a hold-one-out cross validation process where we view an individual domain as an instance to "hold out". It maximizes the training data available for a classifier and provides a metric that does not vary with the randomness of a dataset split. In Table 8 we measure the classification accuracy for both the in-domain and the hold-one-out experiments. We also measure the average accuracy throughout datasets (Δ_a) and their standard deviation (σ_a). An accuracy value on column i corresponds to the case when the classifier was trained on all domains except i and validated as on i. We are also interested in the relative loss (ζ) of cross-domain testing, its average (Δ_ζ) and standard deviation (σ_ζ). We use the same notations for the domains as Table 7. The last two columns contain the average and standard deviation for accuracy (the first two rows) and relative loss of hold-one-out compared with in-domain (the last row).

The results in Table 8 reveals the encouraging result that the hold-one-out approach manages, on average, to outperform in-domain testing (82 % vs 81.2 % accuracy). This is due to important increases in accuracy for domains like books (+9 %) or video (+8 %). This means that, a model built on the consolidated corpus of k−1 domains performs almost the same as in-domain model. Given a model built on the maximum

Table 7. Cross-domain verification.

Test Train	a	ba	b	c	d	e	h	k	m	MR	mu	s	sp	t	v	Δζ	σζ
a	86	-7	-18	-6	-16	-8	-8	-5	-10	-32	-14	-8	-8	-6	-13	-11.35	7.15
ba	0	84	-14	-2	-11	-4	-7	-2	-5	-23	-10	-3	-5	-1	-10	-6.92	6.24
b	-13	-9	70	-9	-1	-10	-10	-10	-7	-12	-4	-4	-8	-6	-7	-7.85	3.30
c	-4	-5	-14	86	-11	-5	-8	-3	-9	-21	-12	-3	-8	-6	-11	-8.57	4.98
d	-5	-7	-4	-3	80	-6	-7	-6	-4	-1	-2	-3	-7	-3	-3	-4.35	1.98
e	-2	-2	-12	1	-6	80	-5	-1	-5	-22	-7	3	-3	-1	-8	-5	6.23
h	+1	-2	-10	-1	-9	-3	83	-1	-6	-21	-10	-2	-4	-1	-11	-5.71	5.94
k	-1	-4	-16	-2	-10	-5	-5	85	-7	-19	-11	-3	-3	-2	-10	-7	5.50
m	-7	-12	-13	-7	-10	-9	-10	-10	85	-24	-11	-3	-10	-8	-12	-10.42	4.66
MR	-32	-33	-29	-30	-27	-32	-32	-32	-30	82	-28	-31	-32	-32	-26	-30.42	2.17
mu	-3	-7	-5	-4	-4	-9	-9	-4	-2	-6	79	-7	-7	-4	-2	-5.21	2.32
s	-17	-11	-14	-8	-11	-10	-14	-13	-12	-16	-14	81	-11	-10	-15	-12.57	2.56
sp	+2	0	-9	-1	-8	-3	-2	0	-3	-18	-7	-1	81	-1	-7	-4.14	5.23
t	-4	-3	-12	-2	-9	-6	-6	-5	-9	-22	-9	-1	-8	83	-12	-7.71	5.35
v	-7	-6	-6	-5	-4	-9	-11	-8	-6	-6	-3	-8	-9	-8	73	-6.85	2.14

Table 8. Cross-domain verification for hold-one-out training.

	a	ba	b	c	d	e	h	k	m	MR	mu	s	sp	t	v	Δ	σ
In-domain	86	84	70	86	80	80	83	85	85	82	79	81	81	83	73	81.2	4.5
Hold-one-out	84	82	79	84	82	80	81	84	84	80	79	84	82	84	81	82	1.9
ζ		-2	-2	+9	-2	+2	0	-2	-1	-1	-2	0	+3	+1	+1	+8	+0.8	3.5

amount of available data (the k−1 domains) and a completely new domain to be processed the performance is to be consistent with in-domain verification.

Compared with the results reported in Table 7, the proposed meta-feature representation exhibits a reduction in the relative loss due to cross-domain validation as the amount of data available for training increases. The average relative loss improved from −7.66 % for training on a single domain to +0.8 % for training on k-1 domains. The reduction in relative loss suggests that the proposed model handles new domains increasingly better as its training corpus grows, which is the goal of our domain independent sentiment polarity identification approach.

6 Conclusions

This paper proposes a document sentiment polarity identification approach based on an ensemble of meta-features. We propose the use of three meta-feature classes that boost domain-independence increasing the degree of generality. Sentiment lexicons provide a basis for the analysis. Part-of-speech patterns reflect syntactic constructs that are a good indicator of polarity. Finally, polarity histograms provide an insight in the distribution of polarized words within the document. All three interact in order to associate sentiment polarity to a document and boost the classification performance while learning from different domains than the target domain.

We are currently integrating a more advanced approach for negation detection leveraging typed dependencies [15]. We also consider exploring objectivity with the help of undistinguishable sentiment lexicons and a third set of part-of-speech patterns. Further efforts will be focused on adapting our meta-feature approach to an optimal dataset size for the problem of cross-domain sentiment identification. We eventually aim to shift towards an unsupervised approach for sentiment detection.

References

1. The Economist: Fair comment (2009). http://www.economist.com/node/13174365. Accessed 22 June 2014
2. Liu, B.: Sentiment Analysis and Opinion Mining. Morgan & Claypool, San Rafael (2012)
3. Socher, R., Perelygin, A., Wu, J.Y., Chuang, J., Manning, C.D., Ng, A.Y., Potts, C.: Recursive deep models for semantic compositionality over a sentiment treebank. In: Proceedings of EMNLP (2013)
4. Liu, S., Agam, G., Grossman, D.A.: Generalized sentiment-bearing expression features for sentiment analysis. In: Proceedings of COLING (Posters) (2012)
5. Xia, R., Zong, C., Li, S.: Ensemble of feature sets and classification algorithms for sentiment classification. Inf. Sci. **181**(6), 1138–1152 (2011)
6. Ohana, B., Tierney, B., Delany, S.-J.: Domain independent sentiment classification with many lexicons. In: Proceedings of the 2011 IEEE Workshops of AINA (2011)
7. Blitzer, J., Dredze, M., Pereira, F.: Biographies, bollywood, boom-boxes and blenders: domain adaptation for sentiment classification. In: Proceedings of the 45th Annual Meeting of ACL (2007)
8. Raaijmakers, S., Kraaij, W.: Classifier calibration for multi-domain sentiment classification. In: Proceedings of the 4th ICWSM (2010)
9. Hu, M., Liu, B.: Mining and summarizing customer reviews. In: Proceedings of the ACM SIGKDD International Conference (2004)
10. Baccianella, S., Esuli, A., Sebastiani, F.: SentiWordNet 3.0: an enhanced lexical resource for sentiment analysis and opinion mining. In: Proceedings of LREC (2010)
11. Turney, P.: Thumbs up or thumbs down? Semantic orientation applied to unsupervised classification of reviews. In: Proceedings of the 40th Annual Meeting of ACL (2002)
12. Witten, I.H., Frank, E., Hall, M.A.: Data Mining: Practical Machine Learning Tools and Techniques, 3rd edn. Morgan Kaufmann Publishers Inc., San Francisco (2011)
13. Pang, B., Lee, L.: A sentimental education: sentiment analysis using subjectivity summarization based on minimum cuts. In: Proceedings of the 42nd Annual Meeting of ACL (2004)
14. Lin, C., He, Y., Everson, R., Rüger, S.: Weakly-supervised joint sentiment topic. IEEE Trans. Knowl. Data Eng. **15**(6), 1134–1145 (2012)
15. Marneffe, M.-C., MacCartney, B., Manning, C.D.: Generating typed dependency parses from phrase structure parses. In: Proceedings of LREC (2006)

An Improved String Similarity Measure Based on Combining Information-Theoretic and Edit Distance Methods

Thi Thuy Anh Nguyen[✉] and Stefan Conrad

Institute of Computer Science, Heinrich-Heine-University Düsseldorf,
Universitätsstr. 1, 40225 Düsseldorf, Germany
{thuyanh,conrad}@cs.uni-duesseldorf.de
http://dbs.cs.uni-duesseldorf.de

Abstract. The lexical similarity measure is used for calculating the similarities between strings. Existing lexical-based methods usually base on either n-grams or Dice's approaches. These measures have a good performance and could be extended by adjusting the parameter. However, they do not return reasonable results in some situations where strings are quite similar or the sets of characters are the same but their positions are different. To deal with this problem, our paper presents an approach to improve a lexical-based measure based on both information-theoretic and edit distance measures. The proposed method is tested on a partial OAEI benchmark 2008. The results show that our approach has some prominent features compared to other lexical-based methods. It is also flexible clearly and convenient in implementation. Moreover, we chose a range of good parameters can be applied in different domains.

Keywords: Information-theoretic model · Feature-based measure · String-based measure · Similarity

1 Introduction

Measurement of similarity plays an important role and is applied in many well-known areas, such as data mining and information retrieval. Several techniques for calculating the similarities between objects have been proposed so far, for example, lexical-based, structure-based and instance-based measures. Among these, lexical similarity metrics find correspondences between given strings. These measures are usually applied in ontology matching systems, information integration, bioinformatics, plagiarism detection, pattern recognition and spell checkers. The lexical techniques are based on the fact that the more the characters in strings are similar, the more the similarity values increase. Existing lexical similarity measures usually based on either n-grams or Dice's approaches to obtain correspondences between strings. Although these measures are efficient, they are inadequate in situations where strings are quite similar or the sets of characters are the same but their positions are different in strings. In this paper, a lexical similarity approach based on combining features-based and element-based

© Springer International Publishing Switzerland 2015
A. Fred et al. (Eds.): IC3K 2014, CCIS 553, pp. 228–239, 2015.
DOI: 10.1007/978-3-319-25840-9_15

measures to determine correspondences among the concept labels is proposed. In particular, it is combined from information-theoretic model and edit-distance measure. Consequently, common and different properties with respect to characters in strings as well as editing and non-editing operations are considered.

The remainder of this paper is organized as follows. Section 2 overviews the related lexical measures. In Sect. 3, a similarity measure taking into account text strings is proposed. In Sect. 4, we describe our experimental results, give an evaluation as well as a discussion of our measure and compare it with other approaches applying Precision, Recall and F-measure. Finally, conclusions and future work are presented in Sect. 5.

2 Related Work

The lexical similarity measures are usually used to match short strings such as entity names in ontologies, protein sequences and letter strings. In the following subsections, a brief description of these measures is presented.

2.1 Dice Coefficient

Dice coefficient (also called coincidence index) computes the similarity of two species A and B as the ratio of two times the size of the intersection divided by the total number of samples in these sets and is given as [3]:

$$sim(A, B) = \frac{2h}{a + b},\qquad(1)$$

where A and B are distinctive species, h is the number of common samples in A and B, and a, b are the numbers of samples in A and B, respectively. Accordingly, the higher the number of common samples in A and B, the more their similarity increases.

Dice's measure can be described as

$$sim(A, B) = \frac{2|A \cap B|}{|A| + |B|}$$
$$= \frac{2|A \cap B|}{2|A \cap B| + |A \backslash B| + |B \backslash A|}.\qquad(2)$$

2.2 N-grams Approach

N-grams of a sequence are all subsequences with a length equals to n. The items in these subsequences can be characters, tokens in contexts or signals in speech corpus. For example, n-grams of the string *ontology* with $n = 3$ consist of {*ont, nto, tol, olo, log, ogy*}. In case of n-grams of size 1, 2 or 3 they are also known as unigram, bigram or trigram, respectively. Let $|c_1|$, $|c_2|$ be lengths of strings c_1 and c_2, respectively, the similarity between these strings can be presented as [4]:

$$sim(c_1, c_2) = \frac{|ngram(c_1) \cap ngram(c_2)|}{\min(|c_1|, |c_2|) - n + 1}.\qquad(3)$$

The Eq. (3) can be reformulated as follows:

$$sim(c_1, c_2) = \frac{|ngram(c_1) \cap ngram(c_2)|}{\min(|ngram(c_1)|, |ngram(c_2)|)}. \tag{4}$$

N-grams method is widely used in natural language processing, approximate matching, plagiarism detection, bioinformatics and so on. Some measures applied n-grams approach to calculate the similarity between two objects [1,6,9]. The combination of Dice and n-grams methods in [1,9] to match two given concepts in ontologies is shown below.

2.3   Kondrak's Methods

Kondrak [9] develops and uses a notion of n-grams similarity for calculating the similarities between words. In this method, the similarity can be written as

$$sim(c_1, c_2) = \frac{2|ngram(c_1) \cap ngram(c_2)|}{|ngram(c_1)| + |ngram(c_2)|}. \tag{5}$$

As can be seen in Eqs. (2) and (5), Kondrak's method is a specific case for Dice's metric in which the samples correspond to n-grams.

2.4   Algergawy's Methods

Matching two ontologies is presented by Algergawy et al. [1], in which three similarity methods are combined in a name matcher phase. Furthermore, Dice's expression is implemented to obtain similarities between concepts by using trigrams. Particularly, this measure applies the set of trigrams in compared strings c_1 and c_2 instead of using the number of samples in datasets:

$$sim(c_1, c_2) = \frac{2|tri(c_1) \cap tri(c_2)|}{|tri(c_1)| + |tri(c_2)|}, \tag{6}$$

where $tri(c_1)$ and $tri(c_2)$ are the sets of trigrams in c_1 and c_2, respectively.

2.5   Jaccard Similarity Coefficient

Jaccard measure [7] is developed to find out the distribution of the flora in areas. The similarity related to frequency of occurrence of the flora is the number of species in common to both sets with regard to the total number of species.

Let A and B be arbitrary sets. Jaccard's metric can be normalized and is presented as [7]

$$sim(A, B) = \frac{|A \cap B|}{|A \cup B|}$$

$$= \frac{|A \cap B|}{|A \cap B| + |A \backslash B| + |B \backslash A|}. \tag{7}$$

Applying n-grams approach to Jaccard's measure leads to the following expression:

$$sim(c_1, c_2) = \frac{|ngram(c_1) \cap ngram(c_2)|}{|ngram(c_1) \cup ngram(c_2)|}. \tag{8}$$

As can be seen in Eqs. 5 and 8, Kondrak and Jaccard measures are quite similar.

2.6 Needleman-Wunsch Measure

The Needleman-Wunsch measure [13] is proposed to determine the similarities of the amino acids in two proteins. This measure pays attention to maximum of amino acids of one sequence that can be matched with another. Therefore, it is used to achieve the best alignment. A maximum score matrix $M(i,j)$ is built recursively, such that

$$M(i,j) = \max \begin{cases} M(i-1, j-1) + s(i,j) & \text{Aligned} \\ M(i-1, j) + g & \text{Deletion} \\ M(i, j-1) + g & \text{Insertion} \end{cases}, \tag{9}$$

where $s(i,j)$ is the substitution score for residues i and j, and g is the gap penalty.

2.7 Hamming Distance

Hamming distance [5] only applies to strings of the same sizes. With this measure, the difference between two input strings is the minimum number of substitutions that could have changed one string into the other. In case of different string lengths, Hamming distance $dis(c_1, c_2)$ is modified as [4]:

$$dis(c_1, c_2) = \frac{\left(\sum_{i=1}^{\min(|c_1|, |c_2|)} [c_1[i] \neq c_2[i]]\right) + ||c_1| - |c_2||}{\max(|c_1|, |c_2|)}, \tag{10}$$

where $|c_1|$, $|c_2|$ are string lengths, and $c_1[i]$, $c_2[i]$ are the ith characters in two strings c_1 and c_2, respectively.

Besides using only the operation of substitutions, the Levenshtein distance applying insertions or deletions for comparing strings of different lengths is presented in the succeeding section.

2.8 Levenshtein Distance

The Levenshtein distance (also called Edit distance) [10] is a well-know string metric calculating the amount of differences between two given strings and then returning a value. This value is the total cost of the minimum number of operations needed to transform one string into another. Three types of operations are used including the substitution of a character of the first string by a character of the second string, the deletion or the insertion of a character of one string into other. The total cost of the used operations is equal to the sum of the costs of each of the operations [14].

Let c_1 and c_2 are two arbitrary strings. The similarity measure for two strings $sim(c_1, c_2)$ is described as [12]:

$$sim(c_1, c_2) = \max\left(0, \frac{\min(|c_1|, |c_2|) - ed(c_1, c_2)}{\min(|c_1|, |c_2|)}\right), \tag{11}$$

where $|c_1|$, $|c_2|$ are lengths of strings c_1 and c_2, respectively, and $ed(c_1, c_2)$ is Levenshtein measure. Note that the cost assigned to each operation here equals to 1.

2.9   Jaro-Winkler Measure

The Jaro-Winkler measure [21] is based on the Jaro distance metric [8] to compute the similarity between two strings. The Jaro-Winkler measure $sim(c_1, c_2)$ between c_1 and c_2 strings can be defined as follows:

$$sim(c_1, c_2) = sim_{Jaro}(c_1, c_2) + ip(1 - sim_{Jaro}(c_1, c_2)), \tag{12}$$

where i is the number of the first common characters (also known as the length of the common prefix), p is a constant and is assigned to 0.1 in Winkler's work [21] and $sim_{Jaro}(c_1, c_2)$ is the Jaro metric, defined as

$$sim_{Jaro}(c_1, c_2) = \begin{cases} 0 & \text{if } m = 0 \\ \frac{1}{3}\left(\frac{m}{|c_1|} + \frac{m}{|c_2|} + \frac{m-t}{m}\right) & \text{otherwise} \end{cases}. \tag{13}$$

In Eq. (13), m is the number of matching characters and t is the number of transpositions.

2.10   Tversky's Model

In Tversky's ratio model [19], determination of the similarity among objects is related to features of these objects. In particular, the similarity value of object o_1 to object o_2 depends on their shared and different features, so that

$$sim(o_1, o_2) = \frac{\phi(o_1) \cap \phi(o_2)}{(\phi(o_1) \cap \phi(o_2)) + \beta(\phi(o_1) \setminus \phi(o_2)) + \gamma(\phi(o_2) \setminus \phi(o_1))}, \tag{14}$$

where ϕ represents the set of features, $\phi(o_1) \cap \phi(o_2)$ presents common features of both o_1 and o_2, $\phi(o_i) \setminus \phi(o_j)$ describes features being held by o_i but not in o_j, $(i, j = 1, 2)$. The parameters β and γ are adjusted and depend on which features are taken into account. Therefore, in general this model is asymmetric, it means, $sim(o_1, o_2) \neq sim(o_2, o_1)$. This model is also a general approach applied in many matching functions in the literature as well as domains [16, 18].

3   Combining Information-Theoretic and Edit Distance Measures

3.1   Our Lexical Similarity Measure

In this section, a lexical similarity measure is proposed. Our approach is motivated on Tversky's set-theoretical model [19] and Levenshtein measure [10]. We agree that the similarities among entities depend on their commonalities and differences based on the intuitions in [11]. The well-known metrics applying Tversky's model take into account features of compared objects such as intrinsic information content [16,17], the number of shared superconcepts [2], the number of common attributes, instances and relational classes [20] in ontologies. In contrast with existing approaches, the objective of our metric is to focus on the features in terms of the contents of the characters and their positions in strings. In particularly, our measure is related to editing and non-editing operations.

As mentioned earlier, Tversky's model is a general approach considering the common and different features of objects in which the different features are represented by their proportions through parameters β and γ. In our method, a parameter α is added to the common feature of Eq. (14). Consequently, the similarity is given by

$$sim(c_1, c_2) = \frac{\alpha(\phi(c_1) \cap \phi(c_2))}{\alpha(\phi(c_1) \cap \phi(c_2)) + \beta(\phi(c_1) \backslash \phi(c_2)) + \gamma(\phi(c_2) \backslash \phi(c_1))}, \quad (15)$$

where the parameters α, β and γ are subjected to a constraint: $\alpha + \beta + \gamma = 1$. Additionally, the similarity of two strings should be a symmetric function and the differences between these strings have the same contribution, the parameters β and γ can be considered to be equal. Therefore, our measure $Lex\_sim(c_1, c_2)$ can be rewritten as

$$Lex\_sim(c_1, c_2) = \frac{\alpha(\phi(c_1) \cap \phi(c_2))}{\alpha(\phi(c_1) \cap \phi(c_2)) + \beta((\phi(c_1) \backslash \phi(c_2)) + (\phi(c_2) \backslash \phi(c_1)))}, \quad (16)$$

where $\alpha + 2\beta = 1$ and $\alpha, \beta \neq 0$.

In case $\alpha = \beta = \gamma = \frac{1}{3}$, our measure can be written as

$$\begin{aligned} Lex\_sim(c_1, c_2) &= \frac{\alpha(\phi(c_1) \cap \phi(c_2))}{\alpha((\phi(c_1) \cap \phi(c_2)) + (\phi(c_1) \backslash \phi(c_2)) + (\phi(c_2) \backslash \phi(c_1)))} \\ &= \frac{\phi(c_1) \cap \phi(c_2)}{\phi(c_1) \cup \phi(c_2)}, \end{aligned} \quad (17)$$

which coincides with the Jaccard's measure.

The representation of the Dice's approach can be obtained by setting $\beta = \gamma = \frac{1}{2}\alpha$. Indeed,

$$\begin{aligned} Lex\_sim(c_1, c_2) &= \frac{\alpha(\phi(c_1) \cap \phi(c_2))}{\alpha(\phi(c_1) \cap \phi(c_2)) + \frac{1}{2}\alpha((\phi(c_1) \backslash \phi(c_2)) + (\phi(c_2) \backslash \phi(c_1)))} \\ &= \frac{2(\phi(c_1) \cap \phi(c_2))}{\phi(c_1) + \phi(c_2)}. \end{aligned} \quad (18)$$

In this work, features of strings are chosen as the contents and positions of characters. Moreover, our measure uses deletions, insertions and substitutions mentioned in Levenshtein approach [10] to achieve common and different values between two strings. The editing operations can be regarded as the difference, while non-editing can be reflected on commonalities. These values are then applied to Tversky's model.

Accordingly, common features between two strings are obtained by subtracting the total cost of the operations needed to transform one string into another from the maximum length of these strings and is represented as

$$\phi(c_1) \cap \phi(c_2) = \max(|c_1|, |c_2|) - ed(c_1, c_2). \tag{19}$$

The differences between two strings are:

$$\phi(c_1) \setminus \phi(c_2) = |c_1| - \max(|c_1|, |c_2|) + ed(c_1, c_2), \tag{20}$$

and

$$\phi(c_2) \setminus \phi(c_1) = |c_2| - \max(|c_1|, |c_2|) + ed(c_1, c_2), \tag{21}$$

respectively.

Our similarity measure for two strings (c_1, c_2) based on Levenshtein measure becomes:

$$Lex\_sim(c_1, c_2) = \tag{22}$$

$$\frac{\alpha(\max(|c_1|, |c_2|) - ed(c_1, c_2))}{\alpha(\max(|c_1|, |c_2|) - ed(c_1, c_2)) + \beta(|c_1| + |c_2| - 2\max(|c_1|, |c_2|) + 2ed(c_1, c_2))},$$

where $|c_1|, |c_2|$ are lengths of strings c_1 and c_2, respectively; $ed(c_1, c_2)$ is Levenshtein measure and $\alpha + 2\beta = 1$.

In case $\beta = \gamma = \frac{1}{2}\alpha$, substitution in Eq. (22) yields

$$Lex\_sim(c_1, c_2) = \frac{2\alpha(\max(|c_1|, |c_2|) - ed(c_1, c_2))}{\alpha(|c_1| + |c_2|)}. \tag{23}$$

When the lengths of two strings are the same, we have $\max(|c_1|, |c_2|) = \min(|c_1|, |c_2|) = |c_1| = |c_2|$, substitution in Eq. (23) yields

$$Lex\_sim(c_1, c_2) = \frac{\min(|c_1|, |c_2|) - ed(c_1, c_2)}{\min(|c_1|, |c_2|)}, \tag{24}$$

which is similar to the Levenshtein's measure.

3.2   Properties of Proposed Similarity Approach

In this section, the properties of the proposed similarity measure are discussed. As can be seen in Eq. (22), our measure satisfies three properties of a similarity function as follows [4]:

- Positiveness: $\forall c_1, c_2 : Lex\_sim(c_1, c_2) \geq 0$
  $Lex\_sim(c_1, c_2) = 0$ if and only if $(\max(|c_1|, |c_2|) - ed(c_1, c_2)) = 0$; consequently, c_1 and c_2 are totally different.
- Maximality:
  $\forall c_1, c_2, c_3 : Lex\_sim(c_1, c_1) \geq Lex\_sim(c_2, c_3)$
  In fact, the values of our measure were taken in the range of $[0, 1]$, $Lex\_sim(c_1, c_1) = 1$, $Lex\_sim(c_2, c_3) \leq 1$. $Lex\_sim(c_2, c_3) = 1$ if and only if $(|c_2| + |c_3| - 2\max(|c_2|, |c_3|) + 2ed(c_2, c_3)) = 0$, it means c_2 and c_3 are similar.
- Symmetry: $\forall c_1, c_2 : sim(c_1, c_2) = sim(c_2, c_1)$

In order to evaluate the performance of our lexical similarity measure, experiments and results are shown in the following section.

4   Experiments and Discussions

We use ontologies taken from the OAEI benchmark 2008[1] to test and evaluate the performance of our measure and other ones through comparing between their output and reference alignments. This benchmark consists of ontologies modified from the reference ontology 101 by changing properties, using synonyms, extending structures and so on. Since the measures here concentrate on calculating the string-based similarity, only ontologies relating to modified labels and the real bibliographic ontologies are chosen to evaluate. Consequently, the considered ontologies consist of 101, 204, 301, 302, 303 and 304. Actually, these chosen ontologies are quite suitable for the validation and comparison among Needleman-Wunsch, Jaro-Winkler, Levenshtein, normalized Kondrak's method combining Dice and n-grams approaches, with using the same classical metrics. These classical metrics are Precision, Recall and F-measure and can be shown as in Eq. (25).

$$Precision = \frac{No.\_correct\_found\_correspondences}{No.\_found\_correspondences},$$

$$Recall = \frac{No.\_correct\_found\_correspondences}{No.\_existing\_correspondences}, \qquad (25)$$

$$F-measure = \frac{2 * Precision * Recall}{Precision + Recall}.$$

Average Precision, Recall and F-measure of measures for six pairs of ontologies with thresholds changed are introduced in [15].

In Fig. 1, Recall values remain stable quantities with all measures. Therefore, the F-measure only changes when Precision values vary. Note that these results in Fig. 1 are obtained by means of thresholds changed for nine different values from 0.5 to 0.9 with the increment of 0.05; in addition, two parameters including $\alpha = 0.2$ and $\beta = 0.4$ were applied. Based on each threshold value, the alignments are achieved for five participants. Then average Precision, Recall and F-measure for all these thresholds are calculated.

---

[1] http://oaei.ontologymatching.org/.

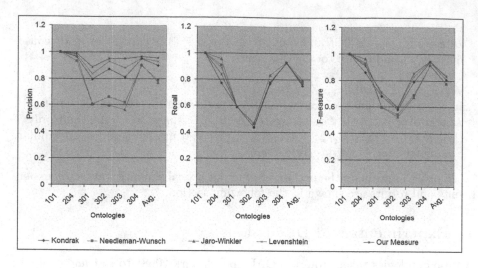

Fig. 1. Average Precision, Recall and F-measure values of different methods.

In addition, average F-measure of our approach gives premier value compared to those of other methods, which indicates that our approach is more effective than the others. Moreover, average F-measure of both our measure and Levenshtein's seem slightly better than Kondrak's metric for each pair of ontologies. Besides, Precision values of Jaro-Winkler and Needleman-Wunsch measures are quite similar and lower than those of others while Precision of our measure is dominantly the best one. The Recall value is quite important because it lets us estimate the number of true positives compared to the number of existing correspondences in the reference alignment. In general, when Recall value is similar, the measure is considered better if it provides higher Precision. For the ontology 101, when compared to itself, all methods above produce the values of Precision, Recall and F-measure to be 1.0. Since ontology 301 consists of concepts which are slightly or completely modified from reference ontology, the number of obtained true positive concepts are the same for string-based metrics mentioned before. Thus, in this case Recall measures have the same values in all methods. In addition, our measure gives better Precision values than those of these measures. Consequently, our approach is better than existing methods for ontology 301. Because ontology 204 only contains concepts modified from the reference one by adding underscores, abbreviations and so on, the measures achieve the rather high results of F-measure. For ontology 204, Recall values of Jaro-Winkler and Needleman-Wunsch methods are a little bit higher than others. Ontology 304 has similar vocabularies to the ontology 101, so Precision and Recall values which are achieved for this pair of ontologies are also good. Jaro-Winkler measure is also known as a good approach because its average Recall value is slightly higher than others. However its average Precision is significantly lower than others. Therefore, the number of obtained false positive concepts of Jaro-Winkler is higher than other measures. This phenomenon occurs in the same manner in the pairs of ontologies 302 and 303.

Precision and F-measure obtained by Levenshtein and our approach are higher than those of other measures in general. Therefore, two methods are used for the comparison hereafter. It is clear that our measure depends on parameters α and β due to deriving from information-theoretic approach. To determine the range of parameters in our measure which could obtain good results, parameter α changed for six different values from 0.2 to 0.7 with an increment of 0.1. The average Precision, Recall and F-measure of two measures for six pairs of ontologies with thresholds and parameters are presented in Table 1.

The results show that increasing parameter α leads to our Precision value decreasing and our Recall value increasing. When $\alpha = 0.5$, our measure is similar to Dice's measure. However, our F-measure is lower than Levenshtein's. In the following experiment, parameter α is chosen the values from the interval [0.2, 0.4] with an increment of 0.05. The results of average Precision, Recall and F-measure of our measure for six pairs of ontologies are described in Table 2, which shows that to obtain good F-measure values, parameter α should be chosen between 0.2 and 0.35. Consequently, β is in range from 0.4 to 0.325.

Besides the above evaluation, our measure is also more rational in several cases. For example, given two strings $c_1 =$ 'glass' and $c_2 =$ 'grass'. There is only one edit transforming c_1 into c_2: the substitution of 'l' with 'r'. Therefore, the Levenshtein distance between two strings 'glass' and 'grass' is 1. Applying Eqs. (11) and (22) to our measure in which the parameter α is assigned to 0.2, the similarity between two strings 'glass' and 'grass' is 0.8 while the similarity degree of our measure yields 0.5. In fact, the two strings 'glass' and 'grass' describe different objects. While the Levenshtein measure returns the high similarity score value (0.8), the result 0.5 of our measure is quite reasonable. In another example, if

Table 1. Average Precision, Recall and F-measure values of two measures for six pairs of ontologies with an increment of parameters of 0.1.

Average	Levenshtein and our measure						
	Levenshtein	$\alpha = 0.2$	$\alpha = 0.3$	$\alpha = 0.4$	$\alpha = 0.5$	$\alpha = 0.6$	$\alpha = 0.7$
Precision	0.930	0.957	0.927	0.896	0.863	0.828	0.787
Recall	0.771	0.762	0.773	0.782	0.787	0.793	0.804
F-measure	0.832	0.836	0.833	0.825	0.815	0.803	0.790

Table 2. Average Precision, Recall and F-measure values of two measures for six pairs of ontologies with an increment of parameters of 0.05.

Average	Levenshtein and our measure					
	Levenshtein	$\alpha = 0.2$	$\alpha = 0.25$	$\alpha = 0.3$	$\alpha = 0.35$	$\alpha = 0.4$
Precision	0.930	0.957	0.944	0.927	0.913	0.896
Recall	0.771	0.762	0.767	0.773	0.780	0.782
F-measure	0.832	0.836	0.846	0.833	0.841	0.825

$n \geq 2$ then two strings *Rep* and *Rap* have no n-grams in common. In this case, applying Dice's measure to these strings brings the dissimilarity. Additionally, the family of Dice's methods has a characteristic which relies on the set of samples but not on their positions. Because the sets of bigrams of two strings *Label* and *Belab* including $\{la,\ ab,\ be,\ el\}$ are the same, the similarity value of these strings equal to 1, which seems inappropriate. In short, our approach overcomes the limits of these cases.

5   Conclusions and Future Work

In this paper, we propose a new lexical-based approach, which considers the similarity of sequences by combining features-based and element-based measures. This approach is motivated by Tversky's and Levenshtein's measures; however, it is completely different from original lexical methods previously presented. The main idea of our approach is that the similarity value of two given concepts depends not only on the contents but also on the editing operations of these concepts in strings. For Levenshtein's measure, it focus on the number of editing operations in order to change one string into another string; whereas, the characteristic of the Tversky's model contains the more common features and the less different features with an increasing in similarity between objects. For this reason, the combination of the two above models reduces the limitations of other methods. The experimental validation of the proposed metric has been conducted through six pairs of ontologies in OAEI benchmark 2008, and compared to four of the common similarity metrics including Jaro-Winkler, Needleman-Wunsch, Kondrak and Levenshtein metrics. The results show that our proposed sequence similarity metric provides good values compared to other existing metrics. Moreover, our metric can be considered as flexible and general lexical approach. In particular, adjusting the parameters α and β produces the popular measures making convenient experiments. In addition, parameters α and β have been changed based on that we can choose the range for good similarity values. Our measure can also be implemented in many domains in which strings are short as labels of concepts in ontologies, proteins and so on.

In this work, strings are considered as a set of characters. However, they can be extended to the set of tokens in which the similarity between chunks in plagiarism detection is calculated. In the future work, our string-based similarity metric might also be combined with relations between entities in ontologies using Wordnet dictionary to improve the semantic similarity of pairs of these entities.

References

1. Algergawy, A., Schallehn, E., Saake, G.: A sequence-based ontology matching approach. In: The 10th International Conference on Information Integration and Web-Based Applications & Services, pp. 131–136. ACM (2008)
2. Batet, M., Sánchez, D., Valls, A.: An ontology-based measure to compute semantic similarity in biomedicine. Biomed. Inf. **44**(1), 118–125 (2011)

3. Dice, L.R.: Measures of the amount of ecologic association between species. Ecology **26**(3), 297–302 (1945)
4. Euzenat, J., Shvaiko, P.: Ontology Matching. Springer, Heidelberg (2013)
5. Hamming, R.W.: Error detecting and error correcting codes. Bell Syst. Techn. J. **29**(2), 147–160 (1950)
6. Ichise, R.: Machine learning approach for ontology mapping using multiple concept similarity measures. In: The 7th IEEE/ACIS International Conference on Computer and Information Science, pp. 340–346. IEEE (2008)
7. Jaccard, P.: The distribution of the flora in the alpine zone. New Phytol. **11**(2), 37–50 (1912)
8. Jaro, M.A.: Advances in record-linkage methodology as applied to matching the 1985 census of Tampa, Florida. Am. Stat. Assoc. **84**(406), 414–420 (1989)
9. Kondrak, G.: N-gram similarity and distance. In: Consens, M.P., Navarro, G. (eds.) SPIRE 2005. LNCS, vol. 3772, pp. 115–126. Springer, Heidelberg (2005)
10. Levenshtein, V.I.: Binary codes capable of correcting deletions, insertions, and reversals. Sov. Phys. Doklady **10**, 707–710 (1966)
11. Lin, D.: An information-theoretic definition of similarity. In: The 15th International Conference on Machine Learning, pp. 296–304. Morgan Kaufmann (1998)
12. Maedche, A., Staab, S.: Measuring similarity between ontologies. In: Gómez-Pérez, A., Benjamins, V.R. (eds.) EKAW 2002. LNCS (LNAI), vol. 2473, pp. 251–263. Springer, Heidelberg (2002)
13. Needleman, S.B., Wunsch, C.D.: A general method applicable to the search for similarities in the amino acid sequence of two proteins. Mol. Biol. **48**, 443–453 (1970)
14. Nguyen, T.T.A., Conrad, S.: Combination of lexical and structure-based similarity measures to match ontologies automatically. In: Fred, A., Dietz, J.L.G., Liu, K., Filipe, J. (eds.) IC3K 2012. CCIS, vol. 415, pp. 101–112. Springer, Heidelberg (2013)
15. Nguyen, T.T.A., Conrad, S.: Applying information-theoretic and edit distance approaches to flexibly measure lexical similarity. In: The 6th International Conference on Knowledge Discovery and Information Retrieval, pp. 505–511. SciTePress (2014)
16. Pirró, G., Euzenat, J.: A feature and information theoretic framework for semantic similarity and relatedness. In: Patel-Schneider, P.F., Pan, Y., Hitzler, P., Mika, P., Zhang, L., Pan, J.Z., Horrocks, I., Glimm, B. (eds.) ISWC 2010, Part I. LNCS, vol. 6496, pp. 615–630. Springer, Heidelberg (2010)
17. Pirró, G., Seco, N.: Design, implementation and evaluation of a new semantic similarity metric combining features and intrinsic information content. In: Meersman, R., Tari, Z. (eds.) OTM 2008, Part II. LNCS, vol. 5332, pp. 1271–1288. Springer, Heidelberg (2008)
18. Sánchez, D., Batet, M., Isern, D., Valls, A.: Ontology-based semantic similarity: a new feature-based approach. Expert Syst. Appl. **39**(9), 7718–7728 (2012)
19. Tversky, A.: Features of similarity. Psychol. Rev. **84**, 327–352 (1997)
20. Wang, X., Ding, Y., Zhao, Y.: Similarity measurement about ontology-based semantic web services. In: The Workshop on Semantics for Web Services (2006)
21. Winkler, W.E.: String comparator metrics and enhanced decision rules in the fellegi-sunter model of record linkage. In: The Section on Survey Research, pp. 354–359 (1990)

Named Entity Recognition from Financial Press Releases

Rebecca J. Passonneau[1], Tifara Ramelson[2]([⊠]), and Boyi Xie[3]

[1] Center for Computational Learning Systems,
Columbia University,
New York, NY, USA
ecky@ccls.columbia.edu
[2] Brandeis University, Waltham, MA, USA
tifarar@brandeis.edu
[3] Department of Computer Science, Columbia University, New York, NY, USA
xie@cs.columbia.edu

Abstract. This paper explores a previous model's use of named entity recognition to predict the changes in stock prices from financial news. Detecting company mentions in the articles is crucial for this task, and we modified these methods to gain additional mentions. We first expanded upon the rules of the named entity recognition from the original model. We also incorporated coreference resolution and modified an existing toolkit to be compatible with our specific domain. After these two adjustments, the number of instances captured increased significantly. Although this did not necessarily improve the overall prediction performance, the results give us an opportunity to explore reasons why the scores stayed around the same, and a full analysis will allow us to achieve our goals.

Keywords: Text mining for financial news · Financial analytics · Coreference resolution · Named entity recognition

1 Introduction

Given that accessing vast amounts of online news is easy, it becomes possible to mine news on a large scale to automatically discover information relevant for human decision making. For example, the ability to detect conflict among political entities, as in [1], could potentially inform policy decisions. Text mining could also affect decisions by individual analysts who track entities of other sorts, such as corporate entities. Our work investigates the problem of mining online financial news sources in order to learn about the fundamental market value of publicly traded companies. The ability to automatically discover aspects of the market through news has broad significance. Such information could be used by individual investors who want to make informed investment decisions, by corporations that want to understand public perception of the market, by government entities that regulate markets, or by intelligence agencies that monitor the market for unusual events. Given a company whose stock price has changed on a

© Springer International Publishing Switzerland 2015
A. Fred et al. (Eds.): IC3K 2014, CCIS 553, pp. 240–254, 2015.
DOI: 10.1007/978-3-319-25840-9_16

given day, the general task we address is to predict whether the price went up or down, based on the news. The specific focus of this paper is to test the benefit of finding additional mentions of companies in the news through Named Entity detection and coreference resolution.

Our goal is to study the impact of a high precision, high recall approach to mining news for mentions of entities of interest. In the financial domain, we currently restrict our attention to publicly traded companies. The two issues we address are (1) to resolve variant names to the same company (e.g., *Eli Lilly and Company, Eli Lilly, Eli Lilly & Co., Lilly & Co.*), and (2) to resolve coreferent expressions consisting of noun phrases and pronouns (e.g., *Eli Lilly and Company is an American global pharmaceutical company with headquarters in Indianapolis, Indiana. The company also has offices in Puerto Rico and 17 other countries. Their products are sold in 125 countries. It was founded in 1876.*). We refer to this task as *company mention detection.*

Our research shows that improved company mention detection will not necessarily improve price prediction from news. This is an extremely challenging prediction problem with many confounding factors. For example, news items that provide novel information about a company potentially have more impact on price than news items that provide old information. Accurate company mention detection might incorporate a higher proportion of sentences that provide old information, which could hurt rather than benefit prediction of price change. Given the complexity of factors involved in testing whether more accurate company mention detection improves prediction of stock price change, it is possible that results would vary, depending on the type of feature representation used. To make our test more general, we use an existing framework that compares alternative document representations in this domain [2]. Because this framework compares several kinds of vector and tree space representations, it serves as a more general test of the impact of improved company mention detection.

One of the challenges in mining financial information from news is that the domain of publicly traded corporate entities is extremely heterogeneous. For example, the features that prove predictive in [2] vary markedly across sectors, and can even predict opposite direction of price change in different sectors, such as retail versus industrials. It is also well known that the performance of NLP techniques varies across domains. Domain adaptation has been addressed in parsing [3–5] and language modeling [6,7]. Sensitivity to domain is undoubtedly true as well of NER and coreference. This suggested to us that to evaluate the effect on performance of existing NLP tools for improving company mention detection, it is important to assess performance sector by sector. We find that the extension of the NER component of the framework in [2] and integration of a coreference toolkit dramatically improves recall, but much more so for one sector in particular. Manual assessment of samples of the data suggests that precision remains high. The impact on prediction, however, is not uniform. Predictive accuracy improves primarily for one of the three sectors, using the more expressive tree space representation. Improving predictions is not necessarily dependent on the number of mentions captured, but rather on the quality of the content surrounding company mentions.

2   Motivation

Company mention detection is certainly a challenging task. Consider the example in Fig. 1. *Baker Hughes Inc.* is a company that provides oil and gas services in the *Energy* sector. Example sentence 1 mentions the full name of the company and an exact match can identify it. The challenge that occurs is when companies mentioned in the articles are referred to by a more abbreviated version of their full name, such as *Baker Hughes* or *Baker*, as in example sentence 2. Further problems lie in the fact that some of these abbreviated mentions name other entities, such as a person, or are generic words, such as the word *baker*, when it occurs at the beginning of a sentence introducing a person of that occupation. We had to consider if increasing the recall to capture these cases would outweigh the negative effect of a decrease in precision. Accordingly, we looked at how frequently abbreviated name strings are in fact used to refer to a company versus a different entity. In addition, there are instances where sub-branches of a company are mentioned, and it is questionable as to whether these are important instances to capture. *Baker Hughes*, in example sentence 2, has divisions *Baker Oil Tools* and *Baker Petrolite*, which are mentioned in the same news article, but an exact match by full name cannot capture these mentions. The question of whether news reports about subsidiary units affect the main company's price requires is a complex one that we do not address here.

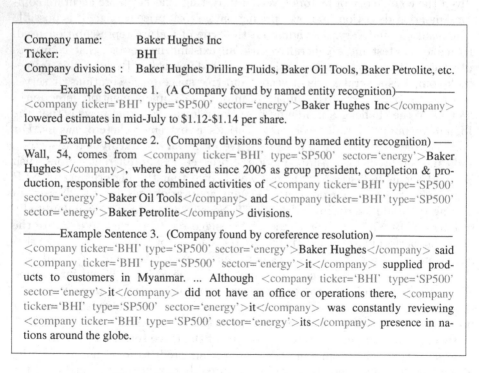

Fig. 1. Example company and news sentences.

Further improvement of company mention detection requires coreference resolution, especially to detect mentions in different sentences, as shown in example sentence 3 of Fig. 1. Coreference resolution was not used in many previous studies on financial news analytics, including [2,8,9]. We found that the Stanford CoreNLP coreference parser [10], a state-of-art coreference resolution toolkit that works well on the CoNLL Shared Task, does not lead to good results when directly applied. It introduces many mention chains that are irrelevant to the company entities, and some chains contain heterogeneous noun phrases that are not appropriate for our company mention annotation task. However, it has a modular design that supports relatively easy re-design, as described in Sect. 6.

3 Related Work

Text mining in the financial domain with shallow techniques has shown some success [11–13]. Recent work has applied NLP techniques to various financial media (conventional news, tweets) to detect sentiment in conventional news [14,15] or message boards [16], or to discriminate expert from non-expert investors in financial tweets [17]. Kogan et al. [18] analyzed quarterly earning reports to predict stock return volatility and to predict whether a company will be delisted. Luss and d'Aspremont [19] used text classification to model daily price movements that are above a predefined threshold. Xie et al. [2] proposed a semantic tree structure for data representation and used tree kernel for support vector learning to predict the price change based on financial news. This work is related to the above studies that explore richer NLP techniques for company driven financial news analysis. However, most of the existing research focuses on task specific modeling, such as price prediction or fraud detection. Little attention is paid to the best ways to integrate fundamental text processing methodologies such as named entity recognition and coreference resolution.

This paper focuses on improving the text processing pipeline to improve the overall financial news knowledge discovery framework. Capturing named entities is essential for making accurate predictions because we rely on named entity recognition to select company relevant news information for price modeling. Named entity recognition is a major area of interest in text mining. A large resource that supports this task is the Heidelberg Named Entity Resource, a lexicon that links many proper names to named entities [20]. It is not used in our study because its coverage is limited: it fails to capture enough mentions for our targeted company list, which is based on the S&P 500.[1] As a result, we require a more general and comprehensive method. In addition to named entity recognition, we also incorporated a coreference resolution step to further improve the performance of text mining procedure. There are coreference parsers that use various approaches in attempts to attain optimal performance. The coreference resolution model that our method builds on is the Stanford CoreNLP parser [21]. Named entity recognition and coreference resolution are the two key

---

[1] S&P 500 is an equity market index that includes 500 publicly traded companies in leading industries.

components in our company mention detection task. We leverage state-of-art tools to maximize compatibility and stock market predictability for the financial news domain.

4   Data

We work with a large dataset for doing extensive news analysis, where publicly available Reuters news data from the year 2007 are used for this study. We concentrated on the first three sectors in GICS: 40 companies in GICS 10 of *Energy* such as *Hess* and *Exxon Mobile*, 26 companies in GICS 15 of *Materials* such as *Du Point*, and 58 companies in GICS 20 of *Industrials* such as *Boeing* and *General Electric*. Table 1 describes our data. C is number of companies in each sector; N is the number of news items; S is the number of sentences; and T is the number of words.

Table 1. Description of news data.

GICS	C	N	S	T
10	40	5,373	109,277	2,014,085
15	26	2,295	53,595	953,133
20	58	8,325	238,570	3,780,129

5   Framework

Our framework to capture news impact on the financial market consists of three main components, as shown in Fig. 2: (1) text processing, (2) data instance formation, and (3) model learning and evaluation. In the text processing component, a four-stage NLP pipeline is used. The title and full text of the news article are first extracted from the HTML documents from Reuters News Web Archive. In the sentence segmentation stage, the full text is split into sentences. The company mention detection stage then identifies if any company of interest is mentioned in the sentence. In this study, we focus on a finite list of companies in the S&P 500. The sentences containing at least one S&P 500 company mention are parsed and used in the following stages for text mining. Therefore, the company mention detection task provides the data foundation for the whole framework. How to improve the coverage of the company mention detection in a way that improves prediction is the main focus in this study.

The remaining framework relies on the implementation described in [2].

After text processing, we align publicly available daily stock price data from Yahoo Finance with the textual news data following the method in [2]. Recall that the task is to predict the change in price of a company on a date based on the analysis of the preceding day's news. A data instance is all the news associated with a company on a given day, and consists of the companies whose

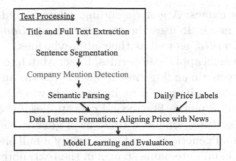

Fig. 2. Framework of the text mining on financial news for stock market price prediction.

price changed above a threshold between the closing price on the day of the news and the closing price on the following day.

In the learning and evaluation component, rich vector space models are used to test the price prediction performance. These vector space models include bag-of-words models, semantic frame features, and part-of-speech based word affective features. A model that encodes rich structured semantic information, *SemTreeFWD* of [2], is also used for model learning and evaluation. It is an enriched hybrid of vector and tree space models that contains semantic frames, lexical items, and part-of-speech-specific affective features trained with Tree Kernel SVM [22].

6   Company Mention Detection

Our Company Mention Detection module attempts to identify all named entities, variants of these names, and coreferential expressions, then replaces the original strings in the text with a unique identifier. For the identifiers, we use the company tickers, character codes between length of two to five, to identify publicly traded companies. The NLP pipeline in [2] used a rule-based method for partial matching on the full company names that only recognized a limited number of the variant names for a company. We have expanded its NER (Named Entity Recognition) rules to capture a much wider range of name variants. We also tested the Stanford CoreNLP coreference parser, and modified it to achieve optimal performance for our domain. This section describes the original and our new NER module, and the changes we made to Stanford coreference parser.

To obtain a lower bound for NER, we used an Exact Match method, defined as matching the exact string to the official names of the S&P 500 companies. This ensures 100 % precision, but recall is low. The approach in [2] for NER relies on a few conservative rules. These rules focus on the structure of the company names, which can consist of two types of tokens. The words that make up the unique name of the company are the general name elements. The second type are the generic endings, a predefined set of possible suffixes that are optionally included in company names. A generic ending, when included, will be the last token of a company name. It uses the generic endings *Company, Corporation, Incorporation*, and *Limited*, as well as their abbreviations.

The NER module in [2] applies three rules, Exact Match to the S&P Wikipedia name, a rule for the generic endings in the Exact Match, and one for the name elements. The first rule, as indicated by its name, is when the name in the text exactly matches the S&P name. The second rule applies if there is a generic ending: the program substitutes, one at a time, each generic element in our predefined list for the original generic element, and finally a null element, and searches for each of these new candidate name strings in the text; note that if the null element is substituted, then the new search string consists only of a sequence of name elements with no generic ending. The third rule, which triggers after the second, truncates the sequence of name elements by iteratively removing the last name element unless the sequence of name elements is length two. After each truncation step, the second rule is re-applied. The process terminates at the first word of a company name.

Our Company Mention Detection module incorporates the NER from [2] described above, but extends the rules so that it does not terminate when the sequence of name elements is length one. Through random sampling and visual inspection, we found that it would be beneficial to include the first word. To maintain high precision, we hard-coded rules for companies where there was a strong possibility that the first token of their names could be mistaken for another entity.

Our Company Mention Detection module also incorporates the Stanford CoreNLP parser, which outputs lists of entities that corefer, called coreference chains [21]. The Stanford parser was trained on various copora where the average F-measure was about 60 %, which is considered a high score for this task. Furthermore, this parser was intended to be easy for others to modify, either by removing or adding methods to capture coreference patterns. Initially, the Stanford parser seemed ineffective for our dataset due to some inaccuracies in the results. It captured many more instances in some of its coreference chains than it should have, thus decreasing precision. By observing the list of entities in these coreference chains, we noticed that there were some incorrect linkings. First, distinct companies were sometimes linked with each other, such that an incorrect ticker was assigned to one of the companies. Second, the parser captured predicate nominative instances, which are not relevant for our purposes. Third, there were general incorrect linkings between company names and other words in the text.

To address these issues, we re-structured the components of the Stanford CoreNLP coreference parser. The original algorithm goes through ten passes, or sieves, to capture different kinds of coreference phenomena for each iteration [10]. By exploring the sieves in the coreference toolkit, we were able to identify the ones causing problems in our data, and to manually tune the parser to meet our needs. The three passes that decreased the accuracy of the mention detection algorithm are called Precise Constructs, Strict Head Match 3 and Relaxed

Head Match. There are a few rules incorporated into Precise Constructs, but the main one causing issues in our data was the predicate nominative condition, which, when capturing an entity, also captures the text following a linking verb [10]. For example, a sentence that mentions the *ConocoPhillips* company says, *ConocoPhillips is an international, integrated petroleum company with interests around the world*. Precise Constructs gives the output *ConocoPhillips is ConocoPhillips*.

Strict Head Match 3 removes a word inclusion constraint used in Strict Head Match 1, where all the non-stop words of one entity must match the non-stop words that appear in the previous one. By removing this sieve and thereby imposing this constraint, our program avoids generating incorrect linkages between entities. Strict Head Match 3 removes this constraint since the score for the dataset the Stanford team tested it on improved. Relaxed Head Match allows any word in the main entity to match with entities in other coreference chains. As a result, for the company *Air Products*, the original algorithm incorrectly recognized *these products* to be the company entity. Once these three sieves were eliminated, we observed a significant improvement.

The passes that remained in the coreference parser include Speaker Identification, Exact String Match, Relaxed String Match, Strict Head Match 1, Strict Head Match 2, Proper Head Word Match and Pronoun Match. The Speaker Identification sieve detects the speakers in the text and captures any pronouns that refer to them. In Exact String Match, the parser captures the exact string of entities, similar to the idea of our Exact Match method, but with the additional property of including modifiers and determiners. Relaxed String Match removes the text following the head words of two entities, and links them together if the remaining strings match. Strict Head Match 1 uses the heads of the entities and imposes constraints to determine if the mentions are coreferent. Strict Head Match 2 eliminates a restriction used in Strict Head Match 1, where in this property, modifiers in one entity must match the modifiers in the previous entity in order to be linked together. Proper Head Word Match links proper nouns that have the same head word, but also has specific restrictions imposed on these entities. Pronoun Match focuses on pronominal rules and imposes agreement constraints to capture the entities that are compatible. These seven sieves [10] provided the results we needed for capturing additional correct instances.

7   Experiment

Before conducting our experiment with the Company Mention Detection module, we did some probes on the data to shape our expectations for performance gains. Taking a randomly selected company, and ten randomly selected documents that mention the company, we counted how many company mentions were captured by each of the three methods: *Exact Match*, the *Initial NER* from [2] and our *Company Mention Detection* (*CMD*). Percentage results for the 54 mentions this yielded are displayed in Table 2. As shown, *CMD* yielded greatly improved recall at a reasonable sacrifice in precision, and an overall increase in F-measure of 13.6 %, compared to the *Initial NER*. Interestingly, the incorrect instances for *CMD* were not entirely wrong: they all referred to units within the company. We count them as incorrect, however, because of our focus on predicting stock price for the S&P 500 (parent) companies. As noted above, what happens to one unit of a company may not necessarily affect public perception of the company as a whole. We, therefore, do not regard sub-companies as correct instances for the purposes of our experiment.

Table 2. A manual evaluation for company detection in a preliminary experiment.

Methods	Precision	Recall	F-measure
Exact match	100.0 %	17.0 %	29.0 %
Initial NER	100.0 %	57.4 %	72.9 %
CMD	90.0 %	76.6 %	82.8 %

Table 3. Counts of company mentions by sentence.

GICS	Initial NER	CMD	Increase
10	8,646	11,252	30.14 %
15	5,445	6,336	16.36 %
20	15,286	17,865	16.87 %
Total	29,377	35,453	20.68 %

The Exact Match method has a very low F-measure since it only captures the full name of a company. Except the first time mentioned in a news article, a company is usually not referred to by its full name. Instead, variations of company names are frequently used. Clearly, the initial NER method from [2] far outperforms this baseline, yet leaves much room for improvement in recall.

As described in Sect. 6, *CMD* further expanded the NER so as to search for abbreviated name strings that include only the first word of the full named entity string of the companies. For the company *Baker Hughes Inc.*, this would lead to the inclusion of mentions by the single name *Baker*. Although in the general case, this could introduce imprecision, if a document already contains the full company name, it is likely that use of the first name token in the full name (e.g., *Baker*) would be a company mention. In addition, *CMD* also captures many coreferential expressions for company mentions. For example, one article says, *Baker Hughes said it supplied products to customers*; where the original NER rules capture *Baker Hughes*. *CMD* also captures *it*. As shown in Table 3, *CMD* captures many additional instances of company mentions. This also leads to some gains in stock price prediction, as will be reported in the next section.

The full experiment uses as input the data described in Table 1 consisting of all the news in three market sectors from Reuters news archive for 2007. Recall, we use the framework described in Sect. 5 because it allows us to test the impact of improved F-measure for *CMD* across multiple document representations. The five document representations we test in the experiment are: (1) *BOW*, which refers to bag-of-words with unigram counts; (2) *BOW (n-gram)*, for BOW with unigram, bigram and trigram counts; (3) *FW* which is like *BOW (n-gram)* but also includes Frame Semantic elements (see next paragraph); (4) *FWD* consists of *FW* plus a prior polarity on words from the Dictionary of Affect in Language (DAL score; see next paragraph); (5) and lastly, *SemTreeFWD*, which is a tree structure that uses the FWD features combined with a tree kernel.

Three of the five document representations make use of the features from frame semantics [23]. Frame semantics aims for a conceptual representation that generalizes from words and phrases to abstract scenarios, or frames, that capture explicit and implicit meanings of sentences. The three basic feature types from frame semantics are frame name, frame target, and frame element. Each frame is evoked by a *frame target*, or *lexical unit*, for example, *sue* or *accuse* evoke the *Judgement Communication* frame, which describes a lawsuit scenario. Its frame elements, or semantic roles, are *Communicator*, *Evaluee*, and *Reason*. *FW* and *FWD* uses bag-of-frames (including frame names, frame targets, and frame elements) features in a vector space representation, while *SemTreeFWD* encodes relational structures between the company entity and the semantic frame features in a tree representation, in addition to *FWD*. The semantic parsing we use to extract frame features is SEMAFOR[2] [24,25], a statistical parser that uses a rule-based frame target identification, a semi-supervised model that expands the predicate lexicon of FrameNet for semantic frame classification, and a supervised model for argument identification.

FWD and *SemTreeFWD* contain word affective features based on DAL, the Dictionary of Affect in Language [26]. It is a psycholinguistic resource designed to quantify the undertones of emotional words that includes 8,742 words annotated for three dimensions: pleasantness, activation, and imagery. We use the average scores, in terms of the three dimensions, for all words, verbs, adjectives, and adverbs in a vector space for feature representation.

The experiments assess the performance of predicting the direction of price change across companies in a sector. Recall that a data instance in our experiment is all the news associated with a company on a given day, and consists of the companies whose price changed above a threshold between the closing price on the day of the news and the closing price on the following day. In this experiment, we use the threshold of 2 % that corresponds to a moderate fluctuation. A binary class label $\{-1, +1\}$ indicates the direction of price change on the next day after the data instance was generated from the news. For each company, 80 % of the data is used for training and 20 % for testing. We report the averaged accuracy and standard deviation of the test data for both the *Initial NER*, as a benchmark, and our *CMD* on a sector-by-sector basis.

---

[2] http://www.ark.cs.cmu.edu/SEMAFOR.

Table 4. Averaged test accuracy for each company by sector that uses 80 % of the data for training 20 % for testing. Boldface identifies a higher *CMD* mean and * identifies the *CMD* that is significantly better than the *Initial NER* with *p-value* < 0.05.

GICS	Sector	type	BOW	BOW (n-gram)	FW	FWD	SemTreeFWD
10	Energy	Initial NER	59.94 ± 16.38	61.18 ± 15.43	59.99 ± 14.46	59.05 ± 16.58	64.26 ± 14.95
		CMD	58.54 ± 17.32	61.11 ± 15.34	58.67 ± 15.76	58.44 ± 18.40	**64.87** ± 15.04
15	Materials	Initial NER	58.23 ± 15.53	59.74 ± 14.33	62.10 ± 14.24	62.69 ± 15.28	68.62 ± 14.72
		CMD	**61.82** ± 15.18	**60.63** ± 15.33	**63.23** ± 13.71	**63.12** ± 15.01	67.18 ± 13.37
20	Industrials	Initial NER	56.70 ± 14.81	55.47 ± 13.86	53.86 ± 13.43	54.29 ± 14.31	57.25 ± 16.88
		CMD	**60.13** ± 14.04*	**58.19** ± 13.44*	**55.37** ± 13.31	**55.75** ± 13.54	56.36 ± 18.38

8   Results

The experiment addresses two questions: (1) Does CMD improve the coverage of company mentions in the domain of interest? (2) Does our *Company Mention Detection* improve accuracy of prediction on the task to identify the direction of price change? Based on our probe of the data where we could manually assess precision (Table 2 in Sect. 7), we expected a large increase in coverage. Projecting from the results of this manual probe, we assume that an increase in recall comes with an acceptable (small) degradation in precision. Yet, because there is no gold standard data set, we cannot assess precision of CMD for the full dataset. Prediction accuracy is the true test of performance on the benefit of increased coverage of company mentions using CMD, but is only a very indirect measure of precision. As noted above, stock price prediction from news is a challenging task with a great deal of noise in the input. Results presented here show a substantial increase in coverage, and statistically significant increases in prediction accuracy for some but not all of the experimental conditions.

As background to interpret the results, it is important to consider the relation between the increased number of mentions versus the number of data instances per company, and the differences across sectors in the average number of data instances per company. Again, each data instance consists of all the news for a given company on a given day. Therefore, new <u>data instances</u> will be added only if CMD identifies news for a given company on a day that was not identified before. If new <u>sentences</u> for a given day are identified, however, then we expect that *BOW* and *BOW (n-gram)* are very likely to be enriched, and prediction could improve in these two cases. If new mentions in an existing sentence are identified, this should not improve *BOW* and *BOW (n-gram)* because all the relevant feature positions in the vector (unigram, n-gram) will already have had values, and the values will not change. In contrast, if new mentions occur not in the same sentence but in new clauses within or across sentences, the representations that use semantic frame parsing (FW, FWD, SemTreeFWD) could be enriched if the new clauses contain words that trigger new frames, and the new mentions fill their roles.

We found that CMD did not increase the number of data instances. This result suggests that if a news item mentions a relevant company, at least one mention will be either an exact match to the full name string, or a near match based on the conservative NER rules in [2]. On the other hand, there were substantial gains in the total number of sentences. Table 3 reports the absolute numbers of sentences with company mentions from the original NER module in [2] compared with those for the Company Mention Detection module. At increases of between 16 % and 17 %, the Materials and Industrials sectors already show large increases; the increase for the energy sector is nearly double that of the two other sectors. This difference between the GICS 15 and 20 versus GICS 10 reflects the underlying domain differences from sector to sector, which accounts to some degree for the difficulty of the prediction task. We further note that the number of data instances per company differs substantially across the three sectors. The mean and standard deviation for each sector are as follows, respectively: GICS 10, $\mu = 24.37, \sigma = 15.80$; GICS 15 $\mu = 20.80, \sigma = 15.52$; GICS 20: $\mu = 16.16, \sigma = 18.96$. Based on these figures, we expect the gains for GICS 15 and 20 to be similar, and the gains for GICS 10 to be larger for the semantic frame representations.

Table 4 gives the average accuracy per sector of the CMD combined with the five document representation methods introduced in the previous section. (Note: None of these results significantly beat the baseline accuracy given by the average over the majority class for each company, but the standard deviations for this baseline (as for the results in Table 4) are quite high. This does not diminish the comparison of the different representations, and the question of whether CMD can improve performance.) Prediction accuracy improved for the BOW representations. The numbers in boldface are the cases where the average accuracy for CMD is higher than for the original NER, and the cells with an asterisk indicate cases where a t-test of the difference is statistically significant. As shown, the two cases where there is a statistically significant improvement are for the two BOW representations for the sector with the fewest average data instances per company, namely Industrials. When using NER, the BOW representations already had very competitive performance, and CMD increases their performance. This suggests that the new sentences that are identified with CMD add new vocabulary that is predictive. The two vector-based representations with frames also have higher accuracy, but the increase is not statistically significant. For the tree-based representation (SemTreeFWD), the performance degrades somewhat. The performance of the frame-based representations suggests that the new sentences for Industrials do not add new frames, or possibly add new frames that have semantic conflicts with the frames that were found earlier. The same general pattern holds for the Materials sector.

The one case where the SemTreeFWD performance improves is for the Energy sector, but the improvement is not statistically significant. We can only speculate that this sector is the only one where SemTreeFWD shows greater accuracy because this is the sector where the number of additional sentences is substantially larger.

The two questions posed by our experiment can be answered briefly as follows: (1) CMD improves the coverage of company mentions dramatically at the sentence level: the number of additional sentences per sector increases on average by over 20 %. This does not, however, increase the number of data instances; (2) CMD has a statistically significant impact on predictive accuracy only for the Industrials sector, for the two BOW representations. In the next section we discuss the ramifications of these results.

9   Conclusion

Evaluation of coreference performance generally involves assessment of the accuracy of coreference as an independent module. Here we provide an evaluation of coreference as an independent module (intrinsic), and as part of an end-to-end system that aims at a real world prediction task (extrinsic). The results presented in the preceding section provide a very dramatic and concrete demonstration that large gains for coreference as a stand-alone module do not necessarily result in system gains. They also demonstrate the importance of considering the overall integration of information for data representation.

Of the fifteen conditions in Table 4, the two conditions where we find statistically significant improvements from CMD pertain to the two data representations that are relatively less rich, *BOW* and *BOW (n-gram)*, for the sector with the fewest data instances. There are marginal improvements that are not statistically significant for FW and FWD, and a degradation for SemTreeFWD. This indicates to us that the new sentences added for the Industrials sector add new features to the BOW feature vector, but do not add as much in the way of frame features. Continuing with this sector, the differences between the five document representations are not as great for NER as they are with CMD, and the unigram BOW representation in the CMD condition ends up with the highest accuracy for the ten conditions. The same general trend for the vector representations holds in Materials as for Industrials, but without statistical significance. For Materials, however, SemTreeFWD remains the representation with the highest accuracy among all five.

Energy, which had a much more substantial gain in number of sentences, has a completely different pattern. There are no gains for the vector based representations. Energy is also the sector with the greatest number of data instances per company. Here we speculate that the addition of new sentences does not add new vocabulary: with such a large number of data instances per company already, vocabulary coverage was perhaps already high. SemTreeFWD shows a small gain in accuracy that is not statistically significant.

In our view, rich semantic and pragmatic data mining for large scale text mining should aim for information that supports more informed decision making, or in other words, is actionable. To summarize the results of the experiment presented here, a substantial increase in coverage for the task of detecting mentions of relevant entities on a large scale prediction task does not necessarily translate to gains in the actionable value of the information gained. Further, the experiment demonstrates the interdependence of semantic and pragmatic data mining with feature representation, and with the end goals of the data mining task. For future work, a detailed post hoc analysis of results across sectors, and across companies within sectors, should yield insights that could inform a more sophisticated processing architecture, as well as a more effective document representation.

References

1. O'Connor, B., Stewart, B.M., Smith, N.A.: Learning to extract international relations from political context. In: Proceedings of the 51st Annual Meeting of the ACL, Sofia, Bulgaria, pp. 1094–1104 (2013)
2. Xie, B., Passonneau, R.J., Wu, L., Creamer, G.: Semantic frames to predict stock price movement. In: Proceedings of the 51st Annual Meeting of the ACL, Sofia, Bulgaria, pp. 873–883 (2013)
3. Ravi, S., Knight, K., Soricut, R.: Automatic prediction of parser accuracy. In: Proceedings of the 2008 Conference on Empirical Methods in Natural Language Processing, Honolulu, Hawaii, pp. 887–896 (2008)
4. McClosky, D., Charniak, E., Johnson, M.: Automatic domain adaptation for parsing. In: Human Language Technologies: The 2010 Annual Conference of the North American Chapter of the ACL. HLT 2010, Stroudsburg, PA, USA, pp. 28–36 (2010)
5. Roux, J.L., Foster, J., Wagner, J., Samad, R., Kaljahi, Z., Bryl, A.: DUC-Paris13 systems for the SANCL 2012 shared task (2012)
6. Bulyko, I., Ostendorf, M.: Getting more mileage from web text sources for conversational speech language modeling using class-dependent mixtures. In: Proceedings of HLT-NAACL 2003, pp. 7–9 (2003)
7. Sarikaya, R., Gravano, A., Gao, Y.: Rapid language model development using external resources for new spoken dialog domains. In: International Congress of Acoustics. Speech, and Signal Processing (ICASSP), Philadelphia, PA, USA, pp. 573–576. IEEE Signal Processing Society (2005)
8. Rosenfeld, B., Feldman, R.: Using corpus statistics on entities to improve semi-supervised relation extraction from the web. In: Proceedings of the 45th Annual Meeting of the ACL, ACL 2007, 23–30 June, 2007, Prague, Czech Republic (2007)
9. Feldman, R., Rosenfeld, B., Bar-Haim, R., Fresko, M.: The stock sonar - sentiment analysis of stocks based on a hybrid approach. In: Proceedings of the Twenty-Third Conference on Innovative Applications of Artificial Intelligence, 9–11 August 2011, San Francisco, California, USA (2011)
10. Lee, H., Chang, A., Peirsman, Y., Chambers, N., Surdeanu, M., Jurafsky, D.: Deterministic coreference resolution based on entity-centric, precision-ranked rules. Comput. Linguist. **39**, 885–916 (2013)
11. Tetlock, P.C.: Giving content to investor sentiment: the role of media in the stock market. J. Fin. (2007)
12. Gentzkow, M., Shapiro, J.M.: What drives media slant? Evidence from U.S. daily newspapers. Econometrica **78**, 35–71 (2010)
13. Engelberg, J., Parsons, C.A.: The causal impact of media in financial markets. J. Fin. **66**, 67–97 (2011)

14. Devitt, A., Ahmad, K.: Sentiment polarity identification in financial news: a cohesion-based approach. In: Proceedings of the 45th Annual Meeting of the ACL, Prague, Czech Republic, pp. 984–991 (2007)
15. Haider, S.A., Mehrotra, R.: Corporate news classification and valence prediction: a supervised approach. In: Proceedings of the 2nd Workshop on Computational Approaches to Subjectivity and Sentiment Analysis. WASSA 2011, Portland, Oregon, pp. 175–181 (2011)
16. Chua, C., Milosavljevic, M., Curran, J.R.: A sentiment detection engine for internet stock message boards. In: Proceedings of the Australasian Language Technology Association Workshop 2009, Sydney, Australia, pp. 89–93 (2009)
17. Bar-Haim, R., Dinur, E., Feldman, R., Fresko, M., Goldstein, G.: Identifying and following expert investors in stock microblogs. In: Proceedings of the 2011 Conference on Empirical Methods in Natural Language Processing, Edinburgh, Scotland, UK, pp. 1310–1319 (2011)
18. Kogan, S., Levin, D., Routledge, B.R., Sagi, J.S., Smith, N.A.: Predicting risk from financial reports with regression. In: Proceedings of Human Language Technologies: The 2009 Annual Conference of the North American Chapter of the ACL. NAACL 2009, Stroudsburg, PA, USA, pp. 272–280 (2009)
19. Luss, R., d'Aspremont, A.: Predicting abnormal returns from news using text classification (2008). CoRR, abs/0809.2792
20. Wentland, W., Johannes Knopp, C.S., Hartung, M.: Building a multilingual lexical resource for named entity disambiguation, translation and transliteration. In: (ELRA), E.L.R.A., ed.: Proceedings of the Sixth International Language Resources and Evaluation (LREC 2008), Marrakech, Morocco (2008)
21. Manning, C.D., Surdeanu, M., Bauer, J., Finkel, J., Bethard, S.J., McClosky, D.: The Stanford CoreNLP natural language processing toolkit. In: Proceedings of 52nd Annual Meeting of the ACL, pp. 55–60 (2014)
22. Moschitti, A.: Making tree kernels practical for natural language learning. In: Proceedings of the 11th Conference of the European Chapter of the ACL (2006)
23. Fillmore, C.J.: Frame semantics and the nature of language. Ann. N.Y. Acad. Sci. **280**, 20–32 (1976)
24. Das, D., Smith, N.A.: Semi-supervised frame-semantic parsing for unknown predicates. In: Proceedings of the 49th Annual Meeting of the ACL. HLT 2011, Stroudsburg, PA, USA, pp. 1435–1444 (2011)
25. Das, D., Smith, N.A.: Graph-based lexicon expansion with sparsity-inducing penalties. In: HLT-NAACL, pp. 677–687 (2012)
26. Whissel, C.M.: The dictionary of affect in language. Emot. Theory Res. Experience **39**, 113–131 (1989)

Knowledge Engineering
and Ontology Development

A New Similarity Measure for an Ontology Matching System

Lorena Otero-Cerdeira[✉], Francisco J. Rodríguez-Martínez,
Tito Valencia-Requejo, and Alma Gómez-Rodríguez

Laboratorio de Informática Aplicada 2, Computer Science Department,
University of Vigo, Ourense, Spain
{locerdeira,franjrm,tvalencia,alma}@uvigo.es

Abstract. The purpose of this paper is twofold. It describes a new similarity measure which is applied in a new ontology matching algorithm, *OntoPhil* that exploits both the lexical and structural information of the input ontologies. The different steps of the algorithm as well as some clarifying examples are also provided. In addition, *OntoPhil* is also compared and evaluated using the datasets from well-known *Ontology Alignment Evaluation Initiative*.

Keywords: Ontology matching · Similarity measure · Lexical measure

1 Introduction

Guaranteeing the interoperability between new and legacy systems is a key point in systems' development. In this environment, the fields of *ontology* and *ontology matching* are fundamental lines of research to approach the issues related to systems interoperability.

An ontology provides a vocabulary to describe a domain of interest and a specification of the meaning of the terms in that context [1]. This definition typically includes a wide range of computer science objects, such as thesauri, XML schemas, directories, etc. These different objects identified as ontologies, belong to various fields of information systems such as web technologies [2], database integration [3], multi agent systems [4] or natural language processing. Therefore, ontologies have been used anywhere as a way of reducing the heterogeneity among open systems. Ontologies are expressed in an ontology language. Although there is a large variety of ontology languages [5], in this paper we focus on OWL [6–8].

Nevertheless, the sole use of ontologies does not solve the heterogeneity problem, since different parties in a cooperative system may adopt different ontologies. Thus, in such case, there is the need to perform an *ontology matching* process to guarantee the interoperability. Ontology matching is a solution to semantic heterogeneity since it finds correspondences between semantically related entities of the ontologies [9]. This set of correspondences is known as an *alignment*. A more formal definition of an alignment was provided in [10].

© Springer International Publishing Switzerland 2015
A. Fred et al. (Eds.): IC3K 2014, CCIS 553, pp. 257–272, 2015.
DOI: 10.1007/978-3-319-25840-9_17

In this paper, we describe our combinational approach to ontology matching, the *OntoPhil* algorithm.

In literature there are general approaches to ontology matching that also exploit the lexical and structural features of the ontologies to align. A detailed description of such algorithms, related to the one proposed in this work can be found in [11].

The rest of the paper is organized as follows. Section 2 we describe our algorithm in detail, and in Sect. 3 we provide the evaluation and comparative results of the performance of our algorithm. Finally, Sect. 4 includes the main conclusions and remarks.

2   A New Algorithm for Ontology Matching

This section presents our approach to ontology matching. It relies on the exploitation of some initial correspondences or *binding points* that connect both ontologies.

The process designed takes a couple of ontologies as input and consecutively applies, first some lexical matchers and later some structural matchers to obtain the final result. Therefore the matching process followed by the algorithm is *sequential* as defined by Euzenat and Shvaiko in [1].

We aim at discovering several *binding points* between the input ontologies by using some new lexical matchers. Thereafter, taking these *binding points* as pivots, an structural matcher is applied to the initial *binding points*. This matcher exploits particular features and properties of the ontologies to discover new *binding points*. Finally, by selecting and refining these *binding points*, we obtain the alignment between the input ontologies. This final output of *OntoPhil* is composed by the entities matched, the relation that holds between them and a confidence value in the interval [0, 1].

The structural matchers used in the second phase are defined as independent modules so that the output of a structural matcher is not necessary for the run of any of the others and it does not affect their outputs. The fact that the different structural matchers are independent allows selecting just the most suitable ones for each task.

In the following sections we present a detailed view of each one of the steps and sub-steps of the algorithm.

2.1   *Step 1:* Obtaining Initial Binding Points

Obtaining the initial *binding points* is a crucial part of the matching process since these binding points are the basis on which the rest of the algorithm is built. To identify these initial *binding points*, we use two different lexical matchers.

These lexical matchers use a new distance measure which combines several string-based distance measures, with language-based methods. The language-based methods used, involve the preprocessing of the words and the exploitation of external resources. In our case we use a WordNet matcher that exploits such database as external resource to solve the synonym problem.

The quality of the *initial binding points* is crucial as it affects the goodness of the final output, therefore in this step the amount of correspondences obtained is sacrificed for their quality by setting a high threshold for pruning the results of this step, i.e., the initial binding points.

Sub-step 1.1: **Lexical Matchers.** This lexical matching phase will produce as result two distinct lexical matrices since classes and properties are separately aligned by means of a *ClassesLexicalMatcher (CLM)* and a *PropertiesLexical-Matcher (PLM)* respectively.

The $CLM = \{LexicalValue(c, c')\}$ denotes the matrix containing the set of similarity values between each $c \in C$, $c' \in C'$ and $PLM = \{LexicalValue(p, p')\}$ denotes the matrix containing the set of similarity values between each $p \in P$, $p' \in P'$ being, C and C' be the set of classes from ontologies o and o', and P and P' the set of properties from ontologies o and o' respectively.

Both these matching approaches use string and language-based methods to identify similar entities in the given ontologies. To do so a new distance measure is introduced.

This new measure is defined according to the following considerations.
[**Step 1**]. If the lexical similarity between two strings s_1 and s_2 is to be computed, the first step is to remove the numbers that may appear in the strings. Numbers are omitted since they would only interfere in the final result and would not provide relevant matching information.

[**Step 2**]. After doing so, each string is tokenized by using as delimiters any non-alphabetical character, blank spaces and the uppercasing - lowercasing changes in the word. This way, for each one of the original strings a bag of words is obtained, s_1' and s_2' respectively.

[**Step 3**]. These bags of words are then compared with the following procedure.

[**Step 3.1**]. First, the words shared by the two bags are removed, hence obtaining s_1'' and s_2''. In this step, in the *PropertiesLexicalMatcher (PLM)*, the *stop words* are also dismissed.

[**Step 3.2**]. If after doing so both bags are empty then the similarity measure with the input strings is 1.0. Otherwise, all the words left in the first bag s_1'' are compared with the words left in the second bag s_2'' by using the **Jaro-Winkler Distance** [12], considering the **Levenshtein Distance** [12] as a complementary measure and exploiting WordNet to improve the results obtained by these similarity measures. For each pair of bags compared a temporary matrix (*pairEvaluation*) is built which stores the similarity results obtained from the pairwise comparison of the words left in the bags. In the following steps, we describe the procedure applied for the words left in bags s_1'' and s_2''.

[Step 3.2.1]. Let $a \in s_1''$ and $b \in s_2''$. If the WordNet matcher for words (a, b) indicates that there is an overlapping in the synonyms sets, obtained from both words, then the first similarity result, sim_1, for these words (1.0) is stored in $pairEvaluation$ (s_1'', s_2''). Otherwise no WordNet similarity value is stored.

In any case, it continues to *Step 3.2.2*.

$$synonyms(a) \cap synonyms(b) \neq \emptyset \Rightarrow sim_1(a, b) = 1.0 \Rightarrow$$
$$put(pairEvaluation(s_1'', s_2''), sim_1(a, b)) \tag{1}$$

[Step 3.2.2]. To compute the second similarity value, sim_2, it starts by calculating the Jaro-Winkler distance of the words (a, b). If *Jaro-Winkler(a,b)* returns a number greater than α, and for these words, the Levenshtein distance measure, *Levenshtein(a, b)*, returns a number lower or equal than β, then the combined similarity value of these words, sim_2, is 1.0 and it is stored in *pairEvaluation*. Otherwise it continues to *Step 3.2.3*. These thresholds, α and β, have been empirically set to 0.90 and 1.0 respectively.

$$(JaroWinkler(a, b) > \alpha) \wedge (Levenshtein(a, b) < \beta) \Rightarrow$$
$$sim_2(a, b) = 1.0 \Rightarrow put(pairEvaluation(s_1'', s_2''), sim_2(a, b)) \tag{2}$$

[Step 3.2.3]. In case Levenshtein distance measure indicates that the shortest word must be completely modified to be matched to the second one, this causes a proportional forfeit in the combined result. This penalized value is stored in *pairEvaluation* as similarity value sim_2 for (a, b).

$$Levenshtein(a, b) \geq minLength(a, b) \Rightarrow$$
$$sim_2(a, b) = JaroWinkler(a, b) * \left(1 - \frac{Levenshtein(a, b)}{maxLength(a, b)}\right) \tag{3}$$
$$\Rightarrow put(pairEvaluation(s_1'', s_2''), sim_2(a, b))$$

Otherwise, the result of the Jaro-Winkler distance measure is stored in *pairEvaluation* as similarity result, sim_2, for that pair of words.

$$Levenshtein(a, b) < minLength(a, b) \Rightarrow sim_2(a, b) = JaroWinkler(a, b)$$
$$\Rightarrow put(pairEvaluation(s_1'', s_2''), sim_2(a, b))$$
$$\tag{4}$$

[Step 4]. Once every possible pair of words of the bags is assigned a value, the final result for the bags is computed. To do so, all partial results stored in the temporary matrix *pairEvaluation* are added up considering an improvement factor ϕ which is used to strengthen the similarity of the bags that share several words in common.

$$bagEvaluation(s_1, s_2) = \frac{\left(\frac{\sum pairEvaluation(s_1'', s_2'')}{length(s_1'') * length(s_2'')}\right) + repeatedWords * \phi}{1 + repeatedWords} \quad (5)$$

[Step 5]. The original strings, s_1 and s_2 are also compared using the Jaro-Winkler and Levenshtein distances, as described in (Step 3.2.2) and (Step 3.2.3). The obtained value is stored in $stringEvaluation(s_1, s_2)$.

[Step 6]. As final result, the best score between the string comparison (Step 5) and the bags comparison (Step 4) is returned.

$$LexicalValue(s_1, s_2) = max[bagEvaluation(s_1, s_2), stringEvaluation(s_1, s_2)] \quad (6)$$

Next, we revisit this process by means of an example.

EXAMPLE:
Let s_1 and s_2 be, s_1 = "isCorrespondingAuthor" and s_2 = "is_author_of_paper" respectively.

[Step 1] # Remove numbers
s_1 = "isCorrespondingAuthor"
s_2 =" is_author_of_paper"

[Step 2] # Create bags using intrinsic language-based methods
s_1' = [is, corresponding, author]
s_2' = [is, author, of, paper]

[Step 3] # Bag comparison procedure
[Step 3.1] # Removing shared words & Stopwords
The stopwords are removed from both bags. So the bags remain as follows:
s_1'' = [corresponding]
s_2'' = [paper]

[Step 3.2] # Bags empty
The bags are not empty as each one of them still retains a word, i.e., each one of them has a cardinality of 1. The remaining words are pairwise compared.
$\#s_1'' = 1$
$\#s_2'' = 1$

[Step 3.2.1] # WordNet synonyms
These words do not have any common synonyms, therefore no WordNet value is added to $pairEvaluation$.

$synonyms(corresponding) \cap synonyms(paper) = \emptyset$

[Step 3.2.2] # Jaro Winkler & Levenshtein

$$\left.\begin{array}{l} JaroWinkler(corresponding, paper) = 0.5 \\ \Rightarrow JaroWinkler(corresponding, paper) < \alpha \\ Levenshtein(corresponding, paper) = 12 \\ \Rightarrow Levenshtein(corresponding, paper) < \beta \\ \quad sim(corresponding,\ paper) \neq 1.0 \end{array}\right\} \Rightarrow$$

As these conditions were not satisfied, it continues to *Step 3.2.3*.

[Step 3.2.3] # Check forfeit
No forfeit is applied so the result of JaroWinkler is stored in *pair Evaluation*.
Levenshtein (corresponding, paper) > 5 (forfeit applied)
$forfeit = 0.5 * (1 - 12/13) = 0.038$
$put(pairEvaluation(s_1'', s_2''), forfeit)$

[Step 4] # Final bags result
Since the word "author" is common to both bags, the improvement factor ϕ is applied.
$bagEvaluation(s_1, s_2) = (0.038 + 2 * \phi)/3 = 0.68$

[Step 5] # Check original strings
$JaroWinkler(isCorrespondingAuthor, is\_author\_of\_paper) = 0.56$
$Levenshtein(isCorrespondingAuthor, is\_author\_of\_paper) = 17$
The procedures described in *Step 3.2.2* and *Step 3.2.3* are now applied to the original strings.

Step 3.2.2
$$\left.\begin{array}{l} JaroWinkler(isCorrespondingAuthor, is\_author\_of\_paper) = 0.56 \\ \Rightarrow JaroWinkler(isCorrespondingAuthor, is\_author\_of\_paper) < \alpha \\ Levenshtein(isCorrespondingAuthor, is\_author\_of\_paper) = 17 \\ \Rightarrow Levenshtein(isCorrespondingAuthor, is\_author\_of\_paper) < \beta \\ \quad sim(isCorrespondingAuthor, is\_author\_of\_paper) \neq 1.0 \end{array}\right\} \Rightarrow$$
Step 3.2.3
$Levenshtein(ConferenceDinner, Conference\_Banquet) < 18 \ \} \Rightarrow$
$\quad put(stringEvaluation, 0.56)$

[Step 6] # Obtain final value
LexicalValue(isCorrespondingAuthor,is_author_of_paper) = max[0.68, 0.56]
= 0.68

Considering these results the final evaluation of the similarity for the strings "*isCorrespondingAuthor*" and "*is_author_of_paper*" is 0.68.

We compute separately the similarity of the different types of entities from the input ontologies, namely, classes, object properties and data properties. By following this procedure we obtain a lexical similarity matrix for each type of entity. The lexical similarity value obtained for each pair of entities is used as confidence value for such entities.

Sub-step 1.2: **Combine and Select Results.** The matrices resulting from the previous steps are combined in a new structure, where all the candidate correspondences generated by the properties lexical matcher and the classes lexical matcher, are joined in a set.

This set is composed by tuples of the form: $(e_1, e_2, LexicalValue(e_1, e_2), WordNetValue(e_1, e_2))$, where e_1 and e_2 are the entities linked (classes or properties), $LexicalValue(e_1, e_2)$ is the value obtained from running the *Classes Lexical Matcher*(e_1, e_2) or the *Properties Lexical Matcher*(e_1, e_2) and *Word Net Value*(e_1, e_2) is the value obtained from running the *WordNetMatcher* (e_1, e_2).

Next, all the initial *binding points* or candidate correspondences are sifted out. The sifting out is fundamental for the algorithm since it reduces the number of *binding points* that are used as the starting point for the next step. For this purpose, only the most accurate points are used, so a sifted set, (*Sifted_CS*), containing just the elements with a lexical value of 1.0 is constructed.

The better these initial binding points the higher the chances that the correspondences obtained in *Step 2* (see Sect. 2.2) are valid ones, therefore only those binding points with the highest lexical similarity (1.0) are chosen to be the start point of the *set expansion procedures*.

These expansion procedures aim at discovering new *binding points* between the two input ontologies by exploiting structural features of the ontologies.

2.2   *Step 2:* Discovering New Binding Points

As stated before, the sifted set (*Sifted_CS*), obtained in the previous step, becomes the *base set* on which the *expansion procedures* are built. The candidate *binding points* obtained in this step, can link either properties or classes and therefore there are *properties candidate correspondences* and *classes candidate correspondences*.

Depending on the structural feature that is exploited in each procedure, there is a different likelihood that the discovered candidate *binding points* are promising, therefore some of these procedures will directly update the *base set* by modifying or inserting new *binding points* while others will insert them in a *candidate set*.

Each one of the new discovered *binding points* is placed in a bag that identifies the procedure and sub-procedure that led to its discovery. The procedures are sequentially applied, and each one exploits a different feature of the ontologies. In case a binding point is reached by several procedures, it is saved in all the corresponding bags. Initially those class and property candidate correspondences which are *binding points* in the *base set*, are put in the bags *CC_BASE_SET* and *CP_BASE_SET* respectively.

The different procedures applied are detailed in the following subsections.

Sub-step 2.1: Properties Inverse Procedure. The aim of this procedure is to discover new property candidate correspondences. For every properties correspondence in the *base set*, the procedure retrieves its *inverse properties correspondence*. This is only feasible if the properties from the original correspondence have defined an inverse property using the construct $owl : inverseOf$.

$$\forall(q_1, q_2, PLM(q_1, q_2), WNM(q_1, q_2)) \in BASE\_SET \Longleftrightarrow$$
$$\exists[((p_1, p_2, PLM(p_1, p_2), WNM(p_1, p_2)) \in BASE\_SET) \land (\exists q_1 \equiv \neg p_1) \land (\exists q_2 \equiv \neg p_2)] \tag{7}$$

If the inverse correspondence already exists in the *base set* the counter of its occurrences is increased, otherwise the new correspondence is inserted in the *CP_BASE_SET_INVERSE* bag.

Sub-step 2.2: Properties Domain Range Procedure. This procedure allows the identification of new class candidate correspondences. The first step is to retrieve all the property candidate correspondences in the *base set*. Thereafter the domain and range for each property in the correspondence is recovered from their corresponding ontologies.

Shall we consider first the domain sets. If both of them have only one class each, then these two classes constitute a new class candidate correspondence (see Eq. (8)). In case this new correspondence already exists in the *base set* or *candidate set* the number of its occurrences is increased, otherwise it is inserted in the *candidate set* in the bag *CC_DIRECT_DERIVED*.

$$\exists c_1 \in dom(p_1) \land \exists c_2 \in dom(p_2) \land \#dom(p_1) = \#dom(p_2) = 1$$
$$\Rightarrow (c_1, c_2, CLM(c_1, c_2), WNM(c_1, c_2)) \in BASE\_SET \tag{8}$$

If the cardinality of any of the domain sets is higher than 1, the *superclasses approach*, described next, is followed. The correspondences discovered with this procedure are inserted in the bag *CC_DIRECT_DERIVED_WITH_SET*.

For each of the classes in both domain sets, its superclasses are retrieved recursively until the higher level in the ontology hierarchy is reached. This way, for every class in every domain a temporary set containing all its superclasses is obtained. If there is a common superclass to several classes in the same domain set, then the intermediate classes are dismissed and only the initial class and the common superclass are considered. All the selected classes are integrated in the same set, so we will have only one final set associated to each domain. These sets are represented as $sup(dom(p_1))$ and $sup(dom(p_2))$ respectively, in Eq. (9). The classes from these final sets are combined to obtain new candidate correspondences.

$$sup(dom(p_1)) \times sup(dom(p_2)) = \{(c_1, c_2, CLM(c_1, c_2), WNM(c_1, c_2))\}$$
$$where \tag{9}$$
$$(c_1 \in sup(dom(p_1)) \land c_2 \in sup(dom(p_2)))$$

If any of these new correspondences already exists in the *base set* or *candidate set* their number of occurrences are properly modified. For the rest of the new correspondences, their lexical and WordNet values are retrieved from the matrices resulting from the lexical and WordNet matcher respectively, to create a new tuple. If the lexical value surpasses the threshold δ then the tuple is inserted in the *candidate set*, otherwise it is dismissed (see Eq. (10)). The δ threshold limits the amount of new correspondences that are inserted in the set by dismissing the less promising ones. We have experimentally determined, in previous works, that a value of 0.9 for δ threshold achieves the best results.

$$\forall t \in \{(c_1, c_2, CLM(c_1, c_2), WNM(c_1, c_2))\}, t \in CANDIDATE\_SET \iff t.s^{lex} > \delta \tag{10}$$

This procedure that we have just described for the domain sets is also applied for the range sets.

Sub-step 2.3: **Classes Properties Procedure.** This procedure discovers not only class candidate correspondences but also property candidate correspondences. To do so, the first step is to retrieve all the existing class candidate correspondences from the *base set* and then, for every pair of classes, their *correlated properties set* is put together.

The set of correlated properties is the result of the cartesian product of the properties from the source ontology whose domain or range includes the first class of the explored correspondence, with the properties from the target ontology, whose domain or range includes the second class in that correspondence.

To choose a pair of properties for the correlated set, if the first class in the explored pair belongs to the domain of the first property in the candidate pair, then the second property in the candidate pair must also have the second class in the explored pair as part of its domain, otherwise the pair of properties is dismissed (see Eq. (11)).

$$Correlated\_Properties(c_1, c_2) = \{p_1\} \times \{p_2\}$$
$$where \tag{11}$$
$$(c_1 \in dom(p_1) \wedge c_2 \in dom(p_2)) \vee (c_1 \in ran(p_1) \wedge c_2 \in ran(p_2))$$

Once all the correlated properties are identified, the procedure continues assessing their domains and ranges to determine whether there can be found new correspondences or not.

For each pair of properties belonging to the correlated properties set, their domains and ranges are retrieved. If a pair of properties is inserted in the correlated properties set because their respective domains has a pair of classes already aligned in the *base set*, then the domains receive the *aligned sets sub-procedure*, and the ranges receive the *non-aligned set sub-procedure*.

- ## Aligned Set Sub-procedure

The initial classes whose correlated properties are under assessment are removed from the aligned sets, since these classes are already part of the *base set*. In case that, after doing so, there is only one class left in each one of the aligned sets, then a new class candidate correspondence has been identified. If this new identified correspondence already exists in the *base set* or in the *candidate set* its number of occurrences is accordingly modified. Otherwise this new correspondence is added to the *candidate set*. This correspondence of classes is inserted as a new class candidate correspondence in the bag *CC _ SOURCE _ INC _ WITHOUT _ SET*.

If the cardinality of the sets is bigger than 1 then the superclasses approach previously described in Sect. 2.2 is followed, although the new correspondences are placed in the bag *CC _ SOURCE _ INC _ WITH _ SET*.

- ## Non-aligned Set Sub-procedure

In the non-aligned sets, the first thing to check is the cardinality, if both sets have just one class then a new class candidate correspondence is created and put in the bag *CC _ DIRECT _CLASSES*. This new correspondence is confronted separately with the *base set* and *candidate set*.

If the class candidate correspondence already exists in the *base set*, then its number of occurrences is increased and the properties that link these classes become a new property candidate correspondence. This new correspondence is evaluated against the *base set* too in order to check for previous occurrences. Its inverse candidate correspondence is also assessed.

These new property candidate correspondences are respectively set in bags *CP _ DIRECT _ CLASSES* and *CP _ DIRECT _ CLASSES _ INVERSE*. If the class candidate correspondence is identified as new, then it is inserted in the *base set*.

If the class candidate correspondence exists in the *candidate set* its number of occurrences is increased, and the corresponding properties are evaluated against the *candidate set* to become a new property candidate correspondence. Its inverse correspondence is also assessed. If the class candidate correspondence is identified as new it is placed in the bag *CC _ PROPERTIES _ WITH-OUT _ SET* and then it is inserted in the *candidate set* as well as the new property candidate correspondences identified in this step. The property candidate correspondence and its inverse, in case it exists, are placed in *CP _ DIRECT _ CLASSES* and *CP _ DIRECT _ CLASSES _ INVERSE* respectively.

In case the non-aligned sets cardinality is different from 1, then the superclasses approach is followed, as described in Sect. 2.2. In this case the bag for the new class candidate correspondences is *CC _ PROPERTIES _ WITH _ SET* and for the new property candidate correspondences *CP _ DIRECT _ CLASSES _ WITH _ SET* and *CP _ DIRECT _ CLASSES _ WITH _ SET _ INVERSE*.

Sub-step 2.4: **Classes Family Procedure.** The Classes Family Procedure is the family approach to identifying new class candidate correspondences. Following this approach class candidate correspondences from the *base set* are retrieved and their familiar relations are exploited.

For each one of the classes in the class candidate correspondences, its superclasses, subclasses and sibling classes are recovered, identified in Eq. (12) as $sup(class)$, $sub(class)$ and $sib(class)$ respectively. Then, a cartesian product is applied between the two sets of classes recovered for each correspondences, dividing the results into the three identified levels. The new identified pairs are compared with the existing ones in both *base set* and *candidate set*.

$$FamiliarCorrespondences = \{(sup(c_1) \times sup(c_2)) \cup \qquad (12)$$
$$(sub(c_1) \times sub(c_2)) \cup (sib(c_1) \times sib(c_2))\}$$

As it has been done in previous steps, if a correspondence already exists in any set (*base set* or *candidate set*) its number of occurrences is increased, otherwise these new correspondence are inserted in the *candidate set* provided that the lexical valued corresponding to these new tuples surpass the δ threshold. These new tuples are placed in the bag *CC_FAMILY*.

By applying the procedures described in Sect. 2.2 (Sub-steps 2.1–2.4) new *binding points* are discovered and classified in different types of bags.

2.3   Selecting Binding Points

Once all iterations have finished the results are combined and the selecting process begins. This selecting process is essential since it is a way of dismissing false correspondences and therefore of defining the final output of the algorithm. For this process various restrictions have been defined which treat differently class candidate correspondences and property candidate correspondences. These restrictions are based in the idea that the different procedures exploit different features of the ontologies and therefore they outcome correspondences with different levels of accuracy. Hence, the location of the correspondences in different bags facilitates their selection, in order to choose for the final output the best possible ones.

To identify the best restriction rules we have performed an study in the overlapping of the different bags using different test ontologies. The idea behind this study is the following, starting from the representation of the overlapping of the bags in different test cases and given the desired output for the algorithm, it is possible to infer the rules that would allow the identification of such desired output, if doing this with enough training cases, then we obtain a stable set of restriction rules.

In Fig. 1 a few examples of Venn diagrams showing the overlapping of the different bags on training ontologies is presented.

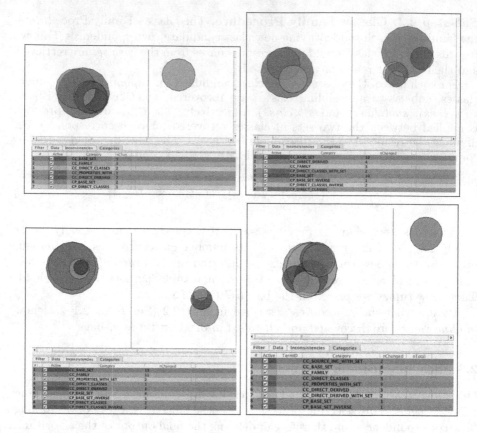

Fig. 1. Venn diagrams showing the overlapping of the different bags.

From the analysis of such diagrams, we have defined a set of restriction rules. As an example, some of the simpler restriction rules taken into account for the algorithm are:

- *Retrieve those correspondences that belong to CC_BASE_SET.*
- *Retrieve those correspondences that belong to CC_DIRECT_DERIVED which have at least α occurrences.*
- *Retrieve those correspondences that belong to CP_DIRECT_CLASSES with at least β occurrences and to CP_DIRECT_CLASSES_WITH_SET with a γ confidence threshold.*
- *For those property correspondences that share the same entity in the source ontology, retrieve those whose confidence value is higher.*
- *Dismiss all classes correspondences with a confidence level lower than δ and all properties correspondences with a confidence level lower than ϵ.*

When the selecting of the binding points has finished, the final output is produced, i.e., the set of correspondences identified between the input ontologies.

For each one of these correspondences, the matched entities are provided together with the relation that holds between them, and a confidence value in $[0, 1]$, obtained from the lexical value computed in *Step 1* Sect. 2.1.

3 Evaluation

The goal of any algorithm for ontology matching is to generate an alignment that discovers the correspondences, and only the correct ones, between two input ontologies. However, the goodness of this alignment, and of the matching process in general, can be evaluated using different *compliance measures*.

These measures evaluate the degree of compliance of matcher alignments with respect to a *reference alignment* [13], i.e., the correctness of the correspondences in a alignment according to a reference matching provided by the human interpretation of the meaning of the input ontologies.

A *reference alignment* is a standard to which the result outputted by the matcher is compared. Even if there are some approaches to the semi-automatic construction of reference alignments [14], most of the reference alignments used in ontology matching evaluation have been defined by users. This fact leads to problems pointed out in [15], since different users may provide significantly different reference alignments in terms of completeness, interpretation and accuracy of certain correspondences, etc. It is then worth noticing that these reference alignments used to evaluate the matchers, may not be consensual, may be partial or may even be useless for the task.

Taking the reference alignment as desired output of an ontology matching algorithm, the most widespread measures used to evaluate the compliance such ontology matching algorithms are the standard information retrieval metrics of *Precision, Recall and F-measure* [16].

The *Precision* measures the ratio of correctly found correspondences over the total number of returned correspondences, which in logical terms reflects the degree of correctness of the alignment computed by the algorithm.

The *Recall* measures the ratio of correctly found correspondences over the total number of expected correspondences which in logical terms measures the degree of completeness of the alignment obtained by the algorithm.

Even though precision and recall are widely used and accepted measures, in some occasions it may be preferable having a single measure to compare different systems or algorithms. Moreover, systems are often not comparable based solely on precision or recall, because the one which has a higher recall may have a lower precision and vice versa [1]. Therefore, the *F-measure* was introduced.

$$f - measure = \frac{precision * recall}{(1 - \alpha) * precision + \alpha * recall}$$

The f-measure combines the precision and recall values by computing their harmonic mean. To do so, the α parameter is set to 0.5, so the outputted value shows no bias towards precision or recall. To compute the f-measure the arithmetic mean is not used as typically a system with 50 % precision and 50 % recall is considered better than another one with 80 % precision and 20 % recall.

We have evaluated the accuracy of our algorithm using these measures and comparing our results with those provided by the algorithm of Akbari and Fathian [17] due to the similarities that our algorithm's structure shares with theirs because both algorithms use lexical and structural matchers applied sequentially to discover the correspondences between the ontologies. Akbari and Fathian's algorithm was tested using the benchmark test set provided in the Ontology Alignment Evaluation Initiative 2008 (OAEI-08) [18].

This is a well known benchmark series of tests that has been used for several years. This allows the comparison with other systems since 2004. This benchmark is built around a seed ontology and variations of it [19] and its purpose is to provide a stable and detailed picture of the algorithms. These tests were organized into *simple tests (1xx)*, *systematic tests (2xx)* and *real-life ontologies (3xx)*. Recently the structure of the benchmark was changed and *real-life ontologies* were removed.

Table 1. Average performance of the algorithm proposed by Akbari and Fathian on the OAEI-08 benchmark test suite.

	1xx	2xx	3xx	Average
Precision	0.98	0.78	0.87	0.88
Recall	0.95	0.74	0.84	0.85
F-measure	0.96	0.75	0.85	0.86

Table 2. Average performance of our algorithm on the OAEI-08 benchmark test suite.

	1xx	2xx	3xx	Average
Precision	0.97	0.96	0.92	0.95
Recall	1	0.71	0.82	0.85
F-measure	0.99	0.77	0.86	0.88

From the results presented in Tables 1 and 2, our algorithm shows better *precision* and *f-measure* values than the system proposed by Akbari and Fathian.

It is specially important for us the results obtained in the set of tests *3xx* (real-life ontologies), since we aim at integrating our system in a real-world application.

Besides, we have compared the results of our algorithm with those provided by the algorithms participating in the Ontology Alignment Evaluation Initiative 2012 (OAEI-12), as shown in Table 3.

From these results we can outline that the proposed algorithm has a better average behavior than most of the competing systems, since even if our approach has not the highest value in any of the three measures *precision, recall and f-measure*, it certainly has the most balanced ones.

4   Conclusions

In this paper we have presented a novel ontology matching algorithm that finds correspondences among entities of input ontologies based on their lexical and

Table 3. Results obtained by the participants in the OAEI-12 compared with our approach.

	edna	AROMA	ASE	AUTOMSV2	GOMMA	Hertuda
P	0.35	0.98	0.49	0.97	0.75	0.90
R	0.41	0.77	0.51	0.69	0.67	0.68
F	0.50	0.64	0.54	0.54	0.61	0.54
	HotMatch	LogMap	LogMapLt	MaasMatch	MapSSS	MEDLEY
P	0.96	0.73	0.71	0.54	0.99	0.60
R	0.66	0.56	0.59	0.56	0.87	0.54
F	0.50	0.46	0.50	0.57	0.77	0.50
	Optima	ServOMap	ServOMapLt	WeSeE	WikiMatch	YAM++
P	0.89	0.88	1.0	0.99	0.74	0.98
R	0.63	0.58	0.33	0.69	0.62	0.83
F	0.49	0.43	0.20	0.53	0.54	0.72
	OntoPhil					
P	0.95					
R	0.83					
F	0.79					

structural information. The algorithm proposed relies on the exploitation of some initial correspondences or *binding points* that connect one ontology to the other. This is an adaptive way of matching two ontologies which allows the identification of new correspondences by exploiting the particular features of the matched ontologies.

Likewise, we have introduced a new lexical measure that determines the lexical similarity among entities by using the terminological information available for each pair of entities. For this lexical similarity measure a full and detailed example is provided to show how the similarity is computed.

The comparison of our algorithm to other similar ones, reflects that it has good and balanced results that encourage the work on this research line.

In spite of the promising start of this line work, there are still several steps that must be taken before it is fully functional. First, the algorithm will continue to be enhanced, for instance, to improve the calculation of the initial binding points, to refine of the restriction rules so the amount of false positives retrieved in the alignment are kept to a minimum or to include other features such as multi-lingual support and instance matching. These improvements are being developed to be included in the algorithm as a way to enhance the results of *OntoPhil*.

References

1. Euzenat, J., Shvaiko, P.: Ontology Matching. Springer, Heidelberg (2007)
2. Antoniou, G., van Harmelen, F.: Semantic Web Primer. The MIT Press, Cambridge (2004)
3. Doan, A., Halevy, A.Y.: Semantic integration research in the database community: a brief survey. Am. Assoc. AI **26**, 83–94 (2005)
4. van Aart, C., Pels, R., Caire, G., Bergenti, F.: Creating and using ontologies in agent communication. In: Proceedings of the Workshop on Ontologies in Agent Systems (AOAS 2002), pp. 1–8 (2002)
5. Suárez-Figuero, C., García-Castro, R., Villazón-Terrazas, B., Gómez-Pérez, A.: Essentials in ontology engineering: methodologies, languages, and tools. In: Proceedings of the 2nd Workshop organized by the EEBuildings Data Models community. CIB conference W078-W012, pp. 9–21 (2011)
6. W3C: OWL 2: Web Ontology Language (2013)
7. W3C: OWL: Web Ontology Language (2013)
8. Cuenca-Grau, B., Horrocks, I., Motik, B., Parsia, B., Patel-Schneider, P., Sattler, U.: OWL 2: the next step for OWL. J. Web Semant. Sci. Serv. Agents World Wide Web **6**, 309–332 (2008)
9. Shvaiko, P., Euzenat, J.: Ontology matching: state of the art and future challenges. IEEE Trans. Knowl. Softw. Eng. **25**(1), 158–176 (2013)
10. Ehrig, M., Euzenat, J.: Relaxed precision and recall for ontology matching. Integrating Ontol. **156**, 8 (2005)
11. Otero-Cerdeira, L., Rodríguez-Martínez, F.J., Gómez-Rodríguez, A.: Definition of an ontology matching algorithm for context integration in smart cities. Sensors **14**, 23581–23619 (2014)
12. Cohen, W.W., Ravikumar, P.D., Fienberg, S.E.: A comparison of string distance metrics for name-matching tasks. In: Proceedings of International Joint Conference on A.I Workshop on Information Integration, pp. 73–78 (2003)
13. Euzenat, J., Meilicke, C., Stuckenschmidt, H., Shvaiko, P., Trojahn, C.: Ontology alignment evaluation initiative: six years of experience. In: Spaccapietra, S. (ed.) Journal on Data Semantics XV. LNCS, vol. 6720, pp. 158–192. Springer, Heidelberg (2011)
14. Giunchiglia, F., Yatskevich, M., Avesani, P., Shvaiko, P.: A large scale dataset for the evaluation of ontology matching systems. Knowl. Eng. Rev. **23**, 1–22 (2008)
15. Tordai, A., van Ossenbruggen, J., Schreiber, G., Wielinga, B.: Let's agree to disagree: on the evaluation of vocabulary alignment. In: Proceedings of the Sixth International Conference on Knowledge Capture (K-CAP 2011), pp. 65–72 (2011)
16. Manning, C.D., Raghavan, P., Schütze, H.: Introduction to Information Retrieval. Cambridge University Press, New York (2008)
17. Akbari, I., Fathian, M.: A novel algorithm for ontology matching. J. Inf. Sci. **36**, 12 (2010)
18. Caracciolo, C., Euzenat, J., Hollink, L., Ichise, R., Isaac, A., Malaisé, V., Meilicke, C., Pane, J., Shvaiko, P., Stuckenschmidt, H., Šváb-Zamazal, O., Svátek, V.: Results of the ontology alignment evaluation initiative 2008. In: Ontology Matching Workshop (2009)
19. Aguirre, J.L., Eckert, K., Euzenat, J., Ferrara, A., van Hage, W. R., Hollink, L., Meilicke, C., Nikolov, A., Ritze, D., Scharffe, F., Shvaiko, P., Svab-Zamazal, O., Trojahn, C., Bernardo, E. J. R., Grau, C., Zapilko, B.: Results of the ontology alignment evaluation initiative 2012. In: The 7th International Workshop on Ontology Matching (2013)

Techniques for Merging Upper Ontologies

Carmen Chui[✉] and Michael Grüninger

Department of Mechanical and Industrial Engineering, University of Toronto,
Toronto, ON M5S 3G8, Canada
{cchui,gruninger}@mie.utoronto.ca

Abstract. In this paper, we examine techniques used to merge the upper ontologies of the Process Specification Language (PSL) and the Descriptive Ontology for Linguistic and Cognitive Engineering (DOLCE) ontologies. In particular, we focus on the parts of these two ontologies relevant to the commonsense notion of participation as a relation between objects, activities, and time. We discuss the obstacles faced to formalize the relationships between these ontologies and provide an overview of the methodology undertaken to bridge the ontologies together. New ontologies are introduced to bring the PSL and DOLCE ontologies together to allow us to specify the mappings between them. We illustrate how ontology verification is used to show faithful interpretations between the two upper ontologies. We also explore applications of ontology transfer between a mathematical ontology and an upper ontology as an additional means of specifying bridges between ontologies.

Keywords: Bridge ontology · COLORE · DOLCE · First-order logic · Ontology merging · Ontology repositories · Ontology verification · Participation · PSL · Upper ontologies

1 Introduction

In order to understand how two ontologies are related to each other, there is a need to explicitly identify the potential relationships between them, and we cannot understand such relationships without analyzing the axioms of the ontologies. In this paper, we explore the relationships between two upper ontologies – the Process Specification Language (PSL) and the Descriptive Ontology for Linguistic and Cognitive Engineering (DOLCE). Our objective is to determine whether one upper ontology can be interpreted in the other by using techniques from ontology verification to examine the model-theoretic properties of both ontologies. In particular, we examine how the theories found in DOLCE can be mapped to the PSL ontology by using existing ontologies found in the COmmon Logic Ontology REpository (COLORE), and then discuss the steps needed to *bridge* these ontologies together.

With respect to ontology mapping, the research community is often interested in determining whether two *contextually equivalent* ontologies contain the same, or similar, axioms and descriptions of concepts. The intent of ontology mapping

© Springer International Publishing Switzerland 2015
A. Fred et al. (Eds.): IC3K 2014, CCIS 553, pp. 273–292, 2015.
DOI: 10.1007/978-3-319-25840-9_18

is to make semantic matches between the ontologies and to utilize these matches to aid us in reasoning tasks [12]. Ontology merging allows the creation of a new ontology from two, possibly overlapping, ontologies [1,7]. Additionally, in this work, we utilize ontology mapping along with ontology merging to identify similarities and conflicts between the ontologies [1].

Our approach to ontology mapping is based on ontology verification, which is concerned with the relationship between the intended models of an ontology and the models of the axiomatization of the ontology. In particular, the models of an ontology are characterized up to isomorphism and then shown to be equivalent to the intended models of the ontology. The objective is the construction of the models of one ontology from the models of another ontology by exploiting the metatheoretic relationships (such as faithful interpretation and definable equivalence) between these ontologies and existing theories in COLORE.

In this paper[1], we begin by examining the fundamental ontological commitments of the DOLCE and PSL ontologies with respect to the notion of participation. We selected these two ontologies because they appeared to have several commonalities in their axiomatizations of time, process, and participation. However, when we examine the fundamental ontological commitments of these ontologies, we can identify two major obstacles to their integration – they differ on their time ontologies (time points vs. time intervals) and they make different assumptions about how objects participate in activity occurrences. We introduce sets of ontologies that combine both time points and intervals, and which can serve as a bridge between the time ontologies of DOLCE and PSL. We also extend the PSL ontology with new axioms regarding the notion of participation. Figure 1 illustrates the bridges that are created to formalize the relationships between the DOLCE and PSL ontologies. Finally, we show that the resulting extensions of PSL can faithfully interpret the DOLCE ontology, with an emphasis on the semi-automatic verification of these ontologies.

2   Ontology Merging Through Verification

Two key techniques used in this paper are ontology mapping and the design of ontologies through the merging of existing ontologies. In this section, we discuss the notion of ontology verification and how it is used to support ontology mapping. We also review existing ontology merging techniques and discuss how these are related to our approach.

2.1   Ontology Verification

Verification is concerned with the relationship between the intended models of an ontology and the models of the axiomatization of the ontology[2]. In particular,

---

[1] The results in this paper are an updated and revised version of the material found in [3], which was presented at the 6th International Conference on Knowledge Engineering and Ontology Development (KEOD).

[2] A first-order ontology is a set of first-order sentences (axioms) that characterize a first-order theory, which is the closure of the ontology's axioms under logical

we want to characterize the models of an ontology up to isomorphism and determine whether or not these models are equivalent to the intended models of the ontology. In practice, the verification of an ontology is achieved by demonstrating that it is equivalent to another logical theory whose models have already been characterized up to isomorphism. We therefore need to understand the different relationships among logical theories, which will also lead us to specify the mappings between ontologies.

A fundamental property of an ontology is the range of concepts and relations that it axiomatizes. Within the syntax, this is captured by the notion of the signature of the ontology. The signature $\Sigma(T)$ of a logical theory T is the set of all constant symbols, function symbols, and relation symbols that are used in the theory. The simplest relationships between logical theories are the different notions of extension, in which the signature of one theory is a subset of the signature of another theory. Let T_1, T_2 be two first-order theories such that $\Sigma(T_1)$ $\subseteq \Sigma(T_2)$. We say that T_2 is an *extension* of T_1 iff for any sentence $\sigma \in \mathcal{L}(T_1)$,

$$\text{if} \quad T_1 \models \sigma \quad \text{then} \quad T_2 \models \sigma.$$

T_2 is a *conservative extension* of T_1 iff for any sentence $\sigma \in \mathcal{L}(T_1)$,

$$T_2 \models \sigma \quad \text{iff} \quad T_1 \models \sigma.$$

T_2 is a *non-conservative extension* of T_1 iff T_2 is an extension of T_1 and there exists a sentence $\sigma \in \Sigma(T_1)$ so that

$$T_1 \nvDash \sigma \quad \text{and} \quad T_2 \models \sigma.$$

Non-conservative extension plays a key role in COLORE. Ontologies within the repository are organized into sets of ontologies with the same signature known as *hierarchies*. The set of ontologies within a hierarchy are ordered by non-conservative extension; an ontology is a root theory of hierarchy if it is not extended by any other ontology within the same hierarchy.

If the logical theories have different signatures, there is a range of fundamental metalogical relationships which are used in ontology verification. All of them consider mappings between the signatures of the theories that preserve entailment and satisfiability. The basic relationship between theories T_A and T_B is the notion of *interpretation*, which is a mapping from the language of T_A to the language of T_B that preserves the theorems of T_A [6]. The interpretation is *faithful* if the mapping also preserves the satisfiable sentences of T_A.

One notion of equivalence among theories is *mutual faithful interpretability*, that is, T_1 faithfully interprets T_2 and T_2 faithfully interprets T_1. An even stronger equivalence relation is that of logical synonymy: Two ontologies T_1 and T_2 are *synonymous* iff there exists an ontology T_3 with the signature $\Sigma(T_1) \cup \Sigma(T_2)$ that is a definitional extension of T_1 and T_2.

---

entailment. In the rest of the paper we will simply drop the term first-order and assume ontologies and theories to be first-order.

If there is an interpretation of T_A in T_B, then there exists a set of sentences (referred to as *translation definitions*) in the language $L_A \cup L_B$ of the following form, where $p_i(\overline{x})$ is a relation symbol in L_A and $\varphi(\overline{x})$ is a formula in L_B:

$$(\forall \overline{x})\, p_i(\overline{x}) \equiv \varphi(\overline{x})$$

Thus, T_B interprets T_A if there exists a set Δ of translation definitions such that

$$T_B \cup \Delta \models T_A$$

T_B faithfully interprets T_A if $T_B \cup \Delta$ is a conservative extension of T_A.

2.2   Existing Merging Techniques

Within the applied ontology community, there exist various terms used to describe the notion of *ontology merging*; some of these terms have slight differences in the notion of merging but all agree that a *new* ontology is created from two ontologies. In [1], the authors indicate that ontology merging is the process of generating a new ontology from two or more existing and different ontologies that contain similar notions of a given subject. In [7], the notion is similar, where the new merged ontology contains the knowledge of the source ontologies. Other notions of merging discussed in [5,19] indicate that ontology merging results in a new ontology that is a union of two source ontologies, and captures all of the knowledge found in the source ontologies.

Bridge axioms are another term used to describe axioms that relate the terms of two or more ontologies together, and serve as the basis for ontology merging when the ontologies are expressed in the same language [7,19]. They are often written in the form of subsumption axioms [17,19]; we make the distinction here that the notion of 'bridging' used in our approach is not restricted to subsumption axioms, but of the creation of an ontology (resulted from the merge) and the usage of translation definitions to illustrate faithful interpretations between theories.

Ontology mappings are used to formalize the correspondences between the entities of one ontology with the entities of another [7]. There are several notions of mappings which will not be discussed in this section; we direct the reader to [1] for such distinctions. In this work, we utilize *translation definitions* to specify the mappings between the theories.

Our approach to ontology merging is distinct from these existing techniques as it utilizes ontology verification in the process to ensure that the source theories, along with the new intermediary theory, are faithfully interpretable with one another. As well, we provide direct mappings between the concepts in the ontologies through the use of translation definitions which are first-order axiomatizations of the interpretations between the ontologies.

3   Background

DOLCE is often known as an ontology of endurants (objects) and perdurants (processes). Similarly, the PSL ontology also axiomatizes classes and properties of objects and activity occurrences. In this section, we briefly review these two upper ontologies, and we review the different ontological commitments that they make. It will be these differences which will become the focus for the design of new bridge ontologies in the remainder of the paper.

3.1 PSL-Core

The Process Specification Language (PSL) is an ontology designed to facilitate the correct and complete exchange of process information among manufacturing systems [8]. These applications include scheduling, process modelling, process planning, production planning, simulation, and project management. The PSL ontology is organized into a core theory, PSL-Core[3] and a set of partially ordered extensions; the core ontology consists of four disjoint classes: *activities* can have zero or more occurrences, *activity occurrences* begin and end at time points, *time points* constitute a linear ordered set with end points at infinity, and *objects* are elements that are not activities, occurrences, or time points [8]. In PSL, the ternary relation, $participates\_in(x, o, t)$, is used to specify that an object x participates in an activity occurrence o at a time point t. In other words, an object can participate in an activity occurrence only at those time points at which both the object exists and the associated activity is occurring.

There are five additional modules within the PSL ontology – $T_{occtree}$ (which is closely related to situation calculus), $T_{subactivity}$ (which axiomatizes the composition relation on activities), T_{atomic} (which axiomatizes concurrent activities), $T_{complex}$ (which axiomatizes complex activities), and T_{actocc} (which axiomatizes the composition relation on occurrences of complex activities). However, none of these notions correspond to concepts within DOLCE, so the only part of the PSL ontology considered in this paper is restricted to PSL-Core.

In PSL, the ternary relation, $participates\_in(x, o, t)$, is used to specify that an object x participates in an activity occurrence o at a time point t. In other words, an object can participate in an activity occurrence only at those time points at which both the object exists and the associated activity is occurring.

3.2 DOLCE

As the first module of the WonderWeb library of foundational ontologies, the Descriptive Ontology for Linguistic and Cognitive Engineering (DOLCE) aims to capture the categories which underlie natural language and human common sense [15]. DOLCE is based on the distinction between enduring and perduring entities, referred to as continuants and occurrents, where the fundamental difference between the two is related to their behaviour in time [15]. *Endurants* are wholly present at any time: they are observed and perceived as a complete concept, regardless of a given snapshot of time. *Perdurants*, on the other hand, extend in time by accumulating different temporal parts, so they are only partially present at any given point in time. In DOLCE[4], endurants are *involved* in

---

[3] http://colore.oor.net/psl_core/psl_core.clif.
[4] http://colore.oor.net/dolce_participation/dolce_participation.clif.

an occurrence, so the notion of participation is *not* considered parthood. Rather, participation is time-indexed in order to account for the varieties of participation in time, such as temporary participation and constant participation.

Based on the distinction between endurants and perdurants, DOLCE has been partially modularized into the following modules in [2]: $T_{dolce\_taxonomy}$ (which axiomatizes the subsumption and disjointness of DOLCE categories), $T_{dolce\_mereology}$ (which axiomatizes parthood for atemporal entities), $T_{dolce\_time\_mereology}$ (which axiomatizes a mereology on time intervals), $T_{dolce\_present}$ (which axiomatizes an entity's existence in time), $T_{dolce\_temporary\_parthood}$ (which axiomatizes the time-indexed parthood of entities), $T_{dolce\_constitution}$ (which axiomatizes the co-location of different entities in the same space-time), and $T_{dolce\_participation}$ (which axiomatizes the participation of entities).

The authors of [13] have shown that the first-order axiomatization of DOLCE is consistent and have provided an alternative modularization of ontology; we have adapted their axioms in the results of this paper and included them in COL-ORE[5]. The concepts found within the $T_{dolce\_participation}$, $T_{dolce\_time\_mereology}$, and $T_{dolce\_present}$ theories are considered in this work as they correspond with concepts found in PSL-Core and other time ontologies in COLORE.

3.3   Relationships Among the Ontologies

In order to understand how the PSL ontology is related to DOLCE, we begin by outlining some observations of both ontologies. As shown on the left-hand side of Fig. 1, the various subtheories of DOLCE are depicted as modules in the ontology. There are no relations in the PSL ontology that intuitively correspond to the concepts of temporary parthood, constitution, or dependence within DOLCE. On the other hand, the PSL ontology focuses on relations between activity occurrences, objects, and time points. In this paper, we therefore focus on the three subtheories of DOLCE that axiomatize relationships between perdurants, endurants, and time intervals: participation $T_{dolce\_participation}$, being present $T_{dolce\_present}$, and time mereology $T_{dolce\_time\_mereology}$. In DOLCE, *time intervals* are used to describe temporal objects in $T_{dolce\_participation}$ and $T_{dolce\_present}$, all of which contain $T_{dolce\_time\_mereology}$. DOLCE does not contain an ordering on time, but has a time mereology. In contrast, the T_{psl_core} PSL-Core ontology uses *time points* to describe the temporal aspects of objects and activity occurrences, as well as uses an ordering on time, but does not contain a time mereology. From this observation, both ontologies appear to have intuitions of perdurants/endurants and activity occurrences/objects being present and participating in some time construct, yet these intuitions seem to be quite different, and the relationship between the two ontologies is not obvious.

The second DOLCE module that we consider is $T_{dolce\_participation}$; this subtheory contains the following axiom (**Ad35** in [15]) which indicates every endurant participates in some perdurant at a given time object:

---

[5] http://colore.oor.net/dolce/.

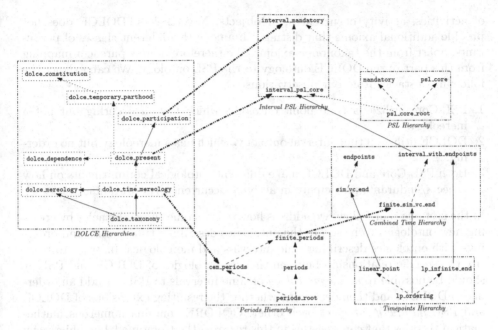

Fig. 1. Relationships between DOLCE modules and theories in COLORE. Solid lines indicate *conservative extensions*, dashed lines indicate *non-conservative extensions*, and the bolded dash-dot-dotted lines indicate *faithful interpretations* between ontologies.

$$\forall x \; ED(x) \supset \exists(y,t) \; PC(y,x,t) \tag{1}$$

A similar axiom is found in T_{psl_core} that indicates activity occurrences require an object to participate in them. From these observations, we can hypothesize that perdurants and endurants from DOLCE are equivalent to activity occurrences and objects in PSL, respectively. We can further conjecture that the notion of participation $PC(x,y,z)$ in DOLCE is equivalent to the *participates_in*(x,y,t) relation in PSL: for any object x, activity occurrence y, and time interval z, there exists a time construct t that is equivalent to the time interval z, where x participates in y during t. We write these equivalences as the following *translation definitions*:

$$\forall x \; PD(x) \equiv activity\_occurrence(x) \tag{2}$$

$$\forall x \; ED(x) \equiv object(x) \tag{3}$$

$$\forall x \; T(x) \equiv timeinterval(x) \tag{4}$$

$$\forall x \forall y \forall z \; (PC(x,y,z) \equiv object(x) \wedge activity\_occurrence(y) \wedge timeinterval(z) \wedge$$
$$(\forall t beforeEq(beginof(z),t) \wedge beforeEq(t,endof(z)) \supset participates\_in(x,y,t))). \tag{5}$$

The DOLCE ontology also contains a taxonomy of classes of perdurants and endurants. The PSL ontology does not contain a corresponding taxonomy of activities, activity occurrences, or objects. Nevertheless, DOLCE does not provide additional axioms that distinguish among the different classes of perdurants, apart from the taxonomic axioms. We therefore do not pursue a mapping from this part of the DOLCE ontology to the PSL ontology. We can extract the following obstacles from our observations:

1. PSL-Core utilizes a time point ontology, which has an ordering but not a mereology.
2. DOLCE utilizes a time interval ontology, which has a mereology but no ordering.
3. Both PSL-Core and DOLCE make different ontological commitments on how objects/endurants participate in activity occurrences/perdurants.

In the sections that follow, we address how to overcome these obstacles by creating new ontologies to integrate the ontologies describing time points and orderings with ontologies describing time intervals and mereologies. In order to identify the specific relationship between the two ontologies of DOLCE and PSL, it should be possible to add a *mereology* of time intervals to PSL, or add an ordering to DOLCE, and then determine whether the resulting extensions of DOLCE and PSL *faithfully interpret* each other. COLORE contains numerous mathematical theories that can assist us in this regard – the Combined Time hierarchy $\mathbb{H}^{combined\_time}$ contains time ontologies that utilize both time point and time interval constructs, and are able to interpret a mereology on time points and time intervals. Figure 1 illustrates the bridges that can be created to formalize the relationships between the ontologies.

It may be noted that there are two different intuitions about bridging ontologies that are being explored here. We first consider new ontologies which are created as either nonconservative extensions of ontologies in existing hierarchies, or as merged ontologies, that is, they are conservative extensions of a set of other ontologies from different hierarchies. We will see this in the role played by the ontologies in the Combined Time and Interval PSL hierarchies. An alternative intuition is that a bridging ontology is strong enough to faithfully interpret one ontology while conservatively extending another, thus providing a way of embedding the two ontologies within the bridging ontologies. We will see this in the role played by ontologies in the Interval PSL hierarchy.

Furthermore, it is interesting to see how this approach to bridging ontologies is related to the notion of bridge axioms. For bridging ontologies which are the merge of other ontologies, there exist sentences whose signature is the union of the signatures of the merged ontologies. For example, we will see that there exist axioms in the ontologies of the Combined Time hierarchy which use relations on both time points and time intervals. Such axioms correspond to the bridge axioms in [7,17]. On the other hand, for bridging ontologies which faithfully interpret an ontology, we do not find such sentences. Instead, the translation definitions that specify the interpretation play the role of bridge axioms between the two ontologies.

3.4 Ontology Transfer

The goal of ontology verification is to characterize the models of an ontology up to isomorphism or elementary equivalence. In practice, this is achieved by proving that the ontology under consideration is synonymous with another ontology whose models are well-understood, and which has a prior characterization. Typically, such ontologies directly axiomatize classes of mathematical structures such as orderings, graphs, and geometries. In the case of participation ontologies, we have used a mathematical structure known as an incidence bundle [3] as part of the verification of the DOLCE and PSL Ontologies.

We can use incidence bundles to illustrate the notion of ontology transfer, in which we specify new ontologies in one hierarchy by identifying synonymous ontologies in another hierarchy. We will focus on the application of this technique to the specification of extensions of $T_{pslcore\_participates}$. On the left side of Fig. 2, we can see the ontologies in the $\mathbb{H}^{in\_bundle}$ Hierarchy which we have used for verification of the participation ontologies. Using the translation definitions from the proof of Theorem 5, we can specify new ontologies which are extensions of $T_{psl\_participates}$[6] and which are synonymous with incidence bundle ontologies. Of particular interest are those ontologies of incidence bundles which were used in the verification of $T_{participates\_owl}$ and $T_{dolce\_participates}$. In this way, the mathematical ontologies for different classes of incidence bundles serve as bridges between the two upper ontologies.

Theorem 1. *$T_{strong\_object\_exist}$ is synonymous with $T_{plane\_flag\_bundle}$.*

Theorem 2. *$T_{strong\_participates}$ is synonymous with T_{flag_bundle}.*

Note that $T_{psl\_participates}$ is the root theory of the $\mathbb{H}^{psl\_participates}$ Hierarchy – there is no weaker theory than $T_{psl\_particpates}$ that axiomatizes intuitions about participation. Since $T_{present\_bundle}$ is synonymous with $T_{pslcore\_participates}$ (which is the root theory of its hierarchy), each theory in the $\mathbb{H}^{in\_bundle}$ Hierarchy which extends it is synonymous with an extension of $T_{pslcore\_participates}$ (see Fig. 2).

We note here that this paper is a revised version of [4] and includes revisions and additions from [3].

4 Merging Time Point and Interval Ontologies

To address the first obstacle, we utilize existing combined time ontologies to integrate time points and time intervals, and mereologies and orderings. In this way, existing temporal ontologies found in COLORE can be used to analyze the interpretations between the DOLCE and PSL. Here we briefly outline the time point and time interval ontologies used. We then examine a set of time ontologies from COLORE which merge the time point ontologies with the time interval ontologies.

---

[6] The new ontologies can be found at http://colore.oor.net/psl_participates/.

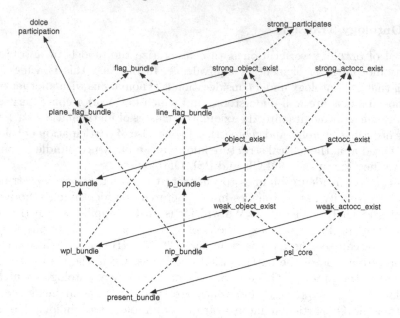

Fig. 2. Relationship between extensions of $T_{psl\_participates}$ and the hierarchy of incidence bundles. Dashed lines denote nonconservative extension, and solid lines denote logical synonymy.

4.1 The Time Points Hierarchy

Within this hierarchy are ontologies that describe time in terms of time points and that introduce a partial ordering on the set of time points using the binary relation $before(x, y)$. We are particularly interested in two of these ontologies, based on the roles they play as modules of other ontologies. The linear point theory, T_{linear_point}[7], derived from axioms found in [11], is a simple ontology whose axioms state that time points infinitely extend a timeline in both forward and backward directions. The linear time points with endpoints at infinity theory, $T_{lp\_infinite\_end}$[8], derived from axioms found in [11], also represents a linear ordering on time points that infinitely extends in both forward and backward directions, but it contains axioms that enforce the existence of endpoints at infinity in both directions.

4.2 The Periods Hierarchy

The axioms for the periods hierarchy, $\mathbb{H}^{periods}$, were proposed in [18], and additional information about other ontologies in this hierarchy can be found in [9]. We are interested in the in the weakest theory of this hierarchy, $T_{periods}$, since it is used by $T_{endpoints}$, which is described in Sect. 4.3. The Minimal Theory of

---

[7] http://colore.oor.net/timepoints/linear_point.clif.
[8] http://colore.oor.net/timepoints/lp_infinite_end.clif.

Periods, $T_{periods}$, constitutes the minimal set of conditions that must be met by any period structure [18]. It contains two relations, $precedence(x, y)$ and $inclusion(x, y)$, and two conservative definitions, $glb(x, y, z)$ and $overlaps(x, y)$, as its non-logical lexicon. Every element in the domain is considered a time interval, and there are transitivity and irreflexivity axioms for the $precedence(x, y)$ relation, making it a strict partial order; similarly, the transitivity, reflexivity, and anti-symmetry axioms for the $inclusion(x, y)$ relation make it a partial order. As well, the axiom, $glb(x, y, z)$, guarantees the existence of greatest lower bounds between overlapping intervals defined by $overlaps(x, y)$.

An interesting property in these ontologies is the convexity of time intervals. Intuitively, convex intervals are those which have no gaps [14][9]. The convexity of time intervals requires an ordering over time intervals and a mereology, and hence it can be defined by ontologies in the $\mathbb{H}^{periods}$ hierarchy. However, convexity is not definable in DOLCE since it lacks an ordering on time intervals.

4.3 The Combined Time Hierarchy

Given that we want to merge ontologies from the time points hierarchy with ontologies from the periods hierarchy, we need to consider ontologies that include both time points and time intervals as primitives, and define a set of functions and relations specifying the interactions between them. These time ontologies[10] are derived from the theories presented in [11], and have been modified and verified in [10]. Depending on the relations and functions used, these theories can represent time in very different ways. For example, the theory of endpoints, $T_{endpoints}$, defines time points only as the boundary of time intervals, where every interval is associated with exactly two time points: the begin of and end of the interval. In contrast, the theory of time point continuum, $T_{point_c ontinuum}$, defines intervals by the set of adjacent time points in which they are contained; another theory, $T_{vector\_continuum}$ introduces the concept of directionality by allowing 'backward intervals' where the end of point is before the begin of point in the timeline.

The theory of endpoints, $T_{endpoints}$[11], combines the language of intervals and points by defining the *beginof*, *endof*, and *between* functions to relate time intervals to time points and vice-versa. In this hierarchy, this theory imports axioms from T_{linear_point} that define a binary $before(x, y)$ relation between time points as transitive and irreflexive, and asserts that all time points are linearly ordered and infinite in both directions. As well, this theory includes axioms that define the $meets\_at(i, x, j)$ relation as one between two intervals and the point at which they meet along, restrict $beginof(i)$ to always come before the $endof(i)$ function, and states that intervals are between two points if they are properly ordered.

---

[9] Additional information about the various relations found in convex and non-convex intervals can be found in [14].

[10] http://colore.oor.net/combined_time/.

[11] http://colore.oor.net/combined_time/endpoints.clif.

The vector continuum theory, $T_{vector\_continuum}$[12], introduces the notion of orientation of intervals, and also imports T_{linear_point}. It contains the same three functions ($beginof(i)$, $endof(i)$, and $between(x,y)$) that transform time intervals into time points and vice-versa, but differs in its definition of $between(x,y)$ by allowing the formation of intervals whose $endof$ point is equal to or before its $beginof$. Thus, every interval in $T_{vector\_continuum}$ has a 'reflection' in the opposite direction via the $back(i)$ function; intervals oriented in the forward direction are defined normally where $beginof(i)$ is before $endof(i)$. As well, single-point intervals, known as *moments*, are defined as intervals whose $beginof(i)$ and $endof(i)$ points are the same.

4.4   Composing the Theory of Intervals with Endpoints

The combined time hierarchy contains ontologies whose models combine structures for time points and time intervals. These ontologies were first proposed in [11], and assume an import of the $T_{endpoints}$ theory, where every time interval is associated with two time points. However, T_{psl_core} contains a *time point* ontology that axiomatizes a linear ordering with endpoints at infinity, whereas $T_{endpoints}$ axiomatizes a time point ontology in which the linear ordering does not have such maximum and minimum time points. Consequently, we need to create a new theory, $T_{interval\_with\_endpoints}$, in $\mathbb{H}^{combined\_time}$ that contains the time interval axioms of $T_{endpoints}$ with a different time point ontology. It is this new time ontology which will be used to extend T_{psl_core} to make it compatible with the existence of time intervals.

The new intervals with endpoints theory, $T_{interval\_with\_endpoints}$, imports axioms from $T_{finite\_sim\_vc\_end}$ from $\mathbb{H}^{combined\_time}$ and $T_{lp\_infinite\_end}$ from $\mathbb{H}^{timepoints}$. The primary difference between the $T_{finite\_sim\_vc\_end}$ and $T_{sim\_vc\_end}$ ontologies within $\mathbb{H}^{combined\_time}$ is that different time point ontologies are used in each theory; while both ontologies share a common set of axioms ($T_{lp\_ordering}$) additional axioms in T_{linear_point} make $T_{sim\_vc\_end}$ different from $T_{finite\_sim\_vc\_end}$, as depicted in Fig. 1. Consequently, $T_{interval\_with\_endpoints}$ non-conservatively extends $T_{finite\_sim\_vc\_end}$ since it contains the same time interval axioms as $T_{finite\_sim\_vc\_end}$, but different time point axioms from $T_{lp\_infinite\_end}$. The axioms of $T_{interval\_with\_endpoints}$ can be found in COLORE[13].

Recall that the DOLCE ontology has a mereology on time intervals, but that there is no ordering relation on time intervals; on the other hand, the combined time ontologies have an ordering over time intervals by which an mereology can be defined. In this way, the periods hierarchy *bridges* DOLCE and combined time hierarchies together since $T_{periods}$ is the common theory between them. We note that the dash-dot-dotted arrows in Fig. 1 outline the faithful interpretations between $\mathbb{H}^{dolce}$ and $\mathbb{H}^{periods}$, and $\mathbb{H}^{periods}$ and $\mathbb{H}^{combined\_time}$; however, these are faithful interpretations are conjectured and proofs will be addressed in future

---

[12] http://colore.oor.net/combined_time/vector_continuum.clif.

[13] http://colore.oor.net/combined_time/interval_with_endpoints.clif.

work. In this paper, we only discuss the composition of theories needed to prove the faithful interpretations between DOLCE and PSL.

5   Extending T_{psl_core}

To address the second obstacle, we extend the PSL-Core theory with time intervals and interpret DOLCE in this new ontology. Within PSL, *activity occurrences* are considered to be occurrents, while *objects* are represented by continuants [8]. The relation $participates\_in(x, o, t)$ is used to specify that an object x participates in activity occurrence o at time point t. Since DOLCE does not utilize time points but time intervals in its time mereology, an extension of T_{psl_core}[14] must be created in order to map the $participates\_in(x, o, t)$ relation to a relation on time intervals.

5.1   Theory of PSL-Core Root

A subset of the axioms in T_{psl_core} were extracted to create the $T_{psl\_core\_root}$ theory. The following *closure* axiom from T_{psl_core} was removed

$$\forall x \left(activity(x) \vee activity\_occurrence(x) \vee timepoint(x) \vee object(x) \right) \quad (6)$$

We need a theory which can incorporate both time points and time intervals, and it is easy to see how such a closure axiom is problematic, since it precludes the existence of time intervals as a distinct class. In other words, there can be no extension of T_{psl_core} that contains an axiomatization of time intervals. It is easy to see that T_{psl_core} is a non-conservative extension of $T_{psl\_core\_root}$. All new theories that incorporate time intervals into the PSL ontology are conservative extensions of $T_{psl\_core\_root}$.

5.2   Theory of Mandatory Participation

The original T_{psl_core} contains a weak axiomatization of the *participates_in* (x, o, t) relation, and does not impose any conditions beyond requiring that an activity is occurring at the same time that the object exists. There are no requirements that an object must participate in some activity occurrence, or that an activity occurrence always have some object participating in it. We can therefore define a new *non-conservative* extension of $T_{psl\_core\_root}$ called $T_{mandatory}$ to take into account the *mandatory* participation of PSL objects in a temporal construct. The axioms found in this extension import $T_{psl\_core\_root}$ and *do not* include the $between(x, y, z)$ and $before(x, y, z)$ relations found in T_{psl_core} since they involve the usage of time points, not time intervals, to describe the participation of objects in activity occurrences and time objects. Figure 3 lists all of the

---

[14] We could not modify the axioms found in T_{psl_core} since the axioms are standardized in ISO 18629-11:2005.

$$\forall x \, (object(x) \supset (\exists o \exists t \, participates\_in(x, o, t))) \tag{7}$$

$$\forall o \forall t \, (activity\_occurrence(o) \land is\_occurring\_at(o, t) \supset (\exists x \, participates\_in(x, o, t))) \tag{8}$$

Fig. 3. Axioms found in $T_{mandatory}$.

axioms found in $T_{mandatory}$, and the axioms can be found in COLORE[15]. Axiom 7 indicates that every object x has to participate in some activity occurrence o at a time object t, and Axiom 8 indicates that, for every activity occurrence o that occurs during the time object t, there exists an object that also participates in that time object.

In $T_{mandatory}$, we *do not* commit to a specific type of temporal object for object participation, but we note that there needs to be a 'bridge' of sorts to connect the DOLCE and PSL ontologies together. Consequently, we are interested in creating a new *bridge ontology* that contains the PSL constructs that are used with time intervals. We discuss this new $\mathbb{H}^{interval\_psl}$ hierarchy in the next section.

5.3 The Interval PSL Hierarchy

Since the PSL ontology only describes object and activity occurrences with respect to time points, we need to create a time interval version of the PSL ontology. This leads to a new hierarchy, called $\mathbb{H}^{interval\_psl}$, with $T_{interval\_psl\_core}$ as its root theory. This hierarchy contains the time interval versions of the T_{psl_core} and $T_{mandatory}$ ontologies which are named $T_{interval\_psl\_core}$ and $T_{interval\_mandatory}$, respectively, and are depicted in Fig. 1. Each of these ontologies is briefly described below, and can be found in COLORE[16].

The ontologies in this hierarchy import axioms from $T_{psl\_core\_root}$ and $T_{interval\_with\_endpoints}$. In order to ensure that the time interval version of $T_{psl\_core\_root}$ contains axioms that describe time intervals, and not time points, $T_{interval\_with\_endpoints}$ is used to describe the time objects found in this compiled ontology.

Three axioms are added to $T_{interval\_psl\_core}$ in addition to the imported ontologies and are outlined in Fig. 4. Axiom 9 indicates that a time interval is not an activity, activity occurrence, object, or time point. In Axiom 10, the relation, $psl\_interval(x, y)$, is introduced to relate a time interval with the begin of and end of an activity occurrence or object. Finally, the $overlay(x, y, z)$

---

[15] http://colore.oor.net/psl_core/mandatory.clif.
[16] http://colore.oor.net/interval_psl/.

$$\forall x \, (timeinterval(x) \supset \neg(activity(x) \lor activity\_occurrence(x) \lor timepoint(x) \lor object(x)))$$
$$(9)$$

$$\forall x \forall y \, (psl\_interval(x, y) \equiv (object(x) \lor activity\_occurrence(x)) \land timeinterval(y) \land$$
$$beginof(x) = beginof(y) \land endof(x) = endof(y))$$
$$(10)$$

$$\forall x \forall y \forall z \, (overlay(x, y, z) \equiv (\exists i_1 \exists i_2 \, (psl\_interval(x, i_1) \land psl\_interval(y, i_2) \land$$
$$beginof(i_2) = beginof(z) \land endof(i_1) = endof(z))))$$
$$(11)$$

Fig. 4. Axioms of $T_{interval\_psl\_core}$.

relation is introduced in Axiom 11 to describe a time interval z that overlays[17] activity occurrences x and y. However, it may not necessarily be the case that both activity occurrence/object y overlays an activity occurrence/object x, or vice versa. This axiom is included in case such overlaying of intervals does occur.

5.4   Theory of Mandatory Intervals

Finally, we have the theory of mandatory intervals which imports axioms from $T_{interval\_psl\_core}$ and $T_{mandatory}$. Since we would like to show that $T_{dolce\_participation}$ can faithfully interpret the time interval versions of PSL ontologies from $T_{interval\_psl\_core}$, we extended $T_{interval\_psl\_core}$ to include the time interval versions of the axioms from $T_{mandatory}$. No additional axioms are included in this ontology and it can be found in COLORE[18]. Essentially this ontology assigns time intervals[19] to $T_{interval\_psl\_core}$ to indicate the *mandatory* participation of PSL over a time interval. The right side of Fig. 1 summarizes the relationships between the Interval PSL, PSL, and Combined Time hierarchies.

6   Interpretations Between DOLCE and PSL

In order to determine whether PSL can faithfully interpret the subtheories of DOLCE, we first need to modify $T_{dolce\_present}$. This module of DOLCE contains class $Q(x)$ of qualities, which is problematic – $T_{psl\_core\_root}$ is unable to define what a quality is because it is unable to discern which $object(x)$ is an endurant $ED(x)$ or a quality $Q(x)$. We must therefore specify a subtheory of $T_{dolce\_present}$

---

[17] The terms *overlap* and *intersect* were not used to describe this relation since they are used in mereology ontologies. To be consistent with PSL, we decided to use the term *overlay* to describe the relationship where time intervals may overlay one another.

[18] http://colore.oor.net/interval_psl/interval_mandatory.clif.

[19] Recall that we did not commit to a particular temporal construct in $T_{mandatory}$.

that does not include qualities for this portion of the interpretation. The axioms of this subtheory $T^*_{dolce\_present}$ can be found in COLORE[20].

The DOLCE subtheories of $T_{dolce\_participation}$ and $T_{dolce\_time\_mereology}$ are able to interpret the $T_{interval\_mandatory}$ and $T_{interval\_psl\_core}$ ontologies in $\mathbb{H}^{interval\_psl}$, respectively. This is graphically depicted in Fig. 1, where the dashed arrows depict the interpretations from the DOLCE ontology to the Interval PSL ontology.

6.1   Interpretations Between $T_{interval\_psl\_core}$ and $T^*_{dolce\_present}$

From our brief discussion of the theories found in COLORE, we make the observation that the concept of *parthood* in DOLCE is equivalent to the *inclusion* of time intervals in $T_{interval\_psl\_core}$:

$$\forall t_1 \forall t_2 \, (P(t_1, t_2) \equiv timeinterval(t_1) \wedge timeinterval(t_2) \wedge$$
$$beforeEq(beginof(t_2), beginof(t_1)) \wedge beforeEq(endof(t_1), endof(t_2))) \quad (12)$$

We can say that the time interval t_1 is *part of* time interval t_2: the beginning of t_2 can either be before or equal to the beginning of t_1, and the end of t_1 can either be before or equal to the end of t_2. Furthermore, we can state that the concept of being present in DOLCE is equivalent to the concept of an object or activity occurrence that exists in a given time interval, where the beginning of the time interval is the time point in which an object or activity occurrence starts, and that the end of the time interval is the time point in which the object or activity occurrence ends.

$$(\forall x \forall y \forall t \, (PRE(x, t) \equiv (object(x) \vee activity\_occurrence(x)) \wedge timeinterval(t) \wedge$$
$$beforeEq(beginof(x), beginof(t)) \wedge beforeEq(endof(t), endof(x)))). \quad (13)$$

We note that, in $psl\_interval(x, y)$, a unique time interval is associated with an object or activity occurrence in PSL. Similarly, the time interval associated in $PRE(x, t)$ in DOLCE need not be a time interval at which an endurant or perdurant is present. The translation definition for the *sum* of time intervals is based on the definition of the overlaps relation in DOLCE:

$$\forall x \forall y \, overlaps(x, y) \equiv (before(beginof(x), beginof(y)) \wedge before(endof(x), endof(y))) \vee$$
$$(before(beginof(y), beginof(x)) \wedge before(endof(y), endof(x))).$$
$$(14)$$

$$\forall x \forall y \forall z \, SUM(z, x, y) \equiv timeinterval(x) \wedge timeinterval(y) \wedge timeinterval(z) \wedge$$
$$(\forall w \, (overlaps(w, z) \equiv (overlaps(w, x) \vee overlaps(w, y)))).$$
$$(15)$$

To show that $T_{interval\_psl\_core}$ interprets $T^*_{dolce\_present}$, we define endurants and qualities in DOLCE to be equivalent to objects in PSL, perdurants to equivalent to activity occurrences, and time intervals in DOLCE to be equivalent to time

---

[20] http://colore.oor.net/dolce_present/dolce_present_star.clif.

intervals in Interval PSL. In regards to mereology, a time intervals t_1 is defined to be a part of a time interval t_2 if the begin of t_2 is before or equal to t_1 and the end of t_1 is before or equal to t_2. The $PRE(x, t)$ relation in DOLCE is defined to be equivalent to an object or activity occurrence that occurs during a time interval. Finally, the $SUM(z, x, y)$ relation in DOLCE is defined to be a time interval z is the sum of the time intervals of two activities x and y in Interval PSL.

Theorem 3. $T_{interval\_psl\_core}$ *faithfully interprets* $T^*_{dolce\_present}$.

Proof. Let Δ_1 be the set of translation definitions found in COLORE[21]. Using Prover9, we show[22] that

$$T_{interval\_psl\_core} \cup \Delta_1 \models T^*_{dolce\_present}$$

It should be noted that, if we consider their entire sets axioms, DOLCE and PSL are not comparable with respect to definable interpretations. Nevertheless, we can identify subtheories of each ontology for which we can specify inter-pretability. For PSL, we needed to weaken the closure axiom and, for DOLCE, we needed to consider the subtheory which omits qualities from the domain.

6.2 Interpretations Between $T_{interval\_mandatory}$ and $T_{dolce\_participation}$

Since the DOLCE ontology contains axioms for participation, we make the observation that the participation relation, $PC(x, y, z)$, is similar to the $participates\_in(x, y, t)$ relation found in PSL. Thus, we can state that any x and y that participate in z in DOLCE is equivalent an object x that partici-pates in an activity occurrence y in a given time interval z and, at *every* time point in that interval, x *participates* in y. We verify this notion of participation with the translation definition found in Eq. 16. As well, we can reuse the set of translation definitions Δ_1 from the previous section (because $T_{interval\_mandatory}$ imports $T_{interval\_psl\_core}$) to show the interpretation of $T_{interval\_mandatory}$ and $T_{dolce\_participation}$ along with the additional translation definition shown below.

$$\forall x \forall y \forall z \forall t \, (PC(x, y, z) \equiv object(x) \land activity\_occurrence(y) \land timeinterval(z) \land$$
$$(beforeEq(beginof(z), t) \land beforeEq(t, endof(z)) \supset participates\_in(x, y, t)))$$
$$(16)$$

Theorem 4. $T_{interval\_mandatory}$ *faithfully interprets* $T_{dolce\_participation}$.

Proof. Let Δ_2 be the set of translation definitions found in COLORE[23]. Using Prover9, we show[24] that

$$T_{interval\_mandatory} \cup \Delta_1 \cup \Delta_2 \models T_{dolce\_participation}$$

---

[21] http://colore.oor.net/interval_psl/mappings/interval_psl_core2dolce_present.clif.

[22] Proofs can be found at http://colore.oor.net/dolce_present/interprets/output/.

[23] http://colore.oor.net/interval_psl/mappings/interval_mandatory2dolce_participat ion.clif.

[24] Proofs can be found at http://colore.oor.net/dolce_participation/interprets/output/.

To show the verification of the notion of participation between PSL and DOLCE, we use the incidence bundle ontologies as the basis for the verification of the participation ontologies.

Theorem 5. $T_{psl\_participates}$ *is synonymous with* $T_{present\_bundle}$.

Proof. If $\Delta_{psl2inbundle}$ is the following set of translation definitions:

$$(\forall p)\, point(p) \equiv timepoint(p) \tag{17}$$

$$(\forall l)\, line(l) \equiv activity\_occurrence(l) \tag{18}$$

$$(\forall q)\, plane(q) \equiv object(q) \tag{19}$$

$$(\forall x, o, t)\, tin(o, x, t) \equiv participates\_in(x, o, t) \tag{20}$$

$$(\forall x, y)\, in(x, y) \equiv is\_occurring\_at(y, x) \vee exists\_at(y, x) \tag{21}$$

and $\Pi_{inbundle2psl}$ is the following set of translation definitions:

$$(\forall t)\, timepoint(t) \equiv point(t) \tag{22}$$

$$(\forall o)\, activity\_occurrence(o) \equiv line(o) \tag{23}$$

$$(\forall x)\, object(x) \equiv plane(x) \tag{24}$$

$$(\forall o, t)\, is\_occurring\_at(o, t) \equiv in(t, o) \wedge line(o) \wedge point(t) \tag{25}$$

$$(\forall x, t)\, exists\_at(x, t) \equiv in(t, x) \wedge plane(x) \wedge point(t) \tag{26}$$

$$(\forall x, o, t)\, participates\_in(x, o, t) \equiv tin(o, x, t) \tag{27}$$

then

$$T_{psl\_participates} \cup \Delta_{psl2inbundle} \models T_{present\_bundle}$$
$$T_{present\_bundle} \cup \Pi_{inbundle2psl} \models T_{psl\_participates}$$

Furthermore, these extensions are conservative. Finally, we have

$$T_{psl\_participates} \cup \Delta_{psl2inbundle} \models \Pi_{inbundle2psl}$$
$$T_{present\_bundle} \cup \Pi_{inbundle2psl} \models \Delta_{psl2inbundle}$$

By [16], $T_{psl\_participates}$ is synonymous with $T_{present\_bundle}$.

Theorems 1 and 2 in [3] characterize the models of T_{in_bundle} up to isomorphism; Theorem 5 shows that there is a one-to-one correspondence between the models of $T_{psl\_participates}$ and the models of $T_{present\_bundle}$.

7 Summary

In cases where direct mappings cannot be specified between ontologies, one can design new ontologies that can serve as bridges between the ontologies, and which then allow mappings to be specified. In this paper we have explored how this can be done with the DOLCE and PSL ontologies. In particular, faithful interpretations specified between the DOLCE ontology and ontologies within the Common Logic Ontology Repository (COLORE) have shown that multiple 'bridges' were needed before any analyses with the $T_{dolce\_participation}$ and $T_{dolce\_present}$ theories could be carried out with theories in COLORE. Firstly, we saw that the Combined Time hierarchy *bridges* the Time Points and Periods hierarchies together to allow us to merge ontologies of time points and time intervals. Secondly, the Interval PSL hierarchy *bridges* both the PSL and DOLCE ontologies together to allow us to do the mapping between them and to identify the faithful interpretations of mereology and orderings in both time points and time intervals. Thirdly, the ontologies that axiomatize the class of mathematical structures known as incidence bundles serve as bridges which can be specified through the technique of ontology transfer. This exercise in bridging ontologies together demonstrates how we can axiomatize the relationships between theories and compose new theories that are required for the bridging task.

The methodology presented in this paper is based on techniques for ontology verification that use an ontology repository to specify faithful interpretations among ontologies. It should therefore be applicable to any set of ontologies which have been verified. Nevertheless, ontologies which have not been explicitly modularized still pose a challenge, and the interplay between ontology merging and decomposition merit further exploration.

References

1. Choi, N., Song, I.-Y., Han, H.: A survey on ontology mapping. SIGMOD Rec. **35**(3), 34–41 (2006)
2. Chui, C.: Axiomatized relationships between ontologies. Master's thesis, University of Toronto (2013)
3. Chui, C., Grüninger, M.: Mathematical foundations for participation ontologies. In: Formal Ontology in Information Systems - Proceedings of the Eighth International Conference, FOIS 2014, 22–25 September, Rio de Janeiro, Brazil, pp. 105–118 (2014)
4. Chui, C., Grüninger, M.: Merging the DOLCE and PSL upper ontologies. In: Filipe, J., Dietz, J.L.G., Aveiro, D. (eds.) KEOD 2014 - Proceedings of the International Conference on Knowledge Engineering and Ontology Development, Rome, Italy, 21–24 October, pp. 16–26. SciTePress (2014)
5. de Bruijn, J., Ehrig, M., Feier, C., Martín-Recuerda, F., Scharffe, F., Weiten, M.: Ontology Mediation, Merging, and Aligning, pp. 95–113. Wiley (2006)
6. Enderton, H.B.: A Mathematical Introduction to Logic. Academic Press, New York (1972)
7. Euzenat, J., Shvaiko, P.: Ontology Matching. Springer-Verlag New York Inc., Secaucus (2007)
8. Grüninger, M.: Using the PSL ontology. In: Staab, S., Studer, R. (eds.) Handbook on Ontologies. International Handbooks on Information Systems, pp. 423–443. Springer, Heidelberg (2009)
9. Grüninger, M., Hahmann, T., Hashemi, A., Ong, D., Özgövde, A.: Modular first-order ontologies via repositories. Appl. Ontol. **7**(2), 169–209 (2012)

10. Gruninger, M., Ong, D.: Verification of time ontologies with points and intervals. In: 2011 Eighteenth International Symposium on Temporal Representation and Reasoning (TIME), pp. 31–38 (2011)

11. Hayes, P.: A catalog of temporal theories. Technical report UIUC-BI-AI-96-01, Beckman Institute and Departments of Philosophy and Computer Science, University of Illinois (1996)

12. Kalfoglou, Y., Schorlemmer, M.: Ontology mapping: the state of the art. In: Kalfoglou, Y., Schorlemmer, M., Sheth, A., Staab, S., Uschold, M. (eds.) Semantic Interoperability and Integration, number 04391 in Dagstuhl Seminar Proceedings, Dagstuhl, Germany, February 2005. Internationales Begegnungs- und Forschungszentrum für Informatik (IBFI), Schloss Dagstuhl, Germany (2005)

13. Kutz, O., Mossakowski, T.: A modular consistency proof for DOLCE. In: Burgard, W., Roth, D. (eds.) Proceedings of the Twenty-Fifth AAAI Conference on Artificial Intelligence (AAAI 2011), 7–11 August, pp. 227–234. AAAI Press, San Francisco (2011)

14. Ladkin, P.B.: Time representation: a taxonomy of internal relations. In: Proceedings of the 5th National Conference on Artificial Intelligence, Philadelphia, PA, vol. 1, pp. 360–366 (1986)

15. Masolo, C., Borgo, S., Gangemi, A., Guarino, N., Oltramari, A.: WonderWeb Deliverable D18 Ontology Library (Final). Technical report, IST Project 2001-33052 WonderWeb: Ontology Infrastructure for the Semantic Web (2003)

16. Pearce, D., Valverde, A.: Synonymous theories and knowledge representations in answer set programming. J. Comput. Syst. Sci. **78**(1), 86–104 (2012)

17. Stuckenschmidt, H., Serafini, L., Wache, H.: Reasoning about Ontology Mappings. Technical report, ITC-IRST, Trento (2005)

18. van Benthem, J.: The Logic of Time: A Model-Theoretic Investigation into the Varieties of Temporal Ontology and Temporal Discourse. Synthese Library. Springer, New York (1991)

19. Zimmermann, A., Krötzsch, M., Euzenat, J., Hitzler, P.: Formalizing ontology alignment and its operations with category theory. In: Proceedings of the Fourth International Conference on Formal Ontology in Information Systems (FOIS 2006), pp. 277–288. IOS Press, Amsterdam (2006)

Ontological Support for Modelling
Planning Knowledge

Gerhard Wickler[1]([✉]), Lukáš Chrpa[2], and Thomas Leo McCluskey[2]

[1] AI Applications Institute, School of Informatics,
University of Edinburgh, Edinburgh, UK
g.wickler@ed.ac.uk
[2] PARK Group, School of Computing and Engineering,
University of Huddersfield, Huddersfield, UK

Abstract. This paper describes the conceptual model underlying the Knowledge Engineering Web Interface (KEWI) which primarily aims to be used for modelling planning tasks in a semi-formal framework. This model consists of three layers: a rich ontology, a model of basic actions, and more complex methods. It is this structured conceptual model based on the rich ontology that facilitates knowledge engineering. The focus of this paper is to show how the *central knowledge model* used in KEWI differs from a model directly encoded in PDDL, the language accepted by most existing planning engines. Specifically, the rich ontology enables a more concise and natural style of representation, including function terms as object references. For operational use, KEWI automatically generates PDDL. Experiments show that the generated PDDL can be processed by a planner without incurring significant drawbacks.

Keywords: Knowledge engineering · Automated planning

1 Introduction

Domain-independent planning has grown significantly in recent years mainly thanks to the International Planning Competition (IPC). Besides many advanced planning engines, PDDL, a de-facto standard language family for describing planning domains and problems, has been developed. However, encoding domain and problem models in PDDL requires a lot of specific expertise and thus it is very challenging for a non-expert to use planning engines in applications.

This paper concerns the use of AI planning technology in an organisation where (i) non-planning experts are required to encode knowledge (ii) the knowledge base is to be used for more than one planning and scheduling task (iii) it is maintained by several personnel over a long period of time, and (iv) it may have a range of potentially unanticipated uses in the future. The first concern has been a major obstacle to using AI-based tools which input formal representations, in that the expertise required to encode such representations has only been possessed by planning experts (e.g. as in NASA's applications [1]). The other concerns are often not covered in the planning literature: in real applications the knowledge encoding is a valuable, general asset, and one that requires a much richer conceptual representation than, for example is accorded by PDDL.

Very few collaborative, domain-expert-usable, knowledge acquisition interfaces are available that are aimed at supporting the harvesting of planning knowledge within a rich language for use in a number of planning-related applications. After initial acquisition, the validation, verification, maintenance and evolution of such knowledge is of prime importance, as the knowledge base is a valuable asset to an organisation.

This paper introduces the Knowledge Engineering Web Interface (KEWI), which aims to enable the acquisition and modelling of knowledge necessary for use with automated plan generation tools. Here we detail the theoretical aspects of KEWI, and evaluate it using a well-understood planning domain.

2   Related Work

A small number of frameworks exist that support the formalisation of planning knowledge in shared web-based systems. Usually, such frameworks build on existing Web 2.0 technologies such as a wiki. A wiki that supports procedural knowledge is available at wikihow.com, but the knowledge remains essentially informal. A system that uses a similar approach, namely, representing procedural knowledge in a wiki is CoScripter [2]. However, their representation is not based on AI planning and thus does not support the automated composition of procedures. More recently, an AI-based representation has been used in OpenVCE [3].

There have been several attempts to create general, user-friendly development environments for planning domain models, but they tend to be limited in the expressiveness of their underlying formalism. The Graphical Interface for Planning with Objects (GIPO) [4] is based on object-centred languages OCL and OCL_h. These formal languages exploit the idea that a set of possible states of objects are defined first, before action (operator) definition [5]. This gives the concept of a *world state* consisting of a set of states of objects, satisfying given constraints. GIPO uses a number of consistency checks such as if the object's class hierarchy is consistent, object state descriptions satisfy invariants, predicate structures and action schema are mutually consistent and task specifications are consistent with the domain model. Such consistency checking guarantees that some types of errors can be prevented, in contrast to ad-hoc methods such as hand crafting.

itSIMPLE [6] provides a graphical environment that enables knowledge engineers to model planning domain models by using the Unified Modelling Language (UML). Object classes, predicates, action schema are modelled by UML diagrams allowing users to 'visualise' domain models which makes the modelling process easier. itSimple incorporates a model checking tool based on Petri Nets that are used to check invariants or analyse dynamic aspects of the domain models such as deadlocks.

The Extensible Universal Remote Operations Planning Architecture (EUROPA) [7], is an integrated platform for AI planning and scheduling, constraint programming and optimisation. This platform is designed to handle complex real-world problems, and the platform has been used in some of NASA's missions. EUROPA supports two representation languages, NDDL and ANML [8], however, PDDL is not supported.

Besides these tools, it is also good to mention VIZ [9], a simplistic tool inspired by itSimple, and PDDL Studio [10], an editor which provides users a support by, for instance, identifying syntax errors or highlighting components of PDDL.

In the field of Knowledge Engineering, methodologies have been developed which centre on the creation of a precise, declarative and detailed model of the area of knowledge to be engineered, in contrast to earlier expert systems approaches which appeared to focus on the "transfer" expertise at a more superficial level. This "expertise model" contains a mix of knowledge about the "problem solving method" needed within the application and the declarative knowledge about the application. Often a key rationale for knowledge engineering is to create declarative representations of an area to act as a formalised part of some requirements, making explicit what hitherto has been implicit in code, or explicit but in documents. Knowledge Engineering modelling frameworks arose out of this, such as CommonKADS [11], which were based on a deep modelling of an area of expertise, and emphasising a life-cycle of this model. The "knowledge model" within CommonKADS, which contains a formal encoding of task knowledge, such as problem statement(s), as well as domain knowledge, is similar to the kind of knowledge captured in KEWI. Unlike KEWI however, this model was expected to be created by knowledge engineers rather than domain experts and users.

AI Planning. The primary aim for KEWI is to ease the formalisation of procedural knowledge, allowing domain experts to encode their knowledge themselves, rather than knowledge engineers having to elicit the knowledge before they formalise it into a representation. We formally describe the conceptual model which consists of three layers: a rich ontology, a model of basic actions, and more complex methods. KEWI is object-centred and allows for a richer representation of knowledge than PDDL. It is more compact and more expressive which means models are easier to maintain, especially for a user who is not an expert in AI planning. KEWI's internal representation can be exported to PDDL and hence standard planning engines can be applied to solve planning problems modelled in KEWI. We demonstrate that PDDL models exported from KEWI are comparable to hand coded ones.

AI planning deals with the problem of finding a sequences of actions transforming the environment from a given initial state to a desired goal state [12]. *Actions* are defined via their preconditions and effects. An action is *applicable* in a given state if and only if its precondition holds in that state. Effects of an action denote how a state where the action is applied will change. A *planning domain model* consists of a set of predicates and/or fluents describing the environment and a set of actions modifying the environment. A *planning problem* consists of a planning domain model, set of objects, an initial state and a set of goal conditions.

3   Conceptual Model of KEWI

KEWI is a tool for encoding domain knowledge mainly by experts in the application area rather than AI planning experts. The idea behind KEWI is to provide a user-friendly environment as well as a language which is easier to follow, especially for users who are not AI planning experts. A high-level architecture of KEWI is depicted in Fig. 1. Encoded knowledge can be exported into the domain and problem description in PDDL on which standard planning engines can be applied. Hence, the user does not have to understand, or even be aware, of any PDDL encodings.

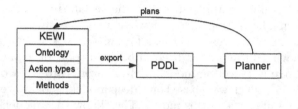

Fig. 1. An architecture of KEWI.

A language in which domain knowledge is encoded in KEWI has three parts, which are explained in the following subsections. First, a rich domain ontology is defined. The domain ontology consists of definitions of classes of objects, hierarchies of classes and relations between objects. Second, action types are defined in terms of their action name, logical preconditions and effects. Third, methods define ways in which high-level task can be broken down into lower-level activities, a so-called task network which includes explicit ordering constraints.

3.1   Ontology: Concepts, Relations and Properties

Ontological elements are usually divided into concepts and instances. Typically, the concepts are defined in a planning domain whereas the instances are defined in a planning problem. Since our focus for KEWI is on planning domains we shall mostly deal with concepts here.

Concepts. A concept is represented by a unique symbol in KEWI. The formal definition of a concept is given by its super-class symbol and by a set of role constraints that define how instances of the concept may be related to other concepts. In KEWI, the definition of a concept also includes other, informal elements that are not used for formal reasoning. However, the knowledge engineering value of such informal elements must not be underestimated, much like the comments in programming often are vital for code to be understandable.

Definition 1 (KEWI Concept). *A concept C in KEWI is a pair $\langle C^{sup}, R \rangle$, where:*

- *C^{sup} is the direct super-concept of C and*
- *R is a set of role constraints of the form $\langle r, n, C' \rangle$ where r is a symbolic role name, C' is a concept (denoting the role filler type), and n is a range $[n_{min}, n_{max}]$ constraining the number of different instances to play that role.*

We assume that there exists a unique root concept often referred to as *object* or *thing* that acts as the implicit super-concept for those concepts that do not have an explicit super-concept defined in the same planing domain. Thus, a concept C may be defined as $\langle \triangle, R \rangle$, meaning its super concept is implicit. This implicit super-concept has no role constraints attached.

For example, in the Dock Worker Robot (DWR) domain [12], the concepts `container` and `pallet` could be defined with the super-concept `stackable`, whereas the concept `crane` could be defined as a root concept with no super-concept (implicitly: $\triangle$). A role constraint can be used to define that a crane can hold at most one container as follows: $\langle \texttt{holds}, [0, 1], \texttt{container} \rangle$.

Since super-concepts are also concepts, we can write a C as $\langle \langle \langle \triangle, R_n \rangle, \ldots, R_2 \rangle, R_1 \rangle$. Then we can refer to all the role constraints associated with C as $R^* = R_n \cup \ldots \cup R_2 \cup R_1$, that is, the role constraints that appear in the definition of C, the role constraints in its direct super-concept, the role constraints in its super-concepts super-concept, etc.

The reason for introducing this simple ontology of concepts is that we can now constrain the set of possible world states based on the role constraints. States are defined as sets of ground, first-order atoms over some function-free language $\mathcal{L}$. This language shall contain symbols to denote each instance of a concept defined in the ontology $(c_1, \ldots, c_L)$ where the type function τ maps each instance c_i to its type C, a concept in the ontology. The relation symbols of $\mathcal{L}$ are defined through the role constraints.

Definition 2 (Relations in $\mathcal{L}$). *Let $\langle r, n, C' \rangle$ be a role constraint of some concept C. Then the first-order language $\mathcal{L}$ that can be used to write ground atoms in a state contains a binary relation $C.r \subseteq C \times C'$.*

In what follows we shall extend the language to include further relation symbols, but for now these relations defined by the ontology are all the relations that may occur in a state. The reason why the relation name is a combination of the concept and the role is simply to disambiguate between roles of the same name but defined in different concepts. Where all role names are unique the concept may be omitted.

We can now define what it means for a state to be valid with respect to an ontology defined as a set of KEWI concepts. Essentially, for a state to be valid, every instance mentioned in the state must respect all the role constraints associated with the concepts to which the instance belongs. Since role constraints are constraints on the number of possible role fillers we need to be able to count these.

Definition 3 (Role Fillers). *Let s be a state, that is, a set of ground atoms over objects $c_1, \ldots, c_L$ using the relations in $\mathcal{L}$. Let $\langle r, n, C' \rangle$ be a role constraint of some concept C. Then we define $vals_s(C.r, c_i) = \{c_f | C.r(c_i, c_f) \in s\}, c_i \in C, c_f \in C'$, that is, the set of all constants that play role r for c_i in s.*

Definition 4 (Valid State). *Let C be a KEWI concept. Then a state s is valid if, for any instance c_i of C and any role constraint $\langle r, n, C' \rangle$ of C or one of its (direct and indirect) super-concepts, the number of ground atoms $a = C.r(c_i, *)$ must be in the range $[n_{min}, n_{max}]$, i.e. $n_{min} \leq |vals_s(C.r, c_i)| \leq n_{max}$.*

Thus, a concept definition defines a set of role constraints which can be interpreted as relations in a world state. The numeric range defines how many ground instances we may find in a valid state. This is the core of the ontological model used in KEWI.

For example, let k1 be a crane and ca be a container. Then a state may contain a ground atom crane.holds(k1,ca). If a state contains this atom, it may not contain another one using the same relation and k1 as the first argument.

Relations. While the relations defined through the concepts in KEWI provide a strong ontological underpinning for the representation, there are often situations where other relations are more natural, e.g. to relate more than two concepts to each other, or where a relation does not belong to a concept. In this case relations can be defined by declaring number and types (concepts) of the expected arguments.

Definition 5 (Relations in $\mathcal{L}$). *A relation may be defined by a role constraint as described above, or it may be a relation symbol followed by an appropriate number of constants. The signature of a relation R is defined as $C_1 \times \ldots \times C_R$ where C_i defines the type of the ith argument.*

A valid state may contain any number of ground instances of these relations. As long as the types of the constants in the ground atoms agree with the signature of the relation, the state that contains this atom may be valid.

Properties. In reality, we distinguish three different types of role constraints: *related classes* for defining arbitrary relations between concepts, *related parts* which can be used to define a "part-of" hierarchy between concepts, and *properties* which relate instances to property values.

The first two are equivalent in the sense that they relate objects to each other. However, properties usually relate objects to values, e.g. an object may be of a given colour. While it often makes sense to distinguish all instances of a concept, this is not true for properties. While the paint that covers one container may not be the same paint that covers another, the colour may be the same. To allow for the representations of properties in KEWI, we allow for the definition of properties with enumerated values.

Definition 6 (Properties). *A property P is defined as a set of constants* $\{p_1, \ldots, p_P\}$.

It is easy to see that the above definitions relating to role constraints and other relations can be extended to allow properties in place of concepts and property values in place of instances. A minor caveat is that property values are usually defined as part of a planning domain, whereas instances are usually given in a planning problem.

3.2   Action Types

Action types in KEWI are specified using an operator name with typed arguments, a set of preconditions, and a set of effects. This high-level conceptualisation of action types is of course very common in AI planning formalisms. KEWI's representation is closely linked with the ontology, however. This will enable a number of features that allow for a more concise representation, allowing to reduce the redundancy contained in many PDDL planning domains.

Object References. In many action representations it is necessary to introduce one variable for each object that is somehow involved in the execution of an action. This variable is declared as one of the typed arguments of the action type. The variable can then be used in the preconditions and effects to consistently refer to the same objects and express conditions on this object.

Sometimes, an action type may need to refer to specific constants in its preconditions or effects. In this case, the unique symbol can be used to identify a specific instance. In the example above, k1 was used to refer to a crane and ca to refer to a container. In most planning domains, operator definitions do not refer to specific objects, but constants may be used as values of properties.

In addition to variables and constants, KEWI also allows a limited set of function terms to be used to refer to objects in an action type's preconditions and effects. Not surprisingly, this is closely linked with the ontology, specifically with the role constraints that specify a maximum of one in their range.

Definition 7 (Function Terms). *Let $\langle r, n, C' \rangle$ be a role constraint of some concept C where $n_{max} = 1$. Then we shall permit the use of function terms of the form $C.r(t)$ in preconditions and effects, where t can again be an arbitrary term (constant, variable, or function term) of type C'.*

Let s be a valid state, that is, a set of ground atoms over objects $c_1, \ldots, c_L$ using the relations in $\mathcal{L}$. Then the constant represented by the function term $C.r(c_i)$ is:

– c_j *if* $vals_s(C.r, c_i) = \{c_j\}$, *or*
– *nothing $(\perp)$ if* $vals_s(C.r, c_i) = \emptyset$.

Note that the set $vals_s(C.r, c_i)$ can contain at most one element in any valid state. If it contains an element, this element is the value of the function term.

Otherwise a new symbol that must not be one of the constants $c_1, \ldots, c_L$ will be used to denote that the function term has no value. This new constant `nothing` may also be used in preconditions as described below.

The basic idea behind function terms is that they allow the knowledge representation to be more concise; it is no longer necessary to introduce a variable for each object. Also, this style of representation may be more natural, e.g. to refer to the container held by a crane as `crane.holds(k1)` meaning "whatever crane `k1 holds`", where the role constraint tells us this must be a container. As a side effect, the generation of a fully ground planning problem could be simpler, given the potentially reduced number of action parameters.

Interestingly, a step in this direction was already proposed in PDDL 1, in which some variables were declared as parameters and others as "local" variables inside an operator. However, with no numeric constraints on role fillers or any other type of relation, it is difficult to make use of such variables in a consistent way. Similarly, state-variable representations [13] exploit the uniqueness of a value. However, this was restricted to the case where n_{min} and n_{max} both must be one.

Condition Types. The atomic expressions that can be used in preconditions and effects can be divided into two categories. Firstly, there are the explicitly defined relations. These are identical in meaning and use to PDDL and thus, there is no need to discuss these further. Secondly, there are the relations based on role constraints which have the same form as such atoms in states, except that they need not be ground.

Definition 8 (Satisfied Atoms). *Let s be a valid state over objects $c_1, \ldots, c_L$. Then a ground atom a is satisfied in s (denoted $s \models a$) if and only if:*

– *a is of the form $C.r(c_i, c_j)$ and $a \in s$, or*
– *a is of the form $R(c_{i_1}, \ldots, c_{i_R})$ and $a \in s$, or*
– *a is of the form $C.r(c_i, \bot)$ and $vals_s(C.r, c_i) = \emptyset$.*

The first two cases are in line with the standard semantics, whereas the the last case is new and lets us express that no role filler for a given instance exists in a given state. Note that the semantics of atoms that use the symbol `nothing` in any other place than as a role filler are never satisfied in any state.

The above definition can now be used to define when an action is applicable in a state.

Definition 9 (Action Applicability). *Let s be a valid state and act be an action, i.e. a ground instance of an action type with atomic preconditions $p_1, \ldots, p_a$. Then act is applicable in s if and only if every precondition is satisfied in s: $\forall p \in p_1, \ldots, p_a : s \models p$.*

This concludes the semantics of atoms used in preconditions. Atoms used in effects describe how the state of the world changes when an action is applied. This is usually described by the state transition function $\gamma : S \times A \to S$, i.e. it maps a state and an applicable action to a new state. Essentially, γ modifies the given state by deleting some atoms and adding some others. Which atoms are deleted and which are added depends on the effects of the action. If the action is not applicable the function is undefined.

Definition 10 (Effect Atoms). *Let s be a valid state and act be an action that is applicable in s. Then the successor state $\gamma(s, a)$ is computed by:*

1. *deleting all the atoms that are declared as negative effects of the action,*
2. *for every positive effect $C.r(c_i, c_j)$ for role constraint $\langle r, n, C' \rangle$ with $n = [n_{min}, 1]$, if $C.r(c_i, c_k) \in s$ delete this atom, and*
3. *add all the atoms that are declared as positive effects of the action.*

Following this definition allows for a declaration of actions using arbitrary relations and state-variables that may have at most one value. The ontology, more specifically the numeric role constraints can be used to distinguish the two cases.

3.3 Methods

The approach adopted in KEWI follows standard HTN planning concepts: a method describes how a larger task can be broken down in into smaller tasks which, together, accomplish the larger task. Technically, a method is defined as an extension of an action type in the object-oriented sense. That is, a method consists of a name, typed arguments, preconditions and effects, which are inherited from action types. In addition, a method must define a task describing *what* is accomplished by the method and the subtask network describing *how* this task is accomplished with this method.

Typical HTN formalisms include all of the above, except for effects. When a method declares that it achieves a (high-level) effect, then every decomposition of this method must result in an action sequence which will achieve the effect after the last action of the sequence has been completed. This allows a flat planner (one that takes PDDL as input) to use a method as if it was an action type. Thus, hierarchical planning is not an alternative to flat planning approaches as it is traditionally viewed, but an extension that may be used to provide optional guidance to a planner.

The task that is accomplished by a method is defined by a task name usually describing what is to be done, and again some parameters which must be a subset of the method parameters. For primitive tasks, the task name will be equal to the name of an action type, in which case no further refinement is required, that is, there will be no subtask network. Note that there may be multiple methods that have the same task.

For non-primitive tasks, a method also includes a set of subtasks. In KEWI, the ordering constraints between subtasks are declared with the subtask, rather than as a separate component of the method. This is simply to aid readability without changing the expressiveness. Subtaks may be specified in one of two forms: as perfomable subtasks or as achievable subgoals.

A performable subtask is defined by a task name and some parameters. The task name may be the name of an action type, in which case the task is considered primitive. Otherwise there must be a method that has a matching task in its definition, and this method may then be used to refine the subtask. This refinement process is typically done by an HTN planner.

An achievable subgoal can be either of the condition types defined above, relations or role constraints. For example, the subgoal "achieve $C.r(c_i, c_j)$" may be used to state that at this point in the subtask network the condition $C.r(c_i, c_j)$ must hold in the corresponding state. Conjunctions of subgoals can be represented my a set of subgoals that are unordered with respect to each other. The refinement process that finds actions to be inserted into the plan which achieve the subgoals is what is typically computed by a flat planner, e.g. using state-space search.

This mixed approach is not new and has been used in practical planners like O-Plan [14].

3.4   Export to PDDL

Given that most modern planners accept planning domains and problems in PDDL syntax as their input, one of the goals for KEWI was to provide a mechanism that exports the knowledge in KEWI to PDDL. Of course, this will not include the HTN methods as PDDL does not support hierarchical planning formalisms.

Function Terms. The first construct that must be removed from KEWI's representation are the function terms that may be used to refer to objects. In PDDL's preconditions and effects only variables (or symbols) may be used to refer to objects. The following function can be used to eliminate a function term of the form $C.r(t)$ that occurs in an action type O's preconditions or effects.

> **function** eliminate-fterms($C.r(t), O$)
> > **if** is-fterm(t) **then**
> > > eliminate-fterms(t, O)
> >
> > $v \leftarrow$ get-variable($C.r(t), O$)
> > replace every $C.r(t)$ in O by v

The function first tests whether the argument to the given function term is itself a function term. If this the case, it has to be eliminated first. This guarantees that, for the remainder of the function t is either a variable or a symbol. We then use the function "get-variable" to identify a suitable variable that can replace the function term. Technically, this function may return a symbol, but the treatment is identical, which is why we shall not distinguish these cases here. The identification of a suitable variable then works as follows.

```
function get-variable(C.r(v), O)
    for every positive precondition p of O do
        if p = C.r(v, v') then
            if is-fterm(v') then
                eliminate-fterms(v', O)
            return v'
    retrieve ⟨r, n, C'⟩ from C
    add new parameter v' of type C' to O
    add new precondition C.r(v, v') to O
    return v'
```

This function first searches for an existing, positive precondition that identifies a value for the function. Since function terms may only be used for constraints that have at most one value, there can only be at most one such precondition. If such a precondition exists, its role filler (v', a variable or a symbol) may be used as the result. If no such precondition can be found, the function will create a new one and add it to the operator. To this end, a new parameter must be added to the action type, and to know the type of the variable we need to retrieve the role filler type from the role constraint. In practise, we also use the type name to generate a suitable variable name. Then a new precondition can be added that effectively binds the function to the role filler. And finally, the new variable may be returned.

(**Handling nothing**). The next construct that needs to be eliminated from the KEWI representation is any precondition that uses the role filler **nothing**. Note that this symbol does not occur in states and thus cannot be bound in traditional PDDL semantics. Simply adding this symbol to the state causes problems since other preconditions that require a specific value could then be unified with this state atom. For example, if we had an explicit atom that stated holds(k1, nothing) in our state, then the precondition holds(?k, ?c) of the load action type would be unifiable with this atom. The approach we have implemented in KEWI is described in the following algorithm.

```
function eliminate-nothing(O)
    for every precondition p = C.r(v, ⊥) do
        replace p with C.r. ⊥ (v)
        if O has an effect e = C.r(v, v')
            add another effect ¬C.r. ⊥ (v)
```

The idea behind this approach is to use a new predicate to keep track of state-variables that have no values in a state. This is the purpose of the new predicate "$C.r. \perp$", indicating the role r of concept C has no filler for the given argument. This is a common approach in knowledge engineering for planning. For example, in the classic blocks world we find a "holds" relation for when a block is being held, and a predicate "hand-empty" for when no block is held.

The algorithm above uses this technique to replace all preconditions that have **nothing** as a role filler with a different precondition that expresses the non-existence of the role filler. To maintain this condition, it will also be necessary to modify the effects accordingly. This is done by adding the negation of this new predicate to corresponding existing effects.

Since this is pseudo code, the algorithm actually omits a few details, e.g. the declaration of the new predicate in the corresponding section of the PDDL domain, and the fact that the planning problem also needs to be modified to account for the new predicate. Both is fairly straight forward to implement.

An alternative approach we have implemented essentially keeps the `nothing` symbol in the representation. To ensure that no action type uses this object a number of inequality preconditions have to be added to the operators. This requires that the planner can correctly handle inequalities. Note that inequalities are static relations that disappear when the domain is grounded by the planner.

Keeping `nothing` literally causes another issue with typed domain, since `nothing` can be an instance of multiple classes in our ontology. To avoid this problem we replace this symbol with different symbols for the different types of which it can be an instance.

This approach is not fundamentally different from the one described above. It trades off a larger number of predicates against larger sets of instances in the domain. The modified problem descriptions are almost identical.

State-Variable Updates. Finally, the cases in which the value of a state-variable is simply changed needs to be handled. The approach we have adopted here is identical to the approach described in [12]. That is, when an effect assigns a new value to a state-variable, e.g. $C.r(v, v_{new})$, we need to add a precondition to get the old value, e.g. $C.r(v, v_{old})$, and then we can use this value in a new negative effect to retract the old value: $\neg C.r(v, v_{old})$.

4   Evaluation: The Dock Worker Robots Domain

In this section we shall describe some experiences gained while re-engineering an existing and well understood planning domain, the dock worker robots (DWR) domain described in [12]. Basically, a problem in this domain consists of a set of locations at which containers are piled into stacks. Cranes at these locations can move the containers around at the same location, and robots can be used to move containers between locations. The current state is a given configuration of containers in piles and the goal is usually to shift the containers to different piles.

4.1   Ontology

The original planning domain specified in PDDL defines a trivial ontology that consists of just the five types of objects that are involved in the actions as shown in Fig. 2. There is hierarchy and concepts are defined by name only. The text following a semicolon are comments and ignored by the reasoning engine.

Apart from the lack of any intensional knowledge about these types, this conceptualisation also does not use a separate type for the pallets that are at

```
(:types
  location      ; there are several connected locations in the harbour
  pile          ; is attached to a location
                ; it holds a pallet and a stack of containers
  robot         ; holds at most 1 container, only 1 robot per location
  crane         ; belongs to a location to pickup containers
  container)
```

Fig. 2. The types declared in the original PDDL domain.

the bottom of each pile. In fact, a single pallet is declared as an instance of type `container` in the problem files and the same pallet is used at the bottom of every pile. This solution works in a planning engine but is clearly unsatisfactory from a knowledge engineering perspective.

The KEWI version of the domain we have developed is shown in Fig. 3 is obviously much richer. There is some hierarchical structure, e.g. the class `stackable` has two sub-classes, `container` and `pallet`. Most classes have associated role constraints that provide an intensional definition of the class. In addition to the types from the original domain, the KEWI version also defines a colour property which is there solely to illustrate the use of properties and should be ignored by a planner. Adjacency between locations is specified as a relation not associated with any concept.

The use of an explicit class for pallets is the only significant difference in the original conceptualisation and the KEWI version of the DWR ontology. What appears as a complication at first is actually a simplification since there is no

```
(:class agent)
(:class crane (:super-class agent)
  (:role at (:min 1) (:max 1) (:class location))
  (:role holds (:max 1) (:class container)))
(:class robot (:super-class agent)
  (:role loaded-with (:max 1) (:class container))
  (:property has-colour (:min 1) (:max 1) (:type colour)))
(:class location
  (:role occupied-by (:max 1) (:class robot)))
(:class stackable)
(:class container (:super-class stackable)
  (:role on (:max 1) (:class stackable))
  (:role piled-on (:max 1) (:class pallet))
  (:property paint (:min 1) (:max 1) (:type colour)))
(:class pallet (:super-class stackable)
  (:role at (:min 1) (:max 1) (:class location))
  (:role top (:min 1) (:max 1) (:class stackable)))
(:property colour
  (:values ( red green blue )))
(:relation adjacent
  (:arguments ( (?loc1 location) (?loc2 location) )))
```

Fig. 3. The ontology of the DWR domain in KEWI.

longer a need for piles. Piles can be identified by the pallets on which they are stacked.

When the PDDL version of the domain is generated from the KEWI environment, most of the ontological information is lost, of course, as PDDL is not sufficiently expressive for the kind of ontology KEWI uses. However, the information is used to generate additional relations and modify the operator specifications that are generated.

4.2 Action Types

The original PDDL specification of the DWR domain specifies five action types:

- move: a robot moves from one location to an adjacent location
- load: a crane loads the container onto a robot
- unload: a crane unloads a container from a robot
- take: a crane takes a container from a pile
- put: a crane puts the container onto a pile

All of these operators were (manually) re-encoded in KEWI, exploiting the richer ontology and other language features described above. The result is a more concise representation that reduces the need for certain explicit, but redundant parameters, preconditions and effects. The generation of PDDL from KEWI results in a specification that cannot make reference to the ontology and therefore is not as concise as the KEWI version. In fact, as shown is Table 1, it even uses some additional parameters, preconditions and effects.

Clearly, the KEWI version of the domain is the most concise. The use of function terms avoids the explicit introduction of parameters. From a knowledge engineering perspective, this means the actions can be specified in terms of the main objects involved. A similar construct is available in PDDL, where certain variables are "local" and not used in the parameter specification. However, this is not widely supported by planners and used in few domain specifications. The KEWI representation also uses fewer preconditions and effects. This is mostly because of the use of role constraints with nothing as their value. Thus, both reasons for a more concise representation are directly related to the richer ontology.

Table 1. Number of parameters, preconditions and effects in different versions of DWR.

	Original PDDL			KEWI			Generated PDDL		
	params	precs	effects	params	precs	effects	params	precs	effects
move	3	3	4	3	3	2	3	3	4
load	4	4	4	3	3	2	4	4	4
unload	4	4	4	3	3	2	4	4	4
take	5	6	6	2	3	4	6	7	8
put	5	4	6	3	4	4	6	7	8

```
(:action put
    :parameters (?k - crane ?c - container ?p - pile)
    :vars (?else - container ?l - location)
    :precondition (and (belong ?k ?l) (attached ?p ?l)
                       (holding ?k ?c) (top ?else ?p))
    :effect (and (in ?c ?p) (top ?c ?p) (on ?c ?else)
                 (not (top ?else ?p)) (not (holding ?k ?c)) (empty ?k))))
```

Fig. 4. The original PDDL version of **put**.

Perhaps the most interesting operator to take a closer look at is the most complex action type specified here, the **put** action. The original PDDL version is shown in Fig. 4. The first local variable, **?else**, is a reference to the container (or pallet) that is at the top of the pile before the action executed. Two of the preconditions are static, two are dynamic. Interestingly, the last of the preconditions, (**top ?else ?p**), is not so much a logical precondition but simply a way to bind the local variable **?else** such that it may be used in the effects. There are no negative preconditions. The effects are a mixture of four positive and two negative effects.

Compare this to the KEWI version of the same operator. In the web interface, the normal view provides many links for navigating the knowledge, whereas the edit view shows a form with fields for different parts of the representation are shown in Fig. 5. The explicit break-down in the edit view is meant to support the knowledge engineer by listing the components available in the language.

The complete formalism specifying the action type in KEWI is shown in Fig. 6. At first glance it appears less concise, but this is simply because the symbols are more verbose, e.g. no single letter variables are used and role constraints make the object type explicit. None of the local variables need to be represented in KEWI as the values they refer to can be described with functional terms.

The first of the preconditions requires the location of the crane and the location of the pile to be equal. This corresponds to the first two preconditions in the PDDL version, where the equality is implicit in the use of the same variable. The main dynamic precondition, that the crane must hold the container, is present in both representations. The remaining preconditions are necessary to make the planner work in both cases, to bind the local variable **?else** or to declare the values of state-variables before execution. Normally this is not necessary in KEWI, unless there is no previous value (i.e. the value is **nothing**) which is the case here. This could be avoided through the use of axioms, which can be declared in the KEWI ontology, but are not currently used for reasoning. Such an axiom would state that, if a container is held by a crane, this container is not on another stackable object (container or pallet) and that this container is not piled on any pallet. Using axioms like this would avoid the last two preconditions shown in the KEWI operator.

The effects of the KEWI version are fairly straight forward. The first effect corresponds to the first effect of the original operator. The second effect

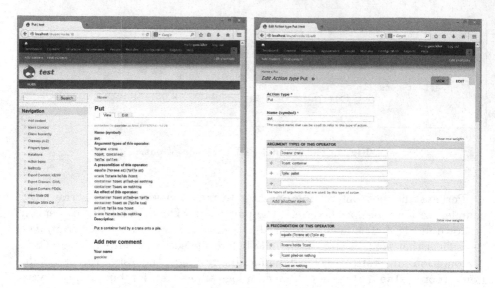

Fig. 5. Different views of put in KEWI.

corresponds to the third original effect, where the previously top-most container is referred to by a function term, rather than a local variable. The third KEWI effect corresponds directly to the second original effect and the first negative effect. The use of a state-variable makes this more concise in KEWI. The final KEWI effect corresponds to the second negative original effect and the final positive effect.

The final version of the put operator is the PDDL version that is generated by KEWI. The full specification is not shown here as the additional parameters, preconditions and effects are just artefacts of the translation process, and they are not very interesting. For example, the equality precondition in KEWI translates into two additional parameters and preconditions, one for each function

```
(:action-type put
  (:arguments ( (?crane crane) (?cont container) (?pile pallet) ))
  (:precondition (:and
    (:relation equals ( (crane.at ?crane) (pallet.at ?pile) ))
    (:constraint crane.holds ( ?crane ?cont ))
    (:constraint container.piled-on ( ?cont nothing ))
    (:constraint container.on ( ?cont nothing ))))
  (:effect (:and
    (:constraint container.piled-on ( ?cont ?pile ))
    (:constraint container.on ( ?cont (pallet.top ?pile) ))
    (:constraint pallet.top ( ?pile ?cont ))
    (:constraint crane.holds ( ?crane nothing )))))
```

Fig. 6. The KEWI version of put.

term. Then the equality itself becomes another precondition on the two new parameters. Clearly, there is optimisation potential, but since this representation is not meant for human consumption and planners should be well capable of compiling out these redundancies, this is not something we are going to dwell on.

4.3 Planning with KEWI

More interesting is to see what a planner actually makes of the PDDL generated by KEWI compared to the original PDDL version. To this end, we have taken two DWR problems[1] and adapted them to the representation used in the generated KEWI. This has to be done manually, which is why only two examples were used. The translation is fairly straightforward, requiring the change of all predicate names and in one case swapping the order of the parameters. The change from a single pallet to multiple pallets but no piles was surprisingly trivial. The only real difference is the occupied predicate used in the original problems. The KEWI-generated PDDL contains a location-no-occupied-by predicate which is effectively the complement of the occupied predicate, indicating when a location is free.

With these two problems translated (manually) it was then possible to run a planner on them. We have chosen the FF planning system, a robust state-space search planner that supports all the features used in the original as well as the generated version of the problem. The result is shown in Table 2.

Table 2. Running FF on different versions of two DWR problems.

	Reachable		States searched	Initial distance	Plan length	Total time
	Facts	Actions				
PDDL pb12	121	362	101	25	34	0.00
KEWI pb12	161	506	96	25	34	0.31
PDDL pb38	1453	15306	94265	104	277	2473.12
KEWI pb38	1889	21642	72565	104	235	3637.81

The first problem is a very simple case with two locations, one robot and six containers to be moved to the other location. The second problem is complex, involving eight locations and three robots that need to shift more than 20 containers around the locations. The initial reachability analysis performed by FF shows that the KEWI generated version has more facts and actions, which means the additional predicates give rise to some redundant information that cannot be compiled away by the planner. Interestingly enough, this redundant information leads to a smaller number of states being explored for both problems. However, processing the additional information incurs an overhead that

---

[1] See http://projects.laas.fr/planning/ for a full definition of these problems.

results in a larger overall search time. While this is not good news, it must also be pointed out that the resulting plan is shorter at least for the more complex problem. Thus, one could argue the performance is roughly equivalent for the original and the KEWI-generated versions.

Finally, we should note that the alternative method for eliminating nothing from the domain is not mentioned here simply because it turns out that FF generates the same ground version of the problem in both cases, thus leading to almost identical performance.

4.4 Further Evaluation

This work is being carried out with an industrial partner with significant experience in control and automation as well as simulation, and we are using a real application of knowledge acquisition and engineering in their area of expertise. The development of KEWI is in fact work in progress, and its evaluation is ongoing, and being done in several ways: (i) An expert engineer from the industrial partner is using KEWI, in parallel with the developers, to build up a knowledge base of knowledge about artefacts, operations, procedures etc. in their domain. (ii) We have created a hand-crafted PDDL domain and problem descriptions of part of the partner's domain and for the same problem area we have generated PDDL automatically from a tool inside KEWI. We are in the process of comparing the two methods and the PDDL produced. An interface to a simulation system is being developed which will help in this aspect. (iii) We are working with another planning project in the same application, which aims to produce natural language explanations and argumentation supporting plans. In the future we believe to combine KEWI with this work, in order that (consistent with involving the user in model creation) the user will be able to better validate the planning operation.

5   Conclusions

In this paper we have introduced KEWI, a knowledge engineering tool for modelling planning tasks, and we have given a formal account of parts of its structure and tools. In contrary to standard literal-centred approach used in PDDL, KEWI represents domain knowledge in an object-centred way. As well as the usual advantages of an object-centred approach, the use of a rich ontology with numeric role constraints enables the use of function terms as object references and explicit non-existence conditions. This allows for a more concise and more natural style of representing planning knowledge. Hence, it is easier, especially for users not being experts in automated planning, to capture and maintain domain knowledge in KEWI. Moreover, KEWI has a user-friendly interface which is simple enough to support domain experts in encoding knowledge and it is designed to enable groups of users to capture, store and maintain knowledge over a period of time, thus facilitating knowledge reuse.

Since PDDL is widespread in the planning community and thus most of the state-of-the-art planning engines supports it, KEWI is able to export domain knowledge into PDDL. We demonstrated that no significant differences are between hand-crafted and automatically generated PDDL models. On the other hand, when running a planner on a more complex problem, results were considerably different.

In future work, we plan to extend KEWI by (i) extending the representation to include numeric fluents, time, and, eventually, continuous processes (ii) developing validation and verification methods which help users to debug and adapt created planning domain and problem descriptions (iii) adding automated acquisition tools which can add to KEWI's knowledge by inputting batch or real time data from process simulations inspired by the real domain KEWI is being used to model.

Acknowledgements. The research was funded by the UK EPSRC Autonomous and Intelligent Systems Programme (grant no. EP/J011991/1). The University of Edinburgh and research sponsors are authorised to reproduce and distribute reprints and online copies for their purposes notwithstanding any copyright annotation hereon.

References

1. Ai-Chang, M., Bresina, J.L., Charest, L., Chase, A., jung Hsu, J.C., Jnsson, A.K., Kanefsky, B., Morris, P.H., Rajan, K., Yglesias, J., Chafin, B.G., Dias, W.C., Maldague, P.F.: Mapgen: mixed-initiative planning and scheduling for the mars exploration rover mission. IEEE Intell. Syst. **19**, 8–12 (2004)
2. Leshed, G., Haber, E. M., Matthews, T., Lau, T. A.: Coscripter: automating and sharing how-to knowledge in the enterprise. In: CHI, pp. 1719–1728 (2008)
3. Wickler, G., Tate, A., Hansberger, J.: Using shared procedural knowledge for virtual collaboration support in emergency response. IEEE Intell. Syst. **28**, 9–17 (2013)
4. Simpson, R., Kitchin, D.E., McCluskey, T.: Planning domain definition using GIPO. Knowl. Eng. Rev. **22**, 117–134 (2007)
5. McCluskey, T. L., Kitchin, D. E.: A tool-supported approach to engineering HTN planning models. In: Proceedings of 10th IEEE International Conference on Tools with Artificial Intelligence (1998)
6. Vaquero, T.S., Tonaco, R., Costa, G., Tonidandel, F., Silva, J.R., Beck, J.C.: itSIM-PLE4.0: Enhancing the modeling experience of planning problems. In: System Demonstration - Proceedings of the 22nd International Conference on Automated Planning and Scheduling (ICAPS-12) (2012)
7. Barreiro, J., Boyce, M., Do, M., Frank, J., Iatauro, M., Kichkaylo, T., Morris, P., Ong, J., Remolina, E., Smith, T.: EUROPA: a platform for AI planning, scheduling, constraint programming, and optimization. In: 4th International Competition on Knowledge Engineering for Planning and Scheduling (ICKEPS) (2012)
8. Smith, D.E., Frank, J., Cushing, W.: The ANML language. In: Proceedings of ICAPS-08 (2008)
9. Vodrka, J., Chrpa, L.: Visual design of planning domains. In: KEPS 2010: Workshop on Knowledge Engineering for Planning and Scheduling (2010)

10. Plch, T., Chomut, M., Brom, C., Barták, R.: Inspect, edit and debug PDDL documents: simply and efficiently with PDDL studio. In: ICAPS12 System Demonstration, p. 4 (2012)
11. Schreiber, G., Akkermans, H., Anjewierden, A., de Hoog, R., Shadbolt, N., de Velde, Wv, Wielinga, B.J.: Knowledge Engineering and Management: The CommonKADS Methodology, 2nd edn. MIT Press, Cambridge (2000)
12. Ghallab, M., Nau, D., Traverso, P.: Automated Planning. Morgan Kaufmann, San Francisco (2004)
13. Jonsson, P., Bäckström, C.: State-variable planning under structural restrictions: algorithms and complexity. Artif. Intell. **100**, 125–176 (1998)
14. Currie, K., Tate, A.: O-Plan: the open planning architeture. Artif. Intell. **52**, 49–86 (1991)

Extending Ontological Categorization Through a Dual Process Conceptual Architecture

Antonio Lieto[1,2(✉)], Daniele P. Radicioni[1], Marcello Frixione[3], and Valentina Rho[1]

[1] Dipartimento di Informatica, Università di Torino, Turin, Italy
{lieto,radicion}@di.unito.it, rho.valentina@gmail.com
[2] ICAR-CNR, Palermo, Italy
[3] DAFIST, Università di Genova, Genova, Italy
frix@dist.unige.it

Abstract. In this work we present a hybrid knowledge representation system aiming at extending the representational and reasoning capabilities of classical ontologies by taking into account the theories of typicality in conceptual processing. The system adopts a categorization process inspired to the *dual process theories* and, from a representational perspective, is equipped with a heterogeneous knowledge base that couples conceptual spaces and ontological formalisms. The system has been experimentally assessed in a conceptual categorization task where common sense linguistic descriptions were given in input, and the corresponding target concepts had to be identified. The results show that the proposed solution substantially improves the representational and reasoning "conceptual" capabilities of standard ontology-based systems.

Keywords: Knowledge representation · Formal ontologies · Conceptual spaces · Common sense reasoning · Dual process theory · Prototypical reasoning

1 Introduction

One of the main open problems in the field of ontology engineering is that formal ontologies do not allow –for technical convenience– neither the representation of concepts in prototypical terms nor forms of approximate, non monotonic, conceptual reasoning. Conversely, in Cognitive Science evidences exist in favor of prototypical representation of concepts, and typicality-based conceptual reasoning has been widely investigated in the field of human cognition. The early work of Rosch [29] showed that ordinary concepts do not obey the classical theory (stating that concepts can be defined in terms of sets of necessary and sufficient conditions). Rather, they exhibit *prototypical* traits: e.g., some members of a category are considered *better instances* than other ones; more *central* instances share certain typical features –such as the ability of flying for birds– that, in general, cannot be thought of as necessary nor sufficient conditions. These results

© Springer International Publishing Switzerland 2015
A. Fred et al. (Eds.): IC3K 2014, CCIS 553, pp. 313–328, 2015.
DOI: 10.1007/978-3-319-25840-9_20

influenced pioneering KR research, where some efforts were invested in trying to take into account the suggestions coming from Cognitive Psychology: artificial systems were designed – e.g., frames [26] and early semantic networks – to represent concepts in "non classical", prototypical terms.

However, these systems were later sacrificed in favor of a class of formalisms stemmed from structured inheritance semantic networks and based in a more rigorous semantics: the first system in this line of research was the KL-ONE system [3]. These formalisms are known today as description logics (DLs) [27]. In this setting, the representation of prototypical information (and therefore the possibility of performing non monotonic reasoning) is not allowed,[1] since the formalisms in this class are primarily intended for deductive, logical inference.

Under a historical perspective, the choice of preferring classical systems, which are based on a well defined – Tarskian-like – semantics left unsolved the problem of representing concepts in prototypical terms. Although in the field of logic oriented KR various fuzzy and non-monotonic extensions of DL formalisms have been designed to deal with some aspects of "non-classical" concepts [15,30], nonetheless various theoretical and practical problems remain unsolved [6].

In this paper a conceptual architecture is presented that, embedded in a larger knowledge-based system, aims at extending the representational and reasoning capabilities available to traditional ontology-based frameworks. The whole system implementing the proposed conceptual architecture is part of a larger software pipeline; it includes the extraction of salient information from the input *stimulus*, the access to the hybrid knowledge base, and the retrieval of the corresponding concept (Fig. 1). The paper is structured as follows: in Sect. 2 we illustrate the theoretical motivations inspiring the proposed system, its general architecture and the main features of the knowledge-base. In Sect. 3 we provide the results of a twofold experimentation to assess the accuracy of the system in a categorization task. Finally, we conclude by presenting some related works (Sect. 4) and outlining future developments (Sect. 5).

2   The System

Two cornerstones inspiring the current proposal are the *dual process theories* of mind and the heterogeneous approach to concepts in Cognitive Science. The theoretical framework known as *dual process theory* postulates the co-existence of two different types of cognitive systems [5,18]. The systems of the first type (*type 1*) are phylogenetically older, unconscious, automatic, associative, parallel and fast. The systems of the second type (*type 2*) are more recent, conscious, sequential and slow, and featured by explicit rule following.

We assume that both systems can be composed in their turn by many subsystems and processes. According to the hypotheses in [6,10], the conceptual representation of our system includes two main sorts of components, based on two sorts of processes. *Type 1* processes are used to perform fast and approximate categorization, and benefit from prototypical information associated to

---

[1] This is the case, for example, of *exceptions* to the inheritance mechanism.

concepts. *Type 2* processes, used in classical inference tasks, do not take advantage from prototypical knowledge. The two sorts of system processes are assumed to interact, since type 1 processes are executed first and their results are then refined by type 2 processes. Another source of inspiration for our work has been the heterogeneous approach to the concepts in Cognitive Science and Philosophy [24]. According to this perspective, concepts do not constitute a unitary phenomenon; rather, concepts can consist of several bodies of knowledge, each one conveying a specific type of information.

Before starting with the description of the system, we introduce the theoretical framework inspiring our present work.

2.1 Theoretical Framework

According to the aforementioned hypotheses (*dual process* theory and *heterogeneous approach* to the concepts), we designed a hybrid conceptual architecture that builds on a classical ontological component, and on a typicality based one. Both components represent a specific conceptual body of knowledge, together with the related reasoning procedures, in a dual process perspective. A DL formalism is the base of the ontological component, which permits to express necessary and/or sufficient conditions to define concepts. For example, if we consider the concept *water*, the classical representation should contain the information that *water* is a natural substance, whose chemical formula is H_2O. On the other hand, the prototypical traits include information about the fact that water usually occurs in liquid state, and it is mostly a tasteless, odorless and colorless fluid.

According to our dual process view, in the implemented system the typical representational and reasoning functions are assigned to the system 1 (hereafter $S1$), that executes processes of type 1, and are associated to the Conceptual Spaces framework [11]. On the other hand, the classical representational and reasoning functions are assigned to the system 2 (hereafter $S2$), to execute processes of type 2, and are associated to a standard DL-based ontological representation. Different from what proposed in [7], the access to the information stored and processed in both components is assumed to proceed from the $S1$ to the $S2$. In the following we introduce the two representational and reasoning frameworks used in our system.

2.2 Conceptual Spaces

Conceptual spaces (CS) are a geometrical representational framework where knowledge is represented as a set of *quality dimensions* [11]. A geometrical structure is associated to each quality dimension. In this framework instances may be represented as points in this multidimensional space, and their similarity can be computed as the intervening distance, based on some suitable metrics. In this setting, concepts correspond to regions, and regions with different geometrical properties correspond to different sorts of concepts.

Conceptual spaces are suitable to represent concepts in "typical" terms, since the regions representing concepts can have soft boundaries. Prototypes have a natural geometrical interpretation, in that they correspond to the geometrical centre of a convex region; conversely, given a convex region we can associate to each point a certain centrality degree, that can be interpreted as a measure of its typicality. Although other forms of typicality-based representation (e.g. the exemplars) are not presently accounted for by our system, this framework can be extended to consider both the exemplar and the prototypical accounts of typicality [9].

Conceptual spaces can be also used to compute the proximity between any two entities, and between entities and prototypes. Concepts, in this framework, are characterized in terms of *domains*; a domain is "a set of integral dimensions that are separable from all other dimensions" [12]. Typical domain examples are color, size, shape, texture. In turn, domain information can be specified along some dimensions: e.g., in the case of the *color* domain, relevant dimensions are hue, chromaticity, and brightness. In order to compute the distance between two points p_1, p_2 we use Euclidean metrics to calculate within-domain distance, while for dimensions from different domains we use the Manhattan distance metrics, as suggested in [1,11]. The weighted Euclidean distance $dist_E$ is computed as follows

$$dist_E(p_1, p_2) = \sqrt{\sum_{i=1}^{n} w_i(p_{1,i} - p_{2,i})^2},$$

where i varies over the n domain dimensions and w_i are dimension weights.

In our implementation of Conceptual Spaces, we represent points as vectors (with as many dimensions as required by the considered domain), whose components correspond to the point coordinates, so that a natural metrics to compute the similarity between them is *cosine similarity*. In this perspective two vectors with same orientation have a cosine similarity 1, while two orthogonal vectors have cosine similarity 0. The normalized version of cosine similarity ($\hat{cs}$), also accounting for the above weights w_i is computed as

$$\hat{cs}(p_1, p_2) = \frac{\sum_{i=1}^{n} w_i(p_{1,i} \times p_{2,i})}{\sqrt{\sum_{i=1}^{n} w_i(p_{1,i})^2} \times \sqrt{\sum_{i=1}^{n} w_i(p_{2,i})^2}}.$$

In the metric space we have defined, the distance between an individual and prototypes is computed with the Manhattan distance. The distance between two concepts can be computed as the distance between two regions: namely, we can compute the distance between their prototypes, or the minimal distance between their *individuals*[2], or we can apply more sophisticated algorithms. Further details about technical issues can be found in [14].

Optionally, a context k can be defined as a set of weights, to grade the relative relevance of the considered dimensions – thus resulting in the following formulas:

---

[2] Individuals can be thought of as *exemplars*, that is elements that belong to the given concept by sharing properties and their related sets of values.

$dist_E(p_1, p_2, k)$ and $\hat{cs}(p_1, p_2, k)-$, and to adapt the computation to a variety of settings, such as, e.g., default values *vs.* known values, explicitly asserted values *vs.* computed values and/or inherited, etc.

We stress that inference in conceptual spaces can be performed on incomplete and/or noisy information: that is, it is frequent the case that only partial information is available to categorize a given input individual, and for some individuals the values of one or more dimensions can be undefined. Conceptual spaces are robust to this sort of lack of information, which is conversely problematic in the context of formal ontologies. In such cases we restrict to considering domains that contain points in the input: if the description for a given individual does not contain values for some domains, the distance for those domains is set to a default value.

2.3 $\mathcal{S}1$: Modeling Domains and Dimensions in Conceptual Spaces

A processing format for the modelling of Conceptual Spaces has been provided and proposed in [23]. Although the format has been developed by attempting to keep it as general as possible (to extend its usage to further domains), the present implementation has been devised based on specific representational needs described in more detail in Sect. 3. The basic format structure processed by the system is named genericDescription; it encodes the salient aspects of the entities being considered. A genericDescription is a super-domain that hosts information about physical and non physical features arranged into nine domains: size, shape, color, location, feeding, locomotion, hasPart, partOf, manRelationship.

The size of entities is expressed through the three Euclidean dimensions; the shape allows expressing that an object has circular, square, spherical, cubic, etc., shape. The color space maps object's features onto the L*a*b* color space. L* $(0 \leq L \leq 100)$ is the correlate of lightness, a* $(-128 \leq a \leq 127)$ is the chromaticity axis ranging from green to red, and b* $(-128 \leq b \leq 127)$ is the chromaticity axis ranging from blue to yellow. The location space indicates the place where the object being modeled can be typically found. It actually results from the combination of five dimensions, and namely: humidity, indicated as a percentage; temperature, ranging in $[-40°, 50°]$; altitude, ranging in $[-11000, 8848]$; vegetation, ranging in $[0, 100]$; time. In turn, time contains a partitioning of the hours of the day into sunrise (4–6 AM), morning (6–12 AM), afternoon (12–5 PM), evening (5–10 PM) and night (10 PM–4 AM). The domain feeding is currently specific to animals, and it allows mapping an element over two dimensions, typeOfFood and amountOfFood. The typeOfFood is associated to an integer indicating 1: herbivore, 2: lectivore, 3: detritivore, 4: necrophage, 5: carnivore. The underlying rationale is that close elements (e.g., necrophage and carnivore, that are one step apart in the proposed scale) are represented as close in this space due to their proximity under an ethological viewpoint, whilst different categories (e.g., herbivore and carnivore) are featured by larger distances in the considered scale [13]. Similar to the previous one, also the locomotion domain combines two dimensions: the former dimension is used to account for the type of movement (1: swim, 2: dig, 3: crawl, 4: walk, 5: run, 6: roll, 7: jump, 8: fly), and the latter one is used to account for the speed, expressed in km/h [2]. The hasPart and partOf domains are used to complement the analogous ontological properties: in particular, we collected information about the following dimensions: name, number, and partSize, partColor that are intended to specialize the above illustrated spaces. Finally, the manRelationship space is used to grasp entities as related to man by function (both a train and a horse can be used as 'transport'), product (chicken produce 'eggs', and 'chicken' themselves are a food product), symbol ('lion' can be used as a symbol for 'strength' and 'royalty').

A simplified example of the *lion* prototype information is reported below.

```
<object name="lion">
  <genericPhysicalDescription>
    <size name="lion_size">
      <x>70</x>
      <y>120</y>
      <z>200</z>
    </size>
    <color name="beige">
      <l>63</l>
      <a>13</a>
      <b>32</b>
    </color>
    <location name="savanna">
      <humidity>50</humidity>
      <temperature>40</temperature>
      <altitude>100</altitude>
      <vegetation>50</vegetation>
    </location>
    <locomotion name="walk">
      <movement>4</movement>
      <speed>10</speed>
    </locomotion>
    <locomotion name="run">
      <movement>5</movement>
      <speed>40</speed>
    </locomotion>
  </genericPhysicalDescription>
</object>
```

2.4 $S2$: Ontological Component of the Knowledge Base

The representation of the classical component $S2$ is implemented through a formal ontology. As already pointed out, the standard ontological formalisms leave unsolved the problem of representing prototypical information. Furthermore, it is not possible to execute non monotonic inference, since classical ontology-based reasoning mechanisms contemplate exclusively deductive processes.

In this setting we cannot represent even simple prototypical information, such as 'A typical rose is red'. This is due to the fact that being red is neither a necessary nor a sufficient condition for being a rose, and therefore it is not possible neither to represent and to automatically identify a prototypical rose (let us call it #$roseP$) nor to describe (and to learn from new cases) the typical features of the prototypical roses. Such aspect have, on the other hand, a natural interpretation in terms of the Conceptual Spaces framework. The ontological component of a given concept, therefore, mainly represents taxonomical necessary information and, as will be described below, is used as control module w.r.t. the output provided by the inferences performed within the conceptual spaces.

2.5   Inference in the Hybrid $S1$-$S2$ System

Categorization (i.e., the process of assigning a given instance to a certain category) is one of the classical inferences automatically performed both by symbolic and sub-symbolic artificial systems. In our system categorization is based on a two-step process involving both the typical and the classical component of the conceptual representation. These components account for different types of categorization: approximate or non monotonic (performed on the conceptual spaces), and classical or monotonic (performed on the ontology). Different from classical ontological inference, in fact, categorization in conceptual spaces proceeds from *prototypical* values. Prototypical values need not be specified for all class members, that vice versa can overwrite them: one typical example is the case of birds that (by default) fly, except for atypical birds, like penguins, that do not fly.

The whole categorization process can be summarized as follows. The system takes in input a textual description d and produces in output a pair of categories $\langle c_0, cc \rangle$, the output of $S1$ and $S2$, respectively (see Algorithm 1). If the $S2$ system classifies it as consistent with the ontology, then the classification succeeded and the category provided by $S2$ (cc) is returned along with c_0, the top scoring class returned by $S1$ (Algorithm 1: line 8). If cc –the class computed by $S2$– is a subclass of one of those identified by $S1$ (c_i), both cc and c_0 are returned (Algorithm 1: line 11). Thus, if $S2$ provides more specific output, we follow a *specificity* heuristics; otherwise, the output of $S2$ is returned, following the rationale that it is *safer*.[3] A pair of results is always returned, including both the output of $S1$ and the output of $S2$, thereby providing typically valid answers (through $S1$) that are checked against a logically valid reasoning conducted on the ontological knowledge base (through $S2$). In so doing, we follow the rationale that despite the $S1$ output can contain errors, it furnishes approximate answers that cannot be obtained by resorting only to classical ontological inference.

If all results in C are inconsistent with those computed by $S2$, a pair of classes is returned including c_0 and the output of $S2$ having for actual parameters d and Thing, the meta-class of all the classes in the ontological formalism.

---

[3] The output of $S2$ cannot be wrong on a purely logical perspective, in that it is the result of a deductive process. The control strategy implements a tradeoff between ontological inference and the output of $S1$, which is more informative but also less reliable from a formal point of view. However, in next future we plan to explore different conciliation mechanisms to ground the overall control strategy.

Algorithm 1. Inference in the hybrid system.

---

input : textual description d
output : a class assignment, as computed by $S1$ and $S2$
 1: $C \leftarrow S1(d)$ /* conceptual spaces output */
 2: **for each** $c_i \in C$ **do**
 3:     $cc \leftarrow S2(\langle d, c_i \rangle)$ /* ontology based output */
 4:     **if** $cc ==$ NULL **then**
 5:         **continue** /* inconsistency detected */
 6:     **end if**
 7:     **if** cc equals c_i **then**
 8:         **return** $\langle c_0, cc \rangle$
 9:     **else**
10:         **if** cc is subclass of c_i **then**
11:             **return** $\langle c_0, cc \rangle$
12:         **end if**
13:     **end if**
14: **end for**
15: $cc \leftarrow S2(\langle d, \mathsf{Thing} \rangle)$
16: **return** $\langle c_0, cc \rangle$

An important function provided by $S2$ regards the explanation of the detected inconsistencies. This function is obtained by recurring to standard DL reasoners.[4] One main problem encountered in the explanation of inconsistencies regards the fact that reasoners' output is usually quite verbose, since it provides the whole chain of all the possible reasons explaining why a given model is not consistent w.r.t. the represented assertions. For example, let us suppose that the ontological KB is provided with an assertion about the fact that *whale isA fish*. Since whales are not fishes (they are in the order of *cetacea* and cetacea are *mammalia*) an inconsistency is detected. The initial results obtained by the reasoner will report all the clauses (i.e. all the models) causing such an incostitency. Not only it will report that — since *whale isA mammal* and mammal and cetacea are disjoint — whales cannot be fishes; it will also provide a huge amount of information about the facts that such an inconsistency has generated. For example, it will provide information about the fact that *whales* cannot be *reptiles*, *birds* and so on. Although factually correct and complete, this explanation is quite long (in an ontological KB with good coverage each class contains many subclasses) and is it not very informative for the punctual explanation of the raised inconsistency. The only relevant information, in this case, regards the fact that the tested class *whale* cannot be classified as *fish* because *mammal* and *fish* are mutually disjoint classes. Therefore we designed and implemented a software layer that runs on top of Jena explanation utilities to extract a *laconic explanation* from the longer one: the main focus of the laconic explanation is to make apparent the cause of the inconsistency. In so doing, we adopted a simple

---

[4] To actually access the KBs we used the Jena framework, https://jena.apache.org.

heuristic according to which the only explanation reported is that focused on the tuples of classes under investigation.

2.6 Categorization Pipeline of the Dual Process Architecture

The whole system embedding the proposed conceptual architecture works as follows. The input to the system is a simple linguistic description, like 'The animal that eats bananas', and the expected output is a given category referred to the description (e.g. the category *monkey* in this case). After the Information Extraction (IE) step, an internal representation is fed into the structure of the hybrid knowledge representation system which is then concerned with the categorization task by adopting the strategies described above.

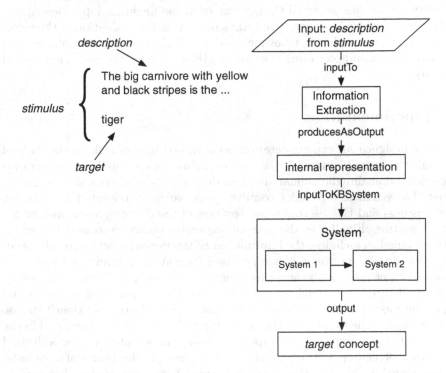

Fig. 1. The software pipeline takes in input the linguistic description, queries the hybrid knowledge base and returns the categorized concept.

A shallow IE approach has been devised: the morphological information computed from input sentences has been used to devise a finite-state automaton to describe the input sentences' structure. In this setting, the POS (Part-Of-Speech) information has been computed through the Stanford POS Tagger [31]. Then POS information has been used to encode the automaton states and transitions, that allow individuating salient information. In this automaton, states contain some kind of salient information required in the internal representation (e.g., the *place* where animals live; the description of their *skin* or *fur*; the *function* of an artifact, etc.; see Sect. 2.3). On the other hand, transitions between any two states encode connectives and prepositions that contain modifications to the the noun phrase they are referred to (like in 'the big carnivore *with* ⌊yellow and black stripes⌋'). This approach makes no use of the sentence dependency structure, and has many known limitations determined from merely using morphological information, and also inherent in finite-state machines (e.g., they cannot deal with parenthetical clauses, like relative clauses). It would not scale to handle more complex sorts of language. We defer to future work the adoption of richer language models; in particular, we will extend to the present context a deep semantic approach developed to perform IE from legal texts [20]. Despite these limitations, this approach allowed us to complete the automatization of the software pipeline going all throughout from the linguistic input description to its final conceptual categorization, thus improving the evaluation of the whole implemented system. In the following we describe the new experimentation and compare current and past results (where the IE step was performed in supervised fashion).

3   Experimentation

We have designed an experimentation on a categorization task to the ends of assessing the overall system. In the past experiments we tested the system over a restricted domain (the animal kingdom domain) [14]. Additionally, in [22] we tested the system in a broader context, since we were interested in assessing its robustness and the discriminative features of the $S1$ component in a multi-domain setting. Finally, in the present experimentation we tested the whole software pipeline, including the Information Extraction step (which, on the other hand, was simply performed in a supervised fashion in experiments 1 and 2).

The dataset used for the new experiment is composed of 87 "common-sense" linguistic descriptions.[5] Each stimulus $st = \langle d, T \rangle$ is a pair of description and target, such as ⟨'The big carnivore with yellow and black stripes', 'tiger'⟩ (please refer to Fig. 1). The target T is the "prototypically correct" category, and in the following it is referred to as the *expected* result. The set of stimuli was devised by a team of neuropsychologists and philosophers in the frame of a broader project aimed at investigating the role of visual load in concepts involved in inferential and referential tasks (further details on neural correlates of lexical processing in [25]). The expected prototypical target category represents a gold standard, since it corresponds to the results provided within a psychological experimentation, where 30 subjects were requested to provide the corresponding target concept for each description. Such input was then used for querying our system as in a typicality based question-answering task.

---

[5] The full list of the stimuli is available at the URL: http://www.di.unito.it/~radicion/datasets/CCIS_2014/stimuli.txt.

In Information Retrieval such queries are known to belong to the class of "informational queries", i.e., queries where the user intends to obtain information regarding a specific information need [16]. Furthermore, this class of queries is characterized by uncertain and/or incomplete information, thereby resulting in the most complex to interpret, if compared to queries where users search for the URL of a given site ('navigational queries'), or look for sites where some task can be performed, like buying music files ('transactional queries'). Additionally, informational queries are by far the most common ones –based on the analysis of 1.5 M user web logs [16]–, and they are therefore of the utmost applicative interest.

3.1   Summary of Previous Experiments

A first experiment was made by using as $S2$ component publicly available common sense domain ontologies. Namely, we selected the Animal in Context Ontology (ACO) developed by the Veterinary Medical Informatics Laboratory at the Virginia-Maryland Regional College, and the BBC WildLife Ontology.[6] They were both retrieved by using a mixed search over Sindice and Swoogle, and they were selected as guaranteeing a granularity of information and a coverage adequate for describing the stimuli being categorized (belonging to the animal domain). In this case we recorded a categorization accuracy of the $S1$-$S2$ about 95 % with $S2$ using the ACO ontology, and over 92 % with $S2$ using the BBC ontology. The accuracy was determined by comparing the results provided by the system with that ones provided by the human subjects in the above mentioned psychological experiment. Full details are provided in [23].

In order to deeply assess the accuracy of the system in a more demanding experimental setting, we devised a second experiment where we used as $S2$ the knowledge base OpenCyc.[7] OpenCyc is one of the largest ontologies publicly available, in that it is an enormous attempt at integrating many diverse semantic resources (such as, e.g., WordNet, DBpedia, Wikicompany, *etc.*). In this case we compared the results obtained by our $S1$-$S2$ systems with the results obtained by the Google and Bing search engines for the same queries [22]. Differently from the previous experiment, we decided to make use of an encyclopedic source of knowledge in order to investigate the differences with results obtained by plugging into $S2$ domain-specific knowledge bases. In this case the hybrid knowledge based $S1$-$S2$ system was able to categorize and retrieve most of the new typicality-based stimuli provided as input and still showed a better performance w.r.t. the general purpose search engines Google and Bing used in question-answering mode. The major problems encountered in this experiment derived from the difficulty of mapping the linguistic structure of stimuli containing very abstract meaning in the representational framework of $S1$ as they are actually encoded according to the conceptual space. For example, it was impossible to map the information contained in the description "the place where kings,

---

[6] Available at the URLs: http://vtsl.vetmed.vt.edu/aco/Ontology/aco.zip and http://www.bbc.co.uk/ontologies/wo.

[7] http://www.cyc.com/platform/opencyc.

princes and princesses live in fairy tales" onto the features used to characterize the prototypical representation of the concept *Castle*. On the other hand, the system shows good performances when dealing with less abstract descriptions based on perceptual and metric-reducible features such as *shape, color, size*. The final categorization accuracy obtained by the system was around 77 %, while Google and Bing successfully categorized 65 % and 57.5 % of linguistic descriptions, respectively.

3.2   New Experimentation

A new experiment to assess the whole pipeline has been devised: specifically, in this case the information available in the descriptions has been extracted through the automatic Information Extraction (IE) process and then used to query the hybrid knowledge base. The $S2$ system has been equipped with the OpenCyc ontology, like in the previous experiment. The dataset included 87 stimuli, 36 related to the artifacts domain, 6 to plants and 45 about animals.

Table 1. Analysis of the POS-tagger and automaton failures and analysis of the correct results, in categorizing artifacts, plants, animals. Overall 36 descriptions of artifacts, 6 of plants and 45 of animals were considered.

	POS-tagger errors	Automaton errors	Successes in the IE step	$\sum$
♯ stimuli (%)	11 (12.6 % of 87)	22 (25.3 % of 87)	54 (62.1 % of 87)	87
$S1$ successes	3 (27.3 % of 11)	6 (27.3 % of 22)	50 (92.6 % of 54)	59

Overall, 59 targets were correctly categorized out of the 87 input descriptions, thereby attaining 67.8 % of correct responses. If we further examine the results, we can *(i)* disaggregate cases where failures in the IE step were determined by the POS-tagger, or by the automaton that did not match the input descriptions structure; *(ii)* compare failures to success cases. Interestingly, this analysis unveils that in some cases also in presence of noisy or lacking information the $S1$ component is able to retrieve the expected category. Detailed figures are reported in Table 1. The first two cells in the first row report the errors committed in the two stages preceding the access to the hybrid knowledge base. The second row illustrates how $S1$ recovers from partial or lacking information provided by the POS tagger and by the automaton: $S1$ obtained a correct categorization in 27 % of cases when wrong and/or incomplete information was received by the POS tagger and by the automaton. On the other hand, the second row shows that even when all the linguistic information is correctly extracted and mapped onto the internal $S1$ format, it does not suffice to provide the expected categorization for all stimuli (the final accuracy for these cases was 92.6 %). This datum is in line with previous results [14,23]; although encouraging, it suggests that the current stage of development of the hybrid system is not yet sufficient to solve the prototypical categorization problem.

Differently from what emerged in past experimentation — where $S2$ detected some inconsistencies in the categories returned by $S1$— in this case $S2$ did not detect an inconsistency of $S1$'s output as expected. In particular, given the description "An intelligent grey fish" (associated to the target concept *Dolphin*), the $S1$ system returned *Dolphin*, but $S2$ did not raise the inconsistency since OpenCyc erroneously represents *Dolphin* as a subclass of *Fish*, rather than a subclass of *Mammal*. Of course, by correcting the information available in the ontological knowledge base, we would obtain a refined, and ontologically correct, result.

A concluding remark is about the advantages of the $S1$-$S2$ system w.r.t. adopting solely the $S2$ for the categorization task. As we have experimentally observed, $S2$ alone is not able to categorize this kind of descriptions [14], which proves to be a challenging task even for state-of-the-art search engines such as Google and Bing [22].

4   Related Work

In the context of a different field of application, a solution similar to the one adopted here has been proposed in [4]. The main difference with their proposal concerns the underlying assumption on which the integration between symbolic and sub-symbolic system is based. In our system the conceptual spaces and the classical component are integrated at the level of the representation of concepts, and such components are assumed to convey different –though complementary– conceptual information. On the other hand, the previous proposal is mainly used to interpret and ground raw data coming from sensors in a high level symbolic system through the mediation of conceptual spaces.

In other respects, our system is also akin to that ones developed in the field of the computational approach to the above mentioned dual process theories. A first example of such "dual based systems" is the *mReasoner* model [19], developed with the aim of providing a computational architecture of reasoning based on the mental models theory proposed by Philip Johnson-Laird [17]. The *mReasoner* architecture is based on three components: a system 0, a system 1 and a system 2. The last two systems correspond to those hypothesized by the dual process approach. System 0 operates at the level of linguistic pre-processing. System 1 uses this intensional representation to build an extensional model, and uses heuristics to provide rapid reasoning conclusions; finally, system 2 carries out more demanding processes to search for alternative models, if the initial conclusion does not hold or if it is not satisfactory.

A second system that is close to our present work has been proposed by [28]. The authors do not explicitly mention the dual process approach; however, they build a system for conversational agents (chatbots) where agents' background knowledge is represented using both a symbolic and a sub-symbolic approach. They also associate different sorts of representation to different types of reasoning. Namely, deterministic reasoning is associated to symbolic (system 2) representations, and associative reasoning is accounted for by the sub-symbolic (system 1) component. Differently from our system, however, the authors do not make any claim about the sequence of activation and the conciliation strategy of the two representational and reasoning processes. It is worth noting that other examples of this type of systems can be considered that are in some sense analogous to the dual process proposal: for example, many hybrid, symbolic-connectionist systems –including cognitive architectures such as, for example, CLARION[8]–, in which the connectionist component is used to model fast, associative processes, while the symbolic component is responsible for explicit, declarative computations. However, at the best of our knowledge, our system is the only one that considers this hybridization with a granularity at the level of individual conceptual representations.

5   Conclusions

In this work we have presented a knowledge-based system relying upon a cognitively inspired architecture for the representation of conceptual knowledge. The system is grounded on a hybrid framework coupling classical and prototypical representation and reasoning, and it aims at extending the representational and reasoning capabilities of classical ontological-based systems towards more realistic and cognitively grounded scenarios, such as those envisioned by the prototype theory. The results obtained in new experimentation reveal that in a broader and composite domain including artifacts, animals and plants (w.r.t. past experimentation, mainly considering the animal kingdom) the proposed architecture provides encouraging results in tasks of prototype-based conceptual categorization.

In next future we plan to test the proposed approach in the area of biomedical domain to assess disease diagnosis tasks by grounding $S2$ on SNOWMED,[9] and $S1$ on conceptual spaces representing the typical symptoms of a given disease. A further development of the current work consists in extending the hybrid knowledge base with an additional layer of typicality-based representation (and reasoning): the exemplars. In particular, exemplar-based representations can be implemented by adopting the conceptual spaces [8] and therefore the proposed approach seems to be useful for the development of a system endowed with an integrated suite of categorization capacities. Such integrated prototype-exemplar based categorization could also be plausibly adopted in the area of cognitive architectures [21].

Acknowledgements. This work has been partly supported by the Project *the Role of the Visual Imagery in Lexical Processing*, grant TO-call03-2012-0046 funded by Università degli Studi di Torino and Compagnia di San Paolo. The authors wish to thank Leo Ghignone, Andrea Minieri and Alberto Piana for working to the early stages of the project.

---

[8] http://www.cogsci.rpi.edu/~rsun/clarion.html.

[9] http://www.b2international.com/portal/snow-owl.

References

1. Adams, B., Raubal, M.: A metric conceptual space algebra. In: Hornsby, K.S., Claramunt, C., Denis, M., Ligozat, G. (eds.) COSIT 2009. LNCS, vol. 5756, pp. 51–68. Springer, Heidelberg (2009). http://dblp.uni-trier.de/db/conf/cosit/cosit2009.html#AdamsR09

2. Bejan, A., Marden, J.H.: Constructing animal locomotion from new thermodynamics theory. Am. Sci. **94**(4), 342 (2006)

3. Brachmann, R.J., Schmolze, J.G.: An overview of the KL-ONE knowledge representation system. Cogn. Sci. **9**(2), 171–202 (1985)

4. Chella, A., Frixione, M., Gaglio, S.: A cognitive architecture for artificial vision. Artif. Intell. **89**(1–2), 73–111 (1997). http://www.sciencedirect.com/science/article/pii/S0004370296000392

5. Evans, J.S.B., Frankish, K.E.: In Two Minds: Dual Processes and Beyond. Oxford University Press, Oxford (2009)

6. Frixione, M., Lieto, A.: Representing concepts in formal ontologies: compositionality vs. typicality effects. Log. Log. Philos. **21**(4), 391–414 (2012)

7. Frixione, M., Lieto, A.: Dealing with concepts: from cognitive psychology to knowledge representation. Front. Psychol. Behav. Sci. **2**(3), 96–106 (2013)

8. Frixione, M., Lieto, A.: Exemplars, prototypes and conceptual spaces. In: Chella, A., Pirrone, R., Sorbello, R., Jóhannsdóttir, K.R. (eds.) Biologically Inspired Cognitive Architectures 2012. AISC, vol. 196, pp. 131–136. Springer, Heidelberg (2013)

9. Frixione, M., Lieto, A.: Representing non classical concepts in formal ontologies: prototypes and exemplars. In: Lai, C., Semeraro, G., Vargiu, E. (eds.) New Challenges in Distributed Information Filtering and Retrieval. SCI, vol. 439, pp. 171–182. Springer, Heidelberg (2013)

10. Frixione, M., Lieto, A.: Towards an extended model of conceptual representations in formal ontologies: a typicality-based proposal. J. Univ. Comput. Sci. **20**(3), 257–276 (2014)

11. Gärdenfors, P.: Conceptual Spaces: The Geometry of Thought. MIT Press, Cambridge (2000)

12. Gärdenfors, P.: The Geometry of Meaning: Semantics Based on Conceptual Spaces. MIT Press, Cambridge (2014)

13. Getz, W.M.: Biomass transformation webs provide a unified approach to consumer-resource modelling. Ecol. Lett. **14**(2), 113–124 (2011)

14. Ghignone, L., Lieto, A., Radicioni, D.P.: Typicality-based inference by plugging conceptual spaces into ontologies. In: Lieto, A., Cruciani, M. (eds.) Proceedings of the International Workshop on Artificial Intelligence and Cognition. CEUR (2013)

15. Giordano, L., Gliozzi, V., Olivetti, N., Pozzato, G.L.: A non-monotonic description logic for reasoning about typicality. Artif. Intell. **195**, 165–202 (2013)

16. Jansen, B.J., Booth, D.L., Spink, A.: Determining the informational, navigational, and transactional intent of web queries. Inf. Process. Manage. **44**(3), 1251–1266 (2008)

17. Johnson-Laird, P.: Mental models in cognitive science. Cogn. Sci. **4**(1), 71–115 (1980)

18. Kahneman, D.: Thinking, Fast and Slow. Macmillan, New York (2011)

19. Khemlani, S., Johnson-Laird, P.: The processes of inference. Argum. Comput. **4**(1), 4–20 (2013)

20. Lesmo, L., Mazzei, A., Palmirani, M., Radicioni, D.P.: TULSI: an NLP system for extracting legal modificatory provisions. Artif. Intell. Law **12**(4), 1–34 (2012)

21. Lieto, A.: A computational framework for concept representation in cognitive systems and architectures: Concepts as heterogeneous proxytypes. Procedia Comput. Sci. **41**, 6–14 (2014). http://www.sciencedirect.com/science/article/pii/S1877050914015233, 5th Annual International Conference on Biologically Inspired Cognitive Architectures (BICA)
22. Lieto, A., Minieri, A., Piana, A., Radicioni, D.P.: A knowledge-based system for prototypical reasoning. Connect. Sci. **27**(2), 137–152 (2015)
23. Lieto, A., Minieri, A., Piana, A., Radicioni, D.P., Frixione, M.: A dual process architecture for ontology-based systems. In: Filipe, J., Dietz, J., Aveiro, D. (eds.) Proceedings of KEOD 2014, 6th International Conference on Knowledge Engineering and Ontology Development, pp. 48–55. SCITEPRESS - Science and Technology Publications (2014)
24. Machery, E.: Doing Without Concepts. OUP, Oxford (2009)
25. Marconi, D., Manenti, R., Catricalà, E., Della Rosa, P.A., Siri, S., Cappa, S.F.: The neural substrates of inferential and referential semantic processing. Cortex **49**(8), 2055–2066 (2013)
26. Minsky, M.: A framework for representing knowledge. In: Winston, P. (ed.) The Psychology of Computer Vision, pp. 211–277. McGraw-Hill, New York (1975). ftp://publications.ai.mit.edu/ai-publications/pdf/AIM-306.pdf
27. Nardi, D., Brachman, R.J.: An introduction to description logics. In: Description Logic Handbook, pp. 1–40 (2003)
28. Pilato, G., Augello, A., Gaglio, S.: A modular system oriented to the design of versatile knowledge bases for chatbots. ISRN Artif. Intell. **2012**, 10 (2012). doi:10.5402/2012/363840
29. Rosch, E.: Cognitive representations of semantic categories. J. Exp. Psychol. Gen. **104**(3), 192–233 (1975)
30. Straccia, U.: Reasoning within fuzzy description logics. arXiv preprint arXiv:1106.0667 (2011)
31. Toutanova, K., Klein, D., Manning, C.D., Singer, Y.: Feature-rich part-of-speech tagging with a cyclic dependency network. In: Proceedings of the 2003 Conference of the North American Chapter of the Association for Computational Linguistics on Human Language Technology, vol. 1, pp. 173–180. Association for Computational Linguistics (2003)

Multiple Dimensions to Data-Driven Ontology Evaluation

Hlomani Hloman[✉] and Deborah A. Stacey

School of Computer Science, University of Guelph, Guelph, Canada
{hhlomani,dastacey}@uoguelph.ca
http://www.uoguelph.ca

Abstract. This chapter explores the multiple dimensions to data-driven ontology evaluation. Theoretically and empirically it suggests two ontology evaluation metrics - temporal bias and category bias, as well as an evaluation approach that are geared towards accounting for bias in data-driven ontology evaluation. Ontologies are a very important technology in the semantic web. They are an approximate representation and formalization of a domain of discourse in a manner that is both machine and human interpretable. Ontology evaluation therefore, concerns itself with measuring the degree to which the ontology approximates the domain. In data-driven ontology evaluation, the correctness of an ontology is measured agains a corpus of documents about the domain. This domain knowledge is dynamic and evolves over several dimensions such as the temporal and categorical. Current research makes an assumption that is contrary to this notion and hence does not account for the existence of bias in ontology evaluation. This chapter addresses this gap through experimentation and statistical evaluation.

Keywords: Ontology · Ontology evaluation · Metric · Measure · Data-drive ontology evaluation · Ontology evaluation framework

1 Introduction

The web we experience today is in fact a fusion of two webs: the hypertext web that we are traditionally accustomed to, also known as the web of documents, and the semantic web, also known as the web of data. The latter is an extension of the former. The semantic web allows for the definition of semantics that enables the exchange and integration of data in communications that takes place over the web and within systems. These semantics are defined through ontologies rendering them the centrepiece for knowledge description. As a result of the important role ontologies play in the semantic web, they have seen increased research interest from both academic and industrial domains. This has lead to the proliferation of ontologies in existence. This proliferation can be a double-edged sword, so to speak. Critical mass is essential for the semantic web to take off, however, in the context of reuse, deciding on which ontology to use presents a

© Springer International Publishing Switzerland 2015
A. Fred et al. (Eds.): IC3K 2014, CCIS 553, pp. 329–346, 2015.
DOI: 10.1007/978-3-319-25840-9_21

big challenge. To that end a varied number of approaches to ontology evaluation have been proposed.

By definition an ontology is a shared conceptualization of a domain of discourse. A conflicting factor is that, while it is a shared conceptualization, it is also created in a specific environmental setting, time, and largely based on the modeller's perception of the domain. Moreover, domain knowledge from which it is based is non-static and changes over different dimensions. These are notions that have been overlooked in current research on data-driven ontology evaluation. The ultimate goal is to answer the question: "How do the domain knowledge dimensions affect the results of data-driven ontology evaluation?" Consequently, this chapter presents a theoretical framework as well as two metrics that account for bias along the dimensions of domain knowledge. To prove and demonstrate the merits of the proposed framework and metrics an experimental procedure that encompasses statistical evaluations is presented in the context of four ontologies in the workflow domain. For the most part the results of the statistical experimentation and evaluation are in support of the hypotheses of this chapter. There are, however, cases where the null hypotheses have been accepted and the alternate rejected.

There exists a question on how data-driven evaluation relates to other requirements based ontology evaluation techniques. There are many ways to look at this question, in the interest of brevity, we will look at it from the point of view of the **Aspects** of an ontology as detailed by Brank et al. [6], and Varendecic [10]. According to them an ontology is a complex structure with different aspects and should thus be evaluated for each of these aspects (vocabulary, syntax, structure, semantics, representation, context). The context aspect is about the features of the ontology when compared with other artifacts in its environment, which may be, e.g. a data source that the ontology describes, a different representation of the data within the ontology, or formalized requirements for the ontology in the form of competency questions or additional semantic constraints. Both requirement-driven and data-driven (including the competency question and the work presented in this chapter) would fall under the context aspect. An evaluation tool will then load the ontology and the additional artifact to perform further evaluation. In this regard, the competency questions or domain corpus would be the artifacts in the ontology's environment and therefore, will be evaluated against them.

2   Related Work: Data-Driven Ontology Evaluation

This evaluation technique typically involves comparing the ontology against existing data about the domain the ontology models. This has been done from different perspectives. For example, Patel et al. [12] considered it from the point of view of determining if an ontology refers to a particular topic(s). Spyns et al. [13] attempted to analyze how appropriate an ontology covers a topic of the corpus through the measurement of the notions of precision and recall. Similarly, Brewster et al. [2] investigated how well a given ontology or a set of ontologies fit the domain knowledge. This is done by comparing ontology concepts and relations to text from documents about a specific domain and further refining the results by employing a probabilistic method to find the best ontology for the corpus. Ontology coverage of a domain was also investigated by Ouyang [5] where coverage is considered from the point of view of both the coverage of the concepts and the coverage of the relations.

The major limitation of current research within the realm of data-driven ontology evaluation is that domain knowledge is implicitly considered to be constant. This is inconsistent with literature's assertions about the nature of domain knowledge. For example, Nonaka [11] asserts that domain knowledge is dynamic. Changes in ontologies have been partially attributed to changes in the domain knowledge. In some circles, ontological representation of the domain has been deemed to be biased towards their temporal, environmental, and spatial setting [2,6]. By extension, the postulation is that domain knowledge would change over these dimensions as well. Hence, it is the intent of this research to succinctly incorporate these salient dimensions of domain knowledge in an ontology evaluation effort with the view of proving their unexplored influence on evaluation measures.

3   General Limitations of Ontology Evaluation

This section discusses *subjectivity* as a common major limitation to current research in ontology evaluation. We demarcate this discussion into: (i) subjectivity in the selection of the criteria for evaluation, (ii) subjectivity in the thresholds for each criterion, and (iii) influences of subjectivity on the results of ontology evaluation.

3.1   Subjectivity in the Criteria for Evaluation

Ontology evaluation can be regarded over several different decision criteria. These criteria can be seen as the desiderata for the evaluation [9,10]. The first level of difficulty has been in deciding the relevant criteria for a given evaluation task. It has largely been the sole responsibility of the evaluator to determine the elements of quality to evaluate [10]. This brings about the issue of subjectivity in deciding which criteria makes the desiderata.

To address this issue, two main approaches have been proposed in literature: (i) induction - empirical testing of ontologies to identify desirable properties of the ontologies in the context of an application, and (ii) deduction - deriving the most suitable properties of the ontologies based on some form of theory (*e.g.* based on software engineering). The advantages of these coincidentally seem to be the disadvantage of the other. For example, inductive approaches are guaranteed to be applicable for at least one context, but their results cannot be generalized to other contexts. Deductive approaches on the other hand, can be generalized to other contexts, but are not guaranteed to be applicable for any specific context. In addition, for deductive approaches, the first level of challenge is in determining the correct theory to base the deduction on. This then spirals back to the problem of subjectivity where the evaluator has to sift through a plethora of theories in order to justify their selection.

3.2 Subjectivity in Thresholds

The issue of thresholds for ontology evaluation criteria has been highlighted by Vrandecic [10]. He puts forward that the goal for ontology evaluation should not be to perform well for all criteria and also suggests that some criteria may even be contradictory. This then defaults to the evaluator to make a decision on the results of the evaluation over the score of each criterion. This leads to subjectivity in deciding the *optimal* thresholds for each criterion. For example, if a number of ontologies were to be evaluated for a specific application, it becomes the responsibility of the evaluator to answer questions like, *"Based on the evaluation criteria, when is Ontology A better than Ontology B?"*.

3.3 Influences of Subjectivity on the Measures/Metrics

The default setting of good science is to exclude subjectivity from a scientific undertaking such as an experiment [11]. This has been typical of ontology evaluation. However, as has been discussed in Sects. 3.1 and 3.2, humans are the objects (typically as actors) of research in most ontology evaluation experiments. The research itself therefore, cannot be free of subjectivity. This expresses bias from the point of view of the evaluator. There exists another form of bias, the kind that is inherent in the design of the ontologies. An ontology (a model of domain knowledge) represents the domain in the context of the time, place, and cultural environment in which it was created as well as the modellers perception of the domain [2,6].

The problem lies in the unexplored potential influence of this subjectivity in the evaluation results. If we take a data-driven approach to ontology evaluation for example, it would be interesting to see how the evaluation results spread over each dimension of the domain knowledge (*i.e.* temporal, categorical, etc.). This is based on equating subjectivity/bias to the different dimensions of domain knowledge. To give a concrete example, let us take the results of Brewster et al. [2]. These are expressed as a vector representation of the similarity score of each ontology showing how closely each ontology represents the domain corpus. This offers a somewhat one dimensional summarization of this score (coverage) where one ontology will be picked ahead of the others based on a high score. It, however, leaves unexplored how this score changes over the years (temporal), for example. This could reveal very important information such as the relevance of the ontology, meaning that the ontology might be aging and needs to be updated as opposed to a rival ontology. The results of Ouyang et al. [5] are a perfect exemple of this need. They reveal that the results of their coverage showed a correlation between the corpus used and the resultant coverage. This revelation is consistent with the notion of dynamic domain knowledge. In fact, a changing domain knowledge has been attributed to the reasons for changes to the ontologies themselves [11]. This offers an avenue to explore and account for bias as well as its influence on the evaluation results. This forms the main research interest of this chapter.

Thus far, to the best of our knowledge, no research in ontology evaluation has been undertaken to account for subjectivity. This has not been especially done to measure subjectivity in the context of a scale as opposed to binary (yes- it is subjective, or no - it is not subjective). Hence, this provides a means to account for the influences of bias (subjectivity) on the individual metrics of evaluation that are being measured.

4   Theoretical Framework

The framework presented in this chapter which is reminiscent of Vrandecic's framework for ontology evaluation [10] is depicted and summarized in Fig. 1. Sections 5 through 7 explain the fundamental components of this framework and provide details on how they relate to each other. An ontology (O) has been defined as a formal specification of a domain of interest through the definition of the concepts in the domain and the relationships that hold between them. An ontology set (S) is a collection of ontologies, $\exists O \in S$. Evaluation methods evaluate an ontology or a set of ontologies. For the purposes of a data-driven approach to ontology evaluation, the evaluation is conducted from the viewpoint of a domain corpus. Put simply, evaluation methods evaluate ontologies against the domain corpus by using metrics and their measures to measure the correctness or quality of the ontologies. In other terms, an ontology evaluation which is the result of the application of an evaluation methodology, is expressed by metrics. In a data-driven ontology evaluation undertaking, the domain corpus is a proxy for the domain of interest. We argue that this proxy is non-static and changes over several dimensions including the temporal, categorical, etc. These dimensions are argued to be the bias factors and this work endeavours to explore their influence on ontology evaluation.

5   The Corpus

Current research in data-driven ontology evaluation assume that domain knowledge is constant. Hence, the premise of this work:

Premise
Literature has suggested that an ontology (a model of domain knowledge) represents the domain in the context of the time, place, and cultural environment in which it was created as well as the modeller's perception of the domain [2,6]. We argue that this extends to domain knowledge. Domain knowledge or concepts are dynamic and change over multiple dimensions including the temporal, spatial and categorical dimensions. There has been recent attempts to formalize this inherent diversity, for example, in the form of a knowledge diversity ontology [7]. We therefore, argue that any evaluation based on a corpus should then do it over these dimensions. This is something that has been overlooked by current research on data-driven ontology evaluation.

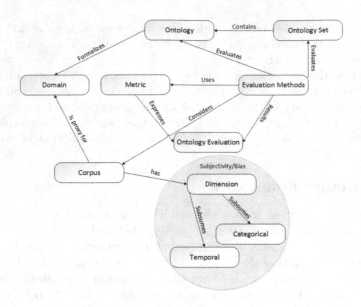

Fig. 1. A Theoretical Framework for Data-driven Ontology Evaluation that identifies and accounts for subjectivity.

5.1 Temporal

As previously mentioned, information about a domain can be discussed on its temporal axis. This is especially true from an academic viewpoint. For example, in the workflow management domain, current provisions are constantly compared in research undertakings with new concepts and languages proposed as solutions to gaps [8]. The word *current* suggest a form of timeline; what was current a decade ago is today considered in historic terms as things evolve over time. For example, in the early years of workflow management, the focus was mostly on office automation [15]. However, from the early 2000s, the focus shifted towards the formalization of business processes in the form of workflow languages. These variabilities would be reflected in the documents about the domain, also referred to as the corpus. Hence, one would be inclined to deduce that there would be a better congruence between a current ontology pitted against a current corpus than there would be for an older ontology. This congruence would suggest that if the ontology requires a lot of revision then the congruence suggests some form of distance between domain corpora and the ontology.

5.2 Categorical

Closely related to the temporal dimension is the categorical dimension. While the temporal would show a diachronic evaluation of an ontology's coverage of the domain, the categorical suggests the partitioning of the domain corpus into several important subject areas. Taking the example of the workflow management domain again, it can be partitioned into many different subjects of interest. At the top level you would consider such topics as workflow in business, scientific workflows, grid workflows all within the umbrella of "workflow" but with differing requirements, environments and operational constraints. At another level of granularity you could consider such topics as business process modelling, workflow patterns, and workflow management tools.

Often ontologies are used not in the applications they were intended for. For example, a workflow ontology created to describe collaborative ontology development [14] could be plugged into a simple workflow management system since it has the notions of task and task decomposition. However, the distance between the ontology as a model of the domain and the different categories of the domain need investigation.

6   The Ontologies and Metric of Interest

An ontology is shared approximate specification of a domain. This implies some sort of distance between the ontology and the domain, hence, the need for evaluation. In the context of the proposed framework, an evaluation undertaking involves one or more ontologies.

In the case of evaluating an ontology or a set of ontologies in the view of a corpus, we put forward that the coverage measure is the most relevant. This may not have been stated explicitly in current research on data-driven ontology evaluation, however, we have observed this to be the case. This is more obvious in the account given by [2] in referencing their creation of the Artequakt application [3]. Their purpose was to evaluate their ARTEQUAKT ontology along side four other ontologies in the view of a corpus by measuring the congruence between the ontologies and the selected corpus. Congruence here is defined as the ontology's level of fitness to the selected corpus [2]. The evaluation consists of (i) drawing a vector space representation of both the domain corpus (documents about the domain) and the ontology corpus (concepts from the ontology), and (ii) calculating the distance between the corpora in their case using Latent Semantic Analysis. The result is a similarity score, which in fact represents the ontology's coverage of the domain. The same can be observed in one of our recent works [4] that instantiates this approach to ontology evaluation.

Coverage is explicitly stated as a measure of interest in [5] with respect to data-driven ontology evaluation. Coverage in this work is partitioned into the coverage of the ontology concepts and the coverage of ontology relations with respect to a corpus. This work also considers the cohesion and coupling metrics, none of which has any bearing on corpus evaluation.

In this regard, if domain knowledge is multi-dimensional and if coverage is the measure that evaluates the congruence between an ontology or set of ontologies and domain knowledge then, coverage should be measured with respect to the dimensions of the corpus. Hence, this work's proposed metrics (temporal bias and category bias).

7   Methods

Methods or methodologies are the particular procedures for evaluating the ontologies within the context of an evaluation framework. With respect to the data-driven ontology evaluation and this work's proposed framework, a method calculates the measure of a given metric in the view of a given corpus. As an example, a methodology will measure the coverage (metric) of a set of workflow ontologies or a single ontology based on a workflow modelling corpus.

One method for evaluating an ontology's coverage of a corpus as suggested by Brewster and his colleagues is that of decomposing both the corpus and ontology into a vector space [2]. This then allows for distances or similarity scores between the two corpora to be calculated. Similar experimentation was conducted by [4] and other variations of these have been documented in the literature, *e.g.* [5]. Latent semantic analysis has been a common technique used for this purpose. Tools have been developed that implement these structures such as the Text Mining Library (TML) by Villalon and Calvo [1] which was employed for the experiments in this chapter. TML is a software library that encapsulates the inner complexities of such techniques as information retrieval, indexing, clustering, part-of-speech tagging and latent semantic analysis.

8   Experimental Design

There are two main hypotheses to this approach each pertaining to a respective dimension. For each hypothesis, there exists the Null Hypothesis (H_0).

Temporal Bias.

1. Null Hypothesis (H_0): If the domain corpus changes over its temporal dimension, then the ontology's coverage of the domain remains the same.
2. Alternate Hypothesis (H_1): If the domain corpus changes over its temporal dimension, then the ontology's coverage of the domain changes along the same temporal dimension.

Category Bias.

1. Null Hypothesis (H_0): If the domain corpus changes over its categorical dimension, then the ontology's coverage of the domain remains the same.
2. Alternate Hypothesis (H_1): If the domain corpus changes over its categorical dimension, then the ontology's coverage of the domain changes.

8.1   Procedure

The main steps of each experiment are outlined in Procedure 1.

Procedure 1: Experimental procedure.
1. *Select the ontologies to be evaluated from the ontology pool*
2. *Select the documents to represent the domain knowledge (corpus)*
3. *Repeat* **step 2** *for each dimension*
4. *Calculate the similarity between the ontologies and the domain corpus*
5. *Perform statistical evaluation*
6. *Repeat* **steps 4 and 5** *for each ontology*

Step 1. Ontologies: The ontologies used for experimentation are listed in Table 1.

Table 1. Profiles of the ontologies in the pool.

Ontology	Size	Focus	Year created
BMO	700+	Business Process Management	2003
Process	70+	Web Services	2007
Workflow	20+	Collaborative workflow	-
Intelleo	40+	Learning and Work related workflows	2011

Step 2. Document Selection: There were three main things that were considered in the selection of the corpus: (i) The source: here we considered three main databases (IEEE, Google Scholar, and ACM); (ii) Search terms: we used the Workflow Management Coalition (WFMC) as a form of authority and used its glossary and terminology as a source for search terms. Ten phrases were randomly selected; (iii) Restrictions: in defining the corpora, bias is simulated by means of restricting desired corpora by date (for the date bias, refer to Table 2) and subject matter (for the category bias, refer to Table 3).

Step 4. Calculate similarity between ontology and corpora. Calculate the cosine similarity between each document vector X_1 and each ontology X_2 as follows:
$$similarity(X_1, X_2) = cos(\theta) = \frac{X_1 \cdot X_2}{\| X_1 \| * \| X_2 \|}$$

Step 5. Perform statistical evaluation. For each dimension we evaluate the ontology coverage measures from two perspectives: (i) multiple ontologies (*e.g.* we take each date bracket and evaluate how coverage of all the ontologies vary for the particular date bracket); (ii) single ontology (*e.g.* we take an ontology and evaluate how its coverage varies across the different date brackets) and thus demarcate the experiments as follows:

Table 2. Corpus definition for experiment #1 showing date brackets and number of documents for each bracket as well as quantity of documents retrieved from each repository.

Bracket	Per repository			
	Google	IEEE	ACM	Sum
[1984 $\cdots$ 1989]	1/3	1/3	1/3	24
[1990 $\cdots$ 1995]	1/3	1/3	1/3	24
[1996 $\cdots$ 2001]	1/3	1/3	1/3	24
[2002 $\cdots$ 2007]	1/3	1/3	1/3	24
[2008 $\cdots$ 2014]	1/3	1/3	1/3	24
			Sum	120

Table 3. Corpus definition for experiment #2 showing key phrases used for each corpus and number of documents for each corpus. C_1 is Business Process Management, C_2 is Grid Workflow, C_3 is Scientific Workflow.

Corpus	Per repository			
	Google	IEEE	ACM	Sum
C_1	1/3	1/3	1/3	24
C_2	1/3	1/3	1/3	24
C_3	1/3	1/3	1/3	24
			Sum	72

Date Bias Part 1: Multiple Ontologies (For each bracket)

1. Compare the ontologies' coverage for each bracket against each other using nonparametric statistics (Kruskal Wallis)
2. Do Post-Hoc analysis where there is significance: $\dfrac{n(n-1)}{2} = 6$ pairwise comparisons (for each date bracket)

Date Bias Part 2: Single Ontology

1. Difference between its coverage across date brackets using nonparametric statistics (Kruskal Wallis)
2. Do Post-Hoc analysis where there is significance: $\dfrac{n(n-1)}{2} = 10$ pairwise comparisons (for each ontology)

The same structure is followed for the Category Bias except instead of date brackets we define corpora for the domain categories or subject areas.

9   Results

9.1   Date Bias - Part 1

Table 4 summarizes the results from the test between the mean coverage of all the ontologies per bracket. The table depicts the results of the statistical significance test of the difference between the mean coverage of the BMO, Process, Workflow, and the Intelleo ontologies per date bracket. The table shows that at the $\alpha = 0.05$ level of significance, there exists enough evidence to conclude that there is a difference in the median coverage (and hence, the mean coverage) among the four ontologies (at least one of them is significantly different). In relating this to our temporal hypotheses, we would reject the Null Hypothesis (H_0) that ontology coverage remains the same if the temporal aspect of domain knowledge changes. This demonstrates the usage of the Temporal bias metric. In contrast to current approaches where definitive answers are given as to whether *OntologyA* is better than *OntologyB*, we see a qualified answer to the same question to the effect that *OntologyA* is better than *OntologyB* only in these defined time intervals.

This test, however, does not indicate which of the ontologies are significantly different from which. Therefore, follow up tests were conducted to evaluate pairwise differences among the different ontologies for each date bracket. This also includes controlling for type 1 error by using the Bonferroni approach.

9.2   Date Bias - Part 1: Post-Hoc

The post-hoc analysis results reveal which ontologies as compared to the others have a significantly different mean coverage for each of the data brackets. Table 5 shows what appears to be a common theme with regards to which ontology performed better than the others. It shows that the BMO ontology's mean coverage is both larger (considering the mean ranks i and j) and significantly different (p value $< \alpha$) from the other ontologies; hence we reject the null hypothesis with regards to the BMO ontology. The table also shows an exception to the earlier sentiments, and that is in the case of the BMO compared to the Intelleo ontology. In this case there is no statistical significance in the difference between the

Table 4. Results for the evaluation of the difference between the means of the four ontologies' coverage of each bracket using the Kruskal Wallis test.

Date bracket	P value	Significant?
[1984 − 1989]	0.008358	yes
[1990 − 1995]	2.743e-12	yes
[1996 − 2001]	3.714e-10	yes
[2002 − 2007]	3.86e-09	yes
[2008 − 2014]	1.335e-07	yes

Table 5. Post-hoc analysis: pairwise comparisons of the ontologies' coverage of the domain between 1984 and 1989.

		i			
		BMO	Process	Workflow	Intelleo
j	Process	p= 0.012 $i = 36.38$ $j = 12.63$			
	Workflow	p= 0.036 $i = 35.63$ $j = 13.38$	p > 0.05 $i = 18.25$ $j = 30.75$		
	Intelleo	p > 0.05 $i = 35.88$ $j = 13.13$	p > 0.05 $i = 18.25$ $j = 30.75$	p > 0.05 $i = 26.92$ $j = 22.08$	

mean coverage of these ontologies. Therefore, at this time interval the ontologies represented the domain similarly. The table also appears to show another trend with regards to the other ontologies as compared to their counterparts. Their P values are greater than the rejection criteria (p value $> \alpha$) and hence the null hypothesis is accepted.

Table 5 shows only one of the date brackets, there are four more of these but in the interest of space and brevity we will only show results where there was statistical significance as depicted in Table 6.

9.3   Date Bias - Part 2

Table 7 shows that at the $alpha = 0.05$ level of significance, there exists enough evidence to conclude that there is a difference in the median coverage (and, hence, the mean coverage) for each of the ontologies coverage across the different date brackets. The difference between the BMO's coverage of at least one of the date brackets is statistically significant. The same applies to the other three ontologies (Process, Workflow, and Intelleo) since their p values are less that the α value (at 0.02007, 0.01781, and 0.03275, respectively). In relating this to the temporal hypotheses, we would reject the Null Hypothesis (H_0) that ontology coverage remains the same if the temporal aspect of domain knowledge changes. This also demonstrates the usage of the Temporal bias metric but only considers each ontology for the different date brackets. This gives perspective to an ontology evaluation of a single ontology.

Like in the case of Experiment #1 Part 1, this test, does not indicate which of the date brackets are significantly different from which. Therefore, follow-up tests were conducted to evaluate pairwise differences among the different date brackets for each ontology. This also includes controlling for type 1 error by using the Bonferroni approach.

Table 6. Pairwise comparisons of the ontologies' coverage for each date bracket.

Date bracket	Ontology	Mean rank	P value
[1990–1995]	BMO	36.38	0.00
	Process	12.63	
	BMO	35.63	0.00
	Workflow	13.38	
	BMO	35.88	0.00
	Intelleo	13.13	
	Process	18.25	0.012
	Workflow	30.75	
	Process	18.25	0.012
	Intelleo	30.75	
[1996–2001]	BMO	35.79	0.00
	Process	13.21	
	BMO	35.58	0.00
	Workflow	13.42	
	BMO	35.75	0.00
	Intelleo	13.25	
[2002–2007]	BMO	33.87	0.00
	Process	13.13	
	BMO	33.09	6e-06
	Workflow	13.91	
	BMO	34.09	0.00
	Intelleo	12.91	
[2008–2014]	BMO	31.24	0.00
	Process	11.76	
	BMO	30.19	2.4e-05
	Workflow	12.8	
	BMO	30.05	3.6e-05
	Intelleo	12.95	

Table 7. Results for the evaluation of the difference between the means of the four ontologies' coverage of each bracket using the Kruskal Wallis test.

Ontology	P_Value	Significant?
BMO	0.01667	yes
Process	0.02007	yes
Workflow	0.01781	yes
Intelleo	0.03275	yes

9.4   Date Bias - Part 2: Post-Hoc

For each ontology, the post-hoc analysis results reveal which date brackets as compared to the others have a significantly different mean coverage. As an example, this would answer questions like "How relevant is a given ontology?" or "How does a given ontology's coverage vary with time?". An answer to these questions would then help in determining how relevant the ontology is to current settings. If we look at the results one ontology at a time, we observe the following:

BMO Ontology (refer to Table 8): In the case of pairwise comparisons of the date brackets, there are only two of the comparisons where there is statistical significance in the difference between the mean coverage. This is the case where the data bracket [1984–1989] is compared to that of [1990–1995] and the comparison between the [1984–1989] and the [1996–2002] brackets. In both these cases, at the $\alpha = 0.05$ we can reject the Null hypothesis (H_0) and conclude that the BMO ontology's coverage of the domain does vary with time at least for those time intervals (with the **p values** $< \alpha$ at 0.04 and 0.02, respectively). In this case we could conclude that BMO was better suited for the domain between 1990 and 1995 as well as between 1996 and 2001 than it was between 1984 and 1989. It does, however, cover the domain at the other time intervals the same.

Table 8. Post-Hoc analysis for the BMO ontology across all date brackets.

		i				
		[84-89]	**[90-95]**	**[96-01]**	**[02-07]**	**[08-14]**
j	**[90-95]**	$p = 0.04$ $i = 15.84$ $j = 26.88$				
	[96-01]	$p = 0.02$ $i = 15.32$ $j = 27.29$	$p > 0.05$ $i = 23.13$ $j = 25.88$			
	[02-07]	$p > 0.05$ $i = 17$ $j = 25.22$	$p > 0.05$ $i = 25.67$ $j = 22.26$	$p > 0.05$ $i = 26.58$ $j = 21.30$		
	[08-14]	$p > 0.05$ $i = 16.89$ $j = 23.76$	$p > 0.05$ $i = 24.67$ $j = 21.10$	$p > 0.05$ $i = 25$ $j = 20.71$	$p > 0.05$ $i = 22.70$ $j = 22.29$	

Table 8 also shows only one of the ontologies there are three more of these but in the interest of space and brevity we will only show results where there was statistical significance as depicted in Table 9.

Table 9. Pairwise comparisons of the date brackets for each ontology.

Ontology	Date bracket	Mean rank	P value
Process	[1984–1989]	15.21	0.02
	[1996–2001]	27.38	
Workflow	[1984–1989]	15.32	0.02
	[1990–1995]	27.29	
Intelleo	[1984–1989]	16.11	0.06
	[1990–1995]	26.67	

Table 10. Results for the evaluation of the difference between the means of the four ontologies' coverage of each Category using the Kruskal Wallis test.

Domain Category	P_Value	Significant?
Business Process Management	5.341e-09	yes
Grid workflow	2.055e-08	yes
Scientific workflow	4.364e-10	yes

9.5   Category Bias - Part 1

Table 10 depicts the results of the statistical significance test of the difference between the mean coverage of the BMO, Process, Workflow, and the Intelleo ontologies per category (Business Process Management, Grid Workflow, and Scientific Workflow). The table shows that at the $\alpha = 0.05$ level of significance, there exists enough evidence to conclude that there is a difference in the median coverage (and, hence, the mean coverage) among the four ontologies (at least one of them is significantly different) for each of the categories. However, this test does not indicate which of the ontologies are significantly different from which (or simply put, where the difference lies). Therefore, follow-up tests were conducted to evaluate pairwise differences among the different ontologies for each domain knowledge category. This also includes controlling for type 1 error by using the Bonferroni approach.

9.6   Category Bias - Part 1: Post-Hoc

At an alpha $(\alpha) = 0.05$, we can conclude that the BMO ontology's mean coverage is both larger (considering the mean ranks) and significantly different (p value $< \alpha$) from the other ontologies across all the categories, hence we reject the Null hypothesis with regards to the BMO ontology. In terms of the category bias metric, it distinguishes the BMO ontology as better representing the Business Process Management Category of the Workflow domain (Table 11).

The same is seen to be true for the Grid Workflow Category and Scientific Workflow Category as depicted in Table 12 which shows the pairwise comparison between the ontologies for the Grid Workflow and Scientific Workflow categories.

Table 11. Post-Hoc analysis for the Business Process Management Category.

		i			
		BMO	Process	Workflow	Intelleo
j	Process	$p < 0.05$ $i = 32.55$ $j = 12.45$			
	Workflow	$p = 6e-06$ $i = 32.18$ $j = 12.82$	$p > 0.05$ $i = 19.32$ $j = 25.68$		
	Intelleo	$p < 0.05$ $i = 32.45$ $j = 12.55$	$p > 0.05$ $i = 20.00$ $j = 25.00$	$p > 0.05$ $i = 23.23$ $j = 21.77$	

Table 12. Pairwise comparisons of the ontologies for each category.

Category	Ontology	Mean rank	P value
Grid workflow	BMO	28.58	0.00
	Process	10.42	
	BMO	28.58	0.00
	Workflow	10.42	
	BMO	28.42	6e-06
	Intelleo	10.58	
Scientific workflow	BMO	34.17	0.00
	Process	12.83	
	BMO	34.39	0.00
	Workflow	12.61	
	BMO	34.83	0.00
	Intelleo	12.17	

Table 13. Results for the evaluation of the difference between the means of each ontology's coverage of the domain categories using the Kruskal Wallis test.

Ontology	P_Value	Significant?
BMO	0.1142	no
Process	0.9869	no
Workflow	0.2025	no
Intelleo	0.4836	no

This was expected for the Business Process Management Category considering that is the ontology's area of focus. The other ontologies, when pitted against each other across the different domain categories seem to cover the domain similarly.

9.7   Category Bias - Part 2

Table 13 summarizes the results from the test between the mean coverage of
each ontology across the five domain categories. This reflects on how each ontology's coverage spreads through the partitions of the domain as defined by the
categories of this chapter. Considering these results we can conclude that for
all the ontologies at an $\alpha = 0.05$ there is no significant statistical evidence to
suggest that the ontologies cover the domain categories differently. For the case
of the BMO ontology, the observed results are contrary to what we had expected
since the ontology was predicated on the Business Process Management category of the workflow domain and therefore, you would have expected a slight bias
towards the same category. We could attribute this observation to the size of the
ontology. You could argue that it contains a large enough number of concepts
to blur the lines between the defined categories.

10   Qualitative Analysis

Section 4 discusses a theoretical framework that advocates for qualifying the
results of data-driven ontology evaluation and thereby accounting for bias. This
has further been demonstrated through experimentation in Sect. 8. When the
results are unqualified as was the case in Brewster et al. [2], important information (e.g. the ontology is aging) remain hidden and its relevance pertaining
to domain knowledge is undiscovered. A diachronic evaluation allows for such
information to be uncovered. For example, between 1984 and 1989 there was
no significant difference in the coverage of the workflow domain by the Process
ontology as compared to the Workflow ontology. However, there was a difference
in the period 1990 to 1995. This would suggest some change to domain knowledge
between those time intervals (e.g. introduction of new concepts). This difference
would not be accounted for if domain knowledge is not partitioned accordingly
during data-driven ontology evaluation.

11   Conclusions

This chapter has discussed an extension to data-driven ontology evaluation where
the main point of discussion was a theoretical framework that accounts for bias
in ontology evaluation. This is a framework that is premised on the notion that
an ontology is a shared conceptualization of a domain with inherent biases as well
as that domain knowledge is non-static and evolves over several dimensions such
as the temporal and categorical. The direct contributions of this work include
the two metrics (temporal bias and categorical bias), the theoretical framework,
as well as an evaluation method that can serve as a template for the definition
of evaluation methods, measures, and metrics.

It is fairly obvious that ontology evaluation constitutes a broad spectrum of
techniques each motivated by several things such as goals and reasons for evaluation as has been show in this chapter. The framework of this chapter is directed

to users and researchers within the data-driven ontology evaluation domain. It serves to fill the gap within this domain where time and category contexts have been overlooked thereby masking their influence of ontology evaluation results.

References

1. Villalon, J., Calvo, A.R.: A decoupled architecture for scalability in text mining applications. J. Univ. Comput. Sci. **19**, 406–427 (2013)
2. Brewster, C., Alani, H., Dasmahapatra, S., Wilks, Y.: Data-driven ontology evaluation. In: 4th International Conference on Language Resources and Evaluation, Lisbon, Portugal (2004)
3. Alani, H., Sanghee, K., Millard, E.D., Weal, J.M., Hall, W., Lewis, H.P., Shadbolt, R.N.: Automatic ontology-based knowledge extraction from Web documents. IEEE Intell. Syst. **18**, 14–21 (2003)
4. Hlomani, H., Stacey, D.A.: Contributing evidence to data-driven ontology evaluation: workflow ontologies perspective. In: 10th International Conference on Knowledge Engineering and Ontology Development, Vilamoura, Portugal (2013)
5. Ouyang, L., Zou, B., Qu, M., Zhang, C.: A method of ontology evaluation based on coverage, cohesion and coupling. In: 8th International Conference on Fuzzy Systems and Knowledge Discovery (FSKD), pp. 2451–2455 (2011)
6. Brank, J., Grobelnik, M., Mladenić, D.: A survey of ontology evaluation techniques. In: Proceedings of the Conference on Data Mining and Data Warehouses (SiKDD 2005), pp. 166–170 (2005)
7. Czajkowski, K., Fitzgerald, S., Foster, I., Kesselman, C.: Thalhammer, A., Toma, L., Hasan, R., Simperl, E., Vrandecic, D.: How to represent knowledge diversity. In: 10th International Semantic Web Conference (2001)
8. Van Der Aalst, W.M.P., Ter Hofstede, A.H.M., Kiepuszewski, B., Barros, A.P.: Workflow patterns. Distrib. Parallel Databases **14**, 5–51 (1981)
9. Burton-Jones, A., Storey, C.V., Sugumaran, V., Ahluwalia, P.: A semiotic metrics suite for assessing the quality of ontologies. Data Knowl. Eng. **55**, 84–102 (2005)
10. Vrandecic, D.: Ontology evaluation. Ph.D. Thesis, Karlsruhe Institute of Technology, Karlsruhe, Germany (2010)
11. Nonaka, I., Toyama, R.: The theory of the knowledge-creating firm: subjectivity, objectivity and synthesis. Ind. Corp. Change **14**, 419–436 (2005)
12. Patel, C., Supekar, K., Lee, Y., Park, E.K.: OntoKhoj: a semantic web portal for ontology searching, ranking and classification. In: 5th ACM International Workshop on Web Information and Data Management, pp. 58–61 (2003)
13. Spyns, P.: EvaLexon: assessing triples mined from texts. Technical report, Star Lab, Brussels, Belgium (2005)
14. Sebastian, A., Noy, N.F., Tudorache, T., Musen, M.A.: A generic ontology for collaborative ontology-development workflows. In: Gangemi, A., Euzenat, J. (eds.) EKAW 2008. LNCS (LNAI), vol. 5268, pp. 318–328. Springer, Heidelberg (2008)
15. Lusk, S., Paley, S., Spanyi, A.: The Evolution of Business Process Management as a Professional Discipline. BPTrends (2005)

A Semi-automatic Approach to Build XML Document Warehouse

Ines Ben Messaoud[1]([⊠]), Jamel Feki[2], and Gilles Zurfluh[3]

[1] Laboratory Mir@cl, University of Sfax, Sfax, Tunisia
Ines.benmessaoud@fsegs.Rnu.tn
[2] Jeddah University, Jeddah, Kingdom of Saudi Arabia
Jamel.feki@fsegs.Rnu.tn
[3] Laboratory IRIT, University of Toulouse 1, Toulouse, France
Zurfluh@univ-tlse1.fr

Abstract. Documents represent an interesting source for decisional analyses. They help decision makers to better understand the evolution of their business activities. Therefore, they merit to be warehoused for decision purposes within organizations. Generally, these documents exist in XML format and are described by multiple structures. In this paper, we present a semi-automatic approach to build the XML Document Warehouse. This approach is made up of two methods namely: *Unification of structures of XML Structures,* and *Multidimensional modeling.* More specifically, this paper focuses on the experiment and evaluation of the proposed approach for warehousing document-centric XML documents.

Keywords: Document warehouse · XML document · Unification of structures of XML documents · Unified tree · Multidimensional modeling · Galaxy model

1 Introduction

The Data Warehouse allows storage and analyses of huge amounts of structured business data [1]. However, these data are generally numeric values. In practice, they represent 20 % of the whole volume of data that could be useful for decision makers. Consequently, decisional data can be found not only as numeric values but also may be presented in documents. Therefore, documents should be integrated into the decision support system [2]. Likewise, [3, 4] advocate that documents should be warehoused. So, the Document Warehouse (DocW) has emerged.

In fact, documents may have different formats. Among these formats, XML (eXtensible Markup Language) is popular for data representation and exchange. It allows the exchange of data in the Web. There are two types of XML documents: *data-centric* and *document-centric* document [5, 6]. These documents are described by heterogeneous structures. Thus, when a decision maker needs to query these documents, he is obliged to write several queries; as the number of queries as the number of structures. To tone down this problem, we expect to provide a global view for the set of documents. In order to define the view, at the beginning we need to unify XML documents structures and, after that, translate documents into a multidimensional model.

© Springer International Publishing Switzerland 2015
A. Fred et al. (Eds.): IC3K 2014, CCIS 553, pp. 347–363, 2015.
DOI: 10.1007/978-3-319-25840-9_22

In [7], we presented a semi-automatic approach to build a DocW. This approach take into account *document-centric* XML documents belonging to the same domain and multidimensional requirements of decision makers. It is composed of two methods: (i) *Unification of structures of XML documents* [8, 9] and *Multidimensional modeling* [10, 11]. In this paper, we focus on the experiments and evaluation of the approach that produces a multidimensional model for the XML DocW. More precisely, we present the tests and assessment of this approach on academic and medical collections, through *USD (Unification of structures of XML Documents)* and *Galaxy-Gen (Galaxy Generation)* software tools. Also, in order to query easily the generated galaxies, we propose graphical notations for the galaxy operators defined in [12].

This paper is organized as follows: Sect. 2 discusses related works addressing unification of XML structures and multidimensional modeling of documents. Section 3 is an overview of our approach to building the XML DocW. In Sect. 4, we present the unification method. Then, Sect. 5 describes our multidimensional modeling method. Section 6 shows experiments and evaluations of the two methods also it presents graphical notations of the galaxy operators. Finally, in Sect. 6, we conclude the paper and addresses future works.

2 Related Works

An XML document is composed of tags that describe its hierarchical structure. According to the content, there are two types of XML documents: *data-centric* documents those are well structured and often used to exchange numeric data between computer applications; and *document-centric* documents those contain text and, therefore are less structured. Whatever its type, an XML document structure is described through one among the two formalisms namely *DTD* (Document Type Definition) and *XSD* (XML Structure Definition). In our work, we are interested in XML *document-centric* documents belonging to the same domain and described in conformity to one of these two formalisms. As our suggested approach aims to build an XML document warehouse relying on two steps: unification of structures of XML documents and Multidimensional modeling of documents, we divided the related work into two subsections, one for each of these two steps.

2.1   Related Works Addressing the Unification

The unification of documents aims to produce a global view for a set of XML documents initially described by heterogeneous structures. In the literature, some works proposed methods to unify DTDs (such as [13, 14 15]) whereas others suggested methods to unify XSDs (like [16]).

The authors of [13] propose an integration strategy called XClust based on clustering DTDs of XML data sources. They use an '*Expansion table*' to handle acronyms, and *Wordnet* as a lexical database to find out synonyms of names. Also, they compute similarity degrees between DTDs. These degrees calculation is based on linguistic and structural information; additionally, they consider semantics. After that, clusters of similar DTDs are determined based on similarity degrees. Finally, an integrated DTD is generated for each cluster. In fact, the similarity calculation depends on the correctness of a tuning phase devoted to set weights of degrees and thresholds [17].

In [14], a unification method was proposed; it translates each DTD into an object-oriented-based canonical schema that represents the conceptual abstractions of DTDs. However, the method does not treat the acronym names of tags used in the DTDs.

On the other hand, [15] developed an algorithm for the unification of DTDs. Their algorithm takes as input a set of DTDs for XML documents having similar structures and belonging to the same domain; the output is a unified DTD. In fact, the resulting DTD plays the role of a global conceptual schema for several subjects to a given domain. This algorithm consists of four steps: pre-processing, DTD representation, uniform DTD generation, and post-processing. Nevertheless, in the pre-processing step, the authors use a data structure called 'Element Name Resolution Table' in order to substitute each group of synonyms with a common and unique term. In practice this table may be incomplete since it cannot cover all synonyms.

In the purpose to integrate XSDs, [16] developed a process that receives a set of XSDs and then produces a global conceptual model. It converts each XSD into an Extended UML class diagram (EUML) and then applies three steps: *Clustering of concepts*, *Unification of concepts*, and *Restructuring of relationships*. The clustering of concepts resolves the naming conflict using *Wordnet*. The unification step resolves data type and structural conflicts. Finally, the third step restructures relationships and removes redundant ones. However, the proposed process does not treat acronyms. Also, the unification does not consider the structural hierarchies between elements of UML diagrams.

In order to summarize we note that in the literature of unification of structures of XML documents some works translate XML structures (i.e., DTD and XSD) into an intermediate model (tree, EUML diagram, etc.) whereas other efforts have focalized on a similarity degree to measure the pertinence of XML structures, and consequently to determine which structures can be merged together. Furthermore, at the design level, we note that decisional users do not participate during the unification of structures of XML documents. Generally, we note that the proposed approaches treat ambiguities (i.e., synonyms and/or acronyms) of names of structures to be unified.

In Sect. 4, we propose a method to unify structures of XML documents that translates each structure into a tree. We elected the formalism of tree because it is graphical and easy to be understood by decision makers, consequently it motivates the designer in order to validate the unification result. Note that our proposed method is based on a similarity degree calculation between each two structures. These degrees are calculated iteratively using a similarity matrix, and interactively in order to involve the stakeholders (mainly decisional designers) during the document warehouse design.

2.2 Related Works Addressing the Multidimensional Modeling of Documents

The multidimensional modeling aims to design multidimensional models that support OLAP (On-Line Analytical Processing) analyses. In the literature, there are two types of works dealing with the multidimensional modeling of XML documents: works related to data-centric XML documents [18, 19], and others related to document-centric XML documents. The remaining of this paper concerns document-centric XML documents. In this section, we present the most pertinent works related to that area where some researchers have proposed methods to model the DocW as a star model (such as [2, 3, 20, 21]) and other researchers adopted the galaxy model (like [12, 22]).

In [3], the authors propose a retrieval method in text collections; to do so, they model the global view of the documents set as a star model. In their star model, they distinguish five types for the *dimension* concept namely: *Localization, Time, Term, Document* and *Category*. The *measure* concept is the number of occurrences of each term within documents.

The author of [20] suggests a process to analyze documents of the DocW. This process relies on this star model. Firstly, the decision maker indicates the analysis components: fact, dimensions, and then selects an aggregate function (e.g. sum, min). Secondly, a document mart is generated and instantiated. Finally, the result is returned as a multidimensional table. However, during the document mart design phase the determination of multidimensional elements (i.e., fact, dimension) is manual. Indeed, the authors do not propose rules or algorithms to identify fact and dimensions to build the document mart.

The authors of [2] adopt the star model in order to analyze documents. This model distinguishes three types of dimensions: *Ordinary* dimension containing keywords extracted from the document, *Metadata* dimension which describes the document with *title, author,* etc.; and *Category* dimension that contains keywords external to the document; *i.e.,* issued from *Wordnet*. The result star model enables counting the number of documents according to these dimensions.

The result star model of [2, 3, 20] accomplish only quantitative analyses because their measures are numeric. Furthermore, analyses are limited since the analyses subject (*i.e.,* the fact) is defined a priori, at the design time of the star model but not at the query time.

The authors of [21] propose to revise the constellation modeling for documents; they suggest adding a new *textual measure* and two new dimensions called *Structure* and *Complementary*. In fact, a *textual measure* can be a word, a paragraph or a whole document. The *Structure dimension* describes the structure of documents whereas the *Complementary dimension* is determined from complementary data sources (*e.g.,* data from the curriculum vitae of authors). Nonetheless, the authors did not propose rules or algorithms to assist the DocW designer elaborating the constellation schema: identification of facts, dimensions and hierarchies.

In [12, 22], the authors propose a hybrid design process to build a document warehouse from document-centric XML documents. Their process combines a top-down approach (*i.e.,* starting from user requirements) and a bottom-up approach (*i.e.,* relying on the source data model). In addition, they suggest a new multidimensional conceptual model called *Galaxy*. This model can be defined as a set of entities, where each entity is modeled according to the dimension concept; several dimensions could be linked by a node when they are semantically compatible (i.e., used within a same query). However, the authors do not define rules to assist the design phase of a galaxy model.

To recapitulate, we have presented relevant works related to multidimensional modeling of documents. We have focused on *star* and *galaxy models*. We stress that a star model is characterized by a predefined subject of analyses (*i.e.*, fact). While, in a galaxy model the fact is not predefined; it will be specified when querying the galaxy. Therefore, analyses expressed on a galaxy model are more flexible than analyses expressed on a star model. Considering these benefits, we are interested in the galaxy for modeling the DocW.

In Sect. 5, we introduce our method for multidimensional modeling of documents. We have elected the galaxy model as a multidimensional model to build the schema of the DocW.

3   Proposed Approach for Warehousing XML Documents

In general, XML documents are described by heterogeneous structures even though they share the same domain. Consequently, when a decision maker wants to query heterogeneous documents, he is constrained to write several queries (*i.e.*, as many queries as the number of different structures in the document set) and then manage their results to obtain the final answer. To alleviate this drawback, our idea is to provide a global view that describes the warehoused XML document set. To do so, we have proposed an approach to build the schema of the DocW. This approach is composed of two methods called *Unification of XML documents structures*, and *Multidimensional modeling of documents* [7]. Figure 1 depicts this approach.

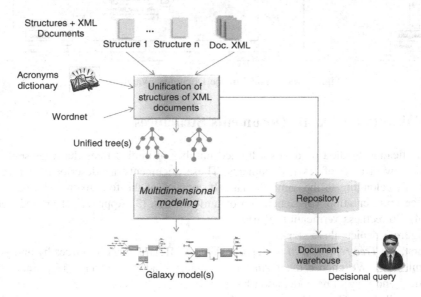

Fig. 1. Approach for building the XML Document Warehouse.

The unification method receives as input a set of XML structures belonging to the same domain and then produces a limited number of unified trees; these trees are validated by the decision makers [8, 9], whereas the multidimensional modeling method accepts the structure of the input XML documents. This structure could be either a unified tree resulting from the previous method (i.e., Unification of XML documents structures) or an XML structure and then produces a galaxy model [7, 11]. We note that the input structures and their output unified trees are saved in a specific repository shown in Fig. 2a; also the output galaxy model is saved according to the meta-model of Fig. 2b.

In the following, we detail the first method (i.e., unification of XML documents structures) and the second method (i.e., multidimensional modeling) in Sects. 4 and 5 respectively.

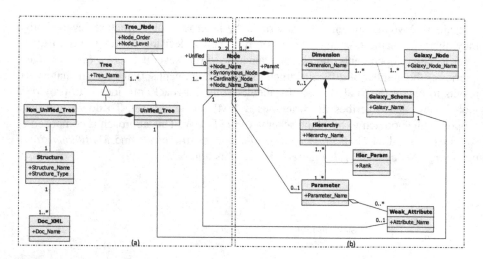

Fig. 2. Meta-model of the Document Warehouse.

4   Unification of XML Documents Structures

The unification method produces a limited number of unified trees that represent the global view of a set of XML documents. These documents are described by several structures belonging to the same domain. It consists of the four main steps namely: (1) Tree representation, (2) Generation of unified trees, (3) Approval of unified trees, and (4) Correctness verification of trees.

Figure 3 depicts these steps.

First, the *Tree representation* step translates XML structures into trees by applying two rules [7]. We choose the formalism of tree, as adopted in [13, 15], since it is graphical and easy to be understood by unskilled stakeholders.

Secondly, the *Generation of unified trees* produces a limited number of *unified trees* by applying a set of operators. It is performed in three sub-steps: *Ambiguity processing of tree nodes*, *Similarity calculation*, and *Unified trees production*.

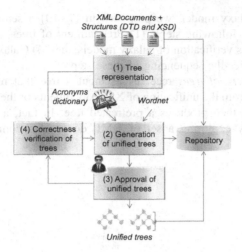

Fig. 3. Unification of XML documents structures.

The *Ambiguity processing* treats both acronym and synonym ambiguities of tree nodes, referring to a dictionary of acronyms and the *Wordnet* lexical database. To build the acronyms dictionary, we have used the Jaro algorithm [23]. Also, this sub-step guarantees the uniqueness of node names by renaming nodes having the same name by prefixing each one with the name of its parent. This is to obtain an accurate similarity factor based on nodes. Then, the *Similarity calculation* computes a triangular *Similarity Matrix* (*SM*) which has n trees in rows and in columns. It facilitates the identification of couples of trees to be merged. Each pair of trees having their similarity factor higher than a given threshold (experimentally determined) is merged in *Unified trees production* sub-step. This fusion is achieved by applying three fusion-operators (*Fusion by inclusion, Fusion by union of sub-trees, or Fusion by merging nodes*) developed in [8].

Thirdly, the *Approval of unified trees* validates trees according to the analytical requirements of decision makers. In fact, they can remove and/or rename nodes.

Finally, the *Correctness verification* checks the syntactic validity of trees according to a set of four constraints called: *Connected nodes, Hierarchy, Uniqueness of the root node* and *Acyclicity* [24].

5 Multidimensional Modeling of Documents

The multidimensional modeling of documents aims to semi-automatically generate a multidimensional model supporting OLAP analyses. Among the existing multidimensional models, we have elected the *Galaxy* model [12]. For readability reasons of the paper, we first introduce the galaxy and then describe our proposed method.

The galaxy model can be seen as a network of entities (*i.e.*, dimensions) connected by nodes (*cf.* Fig. 7). Each node connects compatible entities which could be used together in OLAP queries. In a galaxy, each entity can play a double role: an analysis subject (*i.e.*, fact) or an analysis axis (*i.e.*, dimension). As with the star model [25], an entity is composed of one or more attributes hierarchically organized.

To generate a galaxy model, we proposed in [7, 10] a semi-automatic method composed of the four following steps: (1) Pretreatment of trees, (2) Building galaxy models (3) Correctness verification of galaxy models, and (4) Galaxy models approval.

Figure 4 exemplifies the sequencing of these steps.

Firstly, *Pretreatment of trees* receives as input a tree that represents either the unified tree resulting from the unification of XML documents or the structure of a set of XML documents and then produces a pretreated tree. In fact, a pretreated tree has cardinalities added by exploring a set of XML documents compliant to the input structure(s).

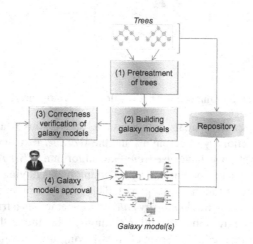

Fig. 4. Multidimensional modeling of documents.

Secondly, *Building galaxy* converts each pretreated tree into a multidimensional galaxy model by applying a set of ten rules and in two sub-steps: *Identification of dimensions (i.e., entities) and galaxy-nodes,* and *Identification of hierarchies.* For the first sub-step, we define three rules to extract dimensions and one rule for galaxy nodes. Since in a multidimensional model, dimensions are composed of parameters, we define four rules to extract parameters and two rules to extract weak attributes (*Identification of hierarchies* sub-step) (cf. [10] for further details). As within a dimensional hierarchy, at the lowest granularity level is the identifier of the dimension. In our work, this identifier is a surrogate key (artificial attribute which values are sequentially generated).

After that, the *Correctness verification of galaxies* step checks the syntactical validity of generated models. As a matter of fact, we find sometimes, among the extracted attributes for a dimension, attributes describing the structure and others relative to the metadata of documents. To tone down this problem, we split the set of attributes into two dimensions. This split relies on the usage of the *Dublin Core Metadata Initiative* (Dublin Core Metadata Initiative: www.dublincore.org.) which facilitates the identification of the metadata attributes.

Furthermore, we define a set of constraints. Eight constraints are adapted from those of the star model and issued from the literature. In addition we define three specific constraints for the galaxy (*Non isolated dimension, Non isolated node,* and *Disjunction of nodes*) (developed in [7]. We classify all these model constraints into three classes according to whether they apply on dimensions (*Identification constraint, Non empty dimension,* and *Non isolated dimension*), nodes (*Non isolated node* and *disjunction of nodes*) or hierarchies (*Hierarchical root, Exclusive hierarchies, Non isolated attribute, Non empty hierarchy, Rollup,* and *Acyclicity*).

Finally, the *Galaxy model approval* displays models issued from the previous step to the decision maker for agreement. In fact, he/she adjusts models according to their analytical requirements. This adjustment consists in removing and/or renaming multidimensional elements (*i.e.,* dimension, parameter). All these changes are saved in the repository of Fig. 2b to be used later in querying the galaxy.

6 Experiment and Evaluation

In order to evaluate our proposed approach, we have implemented two software tools: *USD 'Unification of Structures of XML Documents'* for the generation of the unified tree, and *Galaxy-Gen 'Galaxy Generation'* for the generation of multidimensional galaxy model.

We have carried out two experiments. The first one is applied to a set of twenty XML documents taken from the academic domain and compliant to four complex DTDs [11]. Our second experiment addressed a set of 1691 XML documents issued from the medical collection Clef-2007 and described by three DTDs. Nevertheless, there were some inadequacies in these XML documents; for example, all *keywords* are collected inside a unique textual tag. In order to alleviate this difficulty we have improved the initial DTDs (by adding the + symbol as a cardinality to some elements) in order to obtain more accurate XML documents. These documents are described by three DTDs which have some elements in common, and some different ones. These DTDs have the same root element linked to a set of elements that differs according to the DTD.

In the remaining of this section, we give the results of experiments and the evaluation of the unification and the multidimensional modeling methods on the academic and medical corpuses.

6.1 Experiment and Evaluation of the Unification Method

USD accepts a set of XML structures (i.e., DTDs and/or XSDs) belonging to the same domain and then produces one or more unified trees. It computes a similarity matrix and applies a set of unification operators. Note that in order to generate a DTD/XSD for XML documents not having one, USD invokes XMLSpy.

The first experiment (i.e., academic collection) produces a unified tree after three iterations. Figure 5 shows the obtained unified tree. First, the *Tree representation* produces a tree for each input DTD. Secondly, the ambiguity processing step resolves both synonyms and acronyms. Consequently, *USD* detects three nodes named with acronyms (*Auth, Para* and *Tit*). The *Auth* node is replaced by *Author*, the *Para* node is substituted with *Paragraph*; also, the *Tit* node is changed by *Title*, this is done by referring to the Acronym dictionary. Furthermore, the *Author* node is replaced with its synonym *Writer* which is much more frequent in the remaining trees by using *Wordnet*. In addition, this step treats the ambiguities of tree nodes names; as a result, *USD* renames eleven tree nodes by prefixing each one with the name of its immediate parent node (*Title*, *Paragraph*, *Figure*, *Table*, etc.). After that, the similarity matrix is computed in order to find out trees to be merged. Finally, the result tree is given to the decision-makers/designers in order to adjust it to their analytical requirements.

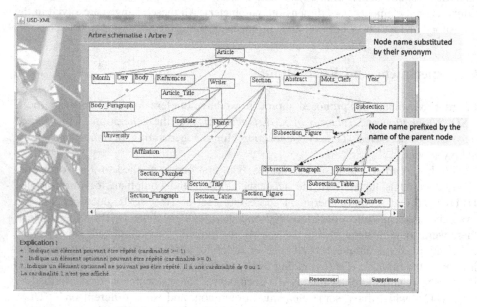

Fig. 5. Result unified tree for the first experiment.

In order to assess the *USD* tool, we have conducted a second experiment on XML documents taken from the medical collection. The processing of DTDs of this collection with *USD* produces a unified tree after two iterations. Figure 6 depicts the result unified tree.

To substantiate the unification method, we check that XML documents of each collection are valid with their unified DTD. To do so, we adopt the following principle: first, we translate the unified tree into a unified DTD and then we use XMLSpy tool to check the validity of the input XML documents with this DTD. By applying this principle on the result tree of the first experiment, we note that academic XML documents are valid with their unified DTD; likewise the second collection.

6.2 Experiment and Evaluation of Multidimensional Method

Let us remember that the multidimensional modeling method aims to design galaxy multidimensional models that support OLAP analyses. Our software tool *Galaxy-Gen* supports this method.

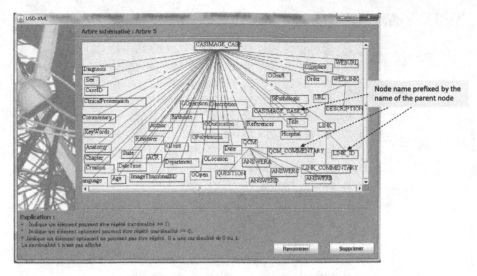

Fig. 6. Result unified tree for the second experiment.

Galaxy-Gen receives as input an XML structure or a unified tree resulting from the unification of a set of DTDs and/or XSDs belonging to the same domain and then produces a multidimensional *Galaxy* model.

The first experiment concerns the unified tree for the four DTDs (cf., Fig. 5). It generates a galaxy model composed of the following five dimensions: *D_Date, D_References, D_Mots_Clefs, D_Article* and *D_Writer* (cf., Fig. 7). The galaxy generation process starts with picking the result unified tree for the academic collection. Secondly, the pretreatment step is launched to produce the pretreated tree. In this tree, cardinalities are added for each parent node (except for the root), one cardinality for each outgoing edge. These cardinalities are determined by exploring the twenty input XML documents conform to the four DTDs of this experiment. We note that the structure of this pretreated tree and the structure of the unified tree are identical. Thirdly, dimensions of the galaxy are extracted by applying a set of four rules. After that, hierarchies of dimensions are determined. In fact, their parameters and weak attributes are extracted by applying rules defined in Sect. 5. Finally, when the decision-maker checks hierarchies (i.e., parameters and weak attributes), the galaxy model is automatically produced.

We have conducted a second experiment; it is performed on documents of the medical collection Clef 2007. After processing the result unified tree (cf., Fig. 6) with the *Galaxy_Gen* tool, we obtained a galaxy composed of five dimensions: D_Casimage_Case, D_Author, D_References, D_Keywords and D_Reviewer (cf., Fig. 8).

In order to justify that the result galaxy models let decision maker to elaborate significant decisional queries, we adopt multidimensional operators defined in [12] to handle galaxies. For readability reasons of this section, we first introduce multidimensional operators and then examples of decisional queries.

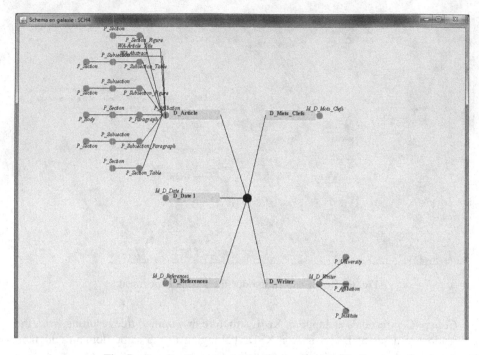

Fig. 7. Result galaxy model for the academic collection.

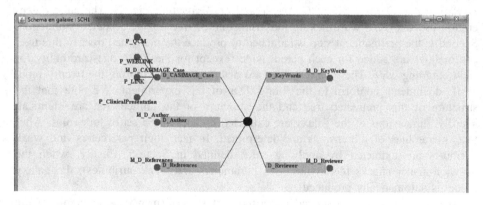

Fig. 8. Result galaxy model for the medical collection.

Language for Galaxy Manipulation. In [12], the author proposes five operators: *FOCUS*, *SELECT*, *DRILL_DOWN*, *ROLL_UP* and *ROTATE*. In order to facilitate their use, we suggest, in this paper, symbolic notations for these operators.

The *FOCUS* operator is used to specify a multidimensional analysis by choosing an analysis subject and to project data on analysis axes.

FOCUS (G, S, ID) = s^G where:

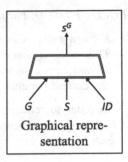

G: is a galaxy model.

S = (D_s, H_s, $Param_s$): an analysis subject where:
- D_s: is a dimension that plays the role of an analysis subject,
- H_s: is a hierarchy of D_s, and
- $Param_s$ = (fagg ($Param_{s\_i}$ [, $Param_{s\_inf}$... $Param_{s\_sup}$])): the data of the dimension are aggregated by the *fagg* function according to a list of optional attributes ($Param_{s\_inf}$... $Param_{s\_sup}$).

Graphical representation

ID = ((D_x, H_x, $Param_x$), (D_y, H_y, $Param_y$) ...): an ordered set of dimensions where:
- D_x is the first selected dimension,
- H_x is the current hierarchy of D_x, and
- $Param_x$ = ($Param_{x\_inf}$... $Param_{x\_sup}$) is an ordered set of parameters of H_x.
- D_y is the second selected dimension,
- H_y is the current hierarchy of D_y, and
- $Param_y$ = ($Param_{y\_inf}$... $Param_{y\_sup}$) is an ordered set of parameters of H_y.

When the analysis subject is determinated, we use the other proposed operators.

SELECT reduces the volume of data, to be analyzed, of a galaxy model s^G by specifying a restriction predicate expressed on a dimension or on a fact of s^G.
SELECT (s^G, pred) = $s^{G'}$ where:

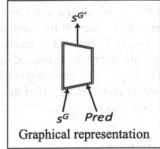

- s^G: is a galaxy model.
- *pred*: is a restriction predicate expressed on attributes of a dimension.

Graphical representation

In order to change the level of detail of data, two drill operators are useful.

DRILL-DOWN displays data at a more detailed granularity level.
DRILL-DOWN (s^G, D, $Param_{inf}$) = $s^{G'}$ where

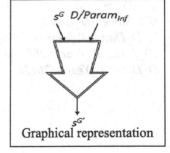

- s^G: is a galaxy model.
- D: is a dimension of s^G.
- $Param_{inf}$ is a parameter with a finer granularity than the parameter already selected.

The *DRILL-UP* operator provides more aggregated view of analyzed data.

Graphical representation

ROLL-UP $(s^G, D, Param_{sup}) = s^{G'}$ where

- s^G: is a galaxy model.
- D: is a dimension of s^G.
- $Param_{sup}$: is a parameter having a higher level than the current parameter.

In order to readdress a multidimensional analysis (i.e., change a dimension or a fact), the ROTATE operator is used.

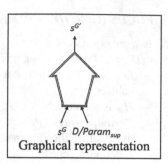

s^G D/Param$_{sup}$

Graphical representation

ROTATE $(s^G, D_{old}, D_{new}, H_{new}, Param_{new}) = s^{G'}$ Where

- s^G: is a galaxy model.
- D_{old}: is the analysis axis to change.
- D_{new}: is the new analysis axis.
- H_{new}: is a hierarchy of D_{new}.
- $Param_{new}$: is a parameter of H_{new}.

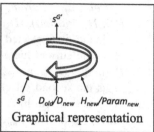

s^G D_{old}/D_{new} $H_{new}/Param_{new}$

Graphical representation

Expression of Multidimensional Queries. To show the usefulness of the generated galaxies, we define decisional queries and use the galaxy model operators. In this section, we present two multidimensional queries; a query for each collection.

Query 1: Number of research papers by author name for the years 2009 and 2012.

This query is expressed on the galaxy generated for the academic collection (cf., Fig. 7). To resolve this query, we use the *FOCUS* operator to select the analysis subject *D-ARTICLE*. After that, we employ the *SELECT* operator to restrict the current analysis to the years 2009 and 2012.

SELECT (
FOCUS (
G1, (D_Article, null,
COUNT(Id_D_Article),
(D_Writer, null,
Wa_Name),(D_Date1,
H_D_Date1, P_Year)),
(D_Date1.P_Year = 2009 or
D_Date1.P_Year = 2012))

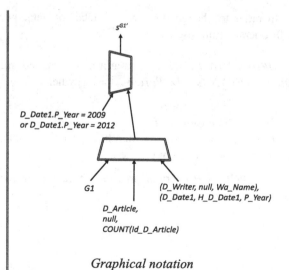

Textual notation   *Graphical notation*

Query 2: Number of casimage case by author.

This query is expressed on the galaxy of Fig. 8. We select the analysis subject *D-CASIMAGE-CASE* and then project data on the dimension *D-Author*. This requires the *FOCUS* operator.

FOCUS (
G2, (D_CASIMAGE_Case,
null, COUNT
(Id_D_CASIMAGE_Case)),
(D_Author, null,
Wa_Author),
)

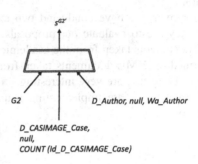

G2   D_Author, null, Wa_Author

D_CASIMAGE_Case,
null,
COUNT (Id_D_CASIMAGE_Case)

Textual notation   *Graphical notation*

7 Conclusion

Documents represent a means to express knowledge and information. Their contents help decision makers to better understand the evolution of their business processes. Consequently, these documents merit to be warehoused for decision purposes. In this paper, our main interest was to enrich the data warehouse with documents in order to improve the quality of decisional analysis. More specifically, we have proposed an approach to build a document warehouse. This approach is composed of two main methods: *Unification of XML document structures*, and *Multidimensional modeling of documents*.

The *Unification* method accepts a set of structures of XML documents and then produces a limited number of unified trees. First, it translates each XML structure into a tree. Secondly, it resolves synonyms and acronyms problems by using respectively *Wordnet* and a dictionary of acronyms. It calculates a similarity degree for each two trees and presents these degrees in a similarity matrix. This matrix helps to decide which trees have to be merged first. Thirdly, selected trees are merged by using three algebraic operators. After that, generated tree(s) are submitted to the designer for validation. Finally, the unification method verifies the correctness of the generated trees by checking a set of four constraints. In order to evaluate this method, we have developed a software tool called *USD 'Unification of Structures of XML Documents'*.

The *Multidimensional modeling* translates each unified tree issued from the unification method into a galaxy model. First, cardinalities are added for each node to obtain a pretreated tree. Secondly, the multidimensional concepts are defined by applying a set of ten rules. After that, the correctness verification of generated galaxy is verified in respect to a set of eleven constraints, in addition the Dublin Core Metadata is used to determine hierarchies that describe the metadata of documents. Finally, the designer validates the resulting galaxy. To substantiate this method, we have developed the *Galaxy-Gen* software tool and we use the galaxy operators defined in [12]. In order to facilitate the use of these operators, we have introduced symbolic notations for these operators.

Moreover, we have conducted two experiments using the USD and Galaxy-Gen tools; they are to evaluate our proposals. The first experiment is applied on a set of XML documents taken from the academic domain. Whereas, the second experiment is performed on XML documents taken from the Clef-2007 collection. For both experiments, the results are very interesting.

Currently, we are implementing a query language for the galaxy based on the galaxy operators.

References

1. Pérez, M.J.M., Berlanga, L.M.R., Aramburu, C.M.J., Pederson, T.B.: Contextualizing data warehouses with documents. In: Decision Support System (DSS), vol. 45, pp. 77–94. Elsevier (2008)
2. Tseng, F.S.C., Chou, A.Y.H.: The concept of document warehousing for multi-dimensional modeling of textual-based business intelligence. In: Decision Support Systems (DSS), vol. 42, pp. 727–744. Elsevier (2006)
3. McCabe, M.C., Lee, J., Chowdhury, A., Grossman, D., Frieder, O.: On the design and evaluation of a multi-dimensional approach to information retrieval. In: Proceedings of the 23rd Annual International ACM SIGIR Conference, pp. 363–365 (2000)
4. Sullivan, D.: Document Warehousing and Text Mining: Techniques for Improving Business Operations. Marketing and Sales. Wiley, New York (2001)
5. Fuhr, N., Grobjohann, K.: XIRQL: a query language for information retrieval in XML documents. In: 24th International ACM Conference on Research and Development in Information Retrieval (SIGIR), pp, 172–180. ACM Press (2001)
6. Kamps, J., Marx, M., De Rijke, M., Sigurbjornsson, B.: Best-match querying from document-centric XML. In: Proceedings of the Seventh International Workshop the Web and Databases, pp. 55–60 (2004)
7. Feki, J., Ben Messaoud, I., Zurfluh, G.: Building an XML document warehouse. J. Decis. Syst. JDS 2013 **22**, 122–148 (2013)
8. Ben Messaoud, I., Feki, J., Khrouf, K., Zurfluh, G.: Unification of XML document structures for Document Warehouse (DocW). In: 13th International Conference on Entreprise Information Systems (ICEIS), pp. 85–94, Beijing (2011a)
9. Ben Messaoud, I., Feki, J., Zurfluh, G.: A first step for building a document warehouse: unification of XML documents. In: Sixth International Conference on Research Challenges in Information Science (RCIS), pp. 59–64, Spain (2012)
10. Ben Messaoud, I., Feki, J., Zurfluh, G.: Modélisation multidimensionnelle des documents XML. Revue des Nouvelles Technologies de l'Information (RNTI) **B-7**, 55–70 (2011b)
11. Ben Messaoud, I., Feki, J., Zurfluh, G.: Galaxy-Gen: a tool for building galaxy model from XML documents. In: 6th International Conference on Knowledge Engineering and Ontology Development KEOD 2014, Rome, Italie (2014)

12. Tournier, R.: Analyse en ligne (OLAP) des documents. Ph.D. thesis, University of Toulouse III, France (2007)
13. Lee, M.L., Yang, L.H., Hsu, W., Yang, X.: XClust: clustering XML schemas for effective integration. In: Proceedings of the ACM International Conference on Information and Knowledge Management (CIKM), pp. 292–299, Virginia (2002)
14. Mello, R.D.S., Castano, S., Heuser, C.A.: A method for the unification of XML schemata. Inf. Softw. Technol. **44**, 241–249 (2002)
15. Yoo, C.-S., Woo, S.-M., Kim, Y.-S.: Unification of XML DTD for XML documents with similar structure. In: Gervasi, O., Gavrilova, M.L., Kumar, V., Laganá, A., Lee, H.P., Mun, Y., Taniar, D., Tan, C.J.K. (eds.) ICCSA 2005. LNCS, vol. 3482, pp. 954–963. Springer, Heidelberg (2005)
16. Zhang, Y., Liu, W.: Semantic integration of XML schema. In: First International Conference on Machine Learning and Cybernetics, Beijing (2002)
17. De Meo, P., Quattrone, G., Terracina, G., Ursino, D.: "Almost Automatic" and semantic integration of XML schemas at various "Severity" levels. In: Meersman, R., Schmidt, D.C. (eds.) CoopIS 2003, DOA 2003, and ODBASE 2003. LNCS, vol. 2888, pp. 4–21. Springer, Heidelberg (2003)
18. Boussaid, O., Ben Messaoud, R., Choquet, R., Anthoard, S., Conception et construction d'entrepôts XML. 2ème journée francophone surles Entrepôts de Données et l'Analyse en ligne EDA 2006, pp. 3–22, Versailles, France (2006)
19. Hachaichi, Y., Feki, J., Ben-Abdallah, H.: Modélisation multidimensionnelle de documents XML centrés-données. J. Decis. Syst. JDS 2010 **19/3**, 313–345 (2010)
20. Khrouf K.: Entrepôts de documents: De l'alimentation à l'exploitation. Thèse de doctorat en Informatique, Université Paul Sabatier, Toulouse, France (2004)
21. Ravat, F., Teste, O., Tournier, R.: Analyse multidimensionnelle de documents via des dimensions OLAP. Document numérique, Hermès, Numéro spécial Entreposage de documents et données semi-structurées, pp. 85–104 (2007)
22. Pujolle, G., Ravat, F., Teste, O., Tournier, R., Zurfluh, G.: Multidimensional database design from document-centric XML documents. In: Cuzzocrea, A., Dayal, U. (eds.) DaWaK 2011. LNCS, vol. 6862, pp. 51–65. Springer, Heidelberg (2011)
23. Jaro, M.A.: Advances in record linking methodology as applied to the 1985 census of Tampa Florida. J. Am. Stat. Soc. **89**, 414–420 (1989)
24. Aouabed, H., Ben Messaoud, I., Feki, J., Zurfluh, G.: USD: Un outil d'unification des structures des documents XML. 6ème Atelier des Systèmes Décisionnels ASD 2012, pp. 83–94, Blida, Algérie (2012)
25. Golfarelli, M., Maio, D., Rizzi, S.: Conceptual design of data warehouses from E/R schema. In: Proceedings of the 31st Annual Hawaii International Conference on System Sciences (HICSS 1998), pp. 334–343. IEEE Computer Society, Washington, D.C., USA (1998)

Improving General Knowledge Sharing via an Ontology of Knowledge Representation Language Ontologies

Jéremy Bénard[1] and Philippe Martin[2,3(✉)]

[1] GTH, Logicells, 3 rue Désiré Barquisseau, 97410 Saint-Pierre, France
jeremy.benard@logicells.com
[2] EA2525 LIM, ESIROI I.T.,
University of La Réunion, 97490 Sainte Clotilde, France
Philippe.Martin@univ-reunion.fr
[3] School of ICT, Griffith University, Queensland, Australia

Abstract. Via a comparison of the currently used language-based components for knowledge sharing, this article first highlights the difficulties caused by the inexistence – and hence absence of exploitation – of a shared core ontology of knowledge representation languages (KRLs), i.e., (i) an ontology of KRL abstract models which represents, aligns and extends standards, and (ii) an ontology of KRL notations. For programmers, these are the difficulties of importing, exporting or translating between KRLs; for end-users, the difficulties of adapting, extending or mixing notations. This article then shows how we have built this shared core ontology plus a tool for exploiting it. We use them for specifying, parsing and translating KRLs, thus allowing their use without additional programming. This ontology can be represented in any KRL that has at least OWL-2 expressiveness. Thus, the results can easily be replicated. A Web address for the tool and the full specifications is given.

Keywords: Language ontology · Meta-modelling · Syntactic translation · Knowledge representation languages · General knowledge sharing

1 Introduction

The term "knowledge representation language" (KRL) may refer (at least) to one of the next notions or their combination: (i) *a semantic model*, i.e., a set of types (alias, an ontology) specifying semantic and/or logical notions, e.g., those of the SHOIN(D) description logic, (ii) *an abstract "data type" (ADT) model*, i.e., an ontology for ADTs such as abstract syntax trees or abstract semantic graphs (e.g., most types in the OWL-DL ontology form an ADT model since they permit to store a graph that only contains binary relations and follows a semantic model for SHOIN(D)), and (iii) a *concrete model*, i.e., a textual/graphical/... KRL *notation*, e.g., Turtle and OWL-Functional-Style. Knowledge sharing (KS) involves many tasks, some of which are notation related: knowledge editing, parsing, importing, exporting, translating, etc. Section 2 compares "approaches implementing these tasks" depending on the language they offer for specifying the ADT model: (i) a notation grammar, e.g., EBNF, (ii) a

© Springer International Publishing Switzerland 2015
A. Fred et al. (Eds.): IC3K 2014, CCIS 553, pp. 364–387, 2015.
DOI: 10.1007/978-3-319-25840-9_23

generic language for creating ADTs, e.g., XML and MOF-HUTN, and (iii) an ADT ontology.

The first approach implies creating a concrete grammar with *actions* associated to its rules, and giving it to a parser generator, e.g., Lex & Yacc. The actions contain functions or rules to build a less syntax-oriented ADT and build a semantic model (directly or not: a parser/translator exploiting rules and a grammar for the semantic model may be used). Even though an "interactive programming environment" generator (e.g., Centaur) may provide parsers, editors, checkers and interpreters or compilers for specified languages, creating these specifications involves – or is akin to – programming. These specifications (grammars and building/translation rules or functions) are difficult to compare, extend and re-use: they cannot be organized by specialization relations. Small changes in the concrete/ADT/semantic models often lead to important changes in the specifications. Translations rules or procedures have to be specified for each pair of languages.

With the second approach, specifying an ADT grammar (e.g., an XML schema) permits to use a concrete parser or editor and specify various presentations (e.g., via CSS and XSLT). However, since the concrete descriptions must then have an explicit structure, they are too bulky to be used *directly* for building or displaying (parts of) programs or knowledge bases, especially with standards in this approach (e.g., XML). Thus, more adapted textual/graphic KRL notations are required and parsers for them are still needed. Furthermore, since these languages are not based on logic, the tools based on them cannot perform logical inferences, hence cannot exploit knowledge representations.

With the third approach – i.e., the use of ontologies on notations, ADTs and semantic elements instead of just grammars on them – the difficulties of the previous two approaches can be reduced. So far however, (i) there were no ontologies for notations, and (ii) the ontologies on KRL abstract models (ADTs and/or semantics) were implicit (i.e., informally or insufficiently described) or about KRLs of insufficient expressiveness for representing many other KRLs (hence for general knowledge representation and sharing). Thus, these last ontologies were also not inter-related. This article shows how (i) we created a core for an "ontology of KRL ontologies" by representing, aligning and extending the major "KRL abstract models", and then extending it with a KRL concrete model ontology, and (ii) we have begun to use/extend it for building an organized library of declarative concise "KRL specifications". This "ontology of KRL ontologies" supports – and also specifies (i.e., provides the declarative code for) – a *generic tool for parsing and exporting KRLs,* hence also performing (many) translations between them. Since programming is avoided and since KRLs or KRL modifications can be specified in a concise way, even the end-users of applications using such a tool can adapt the format of input and output KRLs to their needs or preferences. This ontology, as well as a Web server interface to use this tool, are available from http://www.webkb.org/KRLs/.

Section 3 introduces the notation used in this article for the illustrations.

Section 4 shows some relations between top-level "language elements" and more general concepts, as well as between notation models and abstract models.

Section 5 explains the main primitive concepts and relations which permit to represent and relate the various models in a uniform and concise way.

Section 6 illustrates their use for the general "abstract model" parts of our ontology.

Section 7 illustrates their use for specifying particular KRLs and even grammars. Representing grammars shows the generality of the approach and permits to re-use existing parser generators. However, our tool currently only uses the LALR(1) parser generator "Gold"; thus, it cannot parse KRLs requiring a more expressive grammar.

2   Comparison of the Various Language-Based Approaches for KS

The model-related terminology used in this article is compatible with the terminology used in syntax or semantic related works. To be compatible with the terminology used in Model Driven Engineering (MDE) related works, especially the one related to MOF (the Meta-Object Facility of the OMG: Object Management Group), the prefix "meta-" would need to be systematically added before "model" since in MDE a document or code is a "model" and it follows the specifications of a "meta-model", e.g., an XML schema. For clarity purposes, in this article, a "meta-model" – i.e., a language to specify a language – refers to what is called a "meta-meta-model" in MDE.

The introduction noted that a model may be an "*abstract model*" (e.g., OWL-DL or an XML schema; its *specifies abstract structures* of a certain type) or a "*concrete model*" (e.g., a formal grammar or a CSS script; it *specifies concrete structures* of a certain type). A concrete model, alias a presentation model, specifies either a formal presentation (it is then a notation) or an informal one (e.g., that certain kinds of ADT elements should be presented with bold characters). CSS and EBNF are therefore meta-models for concrete models. MOF is a meta-model for – and a subset of – the UML model. Since UML also refers to a notation, it is both an abstract and concrete model. Even graphical notations implicitly or explicitly follow a grammar [1]. XML is a meta-model for certain ADT models and concrete models. Meta-models are also models.

This terminology permits to compare language-based approaches based on the kind of meta-model they exploit: a notation-dependent one, a structure-dependent one and a logic-based one. In each case, the next subsections show that problems comes from a lack of expressiveness of the meta-model: there are some notions that it allows to *declare* (in a model) but not to *define*. Thus, tools based on this meta-model cannot exploit the semantics of these notions. Other tools that know this semantics are needed for exploiting it. Furthermore, these notions are represented in different and ad-hoc or implicit ways across models, thus making knowledge sharing and re-use more difficult. As can be concluded at the end of the next subsections, the use of an ontology of KRL ontologies, based on higher-order logics, is necessary to avoid problems.

2.1   Exploiting Notation-Dependent Meta-models

The grammar directed parsing of a textual/graphical description leads to a concrete structure – e.g., a concrete syntax tree if a context-free concrete grammar is used – and an ADT – e.g., an "abstract syntax tree" or an "abstract semantic graph". To derive the abstract structure from the concrete one, an "action" – containing a transformation rule or function – is generally associated to all or most concrete grammar rules. Then, static semantic checking (type checking, context-dependent interactions), dynamic semantic checking (interpretation, debugging) and "un-parsing" (pretty-printing of the abstract structures, translation to other notations) can be performed or their code/specifications can be generated. E.g., in the generic "interactive programming environment" generator Centaur [2] – which may generate a parser, a structured editor, a type checker and an interpreter or a compiler for a formal language – the concrete and abstract grammars and building rules can be specified in the Metal formalism (which is then for example transformed into a format on which Yacc is called), while the semantic related specifications are in Typol which is then transformed and executed via Prolog, Lisp and, via the Minotaur system [3], attribute grammars. Centaur has been used for numerous programming languages and one KRL [4].

Tools such as Centaur ease the task of writing parsers, translators, etc. They can be seen as programming focused MDE tools. They could be extended to use ontologies or MOF and XML models. If such a tool allowed the re-use of an ontology of abstract and concrete models (such as the one we propose for KRLs), the various specifications that must be provided by people for each language would be much lighter, comparable and re-usable since the ontology permits to share and categorize them. This is much harder otherwise, even with KRLs such as Prolog which are oriented toward execution rather than modelling. Our KRL ontology includes sufficient information to allow people – programmers as well as application end-users – to specify the peculiarities of their notations for them to be usable as input or output KRLs: programers or users do not need to specify conversions (except for complex ones between abstract elements referring to elements of different logics) since they can be automatically derived.

To avoid the use of multiple structures or models, and thus also allow languages to be directly extended, other (and often earlier) avenues have been proposed. They all imply that extensions can be defined with the language and that it embeds a parser (e.g., via an "eval" function). This is eased when the language is *homo-iconic*, i.e., when its abstract structure can be directly derived from the text (because they have hom same structure). In other words, this is eased when new functions or rules can be built like other data structures and then evaluated, as in Lisp and Prolog. Lithe [5] is a class-based programming language looking like an EBNF grammar with semantic actions containing C-like code; the classes are the non-terminal alphabet of the grammar. Similarly, XBNF [6] is an extension of EBNF that is a KRL since it permits to define some functions, logical relations and sets on each class of objects defined by a rule. However, in all these other avenues, extensions to the language are restricted by its core concrete and abstract models. In other words, these extensions do not allow to represent and follow other models than the predefined ones. Yet, they show that using a unique language to represent abstract and concrete specifications has advantages (concision, ...). A very expressive modelling-oriented KRL has those advantages without the restrictions since it can represent different models (because of it expressiveness, the *defined* models are just specializations of the core models). It also permits to categorize their elements.

2.2   Exploiting Structure-Dependent Meta-models

As explained in the introduction and the previous sub-section, using EBNF and other "languages to write concrete grammars" them implies programming-like tasks to build the abstract structures (ADT and semantic ones) from the concrete ones. The building of the ADT structure can be avoided when homo-iconic ADT meta-models are used, e.g., XML, MOF-HUTN (the Human Usable Textual Notation for MOF) or XMI (the XML-based notation for MOF models). However, the concrete descriptions are then not enough concise and high-level or user-friendly [7] to be used *directly* for developing or displaying (parts of) programs or KBs. Indeed, (i) these descriptions have a very explicit structure, and (ii) current meta-models – and hence their notations – *declare* and allow the use of few *structural* notions (e.g., the notion of object, attribute and – in MOF – association/relation), not logical notions (quantifier, meta-statement, etc.) nor programming notions (parallelism, succession, class inheritance, parameter evaluation, etc.). Although such "structural meta-models" permit to *declare* these additional notions in models, they do not permit to *define* their semantics. In other words, the tools which exploits structural meta-models (e.g., the XML parser and CSS pretty-printer) can only understand the meaning – and hence exploit – the structural elements. This is why, since 1999, for the Semantic Web and, more generally, for knowledge sharing and exploitation purposes, the W3C advocates the use of RDF instead of XML *as meta-model*. Indeed, XML is a tree-based structural meta-model while RDF is a graph-based meta-model which follows a simple logic and can be extended with language ontologies, e.g., those of OWL. However, like KRL models represented with a structural meta-model, extensions by language ontologies are just *declarations* of KRL components (not *definitions*): inference engines must handle them in special ways to take into account their semantics. The OMG followed this approach by proposing ODM (Ontology Definition Metamodel) which, in its version 1.1 [8], *declares* the elements of four KRL models in MOF (with a few relations between them) – RDF, OWL, Common Logic and Topic Maps – with a UML profile for the first two. No integrated ontology is provided. To conclude, since XML and MOF-HUTN cannot be used *directly* (as notations and ways to *define* semantics) for knowledge sharing or programming, tools supporting those tasks include parsers and semantic-handling modules in addition to – or replacement for – XML and MOF related tools.

Some MDE tools are described as having extensible *input notations*, e.g., BAM [9] in process modelling. Actually, they handle an expressive meta-model which includes the primitives for various (already existing) process modelling languages and thus can handle each of them. The authors of these tools also count textual annotations – both formal and informal ones – as a way to extend existing graphical notations.

2.3   Exploiting a Logic-Based Meta-model

As above noted RDF is an ADT meta-model following a simple logic. Its structures can be presented with concrete models specific to RDF (e.g., RDF/XML: RDF linearized with XML) or not (e.g., Turtle). These structures can be used for storing ADTs with more semantics (e.g., SPIN structures are RDF representations of the SPARQL query language). The generic parser and translator specified in this article also works for RDF. No generic parser just for RDF or another model seems to have been undertaken. On the other hand, there have been several works on style-sheet based transformation languages and ontologies for specifying how RDF abstract structures can be *presented*, e.g., in a certain order, in bold, in a pop-up window, etc.: Xenon [10], Fresnel [11], OWL-PL [12] and SPARQL Template [13]. Since these tools do not use a notation ontology, they require a new template or style-sheet for each target notation.

KRLs of low expressiveness ensure good properties for knowledge exploitation, typically speed and completeness. This is why KRL models of the OWL family have different degrees of expressiveness, all inferior to First Order Logic (FOL). For knowledge modeling and sharing purposes, the more expressive the used KRLs the better. Indeed, more expressiveness permit more definitions (instead of declarations) and thus permit to set more relations between different notions (logic ones, programming ones, …) from one or various sources. In other words, a more expressive KRL permits to represent knowledge in more precise (or less biased and ad-hoc) ways and in more generic, high-level, normalized and concise ways, hence in easier to develop and re-use ways. This is clear with the representation of cardinalities (or, more generally, numerical quantifiers; e.g., see the part in italic bold in Table 1), meta-statements and set interpretations. Thus, for knowledge modeling and sharing purposes – and also, as explained below, for knowledge exploitation purposes – a meta-model needs to represent "higher-order logic" (HOL). RIF-FLD [14], the W3C "Framework for Logic Dialects", is based on HOL. To ease its re-use, in our KRL ontology we represented the RIF-FLD elements and organized them via subtype and exclusion links (this organization was left implicit by the W3C, only a grammar and informal descriptions were provided). KIF (Knowledge Interchange Format) [15] is a KRL – with a FOL model but a HOL notation – which reached its purpose: being a de-facto KRL interlingua by allowing KRL authors to *define* elements of their KRLs in KIF and thus ease the translations of these KRLs into KIF.

A HOL model does not necessarily require a HOL inference engine to be handled. One reason is that, interpreted with Henkin semantics, it is equivalent to (many-sorted) FOL. This is how KIF has a HOL notation and a FOL model. Another reason is that conversions to less-expressive models (via losses of information) can be performed for applications, depending on their needs. E.g., a knowledge-retrieval application gain speed and does not loose much precision and completeness by ignoring meta-statements (modalities, …) as long as results are displayed with their associated meta-statements. Since HOL does not restrict possible exploitations – as opposed to knowledge modelled with KRLs of lower expressiveness – it is good for knowledge exploitation purposes too.

For business-to-business KS where the used structural/logic meta-models are well-known and sufficient for both businesses, KRLs of reduced expressiveness may be used and tailored knowledge conversion procedures may then be developed. For general knowledge sharing and re-use, or for making business-to-business KS more efficient, knowledge representation should not be restricted and, to allow the use of various KRLs, a generic parser-translator for KRLs is needed. This requires a shared ontology of KRLs (abstract models and notations) and, more generally, of general concepts. The Ontolingua server [16] was a first step in that direction. It proposed a structured library of interconnected fundamental ontologies, some of which formalized concepts related to KRL models, especially frame-based ones, i.e., concepts similar to those of OWL. It also hosted ontologies of its users in a structured library. However, this server did not have protocols to detect and help avoiding implicit redundancies between the ontologies and hence encouraging numerous relations between the various represented concepts. In other words, the various ontologies did not form an integrated one. The WebKB server [17] provides protocols solving this problem within a KB and between different servers, without restrictions on the content or on the used KRLs, nor forcing the users to agree on terminology or beliefs. The ontology proposed in this article is hosted by a WebKB server and thus can be extended by Web users.

This ontology integrates the main standards for KRL models: RDF + OWL + RIF-FLD from the W3C, Common Logic [18] (a subset of the KIF model) from *ISO/IEC* and the "Semantics of Business Vocabulary and Business Rules" (SBVR [19]) from the OMG. These standards have similar or complementary components which, previously, were not semantically related. Another originality is that our ontology includes a notation ontology.

3   Notation and Conventions Used in this Article for Illustrations

To allow the display and understanding of its numerous required illustrations, this article needs a concise and intuitive notation for KRL models of OWL-2 like expressiveness.

To that end, this article uses the FL notation [20] (it does not advocate the generalized use of FL since it proposes a way for people to use any notation they wish). Indeed, graphical notations are not concise enough and common notations such as those of the W3C are not sufficiently concise and "structured" enough. Here, "structured" means that all direct or indirect relations from an object can be (re-)presented into a unique tree-like statement so that the various inter-relations can readily be seen. Table 1 illustrates this by representing the same statement – or set of statements – in English and then in six formal notations: FE (Formalized-English [21]: it looks like English but it is actually very similar to FL), FL, UML, Turtle (or Notation 3), OWL Manchester notation and OWL Functional-style. This last notation is "positional relation" based. The first five are graph-based notations: they are composed of concept nodes and relation nodes.

The above textual graph-based notations are frame-based. A frame is a statement composed of a first "object" (alias "node": individual or type, quantified or not) and several links associated to it (links from/to other objects). In this article, "link" refers to an instance of a "binary relation type". In OWL, such a type is instance of "owl: Property" ("owl#Property" in FL: the namespace identifier is before the "#"). What is not an individual is a type: relation type or concept type (an instance of owl#Class).

In this article, the default namespace is for the types we introduce via our ontology. Each name for a concept type or individual is a nominal expression beginning by an uppercase letter. The name of a relation type we introduce begins by "r_" (or *"rc_"* if *this is a type of link with destination a concrete term*). Thus, names not following these

Table 1. The same statement – or set of statements – in English and six different KRL notations: FE, FL, UML, Turtle, OWL Manchester, OWL Functional-Style. In all other tables, FL is used.

The type Language_or_Language-element is defined by its subtype partition composed of Language_element and Language. Any instance of Language has for (r_)part at least 2 instances of Language_element. This last type has for subtypes (at least) KRL and Grammar.

/* In **FE:** */ Language_or_Language-element has for partition {Language, Language_element}.
Language has for r_part at least 2 Language_element.
Language has for subtype KRL and has for subtype Grammar.

Language_or_Language-element   //Notes: in **FL**, ">" is an abbreviation for the "subtype" link
= exclusion   // as in some other notations). "<" is its inverse.
  { (Language   // "exclusion{...}" specifies a union of disjoint types :
      r_part: 2..* *Language_element*,   // a real "subtype partition" of T if "T = exclusion{...}",
      >   KRL   Grammar)   // an "incomplete partition" if "T > exclusion{...}".
   Language_element   // A "," separates 2 links of different types. For consecutive links of
  };   // the same type, this type needs not be repeated and there is no ",".

Language_or_Language-element   //In this **UML** representation, no box is drawn
                                                                  // since no attribute or method needs to be
                          △   // represented.
          {disjoint, complete}

Language *r_part*   2..* → *Language-element* // Here, an association/relation of type r_part
△   △   // is used instead of the special arrows used
| | // in UML for aggregations or compositions
KRL   Grammar

:Language_or_Language-element   //**Turtle + OWL** *(which is a low-level KRL model, e.g., the*
   owl:equivalentClass [rdf:type owl:Class;   // *part in italic bold below translates "2..*")*
                           owl:unionOf (:Language *:Language_element*)].
[ ] rdf:type owl:AllDisjointClasses; owl:members (:Language *:Language_element*).
Language *rdfs:subClassOf [a owl:Restriction; owl:onProperty : r_part;*
                           owl:minQualifiedCardinality "2"^^xsd:nonNegativeInteger;
                           owl:onClass Language_element]´
KRL rdfs:subClassOf :Language.   Grammar rdfs:subClassOf :Language.
/ Here is a translation in "RDF/XML with OWL" of the 3rd line in this Turtle example:*
<owl:AllDisjointClasses><owl:members rdf:parseType="Collection">
    <owl:Class rdf:about="Language"/>
    *<owl:Class rdf:about="Language-element"/></owl:members></owl:AllDisjointClasses> */*

Class: Language_or_Language-element EquivalentTo: Language or *Language-element*
DisjointClasses: Language, *Language-element*   //**OWL Manchester**
Class: Language EquivalentTo: r_part min 2 *Language_element*
Class: KRL SubClassOf: Language   Class: Grammar SubClassOf: Language

---

EquivalentClasses(:Language_or_Language-element   //**OWL Functional-Style**
                   ObjectUnionOf(:Language *Language-element*))
DisjointClasses(: Language :*Language_element*)
EquivalentClasses(:Language ObjectMinCardinality (2 :r_part :*Language_element*))
SubClassOf (:KRL : Language)   SubClassOf (:Grammar : Language)

conventions and not prefixed by a namespace are KRL keywords. Within nominal expressions, '_' and '-' are used for separating words. When both are used, '-' connects words that are more closely associated. Since nominal expressions are used for the introduced types, the common convention for reading links in graph-based KRLs can be used, i.e., links of the form "X R: Y" can be read "X has for R Y". If "of" is used for reversing the direction of a link, the form "X R of: Y" can be read "X is the R of Y". The syntactic sugar of FE makes this reading convention explicit. Following this convention reduces the use of verbs and adjectives (which are more difficult to categorize and awkward to use with quantifiers) and thus normalizes knowledge.

In FL, if a link is not a subtype link nor another "link from a type", its first node is quantified and its default quantifier is "any". This one is the "forall" quantifier for definitions (in other words, the type in the first node is defined by this link). FL allows different links with the same first node to quantify this node differently. Indeed, in FL, the quantifiers of the source node and destination node of each link may also be specified in its relation node or in its destination node. This original feature permits FL to gather any number of statements into one visually connected graph. However, in this article, the quantifier for the first node is always "any" and left *implicit*. A destination node can also be source of links if they are delimited by parenthesis.

Given these explanations, the content of the tables in this article can now be read. Every keyword not introduced above will be explained via a comment near it. Comments use the C++ and Java syntax. In these tables, bold and italic characters are only for highlighting some important types and for readability purposes.

4 Situating Top-Level Language Elements in a Top-Level Ontology

Table 2 shows how types for KRL models and notations can be organized and inter-related. E.g., RIF-FLD includes RIF-BLD, both are part of the RIF family of models, and both have a Presentation Syntax ("PS") and an XML linearization.

Table 3 relates Language_element and some of its direct subtypes to important top-level types, thus adding precisions to these subtypes. Such a specification is missing in RIF-FLD but is well detailed in SBVR. This is why Table 3 includes many top-level SBVR types, although indirectly: *the types with names in italics are still types that we introduce* but they *have the same names as types in SBVR* and are equal to them or slight generalizations of them. This approach is for readability reasons and flexibility: if the SBVR authors disagree with our interpretation of their types, only some links to SBVR types will have to be changed, not our ontology. As illustrated by Table 3, to complement and organize types from other ontologies, ours includes many new types.

In RIF-FLD, depending on the context, the word "term" has different meanings. In our ontology, *Gterm* generalizes all these meanings of "term": it is identical to Language_element and sbvr#Expression. In RIF-FLD, an "individual term" is an abstract term that is not a Phrase (see Table 3), although it may refer to one. Individual_gTerm – or, simply "Iterm" – generalizes this notion to concrete terms too. This distinction was very useful to organize types of language elements, especially those from the

Table 2. Examples of relations between KRLs.

KRL r_part: 1..* Language_element,   > exclusion { **KRL_notation KRL_model** },
    r_grammar_head_element_type: Grouped_phrases;

KRL_notation > (*S-expression_based_notation* > LISP_based_KIF)
            (*Function-like_based_notation* > (RIF_PS > RIF-FLD_PS RIF-BLD_PS))
            (*Graph-based_notation* > (Markup_language_based_notation
                            > (XML_based_notation
                                > (RIF_XML > RIF-FLD_XML))
                (Frame_based_notation > FL JSON-LD Turtle)));

KRL_model //KRL abstract model (for an ADT and/or a logic)
 > (*First-order-logic_with_sets_and_meta-statements*
     > (KIF_model r_model_type of: LISP-based_KIF), r_part: 1..* First-order-logic)
   (*First-order-logic* > (CL r_model_type of: CLIF))
   (*RIF* > (RIF-FLD r_model_type of: RIF-FLD_PS, r_part: RIF-BLD)
        (RIF-BLD r_model_type of: RIF-FLD_PS))
   (*Graph-based_model*
     > (JSON-LD_model r_model_type of: JSON-LD)
       (RDF r_part: 1..* JSON-LD_model, r_model_type of: JSON-LD RDF/XML),
       (Frame_model_with_closed_world_assumption > F-Logic_classic_model)
       (Frame_model_with_open_world_assumption
         > (Description_logic_model
           > (**OWL_model**
               > (OWL-1_model > OWL-Lite OWL-DL   OWL-1-Full)
               (OWL-2_model > OWL-2_EL OWL-2-RL OWL-2-Full),
             r_part: 1..* (OWL-1-Full r_part: 1..* RDFS   1..* RDF   1..* OWL-DL)
                 1..* (OWL-2-Full r_part: 1..* OWL-2_EL   1..* OWL-2_RL)))));

---

implicit ontology of RIF-FLD (this framework uses different vocabulary lists, including
one for signatures; in our ontology, all these terms are inter-related). Used in this
context, the informal word "individual" is not equivalent to "something that is not a
type". Indeed, since an Iterm may refer to a Phrase, an Iterm identifier may be a Phrase
identifier. Thus, Table 3 uses the construct "near_exclusion" instead of "exclusion".
This construct has no formal meaning (it does not set exclusion links). It is only useful
for readability purposes. Table 3 also uses it to group and distinguish types for abstract
and concrete terms. Indeed, a (character) string may be seen by some persons as being
both abstract and concrete. Our ontology is – and must be – compatible with many
visions, when this can be achieved without information loss.

5   Ontology of a Core Meta-model for KRL Languages

RIF-FLD distinguishes *three* types of generic structures for a Gterm that is a function
or a phrase. We dropped their RIF-related restrictions and named them Positional_
gTerm, Gterm_with_named_arguments and Frame. Table 1 gave examples for positional and frame terms. A term with named arguments is similar to a frame except that,

Table 3. Situating Language_element w.r.t. other types (note: names in italics come from SBVR).

Thing = owl#Thing,   **r_identifier**: 0..* **Individual_gTerm**,
= exclusion
  { (Situation = exclusion {State Process},   r_description: 0..* **Phrase**)
   (Entity //thing that can be involved in a situation
     > exclusion
       { Spatial_entity //e.g., Square, Physical_Entity
        (Non-spatial_entity //e.g., Justice, Attribute, ...
           > (Description_content = *Meaning*,
             > *Proposition Question*
               (*Concept* > *Noun_concept* /*e.g., types */ (V*erb_concept* = *Fact_type*)))
           (Description_container > (File > RDF_file))
           (Description_instrument
             > (Language_or_Language-element
               = exclusion { (Language > KRL Grammar,
                                     r_part: 1..* Language_element)
                       Language_element //see below
                       })))
       }) };

Language_element = Gterm *Expression*,   r_representation of: 1 *Meaning*,
> near_exclusion   //String is both abstract and concrete
  { (***Representation*** > *Statement*, rc_type: **Concrete_term**)
   (**Concrete_term** > (*Expression* > *Text*) (Concrete_iTerm < Iterm)) //see tables 10 and 11
   }   // (for some subtypes)
  near_exclusion //a reference to a phrase is an Iterm
  { (**Phrase** > *Statement Definition* **Frame**) //+ see Table 6
   (**Individual_gTerm** = Iterm, > *Place_holder*, r_identifier of: 1 Thing) //see tables 10-11
   }
  near_exclusion { **Positional_gTerm Frame Gterm_with_named_arguments** }
  near_exclusion //subtyping these types is KRL dependent
  { Non-referable_gTerm //e.g., a predefined term
   (Referable_gTerm //via constant/variable/function/phrase   //referable→linkable
     r_variable: 0..* Variable,   **r_result of**: 1..* Function, r_annotation: 0..* Annotation,
     > (Gterm_that_cannot_be_annotated_without_link   r_annotation: 0 Annotation)
     Termula) };

---

as in object-oriented languages, local attribute names are used instead of link types (types are global). It could be argued that a same term could be presented in any of these three forms and hence that these three distinctions should rather be syntactic. However, the authors of RIF-FLD have not formalized the equivalence or correspondence between (i) "classes and properties" (or frames "interpreted as sets and binary relations") and (ii) "unary and binary predicates", *in order to have* a "uniform syntax for the RIF component of both RIF-OWL 2 DL and RIF-RDF/OWL 2 Full combinations" [22]. According to this viewpoint, each person re-using ontologies must decide if, for its applications, stating such an equivalence is interesting or not. RIF rules

Table 4. Main links for defining a structure for abstract terms and specifying concrete terms.

Gterm r_identifier_or_description of: 1 Thing, //*these link types are used in tables 6 to 10*
  r_operator: 0..1 Operator , //see Table 10
  r_part: 0..* Gterm, /* object parts or fct/relation arguments */   r_parts: 1 List, //ordered
  r_result: 1 Gterm, /* e.g., a phrase has for r_result a boolean */ **rc_type**: Concrete_term,
  r_variable: 0..* Variable,   r_result of: 1...* Function;

rc_link_to_concrete-term //and from an abstract term but also often from a concrete term
    _[/*from:*/Gterm, /*to*/Concrete_term] //signature of this relation type
 > (rc_begin-mark > rc_operator_begin rc_parts_begin)   (rc_separator > rc_parts_separator)
   (rc_end-mark > rc_operator_end rc_parts_end)   **rc_operator_name**
   rc_infix-operator_position   rc_annotation_position; //-1: before

r_gTerm_part r_type: Transitive_relation_type,
 > (**r_operator** > r_frame_source rdf#predicate) (r_phrase_part > rdf#subject rdf#object)
   (r_parameter = r_part, //"r_part" used for concision
     > (r_link_parameter > (r_link_source   > rdfs#domain)
                             (r_link_destination > rdfs#range)));

//r_parts permits to order the parts, this is sometimes needed for abstract terms and
// this also permits to give a default order for presentation purposes.
r_parts _[?e,?list] :=> [any ^(Thing r_member of: ?list) r_part of: ?e];
r_parts _[?e,?list] :<= [any ^(Thing r_part of: ?e) r_member of: ?list];

/* Notes: **in FL**, ":=' permits to give a full definition,
 ":=>" gives only "necessary conditions", ":<=" gives only "sufficient conditions",
 "^(" and ")" delimit a lambda-abstraction (a construct defining and returning a type),
 "_(" and ")" delimit the parameters of a function call, "_[" and "]" delimit those of a definition,
 ".[" and "]" delimit the elements of a list, ".{" and "}" delimit the elements of a set. */

rc_type _[?t,?rct] := [any ?t rc_: 1..* ?rct]; //people who see concrete terms as specializations
             // of abstract terms can still state: rc_type < subtype;   rc_ r_type: instance;
             //in the next function signature (i.e., in [...]), the variables are untyped

f_link_type _[?operatorName, ?linkType, ?linkSourceType, ?linkDestinationType]
 := ^(Link rc_operator_name: ?operatorName,   r_parts: .[?linkSource ?linkDestination],
            r_operator: ?linkType, r_result: 1 Truth_value,
            r_link_source: 1 ?linkSourceType, r_link_destination: 1 ?linkDestination);

---

or a macro language such as OPPL can certainly be used for such structural translations [23]. However, to avoid imposing this exercise to most users of our KRL ontology, and to avoid limiting its use for specifying KRLs, it formalizes relations between a frame and a Conjunction_of_links_from_a_same_source (this is done in the last 17 lines of Table 9; reminder: a link is – or can also be seen as – a binary relation).

We found that a small number of link types are sufficient for defining a structure for abstract terms and specifying their related concrete terms. Table 4 lists and explains

Table 5. Important structured concrete term types and definition of their default presentation.

Structured_concrete_term_that_is_not_a_string   //the delimiters in comments below permit a
> exclusion   // KRL to have all these structures and still only requires an LALR(1) parser
{ (**List_cTerm** > *Enclosed_list_cTerm* /* e.g., .[A B C] */
                    Fct-like_list_cTerm   /* e.g., A ..[B C] */)
  (**Set_cTerm** > *Enclosed_set_cTerm* /* e.g., .{A B C} */
                    Fct-like_set_cTerm /* e.g., A ..{B C} */)
  (**Positional_cTerm** //e.g., with operator "f" and parts/parameters A, B and C
      rc_operator-name: "", rc_operator_begin: "", rc_operator_end: "", //declared in Table 4
      rc_parts_begin: "(",   rc_parts_separator: "", rc_parts_end: ")",
      rc_infix-operator_position: 0, //if not 0, it indicates the operator position within the parts
      > exclusion
         { (Fct-like-cTerm
              = exclusion { (**Prefix_fct-like-cTerm** rc_parts_begin: "_("")   //e.g., f _(A B C)
                            (*Postfix_fct-like-cTerm* rc_parts_begin: "(_")") }) //e.g., (_A B C)f
           (List-like_fct_cTerm
              = exclusion { (*List-like_prefix-fct_cTerm* rc_parts_begin: ".("") //e.g., .(f A B C)
                            (*List-like_infix-fct_cTerm* rc_parts_begin: "(.",
                                        rc_operator_begin: ".") //e.g., (. A B .f C)
                            (*List-like_postfix-fct_cTerm* rc_parts_begin: "(.."") //e.g., (.. A B C f)
                            }) })
  (**Frame_cTerm** //e.g., with operator type "f" and with parts two half-links of type r1 and r2
      rc_operator-name: "", rc_operator_begin: "", rc_operator_end: "",
      rc_parts_begin: "{", rc_parts_separator: ",", rc_parts_end: "}", //as in JSON-LD
      rc_parts: 1..* Half-link_cTerm,
      > exclusion { (*Prefix_frame_cTerm* rc_parts_begin: "_{"") //e.g., f_{ r1: A, r2: B }
                    (List-like_frame_cTerm rc_parts_begin: "{.",
                        > *List-like_prefix-frame_cTerm*   //e.g., {. f r1: A, r2: B}
                        List-like_infix-frame_cTerm)   //e.g., {. r_id: f, r1: A, r2: B}
                    (*Postfix_frame_cTerm* rc_parts_begin: "{_") //e.g., {_ r1: A, r2: B } f
                    Alternating-XML_cTerm //Frame in the Alternating-XML style where
                    })   // concept nodes alternate with link nodes, as in RDF/XML
  Cterm_with_named_arguments //quite rare in KRLs, hence not detailed in this article
};

fc_prefix-fct-like_type _[?notationSet, ?operator_name, ?begin_mark, ?separator, ?end_mark]
:= ^(**Prefix_fct-like-cTerm** r_direct-or-indirect_part of: ?notationSet, *//uses of this function are*
                            rc_operator-name: ?operator_name,   *// illustrated in Table 13*
        rc_parts_begin: ?begin_mark, rc_parts_separator: ?separator, rc_parts_end: ?end_mark)

Phrase //any phrase has at least these presentations in these 2 kinds of notations (see Table 2),
     // e.g., in RIF-PS and RIF-XML:
  rc_type: ^(fc_prefix-fct-like_type _(.{Function-like_based_notation},"","(","",")")
                rc_annotation-position: -1)
            ^(fc_alternating-XML_type_(.{XML_based_notation},"") rc_annotation-position: 0);

List rc_type: fc_list_type _(.{Notation}, "[", ",", "]"); //by default, in any notation, a list has
     // for representation a comma separated list of element delimited by square brackets;
     // note that fc_list_type has no argument for an operator name

---

the main link types. They can be seen as a representation and *extension* of the signature system of RIF-FLD. The *ideas* are that (1) every composite term can be decomposed into a (possibly implicit) *operator* (e.g., a predicate, a quantifier, a connective, a collection type) and a *list of parameters* (alias, "parts"), and (2) *many non-binary relations can be specified as links to a collection of terms*. Table 6 and the subsequent tables use the link types of Tables 4 and 5 directly or via functions which are shortcuts for specifying such links. This is highlighted via bold characters in those tables. The end of Table 4 specifies one of these functions (f_link_type). In the Tables 6, 7, 8, 9, 10 and 11, which illustrate the organization of subtypes of Phrase and Iterm, f_link_type is used to define some abstract terms *as links* and hence *enables to store them or present them as such when necessary*. To illustrate the way these ADTs are instantiated by knowledge representations, Table 12 shows a *phrase in different notations and then a part of its abstract structure*.

Some links are used for both abstract and concrete terms. E.g., rc_operator_name is often also associated to an abstract term for specifying a default name for its operator. If no such link is specified or if the empty string ("") is given as destination, the operator type name (without its namespace identifier) is used as default operator name.

Table 5 lists major kinds of structured concrete terms and thus also the main presentation possibilities for structured abstract terms (see the 14 names in italics). Based on the five main categories for these concrete terms (see the names in bold and not in italics), it is easy to find the five categories of abstract terms they correspond to, even though such links are not shown in Table 5. We found that each of these concrete term types can be defined with only a few types of links, those that begin by "rc_" and that were listed in Table 4. We defined some functions to provide shortcuts for setting those links when defining a particular concrete term, e.g., fc_prefix-fct-like_type (*its use is illustrated by Table 13 for the definition of RIF-PS and JSON-LD*).

In our ontologies, links from a type do not specify that the given destination is the only one possible (to do so in FL, "=>" would need to be used instead of ":" after the link type name; in OWL-based descriptions, owl#allValuesFrom can be used). Thus, such links represent "default" relationships: if a link from a type T specializes a link from a supertype of T, it overrides this inherited link. This is also true when the link type is functional (i.e., when it can have only one destination) and its destination for T does not specialize the destination for a supertype of T. The links beginning by "rc_" look functional but are not: in FL, multiple destinations can be stated to indicate different presentation possibilities. However, by convention, such links override inherited links of the same types. Table 5 shows how different kinds of "default presentations" can be represented in concise ways.

6   Ontology of KRL Content Models

Table 6. Important top-level types of phrases.

//Note: names in italics come from RIF-FLD, names in bold italics are used in RIF-FLD signatures,
// bold is for highlighting, "cl#' prefixes terms from Common Logics

Phrase < ^(Gterm **r_operator**: 1 (Operator_type > (owl#Property r_instance: r_binary_relation)),
                    r_result: 1 (Truth_value r_instance: True False Indeterminate_truth-value)),
  > (Phrase_not_referable_in_RIF-FLD //→ cannot have an annotation in RIF-FLD
        > (Annotation > cl#Comment (Formal_annotation > RIF_*annotation*)) Module_*directive*
        (Annotating_phrase = **f_link_type**_("",r_annotation,Gterm,Annotation)) Attribute),
  = exclusion

  { (**Modularizing_phrase**   // **Phrases** is the head element of a KRL grammar
    > (**Phrases** = **Grouped_phrases**, r_part: 0..* Phrase, > cl#Text,
        > (**Module** > (*Document* r_part: 0..1 Document_Directive 0..1 Phrases) cl#Module,
            r_part: 0..1 (Module_parts_that_are_directives < Module,
                > (Module_header = **f_link_type**_("",r_header,Module,
                                                        .[0..* **Module_directive**])))
                0..1 (Module_parts_that_are_not_directives =
                                f_link_type_("Group",r_group,Module,.[0..* **Phrases**],
                        < Module, > Module_body ***Group***_of_phrases),
            r_parts: .[0..1 Module_header, 0..1 Module_body]))
      (**Module_directive** = **f_link_type**_("",r_relation,Module,Thing),
        > (Module_name_directive = **f_link_type**_("Name",r_name,**Module**,*Name*))
        (Excluded_Gterm-reference_directive = **f_link_type**_("",r_excluded_gTerm,**Module**,
                                                        .[1..* Gterm_reference]))
            (Document_*directive*
              > (*Dialect_directive* = **f_link_type**_("Dialect",r_*dialect*,**Module**,*Name*))
                (*Base_directive*   = **f_link_type**_("Base",r_base,**Module**,Document_locator))
                (*Prefix_directive* = **f_link_type**_("Prefix",r_prefix,**Module**,
                                                        NamespaceShortcut-DocumentLocator_pair))
                (Import-or-module_directive > cl#Importation,
                  > (*Import_directive* = **f_link_type**_("Import",r_imported-doc,***Document***,
                                                        Imported_document_reference))
                (*Remote_module* = **f_link_type**_("Module",r_imported-module,**Module**,
                                                        Remote_module_reference))
))))
  (**Non-modularizing_phrase** //including non-monotonic phrases: queries, removals, ...
    > (***Formula*** > *Positional_formula Formula_with_named_arguments*
                    Phrase_of_a_grammar cl#Sentence,
        = exclusion //the 3 following distinctions come from KIF
          { (Definition = exclusion { Non_conservative_definition Conservative_definition })
            (Sentence //fact in a world: formula assigned a truth-value in an interpretation
                > Belief //the fact that someone believes in a certain thing
                    Axiom) //sentence assumed to be true, from/by which others are derived
            (Inference_rule> Production_rule) //like an implication but the conclusion is "true"
          }   // only if/when the rule is fired
        near_exclusion{*Composite_formula* **Atomic_formula_or_reference_to_formula**}))
    Termula_phrase /*RIF function/atomic_formula parameter; not detailed in this article*/)
  };

Table 7. A way to restrict this general model for specifying particular KRLs.

r_only_such_part_of_that_type _[?x ?pt] //The source ?x has some parts of type ?pt but no **other**_
< r_part _(?x ?pt), // parts with type the genus of ?pt. The definition in the next line requires that_
:= [?x r_part: 1..* ?pt   0 ^(?t != ?pt, < (?gpt r_genus_supertype of: ?pt))];   // relations of type_
// r_genus_supertype are set by definitions. Thanks to this link type, to our general model for KRLs_
// and to the default presentation associated to its abstract terms, KRLs can be defined in a very concise_
// way. Below are examples for some abstract terms of some KRLs. The next section gives examples_
// for some concrete terms of some KRLs. For the Triplet_notation, nothing else is required._

RIF   r_only_such_part_of_that_type: //RIF models have for part terms defined by these 6 lambdas:
^(Gterm_that_can_be_annotated_without_link > Phrase) //-> only a phrase can be annotated
^(Grouped_phrases r_part: 0..* Document)
^(Quantification > Classic_quantification)   ^(Frame > Minimal_frame)   ^(Collection > List)
^(Delimited_string > Delimited_Unicode_string);

RIF-BLD   r_only_such_part_of_that_type :   //the next two lambdas se are just examples,
^(Rule_conclusion > rif-bld#Formula)   // RIF-BLD has other restrictions
^(Rule_premise > Connective_phrase_on_atomic_formulas Conjunction_phrase);

Triplet_notation = ^(KRL r_only_such_part_of_that_type:
                                        ^(Phrase > Link)   ^(Individual_gTerm > Constant_or_variable));

Table 8. Important types of composite formulas.

//Names in italics come from RIF-FLD, names in bold italics are used in RIF-FLD signatures

Composite_formula = **f_relation_type**_("",r_relation,.[1..* Formula]), //-> r_part: 1..* Formula
> exclusion
  { (Formula_connective **r_operator_type**: 1 connective_operator, > cl#Boolean_sentence,
    > exclusion //e.g., Negative_formula, Production_rule, Logical_implication, ...
      { (**Unary_connective_phrase** = **f_relation_type**_("",r_unary_relation,.[1..* Formula]))
        (**Binary_connective_phrase** = **f_relation_type**_("",r_binary_relation,.[1..* Formula]))
        (**Variable-n-ary_connective_phrase**
          = **f_relation_type**_("",r_variable-ary_relation,.[1..* Formula]),
          > exclusion { (Disjunction_phrase = **f_relation_type**_("Or",r_or,.[1..* Formula]))
                        (Conjunction_phrase = **f_relation_type**_("And",r_and,.[1..* Formula]),
                        > (Conjunction_of_links = **f_relation_type**_("And",r_and,Link),
                          > Frame_as_conjunction_of_links_from_a_same_source)) })
      })
    (Quantification = **f_quantification_type**_("",Quantifier,.[1 Type],Constant-or-variable,Formula),
    > (Classic_quantification = **f_quantification_type**_("",Quantifier,.[],Variable,Formula))
      exclusion   //classic: no guard, no constant
      { (Universal_quantification
        = **f_quantification_type**_("Forall",q_forall,.[1 Type],Constant_or_variable,Formula),
        > (Classic_universal_quantification
          = **f_quantification_type**_("Forall",q_forall,.[],Variable,Formula)))
      Existential_quantification //defined the same way as for Universal_quantification
      }) };

Table 9. Important types of relations between frames, links and positional formulas.

Atomic_formula_or_reference_to_formula > exclusion
{ (Formula_reference /* this is also an Individual_gTerm */
    > exclusion { Variable_for_a_formula
                  Reference_to_formula_in_remote_module //with the same KRL
                  Reference_to_externally_defined_formula
                  })   //external: not in a module and not with the same KRL
  (*Atomic_formula*   //names in bold italics are used in RIF-FLD signatures
    > { Constant_for_a_formula
       (Atomic_formula_that_is_not_a_constant
         > near_exclusion //possible shared subtypes: subclass_or_equal, link
           { (Positional-or-name-based_formula
               r_operator: 1 *Termula*, > cl#Atomic_sentence,
                 > exclusion { (Positional_formula r_part: 1..* Termula)
                               (Name-based_formula r_part: 1..* Name-Termula_pair) })
             (*Equality*_formula = **f_link_type**_("=",r_equal,Termula,Termula), > cl#Equation)
             (*Class-membership* = **f_link_type**_("#",r_type,Termula,Termula))
             (*Subclass*_formula = **f_link_type**_("##",r_supertype,Termula,Termula))
             (*Frame*= (Frame_as_conjunction_of_links_from_a_same_source ?f
                         r_frame_head: 1 Termula ?fh,   r_part: (1..* Link r_link_source: ?fh))
                       (Frame_as_head_and_half-links_from_head ?f
                         r_operator: (1 Termula ?fh r_frame_head of: ?f),
                         r_part: (1..* Half_link r_link_source: ?fh),
                         > (Minimal_frame r_part: 1..* Minimal_half-link)))
           })
       (Binary_atomic_formula_that_is_not_a_constant
         > (Link
             = (Link_as_positional_formula < Positional_formula,
                 < **f_link_type**_("",r_binary_relation_type,Termula,Termula),
                 r_part of: (1 Frame ?f r_frame_head: 1 Termula ?fh), r_link_source: ?fh)
               (Link_as_frame_part r_part of: (1 Frame ?f r_frame_head: 1 Termula ?fh),
                 r_operator: ?fh, r_link_source: ?fh, r_link_destination: 1 Termula ?ld,
                 r_part: (1 Half_link r_link_source: ?fh, **r_operator**: ?fh, r_parts: ?ld)
               )))
  }) };

Table 10. Important top-level types of "individual terms" (not phrases unless referring to one).

Individual_gTerm = near_exclusion   //names in italics come from RIF-FLD
  { (Individual_concrete_term
    > **Concrete-term_for_constant_or_name**   Lexical-grammar_character-set   Character
      Concrete_list-like_term   (String > (Delimited_string > Delimited_Unicode_string)))
  (*Individual*_abstract_term
    > Abstract_individual_gTerm_not_referable_in_RIF-FLD //Quantifier, Half_link, ...
      (**Fterm_or_variable** > (*Functio*nal_term **r_operator**: 1 (Function_type < Type)))
      Individual_abstract_term_of_a_grammar
      (**Operator** > r_relation f_function Operator_not_referable_in_RIF-FLD)
      (Symbol_space > rif#iri rif#local xs#string xs#integer xs#decimal xs#double))   };

Table 11. Important subtypes of one subtype of Individual_gTerm (see Table 10).

Concrete-term_for_constant_or_name   *//examples to show that the same approach*
> (Variable_name   r_identifier of: 1..* Variable, *// also applies for concrete terms*
   < ^(**f_string_type_**("?","","")) r_part: 1 Undelimited_variable-name))
   (Constant_concrete_term r_identifier: 1..* Constant_gTerm,
> (Constant_concrete_term_without_symbol-space
   > (Constant_IRI r_part: 1 IRI_reference)
       (Constant_short-name_via_compact_URI r_part: 1 Compact_URI)
       (Literal_or_datatype_concrete_term r_identifier of: 1..* Literal_or_datatype,
          > (Double_quoted_string
            < ^(**f_string_type_**("","",""")
                r_part: 1..* **f_character_type_with_escape_for_**(Character,"\\",""))))
          (Numeric_literal
             > (Positive_integer < ^(f_string_type ("+","","")) r_part: 1..* Digit))
               (Negative_integer < ^(**f_string_type_**("-","","")) r_part: 1..* Digit))))));

Table 12. One phrase in different notations; part of its abstract structure represented in UML.

In English: "There exists a **car** which is **red** (one shade of red; it may also have other colors)".
In FE: `a **Car** with *color* a **Red**´.   In RIF-PS: Exists ?car ?red (*color*(?car#**Car** ?red#**Red**))
In FL: a **Car** *color :* a **Red** ;   In RIF-PS: Exists ?car ?red ?car#**Car** [color -> ?red#**Red**].
In N-Triples: **Car8** *color* **Red3** .   **Car8** *type* **Car** .   **Red3** *type* **Red** .

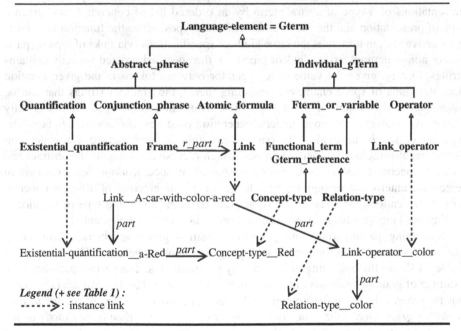

Legend (+ see Table 1) :
- - - - - ->: instance link

7 Ontology of Particular KRLs and Grammars

In a KRL that is "perfectly regular *with respect to a particular kind of abstract/concrete term*" allows all the terms of this kind to be (re-)presented in the same way, e.g., all the terms which in our approach have an operator (the "operator based terms"). A *perfectly regular KRL* is then one which is perfectly regular for all the kinds of terms it allows. The "Triplet notation" is perfectly regular. To be so, a more expressive KRL would have to be fully based on an ontology and be HOL based. Since KIF re-uses the LISP notation, it is perfectly regular with respect to "operator based concrete terms" and "concrete terms for collections". Most KRLs have some *ad hoc* abstract and concrete terms. E.g., in RIF-XML, document directives are presented in different ways: some via links, some via XML attributes. In RIF-PS, they are presented as positional terms but not links. Thanks to the fact that our general model represents the directives both as parts and links (see Table 6), these RIF predefined directives can be represented within/via frames as well as via positional terms. The first part of Table 13 shows how *ad hoc* concrete terms of particular types of KRLs can be specified in a concise way. The approach used to do so for abstract terms (see Table 7) is here re-used. Thus, both abstract and concrete terms of a KRL – or a family of KRLs – can be specified at the same time and in a concise way. Furthermore, since (families of) KRLs and their specifications can be organized by specialization relations, they can be formally and visually compared.

The second part of Table 13 shows how it is possible to declaratively specify all the presentations of a type of abstract term by an ordered list of concrete terms, given a type of presentation and the list of usable notation types. Since the function fc_r_parts is recursive and, in turn, uses the same kinds of specifications (via links of type rc_parts or, for non-structured terms, links of type rc_), the specified ordered list only contains strings. Finally, given the value of rc_separator between tokens in the given notation (i.e., the kinds of space characters separating them), the kinds of strings that can be associated to this collected list are specified. Thus, the whole specification is fully declarative. However, for concrete term generation purposes, choices have to be made, e.g., about space indentation. In our system, this is implemented via generation functions (also included in our ontologies) which recursively navigate the abstract and concrete specifications to find the most precise relevant specifications. Since our system rejects the entering of ambiguous knowledge – e.g., the entering of different concrete term specifications for a same type of abstract term and the same type of notation – finding the most precise relevant specifications was easy to implement.

Specifying parsing rules and generating them – given an abstract term and a grammar notation – can be represented using the same techniques. The first part of Table 14 shows the beginning of an ontology for grammars. The second part shows an example of grammar rule (and its connection to a grammar but this part actually needs not be generated). Once the grammar rules are *generated* – in a way *similar to presentation generation* – the generation of *their presentation is then done exactly as for any other statement*, according to the given grammar notation.

Our ontologies can be represented with KRLs having at least OWL-2 expressiveness. To that end, r_parts links with "lists with cardinalities" (e.g., [0..1 Y, 1..* Z]) as

Table 13. Ways to specify concrete terms for particular kinds of terms in particular notations.

//Thanks to the default values in our specifications for abstract and concrete terms, only the
// following lines are needed for defining the presentation in RIF-PS of the abstract terms shared
// by the KRLs of the RIF family. For instance, the order and operator names of the directives of
// a document can be found in Table 6. Since these directives follow the default presentation for
// phrases in RIF-PS, nothing needs to be specified about them here. The abstract term restrictions
// can be specified here (as illustrated below for "Frame" or separately, as illustrated by the
// second part of Table 6.

RIF   r_only_such_part_of_that_type: //default values make the next lines sufficient
  ^(Phrase rc_type: fc_prefix-fct-like_type_(.{RIF-PS},"","(",",",")")) //default style in RIF_PS
  ^(RIF_annotation rc_type: fc_list_type_(.{RIF-PS},"(*","","*)")) //override for annotations
  ^(Quantification_bound_list rc_type: fc_list_type_(.{RIF-PS},"","",""))
  ^(Rule rc_type: fc-like_infix-fct_type_(.{RIF-PS},":-","","",""))
  ^(Externally_defined_term rc_type: fc_prefix-fct-like_type_(.{RIF-PS},"External","(","",")"))
  ^(Equality_formula rc_type: fc_list-like_infix-fct_type_(.{RIF-PS},"=","","",""))
  ^(Subclass_formula rc_type: fc_list-like_infix-fct_type_(.{RIF-PS},"##","","",""))
                       //e.g., "?t1 ## ?t2"; in FL: "?t1 < ?t2"
  ^(Class-membership_formula rc_type: fc_list-like_infix-fct_type_(.{RIF-PS},"#","","",""))
  ^(Frame > Minimal_frame, rc_type: fc_infix_list-like_frame_type_(.{RIF-PS},"","[","","]"))
  ^(Half_link rc_type: fc_half-link_type_(.{RIF-PS},"","","->","",""))
  ^(Name-Termula_pair rc_type: fc_list_type_(.{RIF-PS},"","->",""))
  ^(Open_list rc_type: fc_prefix-fct-like_type_(.{RIF-PS},"List","(","|",")"))
  ^(Open-list_rest rc_type: fc_list_type_(.{RIF-PS},"","","",""))
  ^(Aggregate_function rc_type: fc_prefix-fct-like_type_(.{RIF-PS},"","{","|","}"))
  ^(Aggregate_function_bound_list rc_type: fc_fct-like_list_cTerm_(.{RIF-PS},"[","","]"));

RIF-FLD r_only_such_part_of_that_type: //only 1 example: a document in RIF-FLD/XML
  ^(Document rc_type: (1 fc_alternating-XML-cTerm_type_(.{RIF-XML},"Document")
                       rc_annotation-position: 0,
                       rc_XML-attribute_type: r_dialect xml#base xml#prefix,
                       rc_XML-link_types: .[rif#directive rif#payload]));

JSON-LD_model r_only_such_part_of_that_type: //specifies also the JSON-LD notation
  ^(Phrase rc_type: fc_list-like_infix-frame_type_(.{JSON-LD},"","{",",","}"))
  ^(Half_link rc_type: fc_half-link_type_(.{JSON-LD},"",":","",""))
  ^(Module_header rc_type: fc_list-like_infix-frame_type_(.{JSON-LD},"@context:","{",",","}"))
  ^(Module_body rc_type: fc_list_type_(.{JSON-LD},"","","",""))
  ^(Formula > ^(Minimal_frame r_operator: 1 Constant_gTerm)) //only 1 destination per link
  ^(Fterm_or_variable > Constant_or_set_or_closed_list)
  ^(Set rc_type: fc_list_type_(.{JSON-LD},"[",",","]")) //in classic JSON, this would be for a list
  ^(Closed_list > ^(Frame r_part: 1 .[r_container, Closed_list],
             rc_type: fc_half-link_type_(.{JSON-LD},"","@container",":","@list",""))
                  ^(Frame r_part: .[r_list, 1 Set], //2nd way to represent a closed list in JSON-LD
             rc_type: fc_half-link_type_(.{JSON-LD},"","@list",":","","")));

//Another kind of specification ("^?" prefixes variables that are implicitly universally quantified):
  ^(Thing ?t rc_: (a Enclosed_list_cTerm ?c r_KRL-set: ^?notationSet))
   rc_parts: f_remove_empty_elements_in_list _(.[(^?cb rc_begin_mark of: ?c),
             fc_r_parts_(?notationSet, (^?tp r_parts of: ?t), (^?cs rc_parts_separator of: ?c))
             (^?cb rc_end_mark of: ?c)]);

Table 14. Important links from Grammar_element, followed by an example of grammar head rule.

Grammar_element   //specifications mainly only for EBNF-like and Lex&Yacc-like grammars
  r_part of: 1..* Grammar,   //and conversely: Grammar r_part: 1..* Grammar_element;
  > exclusion
    { (Phrase_of_a_grammar > Head_grammar-rule,
        = exclusion{**Non-lexical-grammar_rule** Lexical-grammar_rule})
    (Individual_gTerm_of_a_grammar
        = exclusion{ Lexical-grammar_individual-gTerm //what Lex grammars handle
                        Non-lexical-grammar_individual-gTerm }) };

Non-lexical-grammar_rule = NLG_rule,   //this is a beginning but the representation
  r_part: 1 NLG_rule_left-hand-side   1 NLG_expression   // of the whole grammar
          0..1 (Parsing_action_phrase < Phrase),   // is similar
  rc_: (1 fc_list_type_(.{W3C-EBNF,XBNF,Grammar},"","","")
          //fc_list_type is like fc_prefix-fct-like-cTerm_type but without operator as parameter
          rc_parts: .[NLG_rule_left-hand-side "::=" NLG_expression])
                                    //→"A::=B" ("Grammar "→default presentation)
        (1 fc_list_type_(.{ISO-EBNF},"","","")
          rc_parts: .[NLG_rule_left-hand-side "=" NLG_expression])
                                    //→ "A = B" in ISO-EBNF
        (1 fc_list_type_(.{Yacc, Bison},"","","")
          rc_parts: .[NLG_rule_left-hand-side ":" NLG_expression]);
                                    //→ "A : B" in Yacc or Bison (without parsing actions)

Grammar_for_RIF_FLD_in_RIF-PS < Grammar,
  r_description of: 1..* (RIF-FLD < (KRL_model r_part of: 1..* KRL)),
  r_part: 1 (fc_NLG_rule_type_(.{RIF-PS}, "RIF-FLD_document",
                    .[0..1 Annotation "Document" "("   0..1 Dialect_directive
                        0..1 Base_directive   0..* Prefix_directive
                        0..* Import_directive 0..* Remote_module_directive
                        0..1 Group ")"])
              < Head_grammar-rule);

---

destinations can be replaced by lists without cardinalities (e.g., [Y, Z]) as long as r_part links are also used for specifying the cardinalities (e.g., X r_part: 0..1 Y, 1..* Z). The use of functions may also be avoided via macros, i.e., by expanding function definitions.

Replicating our work does not require details on the implementation of our system: our ontologies *are* the required *declarative code*. The used inference engine is irrelevant as long as it can handle the specifications. However, some readers might be interested to know that our translation server exploits the parser available at http://goldparser.org while its inference engine was implemented in Pascal Object (for portability purposes) and exploits "tableaux decision procedures" [24]. This server and its inference engine have recently been designed by GTH (http://www.mitechnologies.net). This work on a generic approach for handling KRLs comes from the many problems encountered to handle various versions of FL and other KRLs in the knowledge sharing servers WebKB-1 [25] and WebKB-2 [17, 20].

Our ontology of KRL ontologies (i.e., its core and the specifications of particular KRLs) and our translation server are accessible from http://www.webkb.org/KRLs/. Its interface is similar to Google Translate except that the input and output languages are KRLs and, instead of KRL names, KRL specifications can also be given by the users.

8 Conclusions

One contribution of this article is a generic model for structured abstract or concrete terms. It is simple: only a few types of links and a few distinctions (Tables 4 and 5). This operator + parameters based model permits to define terms in a concise and flexible way, and thus also their presentation and parsing.

A second contribution is the design of a KRL model ontology by representing, aligning and extending various KRL models, and defining their elements via the above cited few links, as illustrated by Tables 3, 6, 7, 8, 9, 10 and 11. Thus, the merged models are also easier to re-use.

A third one is the design of a KRL notation ontology – to our knowledge, the first one – based on the above two cited contributions, as illustrated by Tables 5, 13 and 14.

These three contributions permit to avoid or reduce the problems listed in the introduction and Sect. 2: those of KRL syntactic translation, KRL parser implementation, dynamic extension of notations, etc. Thus, we provide an ontology-based concise alternative to the use of XML as a meta-language for easily creating KRLs that follow KRL ontologies. Therefore, this also complements GRDDL and can be seen as a new research avenue (GRDDL permits to specify where a software agent can find tools – typically XSLT ones – to convert a given KRL to RDF/XML). This avenue is important given the frequent need for applications to (i) integrate or easily import and export from/to an ever growing number of models and notations (XML-based or not), and (ii) let the users parameter these notations.

Previous attempts (by the second author of this article) based on directly extending EBNF – or directly representing or generating concrete terms in a KRL or transformation language – required much lengthier specifications that were also more difficult to re-use.

Besides its translation server, the GTH company will use this work in its applications for them to (i) collect and aggregate knowledge from knowledge bases, and (ii) enable end-users to adapt the input and output formats they wish to use or see. The goal behind these two points is to make these applications – and the ones they relate – more (re-)usable, flexible, robust and inter-operable.

One theme of our future work on this approach will be the *generation of parsing actions in parsing rules*, given particular ADTs to use. A second theme will be the representation and integration of more abstract models and notations for KRLs as well as *query languages and programming languages*. A third theme will be the extension of our notation ontology into a full *presentation ontology* with concepts from style-sheets and, more generally, user interfaces.

References

1. Golin, E., Reiss, S.: The specification of visual language syntax. J. Vis. Lang. Comput. **1**(2), 141–157 (1990)
2. Borras, P., Clément, D., Despeyrouz, T., Incerpi, J., Kahn, G., Lang, B., Pascual, V.: CENTAUR: the system. In: SIGSOFT 1988, 3rd Annual Symposium on Software Development Environments (SDE3), Boston, USA, pp. 14–24 (1988)
3. Attali, I., Parigot, D.: Integrating Natural Semantics and Attribute Grammars: the Minotaur System. INRIA Research Report no. 2339 (1994)
4. Corby, O., Dieng, R.: Cokace: a centaur-based environment for CommonKADS conceptual modeling language. In: ECAI 1996, Budapest, Hungary, pp. 418–422 (1996)
5. Sandberg, D.: Lithe: a language combining a flexible syntax and classes. In: ACM Sigplan-Sigact Symposium on Principles of Programming Languages, POPL 1982, pp. 142–145 (1982)
6. Botting, R.: How Far Can EBNF Stretch? http://cse.csusb.edu/dick/papers/rjb99g.xbnf.html
7. Lapets, A., Kfoury, A.: A user-friendly interface for a lightweight verification system. Electron. Notes Theoret. Comput. Sci. **285**, 29–41 (2012)
8. ODM: Ontology Definition Metamodel, Version 1.1. OMG document formal/2014-09-02 (2014). http://www.omg.org/spec/ODM/1.1/PDF/
9. Feja, S., Witt, S, Speck, A.: BAM: a requirements validation and verification framework for business process models. In: 11th Quality Software International Conference, QSIC 2011, pp. 186–191 (2011)
10. Quan, D.: Xenon: an RDF stylesheet ontology. In: 14th World Wide Web Conference, WWW 2005, Japan (2005)
11. Pietriga, E., Bizer, C., Karger, D.R., Lee, R.: Fresnel: a browser-independent presentation vocabulary for RDF. In: Cruz, I., Decker, S., Allemang, D., Preist, C., Schwabe, D., Mika, P., Uschold, M., Aroyo, L.M. (eds.) ISWC 2006. LNCS, vol. 4273, pp. 158–171. Springer, Heidelberg (2006)
12. Brophy, M., Heflin, J.: OWL-PL: A Presentation Language for Displaying Semantic Data on the Web. Technical report, Department of Computer Science and Engineering, Lehigh University (2009)
13. Corby, O., Faron-Zucker, C, Gandon, F.: SPARQL template: a transformation language for RDF. In: 25th Journées francophones d'Ingénierie des Connaissances, IC 2014, Clermont-Ferrand, France (2014)
14. RIF-FLD: RIF framework for logic dialects, 2nd edn. In: Boley, H., Kifer, M. (eds.) W3C Recommendation (2013). http://www.w3.org/TR/2013/REC-rif-fld-20130205/
15. Genesereth, M., Fikes, R.: Knowledge Interchange Format, Version 3.0, Reference Manual, Technical Report, Logic-92-1, Stanford University (1992). http://www.cs.umbc.edu/kse/
16. Farquhar, A., Fikes, R., Rice, J.: The ontolingua server: a tool for collaborative ontology construction. Int. J. Hum. Comput. Stud. **46**(6), 707–727 (1997). Academic Press, Inc., MN, USA
17. Martin, P.: Collaborative knowledge sharing and editing. Int. J. Comput. Sci. Inf. Syst. **6**(1), 14–29 (2011)
18. Common Logic: Information technology – Common Logic (CL): a framework for a family of logic-based languages. ISO/IEC 24707:2007(E), JTC1/SC32 (2007)
19. SBVR: Semantics of Business Vocabulary and Business Rules (SBVR), Version 1.0, OMG document formal/08-01-02 (2008). http://www.omg.org/spec/SBVR/1.0/

20. Martin, P.: Towards a collaboratively-built knowledge base of&for scalable knowledge sharing and retrieval. HDR thesis (240 pages; "Habilitation to Direct Research"), University of La Réunion, France (2009)
21. Martin, P.: Knowledge representation in CGLF, CGIF, KIF, Frame-CG and formalized-English. In: Priss, U., Corbett, D.R., Angelova, G. (eds.) ICCS 2002. LNCS (LNAI), vol. 2393, pp. 77–91. Springer, Heidelberg (2002)
22. RIF-FLD-OWL: RIF, RDF and OWL Compatibility, 2nd edn. W3C Recommendation, 5 February 2013. http://www.w3.org/TR/2013/REC-rif-rdf-owl-20130205/
23. Šváb-Zamazal, O., Dudáš, M., Svátek, V.: User-friendly pattern-based transformation of OWL ontologies. In: ten Teije, A., Völker, J., Handschuh, S., Stuckenschmidt, H., d'Acquin, M., Nikolov, A., Aussenac-Gilles, N., Hernandez, N. (eds.) EKAW 2012. LNCS, vol. 7603, pp. 426–429. Springer, Heidelberg (2012)
24. Horrocks I.: Optimising Tableaux Decision Procedures for Description Logics. Ph.D. thesis, University of Manchester (1997)
25. Martin, P., Eklund, P.: Embedding knowledge in web documents. Comput. Netw. Int. J. Comput. Telecommun. Netw. 31(11–16), 1403–1419 (1999)

Towards an Approach for Configuring Ontology Validation

Mounira Harzallah[1(✉)], Giuseppe Berio[2], and Pascale Kuntz[1]

[1] LINA, University of Nantes, Rue Christian Pauc, Nantes, France
{mounira.harzallah,pascale.kuntz}@univ-nantes.fr
[2] IRISA, University of Bretagne Sud, Vannes, France
giuseppe.berio@univ-ubs.fr

Abstract. Ontologies are becoming widely recognised as key components in various types of systems and applications. However, ontology validation remains a critical open issue. In previous work, we have proposed a standard typology of problems that need to be removed for validating one ontology. Indeed, there is no standard vocabulary and definitions for those problems. In this paper, we introduce checking dependencies between standard problems; a validation process needs to satisfy checking dependencies for detecting and removing properly problems mentioned above. Then, we report an experience, based on 2 ontologies automatically generated from textual resources, showing how the typology can practically be deployed for configuring a validation process.

Keywords: Ontology validation · Ontology quality evaluation · Ontology quality problems · Quality problem checking · Automatically generated ontology

1 Introduction

Ontologies are becoming widely recognised as key components in various types of systems and applications: e.g. knowledge management systems, social network analysis, business intelligence and personalised applications. Ontologies have been and still are manually designed by human experts. However, the ever-increasing access to textual sources (as technical documents, web pages and so on) has motivated the development of tools for automating as much as possible ontology design and implementation process (being this process often renamed as "ontology learning"), and further enrichment. As a consequence, human involvement is rather minimised when such tools are used. Promising results have been reached [1, 2]. Unfortunately, experimental studies have then put in evidence limits for real-life applications [2, 3] and recent works recommend a better integration of human involvement [4].

Following this recommendation, we see the process of ontology learning as made of two main processes running in parallel and cooperating: (1) a *generation process* and (2) a *validation process*. The generation process focuses on the extraction of relevant items (such as terms and relations) and the identification and naming of relevant knowledge (such as concepts). The validation process is performed anytime

© Springer International Publishing Switzerland 2015
A. Fred et al. (Eds.): IC3K 2014, CCIS 553, pp. 388–404, 2015.
DOI: 10.1007/978-3-319-25840-9_24

when needed during the generation process and beyond. This is because, according to our experiences, validation should be performed as soon as possible focusing on subparts (such as subset of concepts) of the ontology under construction. Furthermore, validation process can be defined as the process guaranteeing the expected quality of the ontology (while the generation process makes the ontology content available). Thus, the validation process is a process specifically (i) looking for any poor (or bad or lower) quality of the ontology under construction through a sort of "*quality evaluation*" (or "*ontology evaluation*") and then (ii) proposing alternatives and applying selected alternatives for increasing quality by finding and removing recognised defects (i.e. modifying/deleting/adding artefacts to the ontology under construction). Since the pioneering works of Gruber in the 90's [5], *quality evaluation* has been discussed in [6], and, often independently from validation, various procedures and features have been proposed [7–11]: e.g. defining a set of quality measures, comparing ontologies to reference ontologies (also called gold standards), performing assessment of formal correctness, quality qualitative evaluation performed by experts, quality evaluation according to the results of a given application using the ontology, using pre-defined anti-patterns corresponding to defects or to potential defects. Roughly speaking, quality evaluation spans over three major criteria: (1) the *dimensions* which are evaluated (e.g. functional dimension, structural dimension or usability dimension) [7, 12], (2) the *evaluation mode* (manual vs automated) [13] and the *user profile* if any (e.g. knowledge engineers, business analysts, practitioners) [9], (3) and the *phase* in which evaluation is conducted (e.g. during the ontology development, before ontology publication and so on) [9, 14].

An analysis of the state of the art reveals that there are *three facets* when referring to "quality evaluation": (i) *scores of ontology quality* (such as, high, poor, bad or numeric scales) evaluated by using several quality measures, (ii) *quality problems* i.e. symptoms of defects or potential defects impacting ontology quality, and (iii) *defects* in the ontology i.e. the specific ontology artefacts (typically, concepts, relationships, axioms) which are causes of any poor quality and/or problem. The three facets are naturally related: for instance, if an ontology is inconsistent, this is a problem in our terminology i.e. a symptom of a defect in the ontology; the defect is the axioms causing inconsistency; quality scoring may be defined as dependent on the number of axioms causing inconsistency (a quality measure). Even though it seems natural that quality measures being related to defects, this is often not the case (as explained in Sect. 2). For instance, a measure like "ontology depth" can be used for scoring quality saying that "ontology quality" is directly proportional to "ontology depth" without referring to any (potential) defect (in this case, sample defects are the "omitted IS-A relationships"). This situation leads to difficulties for using quality scores in practice for validation purposes. On the contrary, problems are introduced and explained as symptoms of some defects and therefore are closer to defects than scores; thus, using problems and their dependencies for removing defects seems more effective than using quality scores. Therefore, in our previous work [15] we have focused on the problem facet. This previous work has specifically targeted one critical aspect of the problem facet i.e. the standardisation of *problems* and *problems definitions*, currently rather variable. Accordingly, we have proposed a *typology of problems* which: makes a synthesis of the state of the art, is extensible, is easy to understand and based on a well-known quality framework defining quality for conceptual models [16] (ontologies are special cases of conceptual models). However, in that past work, we have not provided details on how, in practice, the proposed typology can be deployed and used in the context of validation process. In this paper, we are going to present one experience (based on 2 ontologies automatically built from textual sources) showing how the proposed typology can concretely be deployed and lessons learned for *configuring a validation process*.

The paper is organised as follow. Section 2 provides a short overview on quality and problems in ontologies. Section 3 introduces the proposed typology of ontology problems and dependencies between them. Section 4 describes the performed experience and provides feedback reporting (discussion and lesson learned). Finally, conclusions summarise key results.

2   State of the Art Insights

As reported in the Introduction, ontology quality evaluation concerns three related facets: scoring quality through measurements based on quality measures, quality problems and defects. Figure 1 below provides a simple picture (as a UML class diagram), for representing the three facets (i.e. quality problem, quality measure and defects) and key relationships according to state of the art. Figure 1 can be explained as follow. Whenever precise and scoped definitions of quality problems are available is possible to select (or to develop) techniques for detecting or warning about those problems (a). Quality measures can be used for detecting or warning about defects/problems, especially when combined with reference values: quality measures are therefore techniques for that purpose (b). Quality measures are also techniques to evaluate the ontology quality (c). Ontology problems can be used for a qualitative assessment of ontology quality (d). When a quality problem is detected, it can trigger the usage of techniques for detecting defects causing that problem (e). Finally, even if a problem may not be associated to some techniques for identifying it, this is not suitable.

Sections 2.1 and 2.2 shortly present relevant state of the art insights on quality measures and problems.

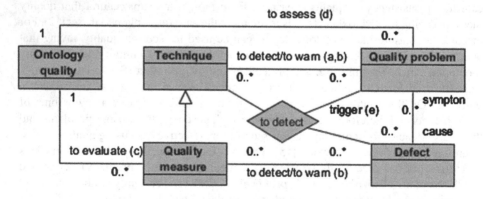

Fig. 1. The three facets of quality evaluation and key relationships.

2.1 Quality Measures

As reported in the Introduction, existing proposals cover various "quality dimensions". In the context of ontologies, dimensions have not been standardised. For instance, they may be referred to as, syntax, semantics, maintenance and ergonomics or, functional, structural and usability.

One of the most complete proposals associating dimensions and measures is probably oQual [12]. In oQual, an ontology is analysed according to three dimensions: (i) structural (syntax and formal semantics of ontologies), (ii) functional (intended meaning of the ontology and its components) and (iii) usability (pragmatics associated with annotations, which contribute to the ontology profiling). A set of measures is associated with each dimension to score the quality. For instance: for structural dimension, depth and breadth of a taxonomy; for functional dimension, precision and recall of the ontology content with respect to its intended meaning; for usability dimension, number of annotations.

More generally, despite the potential interests of measures proposed in literature, some of them remain quite disconnected to defects/problems. For instance, the ratio between number of concepts and number of relations (N°Concepts/N°Relations) is a quality measure for evaluating "cognitive ergonomics", which is in turn closely related to a "easy to use" quality: however, the ontology may not suffer of any problem/defect concerning its artefacts (concepts, relationships, axioms etc.) because representing as it is the targeted domain. Therefore, defining quality measures (possibly organised alongside several dimensions) as entry point does not necessarily make explicit (occurring or potential) defects/problems. On the contrary, defining problems as entry point provides an evidence of occurring defects that, in turn, lead to poor quality.

2.2 Quality Problems

Roughly speaking, the generic notion of "ontological error" covers a wide variety of problems affecting different dimensions. In the relevant literature, it is possible to find precise and less precise definitions for several recognised problems: (1) "taxonomic errors" [8, 10, 17, 18] or "structural errors" [19], (2) "design anomalies" or "deficiencies" [10, 20], (3) "anti-patterns" [11, 19, 21], (4) "pitfalls" or "worst practices [22] and (5) "logical defects" [19]. Additional errors could complete this list: e.g. (6) "syntactic errors" [19].

Hereinafter, we shortly present insights on each of the above mentioned problem cases. Syntactic errors (6) are due to violations of conventions of the language in which the ontology is represented. While interesting in practice for building support tools, they are conceptually less important than others: therefore they will be no longer considered in the remainder.

Taxonomic errors (1) concern the taxonomic structure and are referred to as: inconsistency, incompleteness and redundancy [17]. Three classes of "inconsistency" both logical and semantic have been highlighted: circularity errors (e.g. a concept that is a specialization of itself), partitioning errors (e.g. a concept defined as a specialization of two disjoint concepts), and semantic errors (e.g. a taxonomic relationship in contradiction with the user knowledge). Incompleteness occurs when for instance, relationships or axioms are missing. Finally, redundancy occurs when for instance, a taxonomical relationship can be deduced from the others by logical inference.

Design anomalies (2) concern ontology understanding and maintainability [10, 20]: lazy concepts (leaf concepts without any instance or not considered in any relation or axiom), chains of inheritance (long chains composed of concepts with a unique child), lonely disjoint concepts (superfluous disjunction axioms between distant concepts), over-specific property range and property clumps (duplication of the same properties for a large concept set which can be retrieved by inheritance).

Anti-patterns (3) are known or recognised templates potentially leading to identified problems [11, 19]. Some classes of anti-patterns are: logical anti-patterns (producing conflicts that can be detected by logical reasoning), cognitive anti-patterns (caused by a misunderstanding of the logical consequences of the axioms), and guidelines (complex expressions true from a logical and a cognitive point of view but for which simpler or more accurate alternatives exist).

Pitfalls (4) cover problems for which ontology design patterns (ODPs) are not available. An ODP cover ad-hoc solutions for the conception of recurrent particular cases [21]. Poveda et al. [23] have established a catalogue of pitfalls grouped on 7 classes, them-self classified under the three ontology dimensions cited above [12]. Four pitfalls classes are associated with the structural dimension: modelling decisions (false uses of OWL primitives), wrong inference, no inference (lacks in the ontology which prevent inferences required to produce desirable knowledge), real-world modelling (common sense knowledge missing). One class is associated with the functional dimension: requirement completeness (e.g. uncovered specifications). And, two classes are associated with the usability dimension: ontology understanding and ontology clarity (e.g. variations of writing-rule and typography for the labels). Poveda et al. [22] have also tried to classify these pitfalls according to the three taxonomic error classes [17]; but pitfalls concerning the ontology context do not fit with this classification.

3   Problem Standardisation Overview

Mentioned in the Introduction and made evident in Sect. 2.2, heterogeneity in quality problems and their definitions is due to distinct experiences, communities and perception of ontologies. Standardisation enables a much better understanding of what problems are and to what extent these problems are critical before using the ontology. We have therefore proposed a two-level rigorous problem typology summarised in Table 1. Level 1 distinguishes logical from social ground problems and level 2 distinguishes errors from unsuitable situations. Errors are problems mostly preventing the usage of an ontology. We add "mostly" because in the case of "inconsistency error" (Table 1), some researches focus on how to make usable inconsistent ontologies [24]. On the contrary, unsuitable situations are problems which do not prevent ontology usage (within specific targeted domain and applications). Therefore, while errors need to be solved, unsuitable situations may be maintained as such.

Social ground problems are related to the interpretation and the targeted usage of the ontologies by social actors (both humans and applications). Logical errors and most of logical unsuitable cases can be rigorously formalised within a logical framework; for instance, they can be formally defined by considering key notions synthesised by Guarino et al. [25] i.e.: Interpretation (I) (extensional first order structure), Intended Model, Language (L), Ontology (O) and the two usual relations $\models$ and $\vdash$ provided in any logical language. The relation $\models$ is used to express both that one interpretation I is a model of a logical theory L, written as I $\models$ L (i.e. all the formulas in L are true in I: for each formula $\varphi \in L$, I $\models \varphi$), and also for expressing the logical consequence (i.e. that any model of a logical theory L is also a model of a formula: L $\models \varphi$). The relation $\vdash$ is used to express the logical calculus i.e. the set of rules used to prove a theorem (i.e. any formula) φ starting from a theory L: L $\vdash \varphi$). Accordingly, when needed, problems are formalised by using classical description logic syntax that can also be transformed in FOL or other logics.

Problems in Table 1 are not independent and need to be checked out and removed in meaningful order in the context of a validation process. Table 2 below provides a list of *checking dependencies* that constraint orders for checking and removing problems. Checking dependency $A1,.., An \rightarrow B1,..., Bm$ means that before checking out any Bi, all problems Ai (i.e. A1 to An) need normally to be checked out and removed (i.e. left out) by appropriate techniques. For in practice configuring a validation process, dependencies can be used *forward or backward* as follow: $A1,.., An \rightarrow B1,..., Bm$, is used forward if A1,.., An are checked out and removed before taking into account B1,..., Bm; $A1,.., An \rightarrow B1,..., Bm$ is used backward if B1,..., Bm are checked out before and this is used for checking out and removing A1,..., Am.

It should be noted that removing a logic ground problem from an ontology results in a new ontology which is or is not logically equivalent to the previous one. Depending on the problem, there may or may not be a way for removing it by guaranteeing logical equivalence. It is clear that L1, L2, L3, L10 can only be removed without guaranteeing logical equivalence; L4, L5, L6, L12 can be removed by guaranteeing logical equivalence; for L7, L8, L11 both ways are possible. It is quite natural to consider by default that removing any social ground problem results in a new non-logically equivalent ontology.

In the context of validation process, checking dependencies:

1. Make meaningful or optimize problem checking (for instance, it does not make sense checking if an ontology is unadapted, if the ontology is inconsistent and inconsistency problem has not removed yet);
2. Suggest which problems can be removed without causing injection of additional problems. For instance, checking out if an ontology is not minimal (L12) and removing the problem, does not make sense without having previously checked out and removed social contradictions (S1) within the ontology; as a concrete case, if redundant IS-A relationships (a case for L12) are found, if one of them is removed and then, another IS-A relationship is removed because suffering of S1, the ontology may become incomplete (injection of L3 or S4). However, when a problem is removed without guaranteeing logical equivalence with the new resulting ontology, all the problems need to be checked out again and all dependencies need to be taken into account once again (as, for instance, when L2 is removed).

Table 1. The typology of quality problems.

Logical ground problems

Errors	L1. Logical inconsistency: no I of s.t. I $\models$ O
	L2. Unadapted ontologies: there is a formula φ for some intended models of L, φ is false and O $\models \varphi$
	L3. Incomplete ontologies: there is a formula φ for each intended models of L, φ true and O $\not\models \varphi$
	L4; Incorrect (or unsound) reasoning: when a false formula φ in the intended models O $\not\models \varphi$, can be derived from a suitable reasoning system (O $\vdash \varphi$)
	L5. Incomplete reasoning: when a true formula φ in the intended models O $\models \varphi$, cannot be derived from a reasoning system (O $\not\vdash \varphi$)
Unsuitable cases	L6. Logical equivalence of distinct artefacts: O $\models A_i = A_j$
	L7. Logical indistinguishable artefacts: impossible to prove any of the following statements: (O $\models A_i = A_j$), (O $\models A_i \cap A_j \subseteq \perp$) and (O $\models c \subseteq A_i$ and $c \subseteq A_j$)
	L8. OR artefacts: A_i equivalent to $A_j \cup A_k$, $A_i \neq A_j$, $A_i \neq A_k$, but for which (if applicable) there is neither role R s.t O $\models (A_j \cup A_k) \subseteq \exists R.\top$, nor instance c s.t. O $\models c \subseteq A_j$ and O $\models c \subseteq A_k$
	L9. AND artefacts: A_i equivalent to $A_j \cap A_k$, $A_i \neq A_j$, $A_i \neq A_k$, but for which (if applicable) there is no common (non optional) role/ property for A_j and A_k
	L10. Unsatisfiability: given an artefact A, O $\models A \subseteq \perp$)
	L11. Complex reasoning: unnecessary complex reasoning when a simpler one exists
	L12. Ontology not minimal: unnecessary information

Social ground problems

Errors	S1. Social contradiction: contradiction between the interpretation and the ontology axioms and consequences
	S2. Perception of design errors: e.g. modelling instances as concepts
	S3. Socially meaningless: impossible interpretation
	S4. Social incompleteness: lack of artefacts
Unsuitable cases	S5. Lack of/poor textual explanations: lack of annotations
	S6. Potentially equivalent artefacts: similar artefacts identified as different
	S7. Socially indistinguishable artefacts: difficult to distinguish different artefacts
	S8. Artefacts with polysemic labels
	S9. Flatness of the ontology: unstructured set of artefacts
	S10. Non-standard formalization of the ontology: unreleased specific logical use
	S11. Lack of adapted and certified version of the ontology in various languages
	S12. Socially useless artefacts

Table 2. Checking dependencies between quality problems.

L1 → L2, L3, L4, L5, L6, L7, L8, L9, L10, L11, L12	If an ontology is inconsistent, inferences do not make sense and any other problem can be trivially detected; this dependency should be only used as forward dependency
L2, L3 → L6, L7, L8, L9, L10, L11, L12 L2, L3 → L4, L5	It does not make sense to assess unsuitable situations on one ontology which is not completely finalised; the same is true for unsound/incomplete reasoning; this dependency should be used as forward dependency; however a backward usage is possible (for instance, L6 can be checked out and this may be used for highlighting unintended models)
S12, S3 → S2, S5, S6, S7, S8	Useless and meaningless artefacts should be removed and ontology updated accordingly before checking out any other problem; however, S10 and S11 can be checked independently; this dependency has be used as forward dependency
S1, S2, S3, S12 → L2, L3, L4, L5, L6, L7, L8, L9, L10, L11, L12	Meaningless and useless artefacts, design errors and social contradictions should be removed before checking out any logic ground problem (except L1); this dependency can be also used backward: for instance, L12 is checked out and among redundant artefacts, S1 is then checked out and removed accordingly
S2 → S1, S4, S9	Modelling errors should be removed before checking out social contradiction, incompleteness and flatness; this dependency has be used as forward dependency

4 Experience

This section presents an experience on the deployment of the typology based on two ontologies automatically generated from different corpora by using Text2Onto [1]. We have used Text2Onto in one of our past research projects [26] and realised a full comparison with similar tools. The comparison results made possible to select Text2Onto as the best choice for realising the work. This was also confirmed by successful work performed in the project, making possible to use extracted ontologies as components for interoperating enterprise systems. However, Text2Onto capability for extracting concepts and taxonomic relationships has been shown to significantly outperform its capability for extracting other types of artefact [27, 28].

4.1 Experience Setting

As said above, we have generated two ontologies by using Text2Onto. Generated raw ontology O1 (resp. O2) contains 441 (resp. 965) concepts and 362 (resp. 408) taxonomic relationships. The first ontology (O1) has been generated starting from a scientific article in the domain of "ontology learning from texts" containing 4500 words. The second ontology (O2) has been generated starting from a technical glossary composed of 376 definitions covering the most important terms used in the composite material domain. The glossary contains 9500 words. It has been provided by enterprises involved in the project. It should be noted that although showing quite different content features, the size of the two selected textual resources has been deliberately limited to enable further detailed analysis of the experience results.

4.2   Typology Deployment for the Experience

In the experience, the deployment of the typology for the ontology problem detection is performed in two steps. The first step is about the selection of problems that may occur within the ontology to validate. The second step concerns the identification of appropriate techniques or the development of new ones for the detection of each of selected problems.

Problem Selection (First Step). Appropriate techniques are required for the detection of ontology problems. However, there is no need to possess a technique for each typology problem to identify problems within an ontology, especially because ontologies range from very simple (or light) to very complex (or heavy).

In the experience, ontologies generated with Text2Onto are very simple and basically represented as a list of concepts related by IS-A relationships (i.e. concepts organised as a taxonomy, what is sometimes referred to a *lightweight ontology*).

The logic ground problems L1, L10 and L11 cannot trivially occur. Indeed, they may only occur iff the ontology comprises axioms other than axioms for specifying the taxonomy.

L2 to L5 are not applicable because they can be applied only iff intended models are known in some way, which is not the case within the experience.

The remaining logical ground problems L6, L7, L8, L9 and L12 may occur. Indeed L6, L8 and L9 may occur when two inverse taxonomic relationships exist within the ontology (A IS-A B and B IS-A A). L7 trivially occurs because two concepts that are not equivalents are indistinguishable in this ontology. Finally L12 may occur, for instance, if it exists three taxonomic relationships as A IS-A B, B IS-A C and A IS-A C (possible in the ontologies of the experience).

Concerning the social ground problems, all of them may occur. Specifically, some of them trivially occur. Indeed, S5 (Lack or poor textual explanation) trivially occurs because Text2Onto does not provide any annotation as outcome. Finally, Text2Onto outcome is transformed on OWL but the produced version is not necessarily certified (as rules for making the transformation are proprietary): then, S11 trivially occurs.

The problem selection impacts on general checking dependencies (Table 2). Indeed, whenever a dependency comprises a non-selected problem, this dependency needs to be rewritten by deleting the problem. For instance, in this experience, the dependency: L2, L3 $\rightarrow$ L6, L7, L8, L9, L10, L11, L12

is rewritten as $\rightarrow$ L6, L7, L8, L9, L10, L11, L12.

In this case, the dependency is no longer considered because its premises are "void".

Techniques for Selected Problem Detection/Warning (Second Step). For logical ground problems L6, L7, L8 and L12, the OWL ontology version and a reasoner (Pellet) have been used:

- Concerning L6 (Logical equivalence of distinct artefacts), the reasoner has been able to identify equivalences between concepts(e.g. area = domain = issue = end = section = object, path = shape);
- Because of the special form of the ontology comprising only concepts and IS-A relationships, detecting L7 has been made possible by counting the pairs of non logically equivalent concepts (checked with L6).
- Concerning L8 and L9 problems, the reasoner has not been able to find any concept equivalent to union or intersection of other concepts. So that, L8 and L9 do not occur
- Concerning L12 (Ontology non minimal), the reasoner has been able to detect some IS-A relationships (original) as inferred other ones.

Apart S5 and S11 problems mentioned above, the other social ground problems have required to develop our own techniques. However, because most of these problems can only be detected if stakeholders/users are directly involved (such as end-users, experts and so on), employed techniques do not guarantee unbiased results.

Through *formal inspection*, S1 (Social contradiction) has been detected by specifically inspecting IS-A relationships and pointing the ones contradicting our own IS-A relationships. S2 (Perception of design errors) has been detected by focusing on the ambiguity/vagueness of the dichotomy concept vs. instance.

S3 (meaningless artefacts) has been raised for concepts labelled with artificial labels (e.g. a label such as "tx12").

S4 (social incompleteness) has been detected as follow: whenever a concept is connected only to the root (so that it has no other relationship with other concepts because ontology is lightweight), the ontology is considered to be incomplete because probably lacking of additional IS-A relationships; this technique only warns about the problem.

S6 (Potentially equivalent artefacts) has been detected as a problem occurring when labels for concepts are synonyms according to our domain knowledge (e.g. area = field, human = person, sheet = plane) or according to known domain references.

S7 (socially indistinguishable artefacts) has been detected whenever it was impossible for a pair of concepts to both provide factual raison to made them equivalent and factual raison to made them distinct.

S8 (polysemy in artefact labels) has been detected by looking to the existence of several definitions, within the given domain, for the single concept label (e.g. labels such as cycle, repair).

S9 has been simply detected by calculating the average depth of the ontology as the average of taxonomy leaf depth, and comparing it to an expected typical depth (found in a manually built ontology based on the same documents).

Table 3. Identified quality problems in O1 and O2.

Types of problems	Detected problems	
	Ontology O1 (441 concepts and 362 is-a relationships)	Ontology O2 (965 concepts and 408 relationships)
L1	Trivially non occurring	Trivially non occurring
L2	NA	NA
L3	NA	NA
L4	NA	NA
L5	NA	NA
L6	276 (= 24*23/2, because we found 24 equivalent concepts) pairs of equivalent concepts (detected on the OWL version)	57 pairs of equivalent concepts (detected on the OWL version)
L7	Trivially occurring; all pairs of concepts that are not equivalent are indistinguishable ((441*440/2)-276 indistinguishable pairs)	Trivially occurring; all pairs of concepts that are not equivalent are indistinguishable ((965 *964/2)-57 indistinguishable pairs)
L8	No "OR artefact"	No "OR artefact"
L9	No "AND artefact"	" No "AND artefact"
L10	Trivially non occurring	Trivially non occurring
L11	The ontology does not contain any situation that can make inferences more complicated	The ontology does not contain any situation that can make inferences more complicated
L12	32 redundant taxonomic relations	49 redundant taxonomic relation
S1	130 taxonomic relations contradict the evaluator's knowledge	60 taxonomic relations contradict the evaluator's knowledge
S2	2 instances were identified as concepts according to evaluator's knowledge	5 instances were identified as concepts according to evaluator's knowledge
S3	13 concepts with meaningless labels according to evaluator's knowledge	21 concepts with meaningless labels according to evaluator's knowledge
S4	168 concepts only connected to root	360 concepts only connected to root
S5	Trivially occurring (not counted)	Trivially occurring (not counted)
S6	9 pairs of concepts with synonymous labels	3 pairs of concepts with synonymous labels
S7	No couple of socially indistinguishable artefacts	No couple of socially indistinguishable artefacts
S8	7 concepts with polysemic labels	9 concepts with polysemic labels
S9	Flat ontology, Average depth of leaves = 2.02, Expected depth = at least 5	Flat ontology, Average depth of leaves = 1.99, Expected depth = at least 7
S10	No: a OWL version is available	No: a OWL version is available
S11	The ontology is not certified	The ontology is not certified
S12	121 useless concepts according to evaluator's knowledge	31 useless concepts according to evaluator's knowledge

Finally, useless artefacts (S12) have been considered as such if it is impossible to provide simple and clear raison for including artefacts in the ontology (for instance, 'train', 'cannot' were trivially out of the ontology domain scope).

Table 3 above summarizes the problems detected, by using deployed techniques, in the two generated ontologies. Next section provides a discussion on experience feedback, mostly based on Table 2.

4.3   Discussion

During the experience, we have remarked the interest, when applicable, of keeping in mind "numbers of occurrences" of a given problem (for instance, S1 can be considered occurring several times as many as ontology artefacts suffer of the problem). Indeed, occurrences are a simple way to highlight differences in the two ontologies, then to identify causes of problems (i.e. defects) and potential correlations between problems. However, not all the problems can be counted: for instance, flatness problem (S9) cannot be counted.

The six most occurring problems are the same for O1 and O2, three are social and three are logical problems: S1, S4, S12, L6, L7 and L12. These problems have been checked involving both concepts and relationships. Occurrences of these problems are quite different in O1 and O2: S1 (O1: 130, O2: 45), S12 (O1: 121, O2: 31), L6(O1: 300, O2: 65), These differences may be quite surprising because numbers of concepts and relations in O1 are lower than in O2. We have therefore tried to provide alternatives non-exclusive explanations. Two explanations have been provided.

One alternative explanation concerns the nature of the content of the incoming textual resource. A technical glossary (starting point for O2) naturally providing definitions of terms, is more self-contained and more focused than a scientific paper (starting point for O1). Indeed, few concept labels in O2 can be considered very generic thus loosely related to the domain while this is not the case for O1. This seems to be confirmed by the fact that S12 (useless artefacts) occurs very often in O1 if compared to O2.

A second non-exclusive alternative explanation is traced back to the usage of Wordnet made by Text2Onto. Indeed, generic and rather useless concepts belonging to O1 enable Text2Onto to also introduce IS-A relationships belonging to Wordnet; these IS-A relationships are due to the several meanings associated by Wordnet to terms (for instance, for term "type" in case of O1, Text2Onto extracted: "type" IS-A "case"; "type" IS-A "group" and "type" IS-A "kind"; each IS_A relation concerns one quite specific and distinct meaning of the term "type"). This is confirmed by much higher occurrences of S1 in O1 than in O2 (remember that S1 has been detected by focusing on IS-A relationships only, see Sect. 4.2).

Occurrences can also be fruitful for establishing *potential correlations* between problems. A problem is potentially correlated to another one if the presence of one problem is potentially due to the presence of another one. However, correlations cannot substitute checking dependencies mentioned in Sect. 3: indeed, checking dependencies are by definition independent from the ontology and the technique used to detect problems while correlations are based on occurrences which are consequences of those techniques and the ontology. As a consequence, dependencies are stronger than correlations. Correlations should confirm established checking dependencies (Table 2) and provide insights on using those dependencies in forward or backward direction; however, correlations can also be additional to those dependencies.

We have found the following three interesting correlations.

Correlation 1: S12 seems correlated with S1 (raised from results reported Table 3: O1(S12: 121, S1: 130) and O2 (S12: 31, S1:45)). The correlation, as also explained above, can be justified because in the experience useless concepts are often source of incorrect taxonomic relationships; the correlation confirms the (forward) checking dependency between S12 and S1.

Correlation 2: S9 (ontology flatness) shows similar values for the two ontologies (Table 3 reports O1(S9: 2:02) and O2(S9: 1:99)). S9 seems to be correlated to S4: frequency of S4 is mostly the same for the two ontologies, and S9 also occurs o. The correlation can be justified because S4 is checked by counting the concepts only related to the root. If S4 occurs often, in any case, the average depth tends to depend on number of concepts only related to the root. This correlation is additional to the ones in Table 2.

Correlation 3: S12 seems correlated with L6 (raised from results reported in Table 3: O1(S12: 121, L6: 276) and O2 (S12: 31, L6: 57)) because useless artefacts have generated additional logical equivalences; the correlation confirms the (forward) checking dependency between S12 and L6.

4.4 Lessons Learned: Configuring a Validation Process

Discussion above points out that explanations for quality problems can be traced back to the content features of the incoming textual resources (e.g. technical content vs scientific content) and the usage of external resources. It is therefore suggested that to reduce the complexity of the validation process contents of the incoming textual resources should be evaluated and possibly improved before learning the ontology.

Correlations (Sect. 4.3) and dependencies (as in Table 2 but rewritten as explained in Sect. 4.2) between problems can be merged, when not in contradiction. Once merged, an order for configuring (and then running) a validation process can be identified (Fig. 2 above). It should be noted that, according to what has been said in Sect. 3, whenever a problem is removed, if the new resulting ontology is not logically equivalent to the previous ontology, all dependencies need to be reconsidered: as a consequence, the validation process moves to the initial step (status "Checking S3, S12" in Fig. 2 above). This order configuration is reasonable for any other *lightweight ontology* built and validated in the same context of the reported experience i.e. (i) with the same tool (e.g. Text2Onto) with the same tool parameters, (ii) the same typology deployment (same selected problems, same associated techniques, see Sect. 4.2), and (iii) whenever a problem is removed, the resulting ontology remains lightweight (otherwise additional problems need to be selected when typology is deployed).

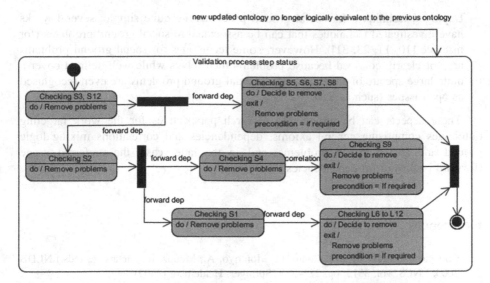

Fig. 2. Identified ordered steps (UML state machine).

5   Conclusions

Through the paper, we have reported a typology of problems impacting the quality of an ontology and introduced checking dependencies for properly detecting and removing problems within a validation process. These dependencies are generally valid and can be reused in any case. Through an experience, we have presented how in practice the typology can be deployed (i.e. problems to be considered and techniques for detecting them). Based on experience results, we have then analysed how a validation process can be configured (i.e. in which order problems need to be checked out and removed). The resulting configuration merges dependencies (independent from the experience) with correlations (dependent from the experience); however, the configuration seems reasonable when other ontologies built with the same tool (Text2Onto) are validated according to the typology deployment made for the experience.

Of course, the experience itself does not cover various important aspects reported below:

- Because ontologies used in the experience are lightweight, typology deployment has only concerned a subset of problems; important problems, especially logic ground errors, are not covered by the deployment; however, specific techniques have been developed for trying to detect most of the logical ground problems; these techniques focus on algorithms for explaining reasoning and supporting users for expressing expected facts; however, some works (through SPARQL queries [10], anti-patterns [21, 29], heuristics [30], tools [23, 30], have undertaken more empirical ways for looking to problems (therefore, often more focusing warning about problems than on detecting occurring problems);

- Deployed techniques for social ground problems are quite simple; several works have investigated techniques that can be associated to social ground problems (for instance [10, 11, 23, 31]). However, some techniques for social ground problems are not clearly confined because problems themselves while well-defined cover a quite large spectre of situations; other social ground problems are even recognised as open issues (such as S10, [32]).

These aspects can be turned into research perspectives for the work targeting ontologies comprising general axioms, dependencies and correlations mixing logic ground and social ground problems, and in the latter case, the preferred usage (forward-backward) of dependencies.

References

1. Cimiano, P., Völker, J.: Text2onto. In: Montoyo, A., Muñoz, R., Métais, E. (eds.) NLDB 2005. LNCS, vol. 3513, pp. 227–238. Springer, Heidelberg (2005)
2. Cimiano, P., Maedche, A., Staab, S., Volker, J.: Ontology learning. In: Studer, R., Staab, S. (eds.) Handbook on ontologies. International Handbooks on Information Systems, pp. 245–267. Springer, Heidelberg (2009)
3. Hirst, G.: Ontology and the lexicon. In: Studer, R., Staab, S. (eds.) Handbook on Ontologies. International Handbooks on Information Systems, 2nd edn, pp. 269–292. Springer, Heidelberg (2009)
4. Simperl, E., Tempich, C.: Exploring the economical aspects of ontology engineering. In: Studer, R., Staab, S. (eds.) Handbook on Ontologies. International Handbooks on Information Systems, 2nd edn, pp. 445–462. Springer, Heidelberg (2009)
5. Gruber, T.R.: A translation approach to portable ontology specifications. Knowl. Acquisition 5(2), 199–220 (1993)
6. Gomez-Perez, A.: Some ideas and examples to evaluate ontologies. In: Proceedings of the 11th Conference on Artificial Intelligence for Applications, pp. 299–305 (1995)
7. Duque-Ramos, A., Fernandez-Breis, J.T., Aussenac-Gilles, N., Stevens, R.: Oquare: asquare based approach for evaluating the quality of ontologies. J. Res. Prac. Inf. Technol. **43**, 159–173 (2011)
8. Gomez-Perez, A.: Ontology evaluation. In: Staab, S., Studer, R. (eds.) Handbook on ontologies. International Handbooks on Information Systems, 1st edn, pp. 251–274. Springer, Heidelberg (2006)
9. Hartmann, J., Spyns, P., Giboin, A., Maynard, D., Cuel, R., Suarez-Figueroa, M.C., Sure, Y.: Methods for ontology evaluation. Technical report. Knowledge Web Deliverable D1.2.3 (2004)
10. Baumeister, J., Seipel, D.: Smelly owls-design anomalies in ontologies. In: Proceedings of 18th International Florida Artificial Intelligence Research Society Conference, pp. 215–220 (2005)
11. Roussey, C., Corcho, O., Blazquez, L.M.V.: A catalogue of owl ontology antipatterns. In: Proceedings of 5th International Conference on Knowledge Capture, pp. 205–206 (2009)
12. Gangemi, A., Catenacci, C., Ciaramita, M., Lehmann, J.: Modelling ontology evaluation and validation. In: Sure, Y., Domingue, J. (eds.) ESWC 2006. LNCS, vol. 4011, pp. 140–154. Springer, Heidelberg (2006)

13. Vrandecic, D.: Ontology evaluation. In: Studer, R., Staab, S. (eds.) Handbook on ontologies. International Handbooks on Information Systems, 2nd edn, pp. 293–314. Springer, Heidelberg (2009)
14. Tartir, S., Arpinar, I.B., Sheth, A.P.: Ontological evaluation and validation. In: Poli, R., Healy, M., Kameas, A. (eds.) Theory and Applications of Ontology: Computer Applications, pp. 115–130. Springer, Netherlands (2010)
15. Gherasim, T., Harzallah, M., Berio, G., Kuntz, P.: Methods and tools for automatic construction of ontologies from textual resources: a framework for comparison and its application. In: Guillet, F., Pinaud, B., Venturini, G., Zighed, D.A. (eds.) Advances in Knowledge Discovery and Management. SCI, vol. 471, pp. 177–201. Springer, Heidelberg (2013)
16. Krogstie, J.: Specialisations of SEQUAL. In: Model-Based Development and Evolution of Information Systems, pp. 281–326. Springer, London (2012)
17. Gomez-Perez, A., Fernandez-Lopez, M., Corcho, O.: Ontological Engineering: With Examples from the Areas of Knowledge Management, e-commerce and the Semantic Web. Advanced Information and Knowledge Processing. Springer, New York (2001)
18. Fahad, M., Qadir, M.A.: A framework for ontology evaluation. In: Proceedings of the 16th International Conference on Conceptual Structures (ICCS 2008), vol. 354, pp. 149–158 (2008)
19. Buhmann, L., Danielczyk, S., Lehmann, J.: D3.4.1 report on relevant automatically detectable modelling errors and problems. Technical report LOD2-Creating Knowledge out of Interlinked Data (2011)
20. Baumeister, J., Seipel, D.: Anomalies in ontologies with rules. Web Seman. Sci. Serv. Agents World Wide Web **8**(1), 55–68 (2010)
21. Corcho, O., Roussey, C., Blazquez, L.M.V.: Catalogue of anti-patterns for formal ontology debugging. In: Atelier construction d'ontologies: vers un guide des bonnes pratiques, afia, pp. 2–12 (2009)
22. Poveda, M., Suarez-Figueroa, M.C., Gomez-Perez, A.: A double classification of commonpitfalls in ontologies. In: Proceedings of the workshop on ontology quality at ekaw, pp. 1–12 (2010)
23. Poveda-Villalón, M., Suárez-Figueroa, M.C., Gómez-Pérez, A.: Validating ontologies with OOPS! In: ten Teije, A., Völker, J., Handschuh, S., Stuckenschmidt, H., d'Acquin, M., Nikolov, A., Aussenac-Gilles, N., Hernandez, N. (eds.) EKAW 2012. LNCS, vol. 7603, pp. 267–281. Springer, Heidelberg (2012)
24. Bertossi, L., Hunter, A., Schaub, T.: Inconsistency Tolerance. Springer, Heidelberg (2005)
25. Guarino, N., Oberle, D., Staab, S.: What is an ontology? In: Studer, R., Staab, S. (eds.) Handbook on ontologies. International Handbooks on Information Systems, 2 edn, pp. 1–17. Springer, Heidelberg (2009)
26. Harzallah, M.: Développement des ontologies pour l'interopérabilité des systèmes hétérogènes, applications aux cas industriels du projet ISTA3. Livrable final de la tâche 2.4 du projet ISTA3. Interop-vlab.eu/workspaces/ISTA%203 (2012)
27. Volker, J., Sure, Y.: Data-driven change discovery-evaluation. Technical report. Deliverable D3.3.2for SEKT Project, Institute AIFB, University of Karlsruhe (2006)
28. Gherasim, T., Harzallah, M., Berio, G., Kuntz, P.: Problems impacting the quality of automatically built ontologies. In: Proceedings of the 8th Workshop on Knowledge Engineering and Software Engineering, held in Conjunction with ECAI 2012, pp. 25–32 (2012)
29. Roussey, C., Scharffe, F., Corcho, O., Zamazal, O.: Une méthode de débogage d'ontologies owl basées sur la détection d'anti-patrons. In: Actes de la 21e conférence en ingénierie des connaissances, pp. 43–54 (2010)

30. Pammer, V.: Automatic support for ontology evaluation-review of entailed statements and assertional effects for owl ontologies. Ph.D. thesis, Graz University of Technology (2010)
31. Burton-Jones, A., Storey, V., Sugumaran, V.: A semiotic metrics suite for assessing the quality of ontologies. Data Knowl. Eng. 55(1), 84–102 (2005)
32. Kalfoglou, Y.: Cases on semantic interoperability for information systems integration: Practices and applications. IGI Global, Hershey (2010)

A Tool for Complete OWL-S Services Annotation by Means of Concept Networks

D. Redavid[✉], S. Ferilli, B. De Carolis, and F. Esposito

Computer Science Department, University of Bari "Aldo Moro",
Via E. Orabona, 4, Bari, Italy
domenico.redavid1@uniba.it

Abstract. Current tools to create OWL-S annotations have been designed starting from the knowledge engineer's point of view. Unfortunately, the formalisms underlying Semantic Web languages are often obscure to the developers of Web services. To bridge this gap, it is desirable that developers are provided with suitable tools that do not necessarily require knowledge of these languages in order to create annotations on Web services. With reference to some characteristics of the involved technologies, this work addresses these issues, proposing guidelines that can improve the annotation activity of Web service developers. Following these guidelines, we also designed a tool that allows a Web service developer to annotate Web services without requiring him to have a deep knowledge of Semantic Web languages. A prototype of such a tool is presented and discussed in this paper.

Keywords: Semantic Web Services · OWL-S annotation · OWL · SWRL · Semantic networks

1 Introduction

Ontologies are written by knowledge engineers and require specific skills for their use [14]. Formalisms such as Description Logics [1], underlying the Web Ontology Language (OWL) (http://www.w3.org/TR/owl2-new-features) standards, are incomprehensible to the developers of Web services [4]. As it happened for the Web, where increasingly powerful browsers have allowed ordinary users to use the Internet without knowing languages such as HTML or XML, in order to promote the technologies related to Semantic Web Services (SWS) [9], it is desirable that developers are provided with tools that do not necessarily require knowledge of languages such as Resource Description Framework (RDF) (http://www.w3.org/RDF), OWL, OWL for services (OWL-S) (http://www.w3.org/Submission/OWL-S) and Semantic Web Rule Language (SWRL) (http://www.w3.org/Submission/SWRL).

The objective of this work is to propose solutions that simplify the process of annotating SWSs in order to facilitate the operation of automatic service composition. Specifically, we propose the use of particular type of Semantic Network to limit the effort to interpret and describe conditions related to the correct invocation of a Web service.

© Springer International Publishing Switzerland 2015
A. Fred et al. (Eds.): IC3K 2014, CCIS 553, pp. 405–420, 2015.
DOI: 10.1007/978-3-319-25840-9_25

2   Technological Background

A Web service, as defined by the World Wide Web Consortium (W3C) (http://www.w3.org), is a software system designed to support interoperability between different machines on the same network by means of an interface that describes the service in a suitable language, through which other machines can interact with it. This interaction takes place through the exchange of SOAP messages, formatted according to the XML standard and typically exchanged using the HTTP protocol. A Web service is an abstract interface that needs to be implemented by a software agent. The agent is the concrete part (software or hardware) that sends and receives messages, while the Web service is a resource that contains the abstract set of functionalities provided by the agent. In a practical context, the agent can vary continuously while still maintaining the same functionalities. For example, its implementation might be modified or rewritten in a different language, while the Web service stays unchanged over time. Thus, a Web service is independent from both implementation language and platform type. For it to be used, in general, a service must be described and advertised [15]. The Web Service Description Language (WSDL) specifications (http://www.w3.org/TR/wsdl) have been designed for this purpose. WSDL provides all the details needed to invoke the service operations. Once the WSDL is created, it can be published in a register called Universal Description Discovery and Integration (UDDI) (http://www.uddi.org/pubs/uddi_v3.htm). Using WSDL and UDDI any requester may seek for an appropriate service (i.e., having the desired characteristics) and understand how to invoke it.

The strengths of Web services can be summarized as follows:

- they allow interoperability between different software applications and on different hardware/software platforms;
- they are specified using a text-based data format, which is more understandable and easier to use for developers;
- as they are based on the HTTP protocol, no change is required to the security rules used as firewall filters;
- once published, they can be combined with each other (no matter who provides them and where they are made available) to form integrated complex services;
- they allow reuse of already developed applications;
- as long as the interface remains constant, the changes made to the service remain transparent.

Despite their many strengths, Web services have limitations for which solutions must be found. The tasks of search, selection, composition and execution of Web services are delegated to the developers, since they know the semantics underlying each service. In this way, they can obtain suitable combinations which result in complex applications. In this practice lies the main limitation of the commonly used Web services. Indeed, since Web services were created to be used by machines, delegating to humans such tasks as discovery, selection and composition is too strict a constraint. To be able to automate these steps, machines should be able to understand the semantics of the services. This is achieved by enriching the existing Web services with a semantic layer that expresses the functionality of the service in a machine-understandable way.

The choice of enriching existing resources is clearly important in the context of the Web, where the information architecture cannot be rewritten. The key to overcoming the syntax limitations is provided by the Semantic Web [2], that is purposely aimed at making the Web machine-understandable. Currently, the machines maintain information without understanding its meaning. When a search engine like Google stores Web pages in its cache, it is not able to distinguish whether the word 'espresso' refers to a train or a type of coffee, or if 'verdi' is a color or the name of a composer. With the Semantic Web this limitation is overcome, since the resources in the network are associated with information that specifies their semantics in a format suitable to be interpreted and processed automatically.

2.1   OWL-S and SWRL Characteristics

OWL-S enables semantic descriptions of Web services using the *Service Model* ontology, that defines the OWL-S process model. Each process is based on an IOPR (Inputs, Outputs, Preconditions, and Results) model. The *Inputs* represent the information that is required for the execution of the process. The *Outputs* represent the information that the process returns to the requester *Preconditions* are conditions that are imposed over the *Inputs* of the process and that must hold for the process to be successfully invoked. Since an OWL-S process may have several results with corresponding outputs, the *Results* entity of the IOPR model provides a means to specify this situation. Each result can be associated to a result condition, called *inCondition*, that specifies when that particular result may occur. It is assumed that such conditions are mutually exclusive, so that only one result can be obtained for each possible situation. When an *inCondition* is satisfied, there are properties associated to this event that specify the corresponding output (*withOutput* property) and, possibly, the *Effects* (*hasEffect* properties) produced by the execution of the process. *Effects* are changes in the state of the world. The OWL-S conditions (*Preconditions*, *inConditions* and *Effects*) are represented as logical formulas. Since OWL-DL offers limited support to formulate constructs such as property compositions without becoming undecidable, a more powerful language is required for the representation of OWL-S conditions. For this reason, in OWL-S these logical formulas are represented as simple string literals or XML literals. The latter allows to use languages whose standard encoding is in XML, such as SWRL [7]. Since these are Semantic Web languages, their use overcomes the loss of semantics. Even though SWRL is undecidable, [10] solves the problem by restricting the application of SWRL rules only to the individuals explicitly introduced in the ontology. This kind of SWRL rules, called DL-safe, makes this language the best candidate for representing OWL-S conditions [11]. Let us now briefly mention the features of SWRL that are relevant to our aims. SWRL extends the set of OWL axioms to include *Horn-like* rules in the form of implications between an antecedent (body) and consequent (head), both consisting of conjunctions of zero or more atoms having one of the following forms:

- $C(x)$, with C an OWL class, $P(x, y)$, with P an OWL property
- $sameAs(x, y)$ or $differentFrom(x, y)$, equivalent to the respective OWL properties
- $builtIn(r, z_1, \ldots, z_n)$, functions over primitive datatypes

where x, y are variables, OWL individuals or OWL data values, and r is a built-in relation between $z_1, \ldots, z_n$ (e.g., $builtIn(greaterThan, z_1, z_2)$). The intended meaning can be read as: whenever the conditions specified in the antecedent hold, then the conditions specified in the consequent also hold. A rule with conjunctive consequent can be transformed into many rules having atomic consequents by means of Lloyd-Topor transformations. atomic consequent.

2.2 Related Works

To the best of our knowledge, there are no recent proposals of tools for OWL-S annotation. The Protégé [8] plugin called OWL-S Editor [3] is the most popular tool for the creation of OWL-S descriptions.

The OWL-S tab can be considered as the main point of user interaction, providing a more direct view of the OWL-S classes and instances than what Protégé provides by default. It contains the necessary panels representing all instances of the main OWL-S classes: Service, Profile, Process, and Grounding. Furthermore, it has the following characteristics: (1) WSDL Support to create a 'skeleton' of OWL-S description; (2) IOPR description and management; (3) Graphical Overview of relationships; (4) an integrated execution environment for OWL-S; (5) Process Modeling to model composite processes. This plugin is available only for an old version of Protégé that provides limited support to OWL.

The OWL-S IDE project (http://projects.semwebcentral.org/projects/owl-s-ide) is also concerned with the development of OWL-S services. The OWL-S IDE is a plugin for Eclipse (www.eclipse.org), which attempts to integrate the semantic markup with the programming environment. Developers can write their Java code in Eclipse, and run an ad hoc tool to generate an OWL-S 'skeleton' directly from the Java sources. The idea of integrating SWSs more closely with the programming environment used to develop the service implementations is good. However, Eclipse does not support ontology editing, and there is no KB from which choosing the domain concepts to which the OWL-S files should relate. Furthermore, it is often more useful to generate the semantic markup before the code, as the semantic descriptions can be seen as a higher level of abstraction of the programming modules. The OWL-S IDE does not provide any graphical visualization of services or processes.

Another OWL-S Editor is presented in [13]. It is a stand-alone program, providing WSDL import as well as a graphical editor and visualization for control flow and data flow definition. It is not integrated with an ontology editor and shares some of the drawbacks of the OWL-S IDE.

ODE SWS is a tool for editing SWSs "at the knowledge level" [5]. It describes services following a Problem-Solving Methods (PSMs) approach. The annotation task plays a subordinate role in this environment, whereas a simplified vision of the OWL-S annotation procedure is the main focus of our work.

The most popular tool to manage SWRL rules is SWRLTab, a development environment integrated in Protégé-OWL (http://protegewiki.stanford.edu/wiki/Protege-OWL). It supports the editing and execution of SWRL rules and includes a set of libraries that can be used in rules, including libraries to interoperate with XML documents, spreadsheets, and libraries with mathematical, string, RDFS, and temporal operators. This plugin was developed for Protégé-OWL version 3. While preserving the SWRL semantics, it has some drawbacks: the OWL's open world assumption is not guaranteed, leading, in some situations, to nonmonotonicity. Unlike OWL and SWRL, SQWRL adopts the unique name assumption when querying.

The main difference between these tool and our proposal lies in the fact that they are oriented to knowledge engineers rather than to WSDL developers.

3 Guidelines for a Correct Annotation

With reference to the OWL-S and SWRL characteristics described in Sect. 2.1, we propose the following guidelines for encoding an OWL-S process into SWRL rules. These guidelines provide precise indications to develop a prototype.

1. For every result of the process, there exists an *inCondition* that expresses the binding between input variables and that particular result's (output or effect) variables.
2. Every *inCondition* related to a particular result will appear in the antecedent of each resulting rule, whilst the *Result* will appear in the consequent. An *inCondition* is valid if it contains all the variables appearing in the *Result*.
3. If the *Result* contains an *Effect* made up of many atoms, the rule will be split into as many rules as the atoms. Each resulting rule will have the same inCondition as the antecedent and a single atom as the consequent.
4. The *Preconditions* are conditions that must be true in order to execute the service. Since these conditions involve only the process *Inputs*, they will appear in the antecedent of each resulting rule together with *inConditions*. In this work we consider all the *Preconditions* as being always true.

The first guideline is necessary because there may be processes where the binding is implicit. For example, consider an atomic process having a single output. Due to the single output, an explicit declaration of the inCondition is not necessary. However, if the inCondition is not specified, the second guideline is not applicable. To overcome this problem, we add a new inCondition that makes explicit the implied binding in atomic processes with one output. This inCondition will be represented as an SWRL atom that is always true, and therefore as an OWL property, that will bind input and output explicitly. In addition, in order to represent SWSs using SWRL rules, one needs to represent in them also preconditions, inConditions and effects explicitly declared in the service, that is, the entire IOPR model. Preconditions, inConditions and effects are represented as SWRL logical formulas (i.e., as an antecedent or a consequent). These logical formulas can be combined in order to obtain SWRL rules as follows:

$$\text{Preconditions} \wedge \text{inCondition} \rightarrow \{\text{Outputs}\} \wedge \text{Effect}$$

Finally, a general problem concerns how the Web services are annotated. Web services must be annotated using ontological classes only, without ever using primitive types (string, int, etc.) as their semantics is too general. For example, consider an SWS for searching for books, with the following input:

```
<process:Input rdf:ID="_CAR">
    <process:parameterType rdf:resource="xsd:#string"/>
</process:Input>

*where xsd = http://www.w3.org/2001/XMLSchema
```

Due to the fact that `process:parameterType` is declared as a datatype, the class of this input is an XML Schema datatype (string) instead of being an entity belonging to the domain ontology. This is critical problem during SWS composition. In fact, suppose that we need to find a service whose output type is the same as the input type defined in the example. Since such an input is of type string, any service returning a string can be composed with that SWS. In this way, the final result might be incorrect due to the semantics of the primitive types. The problem is solved by annotating the parameters of the input and output using ontological classes without ever using primitive types. Referring to the previous example, the input parameter should annotated with the class *Car* of an ontology describing vehicles.

```
<process:Input rdf:ID="_CAR">
    <process:parameterType rdf:resource="&kb;#Car"/>
</process:Input>

*where &kb; is the ontology URI of Car
```

Figure 1 shows the correct placing and relations between Web service and SWS.

4   OWL-S Semantic Annotator

To ensure proper annotation of Web services and a proper division of tasks among different users, we have developed a tool that allows *ad hoc* annotation using only the classes in the domain ontologies, thereby taking into proper consideration the dichotomy between the knowledge engineer and the developer of Web services. Thus, the developer of Web services will be able to annotate Web services without necessarily knowing the Semantic Web ontology and rule languages. Moreover, in a context where there are various professional roles, each in

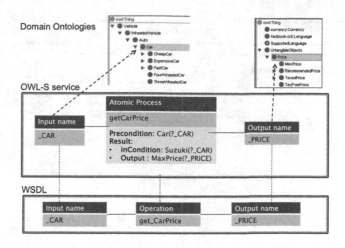

Fig. 1. SWS abstraction layer.

charge of specific tasks, a clear division of tasks must be ensured. For example, there may be users in charge of annotating Web services and others that must release them after checking. Once Web services are annotated, it is necessary to ensure their persistence in such a way that reusability is guaranteed. In this way, it will be possible to search for existing OWL-S descriptions through the Web before creating a new one. At the end of this section an abstract solution to manage OWL-S descriptions will be proposed. To ensure that every user has a defined role that allows him to perform only certain operations, the following subdivision of roles was identified:

1. Annotators: can only create the OWL-S descriptions;
2. Publishers: can also publish the OWL-S descriptions;
3. Group Leaders: can assign permissions (Annotator, Publisher, Group Leader) to existing users, edit their data (username and password) or create new users;
4. Administrators: can perform all the operations of group leaders, and also manage the settings for the connection to the repository for the publication of OWL-S descriptions.

To be published, an OWL-S description requires a common Web server that can potentially be deployed anywhere on the Web. Once it has been created, a mechanism is needed that allows the localization of the resource on a physical machine which may be different from the one where the OWL-S annotator works. Currently, our prototype uses the File Transfer Protocol (FTP) (http://www. w3.org/Protocols/rfc959/) for this purpose.

4.1 A Prototype Implementation

In this section we present the details about the annotation procedure of the OWL-S annotator. Starting from a WSDL document, the user can create an

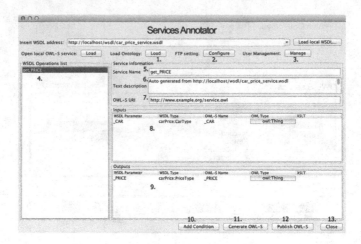

Fig. 2. OWL-S annotator main page.

OWL-S service by simply choosing an OWL domain ontology and annotating every single parameter of the service with an ontology class selected from the ontological class hierarchy (i.e., the taxonomic view of the ontology). Furthermore, once inputs and outputs have been annotated, the user can declare logical conditions corresponding to preconditions, inConditions, and Effects in an intuitive manner. The interface shown in Fig. 2 allows to:

a. enter the URL of the WSDL of the Web service to be annotated;
b. select a local WSDL document to be annotated;
c. load a previously created OWL-S description for editing purposes.

After loading a WSDL or an existing OWL-S description, the functionality of the various numbered items in the graphical interface is as follows:

1. allows to choose the domain ontologies to be used for annotating the parameters;
2. allows to configure the default settings for the publication of OWL-S (this option is available only to the Administrator);
3. is a button to access the panel that allows to manage users, register new ones, change their username and password and assign a set of permissions;
4. contains the list of all operations in the selected WSDL (for each of these operations an OWL-S atomic service can be created);
5. reports the name of the OWL-S service;
6. reports a text description of the OWL-S service;
7. reports the URI of the OWL-S service;
8. is the list of input parameters (for each of them, the name in OWL-S can be changed and an annotation can be made by clicking in the 'OWL class' column);

9. is the list of output parameters (for each of them, the name in OWL-S can be changed and an annotation can be made by clicking in the 'OWL class' column);
10. is a button to access the panel for entering SWRL logical constructs;
11. is a button for generating the OWL-S description;
12. is a button for publishing the OWL-S description on the Web (this option is not permitted to Annotator users);
13. is a button to close the application.

4.2   SWRL Annotation

Once all the service parameters have been annotated, one may create the service or continue with the addition of SWRL logical formulas by clicking on the 'SWRL constructs' panel. More specifically, it:

1. includes a Tab to enter preconditions;
2. includes a Tab to enter output, effects, and under what conditions they occur;
3. can be used to insert an OWL class in preconditions, selecting it from a tree view of the taxonomy (as in the case of record of the parameters of the service);
4. can be used to insert an OWL Property in preconditions.
5. after selecting a class or property, allows to enter the process parameter that will be the argument of the selected class or property;
6. after selecting a class or property, allows to enter the individual who will be the argument of the selected class or property;
7. displays in a comprehensible way all atoms entered for the SWRL construct;
8. allows to cancel the atoms entered for the SWRL construct;
9. allows to add the SWRL construct to the process;
10. contains a complete list of all the SWRL constructs entered in the process;
11. allows to save the changes.

By opening the 'Output/Effects' panel, the window shown in Fig. 3 is displayed. It consists of a panel that includes three tabs. These tabs allow to specify 'Output', 'Effects' and 'Conditions', respectively. In the Output panel the user can select the output to be returned, by clicking on the 'Output' button. Then, if the process returns multiple outputs, he will be asked to select which output must be returned. The elements contained in tabs 'Effects' and 'Conditions' are the same as those appearing in the tab 'Preconditions'. The buttons carry out the same functions as those described for the Preconditions tab, as well. The difference lies in the fact that by selecting some items from the Preconditions tab they will be added as preconditions of the process, while selecting them from the Effects tab they will be added as the effects of the process, and selecting them from the Conditions tab they will be added as necessary conditions to return specific effects and/or specific output.

Let us now show an example of insertion. Our SWRL construct will contain two conditions that are combined in order to form a logical condition. When these

Fig. 3. The "Output/Effects" panel.

conditions are verified, the returned output belongs to a particular OWL class. Specifically, suppose the user wants to define that "the price of Suzuki cheap cars the service will return must be of type RecommendedPrice" (an OWL class). Then he must define the following two conditions on the input parameter _CAR:

- Suzuki(?_CAR),
- CheapCar(?_CAR).

To define these conditions he opens the tab 'Output and Effects', selects the tab 'Conditions' and clicks on the button to select the class 'Suzuki'. Once the class is selected, he clicks on the button 'Variable' and selects the parameter. He repeats the same procedure for the second condition, this time selecting the class 'CheapCar'. Now he must define the fact that the output returned by the service will be of type 'RecommendedPrice(?_CAR)'. To do this, he selects the tab 'Output' and, after clicking on the 'Output' button, he selects the class 'CheapCar'. The entered SWRL construct will be displayed as shown in Fig. 4.

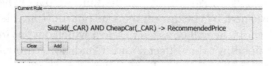

Fig. 4. SWRL construct display panel.

Through this view he may have a summary of the SWRL atoms included in the construct. After entering all the constructs he deems as appropriate, he can go back to the main screen and generate the OWL-S service he has just created by means of the button 'Generate OWL-S'. Later, he will also be able to publish it on the Web using the button 'Publish'.

5   Automatic Elicitation of Pre-condition

Developers might still experience some difficulties to set the conditions of the OWL-S process model using the proposed tool. This is because WSDL does not define the conditions of the OWL-S process model. Furthermore this task requires knowledge of formal languages, such as SWRL, that do not allow intuitive representations like the taxonomy used to create the parameter annotations. To this end, it might be useful to automatically extrapolate and suggest OWL-S process model conditions starting from the text in natural language included in the WSDL. This requires an understanding of which parts of the WSDL contain meaningful. In particular, two types of features are considered: the parameter names and the natural language text used to document the Web service itself. In the next subsection, in addition to the details about these features, three repositories of services are compared in order to give an idea of what happens in practice.

5.1   WSDL and OWL Meaningful Text

As described in the language specification (http://www.w3.org/TR/wsdl), WSDL uses the optional *documentation* element information item as a container for human-readable or machine-processable documentation. The content of the element information item consists of arbitrary character information items and element information items ("mixed" content in XML Schema). The documentation element information item is allowed inside any WSDL element information item, i.e., service, operations, inputs and outputs. Like other element information items in the WSDL namespace, the *documentation* item allows qualified attribute information items with different namespace. Furthermore, the xml:lang attribute can be used to indicate the language used in its content. All the tools to write WSDL intuitively lead the Web service developer to give significant names to the input and output of the service. In this way it is easier to manage and maintain the service. Since the specifications do not impose the use of meaningful names, it is reasonable to perform a check to see how often this good practice is followed.

With this aim, we have performed an analysis to see if and how the element *documentation* and meaningful parameter names are used in practice. To this end, we have considered a sample made up of several WSDL files, possibly of different application domains, from the following repositories available on the Web:

– WebserviceX.NET (http://www.webservicex.net). It is a repository of published Web Services in which the services are divided into categories for easy reference. For each category there is a list of Web Services with their name and a brief description.
– WebServiceList (http://www.webservicelist.com). The site looks like a long list of Web Services with a short description and a rating for each. Web services are organized into categories so as to facilitate their search.

– XMethods (http://www.xmethods.com). It is a repository of Web services that allows also the publication of third-party services. For each service there is a brief description, an indication of the used technology and the publisher info.

From each of them, a sample of WSDL services was taken. From the first repository, due to the division of services by application domain, we have considered a subset of services for each domain. On the contrary, in the second and third repositories services are presented in a single structured list, from which we have drawn at random. For our analysis we have considered 19 documents for each repository. All the parameters have meaningful names, while 78.57 % of *documentation* elements have a text description in WebserviceX.NET, 60.27 % in WebServiceList and 84.31 % in Xmethods.

These results show that in most cases the textual description of the service (or of its individual operations) is given. Since the specific "description" is an XML element that might be prepared to indicate structured information on the operation including the indication of the language in which they are written, we can assume that it is possible to structure the information in order to specify constraints. Furthermore, in all the analyzed cases the parameters have a meaningful name and this suggests that the developer has in mind what is their meaning. For this reason we can suppose that a textual description of parameters can be given by the developer. Then, the corresponding XML structure from which to extract the text can be described as follows:

```
<wsdl:operation name="[operation name]">
    <wsdl:documentation lang=eng>
       [text describing the functionality]
       <conditions>
          [text describing the constraints]
       </conditions>
    </wsdl:documentation >
    <wsdl:input name="[input name]">
       <wsdl:documentation lang=eng>
          [text describing the input]
       </wsdl:documentation>
    </wsdl:input>
    . . .
    <wsdl:output name="[output name]">
       <wsdl:documentation lang=eng>
          [text describing the output]
       </wsdl:documentation>
    </wsdl:output>
    . . .
</wsdl:operation>
```

On the other side, OWL ontologies inherit from RDF a mechanisms to describe in natural language their resources. These textual comments (identified by rdfs:comment) help to clarify the meaning of classes and properties. Such in-line documentation complements the use of both formal techniques (Ontology and rule languages) and informal ones (prose documentation, examples, test cases). A variety of documentation forms can be combined to indicate the intended meaning of the classes and properties described in an RDF vocabulary. Since RDF vocabularies are expressed as RDF graphs, vocabularies defined in other namespaces may be used to provide richer documentation. Furthermore, the tag rdfs:label may be used to provide a human-readable version of a resource's name (i.e., at OWL class "A102" we can associate the label "MicrobioticGene"). It is important to note that OWL properties can have various characteristics that can be useful to interpret their textual descriptions, in particular they *Transitive* (i.e., $P(x,y)$ and $P(y,z)$ implies $P(x,z)$), *Symmetric* (i.e., $P(x,y)$ iff $P(y,x)$), *Functional* (i.e., $P(x,y)$ and $P(x,z)$ implies $y = z$), *Inverse* (i.e., $P1(x,y)$ iff $P2(y,x)$) and *InverseFunctional* (i.e., $P(y,x)$ and $P(z,x)$ implies $y = z$). For our aims, we may assume that ontologies with a rich set of textual comments (e.g., rdfs:comment), meaningful resource names (e.g., rdfs:label), and a complete specification of property characteristics have been used during the manual annotation task.

5.2   ConNeKTion

ConNeKTion (acronym for CONcept NEtwork for Knowledge representaTION) [12] is a tool for concept network learning, consultation and exploitation. As to the learning task, working on plain text it is able to identify the network of concepts and relationships underlying a document collection, and to further structure and enrich this network, also in the presence of missing or partial knowledge (a typical problem in small collections). It also provides functionalities to exploit the learned network in order to partially simulate some human abilities in text understanding and concept formation. It adopts both propositional and relational concept descriptions, that allow to handle different levels of complexity and expressiveness in concept representation. The current prototype works on English, but the underlying methodologies are completely general and applicable to any language.

ConNeKTion can be useful to obtain a lexical ontology when existing resources do not exist, or the existing solutions do not provide sufficient quality or reliability. For instance, consider the state-of-the-art ontology WordNet [6]. It has several shortcomings. It is general, so it does not cover satisfactorily specific branches of knowledge or the topics of specific collections. It is maintained manually, which introduces errors and causes new versions not to be backward compatible. Furthermore, the original WordNet is available only for English; extensions and similar resources for some other languages have been developed (e.g., ItalWordNet for Italian, MultiWordNet for cross-language applications, WordNet Domains for associating words to categories, SentiWordNet for the sentiment domain, etc.), but often do not have the same quality, and for some languages a similar resource is not available at all (at least, not for free).

The main feature of ConNeKTion of interest for the present work is its ability to process a corpus of texts in natural language and to build from it a concept network. Its basic assumption is that nouns express concepts, and that a concept can be defined by (a) the set of other concepts that interact with it in the world described by the corpus, and (b) the set of its associated properties and attributes, expressed by verbs and adjectives. This knowledge can be formalized as a graph, where nodes are the concepts/nouns appearing in the corpus, and edges represent the relationships expressed by the grammatical structure of the text (the direction of edges denoting the role of the nodes in the relationship). The resulting network can be considered as an intensional representation of the corpus. Translating it into a suitable First-Order Logic (FOL) formalism enables the subsequent exploitation of logic inference engines in applications that use that knowledge.

We propose to exploit this feature in order to extract SWRL conditions on the WSDL parameters that can be interpreted as OWL-S pre-conditions, as outlined in the following.

5.3   The Proposed Approach for SWRL Rules Elicitation

As a first step we need to create the conceptual network through ConNeKTion. To do this we need a corpus of texts in natural language. Based on the previous assumptions, we use:

- the WSDL parameter names and all the text associated with a set of services related to a domain;
- the text of the tag *rdfs:comment* associated to OWL classes and properties and their *rdfs:label* to get meaningful names.

After building the conceptual network, ConNeKTion allows to express it in a FOL representation consisting of a conjunction of logical atoms, where each atom denotes a concepts, a property of a concept or a relationships between concepts. For each concept or relationship, a formal definition can be obtained as a portion of this conjunction, including the subset of atoms that lie in a neighborhood of specified radius, centered in the concept or relationship under consideration. For instance, a portion of the definition of concept 'book()' obtained with radius 2 might turn out to be:

```
has\_price(X), write(Y,X), author(Y), read(Z,X), student(Z),
    attend(Z,W), course(W)
```

where 'Book', 'Author', 'Student' and 'Course' might be concepts in the DL ontology, and has_price, write, attend might correspond to properties hasPrice, hasAuthor, hasCourse in the DL ontology. Among all the concepts in the network extracted by ConNeKTion, we are interested in the definition of those concepts corresponding to the parameter names in the WSDL, which in turn correspond to OWL classes in which they were annotated. Therefore, we can say that these FOL representations express relationships between OWL concepts. With reference to the characteristics of the OWL-S conditions (see Sect. 2.1), our interest is focused on those that bind the inputs to the outputs also indirectly (e.g., $R1(Input1, X)$ and $R2(X, Y)$ and $R3(Y, output1)$). So, the last step is to map these relationships with existing OWL properties in the used ontologies. To do this, the concept network learned by ConNeKTion might be exploited again, but with a different perspective such that now verbs (denoting relationships) become the focus, and are described in terms of the nouns (i.e., concepts) that may play some role in the application of those relationships. These relationships are to be associated to corresponding properties in the DL ontology. The match of specific relationships with the selected OWL property names can be obtained with the application of classical NLP techniques (e.g., from "Distribute" one may obtain related word such as "distributed", "distributor" etc.) combined with the indications coming from OWL property characteristics (the inverse property of "isDistributedBy" is "hasDistributor").

In this way we obtain SWRL logical formulas that could be used in the proposed annotation tool in order to complete the generation of the OWL-S service with Precondition, inCondition and Effects. In particular due to the mapping with the terms of ConNeKTion, a prompter in natural language can be integrated in the 'SWRL constructs' panel allowing to web service developer to specify conditions without necessarily knowing SWRL.

6   Conclusions and Future Works

This work aimed at providing a tool for manual annotation of Web services, that would take into account the strengths and weaknesses of Semantic Web technologies. We started from the description of Web services, by specifying their meaning within the Web community and describing the WSDL, the de facto XML-based standard that allows to abstract away from the concrete implementation of the service. Then we analyzed the syntactic limitations of Web Services, and specifically their constraining human experts to search, select, compose and execute the Web Service. The key to overcoming the syntax limitations is provided by the Semantic Web. In particular, OWL allows to specify formal ontologies that can be used to attach a meaning to WSDL inputs and outputs. OWL-S is an OWL ontology that models Web services at the abstract level by simplifying the semantic association to the input parameters and output. In order to specify the process model of a service, rule based languages are necessary. As we have seen, OWL-S allows to specify logical constructs in SWRL, an extension of OWL, which enables the description of preconditions and effects. The implemented tool following the given guidelines for a correct annotation that can be used by a Web service developer that is not an expert in Semantic Web technologies. An interesting approach is proposed to facilitate the formulation of the conditions by means of ConNeKTion. As future works we plan to extend the ontology management tab in order to further facilitate the developer during the annotation procedure, and to introduce the management of OWL-S composite processes, a validator that will allow to check the correctness of the descriptions before generating the OWL-S service, and a graphical tab to visualize the complex services. Furthermore, and most important, we plan to integrate ConNeKTion in order to investigate strengths and weaknesses about SWRL rules elicitation applying the proposed approach.

Acknowledgments. This work was partially funded by the Italian PON 2007-2013 project PON02_00563_3489339 'Puglia@Service'.

References

1. Baader, F., Calvanese, D., McGuinness, D., Nardi, D., Patel-Schneider, P. (eds.): The Description Logic Handbook. Cambridge University Press, Cambridge (2003)
2. Berners-Lee, T., Hendler, J., Lassila, O.: The semantic web. Sci. Am. 29–37 (2001)
3. Elenius, D., Denker, G., Martin, D., Gilham, F., Khouri, J., Sadaati, S., Senanayake, R.: The OWL-S editor – a development tool for semantic web services. In: Gómez-Pérez, A., Euzenat, J. (eds.) ESWC 2005. LNCS, vol. 3532, pp. 78–92. Springer, Heidelberg (2005)
4. Erl, T.: Service-Oriented Architecture: Concepts, Technology, and Design. Prentice Hall PTR, Upper Saddle River (2005)
5. Gómez-Pérez, A., González-Cabero, R., Lama, M.: Development of semantic web services at the knowledge level. In: (LJ) Zhang, L.-J., Jeckle, M. (eds.) ECOWS 2004. LNCS, vol. 3250, pp. 72–86. Springer, Heidelberg (2004)
6. Hirst, G., St-Onge, D.: Lexical chains as representations of context for the detection and correction of malapropisms. In: WordNet: An Electronic Lexical Database, pp. 305–332 (1998)
7. Horrocks, I., Patel-Schneider, P.F., Bechhofer, S., Tsarkov, D.: OWL rules: a proposal and prototype implementation. J. Web Semant. **3**(1), 23–40 (2005)
8. Knublauch, H., Fergerson, R.W., Noy, N.F., Musen, M.A.: The Protégé OWL plugin: an open development environment for semantic web applications. In: McIlraith, S.A., Plexousakis, D., van Harmelen, F. (eds.) ISWC 2004. LNCS, vol. 3298, pp. 229–243. Springer, Heidelberg (2004)
9. McIlraith, S.A., Son, T.C., Zeng, H.: Semantic web Services. IEEE Intell. Syst. **16**(2), 46–53 (2001)
10. Motik, B., Sattler, U., Studer, R.: Query answering for OWL-DL with rules. J. Web Semant. Sci. Serv. Agents World Wide Web **3**(1), 41–60 (2005)
11. Redavid, D., Ferilli, S., Esposito, F.: Towards dynamic orchestration of semantic web services. In: Nguyen, N.-T., Kołodziej, J., Burczyński, T., Biba, M. (eds.) Transactions on Computational Collective Intelligence X. LNCS, vol. 7776, pp. 16–30. Springer, Heidelberg (2013)
12. Rotella, F., Leuzzi, F., Ferilli, S.: Learning and exploiting concept networks with ConNeKTion. Appl. Intell. **42**(1), 87–111 (2015)
13. Scicluna, J., Abela, C., Montebello, M.: Visual modelling of OWL-S services. In: IADIS International Conference WWW/Internet (2004)
14. Staab, S., Studer, R. (eds.): Handbook on Ontologies. International Handbooks on Information Systems. Springer, Heidelberg (2004)
15. Walsh, A.E. (ed.): UDDI, SOAP, and WSDL: The Web Services Specification Reference Book. Prentice Hall Professional Technical Reference, Englewood Cliffs (2002)

Interview Based DEMO Axioms' Benefits Validation: A Regional Government Case Study

David Aveiro[1,2,3](✉) and Duarte Pinto[1]

[1] Exact Sciences and Engineering Centre, University of Madeira,
Caminho da Penteada, 9020-105 Funchal, Portugal
daveiro@uma.pt, duartenfpinto@gmail.com
[2] Madeira Interactive Technologies Institute,
Caminho da Penteada, 9020-105 Funchal, Portugal
[3] Center for Organizational Design and Engineering,
INESC-INOV, Rua Alves Redol 9, 1000-029 Lisbon, Portugal

Abstract. This paper has as its background, a practical enterprise change project where the Design and Engineering Methodology for Organizations (DEMO) was used in the initial stage as to give a neutral and concise but comprehensive view of the organization of a local government administration in the process of implementing an e-government project. The main contribution presented in this paper is an interview based qualitative validation of some of DEMO's axioms and claimed benefits – something that, to our knowledge has never been done up to now. Namely, we were able to validate DEMO's qualities of conciseness and comprehensiveness brought about by the transaction and distinction axioms and also the stabilty of its ontological models which are, by nature, highly abstracted from the human and technological means that implement and operate an organization.

Keywords: Enterprise engineering · Enterprise change · DEMO · Case study · Validation

1 Introduction

This paper has as its background a practical enterprise change project where the Design and Engineering Methodology for Organizations (DEMO) was used in the initial stage with the purpose to give a neutral and concise but comprehensive view of the organization of a local government administration having, itself, the purpose to implement an e-government project. This administration is present in a small island of a European archipelago that is dependent on a main island that has its own autonomous regional government. We chose to apply DEMO in the project, due to its growing use in projects in Europe and purported qualities and benefits given by the method. Such qualities highly fitted our work context which had the need to harness the huge complexity of the government administration target of the project. The fact that, as far as we are aware of, no academic study (qualitative or quantitative) has ever been made to validate DEMO's qualities gave rise to the idea of realizing research presented in this

© Springer International Publishing Switzerland 2015
A. Fred et al. (Eds.): IC3K 2014, CCIS 553, pp. 421–443, 2015.
DOI: 10.1007/978-3-319-25840-9_26

paper. This small local government administration – from now on referred to as SLGA – is a kind of "miniature" replica of almost all government functions from national to regional level and – thanks to having so many functions concentrated in a few persons – was chosen to be a test pilot for the e-government project, later to be extended to all government entities of the main island. This project has three main aspects: (1) the implementation of a work flow system to simplify and automate many operational processes currently paper based and/or – although using Word/Excel documents – lacking in structure and coherence; (2) the development of an online portal to automate as much as possible the interactions and services currently provided at a local physical Citizen Service Desk (CSD), so that the citizens can initiate such interactions in the comfort of their homes; and (3) the development of an IT integration layer with other regional and national government entities that end up executing most of the processes. In this context, our research team was assigned with the responsibility of applying DEMO to model the processes, interactions and information flows occurring in the SLGA, to be used as a base for the production of a strategic roadmap of organizational changes that will have to occur for several alternative scenarios of e-government implementations, according to the possible levels of integration and change in current government entities and/or their IT systems. Our team comprised 4 DEMO experts, 2 working in the project full-time and 2 part-time – one 50 % and the other 25 % – totaling 55 man-days in a month of project execution. Many interviews were made to officials head of each of the SLGA's departments and also to most of the officials responsible for each unit of each department. Interviews were made both for information collection and model validation. A final global workshop with the presence of all interviewees was made for final validation where most models were deemed adequately correct and complete after some small corrections and additions. In the end we specified: 216 transactions – and their associated result types; and 232 fact types – these include classes/categories and fact types and exclude properties. We additionally specified 250 ontological transaction kinds that followed a certain repetitive pattern in certain departments and, because of that, were abstracted into a small subset of generic transactions of the above mentioned 216 transactions set. So, in fact, we specified almost 500 transaction kinds in this project.

Ten months after our main project activities summarized above, we decided to conduct another round of interviews having, as the main purpose, a qualitative evaluation of DEMO's qualities of conciseness and comprehensiveness brought about by the transaction and distinction axioms and also the stability of its ontological models. We took the opportunity to re-validate all previously collected data, and update existing models in case of organizational changes. Very few changes and/or corrections were needed demonstrating the stability of the DEMO models which are, by nature, highly abstracted from the human and technological means implementing and operating an organization. The qualities of conciseness and comprehensiveness were also validated by the vast majority of the interviewees. Regarding the interviews and their analysis, we used a qualitative research method, collecting the data with a previously conceived set of questions specific for this case, most open ended but with short answers. The outcomes in most questions were mostly as expected but there were, however, some peculiar answers.

In the remainder of this paper, Sect. 2 presents our Motivation, problem and research method. In Sect. 3, we present a brief introduction of DEMO - Operation, Transaction and Distinction Axioms. Section 4 has our Case details and Example including some models of this case study. Section 5 explores the Interview questions and results based on our experience and states the intentions behind each question. Section 6 wraps it up with a Results analysis and evaluation, and finally, in Sect. 7, we present our Conclusions.

2   Motivation, Problem and Research Method

We frame our motivation and research method in the Design Science Research paradigm [1, 2] which claims that all design science research should take in account the three cycles presented in Fig. 1.

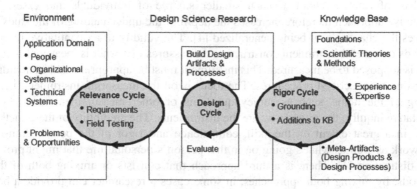

Fig. 1. Design science research cycles.

Regarding the relevance cycle, the motivation of this study is the following problem: it is claimed in [3] that DEMO possesses several qualities but no formal proofs or studies are provided that validate such claims. So our purpose was to validate DEMO's qualities of conciseness, comprehensiveness and stability of the ontological models as to bring more weight and value in practice to this method and associated theories. As for clear definitions of these qualities, we adopt the ones from [3]. Namely by conciseness we mean that no superfluous matters are contained in it, that the whole is compact and succinct [3]. That is, models should provide a view containing the essence that is a global picture of an organization out of which all details can be properly specified. Comprehensiveness implies that all relevant issues are covered, that the whole is complete [3]. That is, all relevant perspectives like the dynamic and static aspects of operation, human responsibilities, operation flow and inter-dependencies should be clearly understandable and covered by the models. Stability of the ontological models is supposedly guaranteed by the implementation independence of DEMO models. And by implementation it is understood the assignment of human and/or computer resources to operationalize an organization [3].

Looking at the rigor cycle we ground our study on the sound formal theories behind DEMO and aim to provide expertise to the Knowledge base while contributing with a validation case study.

In respect to the design cycle, the research reported in this paper aims to apply the DEMO artifact itself and evaluate its claimed qualities by means of interviews with key collaborators on the organization target of study. Regarding such evaluation, qualitative methods can facilitate the study of issues in both depth and detail. They do not have the constraints of predetermined categories of analysis therefore allowing for a bigger depth, openness and detail in the inquiry. On the other end we find that quantitative methods require standardized measures so that the varying perspectives and experiences can fit in a limited number of predetermined response categories to which numbers are assigned [4]. The main advantage of a quantitative approach is the possibility to measure reactions of a large amount of people to a limited set of questions, therefore making comparisons and statistical aggregations of data easier, and allowing for it to be presented in a succinct way [4]. The qualitative approach produces far more detailed information about a much smaller sample of individuals and cases. The qualitative approach therefore increases the depth of the understanding of the study but reduces the chances of it being generalized [4]. The validity of a quantitative research depends on careful instrument construction that assures that what is measured is really what is supposed to be measured. This instrument must be appropriate and standardized according to prescribed procedures. The focus is on the measuring instrument, i.e. the testing of the items, such as survey questions or other measurement tools. In the qualitative inquiry, the researchers are the instruments. The credibility of these methods hinge in a great extent on the skill, competence and rigor of the person doing the fieldwork as well as what's going on in that person's personal life that might prove to be a distraction [4]. There is a third approach that consists on mixing both of these methods, by mixing both approaches, in some cases a researcher can provide a better understanding of the problem not using either the quantitative or qualitative methods alone [5]. A research using this third approach is usually named a mixed methods research and can be defined as the "class of research where the researcher mixes or combines quantitative and qualitative research techniques, methods, approaches, concepts or language into a single study" [6]. By using a mixed methods research, the researchers can provide more comprehensive evidence than either quantitative or qualitative research alone. Thus the researchers are given permission to use all tools of data collection available rather then being restricted to the types of data collection associated with either of the methods alone [5]. Given the dimension of the SGLA target of our project and analysis we have chosen a qualitative method approach as we were limited to a small amount of subjects having the knowledge about the modeled processes.

As previously mentioned to achieve this evaluation we opted to interview the key collaborators involved, using a standardized open-ended format that although lacking flexibility still allowed us the use of open ended questions while facilitating their analyzes furthermore the generalization of the results [4] The interview method used can also be framed in the seven stages of an interview investigation proposed in [7].

1. Thematizing: formulation of a purpose of the investigation and description of the topic being investigated before starting the interviews – in this case the purpose was the validation of the DEMO's axioms in terms of the qualities of conciseness and comprehensiveness and also the stability of its ontological models. We wanted also to evaluate the interview method itself. To achieve this we specified several key points that the interviews should cover, namely: (1) the duration of the interviews, (2) ability by the collaborators to answer the questions in the initial stage of the project both in the terms used by the interviewers and their knowledge of what was being asked for, (3) their opinion on the interview methodology, (4) their current view on the processes and eventual changes, (5) their perception of the modeled workflow and ability to relate to the real workflow in operation, (6) the names used in the models, either in the organizational functions or the transactions, (7) their self knowledge of the organization, (8) the models and their correspondence to current reality after almost a year passed and (9) questions regarding the application of the DEMO methodology and benefits obtained thanks to its axioms.

2. Designing: planing the study taking in consideration all the stages before the interviews take place – to achieve this we devised a set of 43 questions that met the criteria set in the thematizing as to approach all those 9 subjects, being most of them open ended, but with the expectancy of rather short answers considering the extent of the subjects being inquired.

3. Interviewing: conducting the interviews based on a guide and with a reflective approach considering the desired knowledge – the round of interviews was conducted with eleven SLGA collaborators, ten that had been previously interviewed, all either head of a department or chief of a division, and one that, although not previously interviewed, was now the current head of the human resources department. Interviews took place individually and were composed by the previously mentioned set of 43 questions, placed after a re-validation of the models that had been created for the interviewee's department. Meetings were previously scheduled and normally had a duration of approximately one hour for all heads of department, and two hours for the two chiefs of division so that they could give their input on all the several departments that they are responsible for.

4. Transcribing: preparing the interview results for analysis; commonly translating oral speech into written text – all answers were written down and later organized into a spreadsheet containing all participants together with the list of questions.

5. Analyzing: deciding, considering the purpose of the interview and the interview material, what methods are appropriate for analysis – in order to facilitate analysis, our answer data was grouped in sets according to common-theme questions. We then studied the outcomes of each of those sets taking in account the devised goals. All answers were also analyzed individually for particularities and properly considered in the presented results.

6. Verifying: ascertain the generalizability, reliability, and validity of the interview findings i.e. the possibility to apply the results in other contexts, the consistency of the results, and if the study meets the intended purpose – the findings of our research are presented in chapter 6 Results analysis and evaluation as also are the considerations relating to those findings.

7. Reporting: communicate the findings and the methods applied in a form that lives up to scientific criteria, while taking the ethical aspects of the investigation into consideration, and that having the results in a readable and usable product for its audience – in our case, to communicate our findings we are using this paper, presenting the background, contextualization and outcomes as well as a description of the process used.

3   DEMO - Operation, Transaction and Distinction Axioms

In the Ψ-theory [8] – on which DEMO is based – the operation axiom [3] states that, in organizations, subjects perform two kinds of acts: production acts that have an effect in the production world or P-world and coordination acts that have an effect on the coordination world or C-world. Subjects are actors performing an actor role responsible for the execution of these acts. At any moment, these worlds are in a particular state specified by the C-facts and P-facts respectively occurred until that moment in time. When active, actors take the current state of the P-world and the C-world into account. C-facts serve as agenda for actors, which they constantly try to deal with. In other words, actors interact by means of creating and dealing with C-facts. This interaction between the actors and the worlds is illustrated in Fig. 3. It depicts the operational principle of organizations where actors are committed to deal adequately with their agenda. The production acts contribute towards the organization's objectives by bringing about or delivering products and/or services to the organization's environment and coordination acts are the way actors enter into and comply with commitments towards achieving a certain production fact [11].

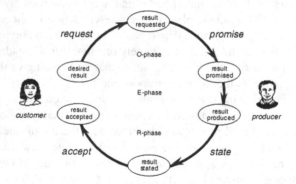

Fig. 2. Basic transaction pattern.

According to the Ψ-theory's transaction axiom the coordination acts follow a certain path along a generic universal pattern called transaction [3]. The transaction pattern has three phases: (1) the order phase, were the initiating actor role of the transaction expresses his wishes in the shape of a request, and the executing actor role promises to produce the desired result; (2) the execution phase where the executing actor role produces in fact the desired result; and (3) the result phase, where the executing actor role states the produced result and the initiating actor role accepts that

COORDINATION   ACTOR ROLES   PRODUCTION

Fig. 3. Actors interaction with production and coordination worlds.

result, thus effectively concluding the transaction. This sequence is known as the basic transaction pattern, illustrated in Fig. 2, and only considers the "happy case" where everything happens according to the expected outcomes. All these five mandatory steps must happen so that a new production fact is realized. In [11] we find the universal transaction pattern that also considers many other coordination acts, including cancellations and rejections that may happen at every step of the "happy path".

Even though all transactions go through the four – social commitment – coordination acts of request, promise, state and accept, these may be performed tacitly, i.e. without any kind of explicit communication happening. This may happen due to the traditional "no news is good news" rule or pure forgetfulness which can lead to severe business breakdown. Thus the importance of always considering the full transaction pattern and the initiator and executor roles when designing organizations [11].

The distinction axiom from the Ψ-theory states that three human abilities play a significant role in an organization's operation: (1) the forma ability that concerns datalogical actions; (2) the informa that concerns infological actions; and (3) the performa that concerns ontological actions [3]. Regarding coordination acts, the performa ability may be considered the essential human ability for doing any kind of business as it concerns being able to engage into commitments either as a performer or as an addressee of a coordination act [11]. When it comes to production, the performa ability concerns the business actors. Those are the actors who perform production acts like deciding or judging or producing new and original (non derivable) things, thus realizing the organization's production facts. The informa ability on the other hand concerns the intellectual actors, the ones who perform infological acts like deriving or computing already existing facts. And finally the forma ability concerns the datalogical actors, the ones who perform datalogical acts like gathering, distributing or storing documents and or data. The organization theorem states that actors in each of these abilities form three kinds of systems whereas the D-organization supports the I-organization with datalogical services and the I-organization supports the B-organization (from Business = Ontological) with informational services [10]. By applying these axioms, DEMO is claimed to be able to produce concise, coherent and complete models with a reduction of around 90 % in complexity, compared to traditional approaches like flowcharts and BPMN [9].

4 Case Details and Example

The SLGA currently has two divisions (had three at the time of the first round of interviews in the beginning of our project) which include ten main departments. The first of those two is the Division of Natural Resources Management (DNRM) that

includes the Veterinary, Fish, Parks and Agriculture departments, all with a collaborator in charge, being that the chief of the DNRM division is also in charge of the
Veterinary department. The second division is the Division of Administration Finances,
Maintenance and Infrastructure Management (DAFMIM) and comprises the departments of Human Resources, Supply, Finance, Fleet, Maintenance and the Citizen
Service Desk, each also with a different collaborator in charge. Each of these departments deals with specific different aspects of the SLGA. For example, the Veterinary
department has the only available veterinarian on the island and deals mostly with farm
animals health, safety and well-being, and food safety issues regarding animal based
food and animal food itself. The Fish department makes the bridge between the local
fisherman and the local commerce but also deals with matters related with fishing boats
diesel oil, the selling of ice to local businesses and cold storage units rental. The
Gardens and Ornamental Plants department deals with the public property gardens of
the island, their planning, execution and maintenance, as well as all the other public
green spaces. Besides this, a small portion of their work is also dedicated to the
propagation of flowers for local events, and when there is surplus the sale of those
flower. Finally in the DNRM the Agricultural Station makes the bridge between local
farmers and the regional authorities in the main island for any needed supplies as well
as supplies them with heavy farming machinery such as tractors. Besides the work with
the local farmers the Agricultural Station has a production of itself mostly based on
grapes for exhibitions either local or in the main island, as well as a small production of
vegetables and fruit for internal consumption in the SLGA canteen.

On the DAFMIM division we find the department of Human Resources that deals
with the allocation of the workers to the different departments, their vacations, their
evaluations, and training programs in their unit as well as their day to day task management realized by the head of each department. The Supply department deals with
the management of most of the equipment of the SLGA and its storage, and partially
deals with the requests for any new item or service before sending it over for the
finance department. Not all requests for an acquisition need to be bought from an
outside supplier, some demands can be met by the regional patrimony, and in those
cases they don't need to be passed to the finance department. As previously mentioned
the Finance Department leads with any acquisition or service that needs to be bought
from an outside source as well as the salary payments to all SLGA collaborators. The
Maintenance department deal with the maintenance of both equipment and infrastructures with the exception of the vehicles. Besides dealing with the maintenance
issues, the Maintenance department also deals with any infractions on the local
infrastructures and with heavy equipment loans such as chainsaws. The Workshop and
Fleet deal with the maintenance of the fleet of cars and machinery as well as their
scheduling and repairs.

The CSD, although included in the DAFMIM is barely connected to the other
departments as it works as a local proxy for services offered by multiple regional
divisions located at the main island, such as employment related issues, housing,
driving related issues and so on. In Figs. 4, 5 and 6 we can see, an excerpt of the Actor
Transaction Diagram, Process Step Diagram and Object Fact diagram produced for one
of the SLGA departments, the CSD, and the description can be found in the following
paragraphs.

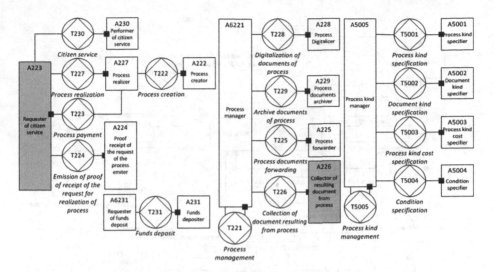

Fig. 4. CSD - actor transaction diagram.

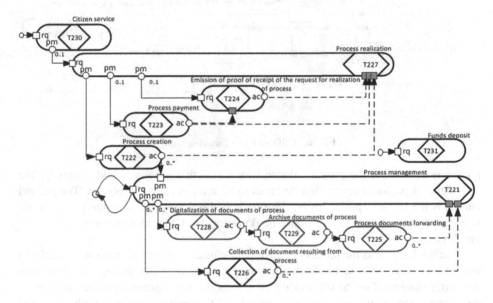

Fig. 5. CSD - process step diagram (Partial).

In Fig. 4 we can notice four clear clusters in the ATD diagram, the first being the transactions initiated by the citizen; it starts with a citizen service that may or may not lead to a process realization – e.g. the case that what the citizen needs is not provided at this desk but in another specific government office. If there is a process realization then there will be a creation of process. The process realization may have an associated cost communicated in the process payment transaction. But there are many processes with no costs associated that may be target of an emission of proof of receipt of the request

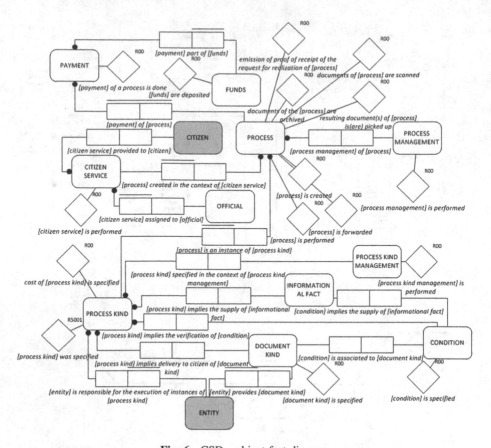

Fig. 6. CSD - object fact diagram.

for the realization of the process. Hence why a step that usually would simply be the state act of a payment transaction deserves to be a transaction on its own. The second cluster is the funds deposit cluster, this one is isolated from the rest due to its nature. It is a daily transaction that can only be done by one CSD coordinator at the end of the day.

The third cluster is the process management cluster, here are the transactions related with the process done in the back-office when the CSD collaborators are free from attending citizens. Two of the datalogical transactions, the scanning of documents of the process and archiving of these documents of the process take place whenever the CSD collaborators have free time, while the forwarding process documents may vary depending on the process forwarding method. If it's sent by fax it takes place at the time of scanning. If it is sent by paper through the ferry boat it takes place every afternoon sometime before the ferry trip. The last transaction in this cluster also takes place when the CSD collaborators have free time, but it does not happen every day as many of the CSD processes have no returning documents, and in most that do, those documents are sent directly to the citizen by postal mail instead of returning to the CSD building. Finally in the fourth cluster we have the process kind management and its related transactions that, as we previously stated, are meant to deal with the constant change in processes and their related documents and conditions.

The diagrams presented and described previously is a typical example of contents presented to the interviewees and already give an idea of the conciseness quality of DEMO, something to which, as we will see just next, almost all interviewees agreed to.

5 Interview Questions and Results

The research pool for this interviews was rather small with only eleven individuals but, considering their positions within the organization and the objectives of these interviews, this was a very significant and useful sample. As previously mentioned, the interview's questions were mostly based on open ended questions, although most with short sentence responses. When a simple yes or no question was made, the usual follow up question consisted in asking the reasons for such answer.

The 43 questions made in the interviews are listed below in Table 1 with a summary of the multiple outcomes. Unfortunately we were unable to obtain enough information to compare our approach with the others, as the key collaborators involved in these other initiatives were no longer working at the SGLA. As many of the placed questions were open ended, we have opted to summarize and group the related answers in order to present them in a more compact and intuitive manner. For each question presented in the following table, we present the number of interviewees that answered with each of the alternative or generally given responses, as well as the total number of collaborators that in fact answered to each specific question. After the table we explain the goals and/or reasoning for each question asked while already providing some comments on the results. In the next section we do a more thorough analysis on the outcome.

Question 1 intended to get some input over the methodology we used in the interviews, described in detail in [12, 13] and relating amount of time each collaborator spent on the interviews. Every answer to this question was centered in the same content, as in it was the appropriate amount of time considering the amount of transactions involved. It would be good to have some other interviews using another process in order to establish a benchmark but unfortunately that was not the case, leaving us with just their perceptions on the matter.

Questions 2 and 3 were in line with the first one, trying to use some concrete surveys that have happened in the SLGA. But unfortunately, the only outcome was to find out that no interviews were used on either cases, as well as many of the collaborators did not even know of the existence of the trainee survey added to the fact that very few departments of the SLGA have in fact a quality manual.

Leaving the interviews aside, Questions 4 and 5 tried to establish the participation of the interviewees in the elaboration of the quality manual, and their experience in doing so, relating to their experience in our first interviews. Only one of the collaborators interviewed, a chief of division, had participated in the execution of the Quality Manual for the Supply department, and qualified that experience as harder, and more so until they found a proper foundation for such process.

Table 1. Interview answers.

1. What did you think of the length of the meetings held for the elaboration of DEMO models? (in terms of the time taken vs quantity of transactions addressed)		2. Were you interviewed for the survey of trainee regarding the organization processes and / or the quality manual (if exists)?	
Appropriate To The Amount of Work	10/10	Yes	0/11
		No	11/11
3. (if yes) In the meetings and / or interviews concerning initiatives: (1) quality manual, (2) survey of the trainee (3) our survey in January can you compare in terms of: (a) the time spent, (b) quality and (c) the scope of the results obtained in models?		4. Have you been involved in the production of quality manual? (if available for the service)	
		Yes	1/11
No Answers		No	10/11
5. (if yes) Can you describe your experience in terms of time spent and difficulties in obtaining the necessary information?		6. Describe how you felt during the meetings in relation to the questions asked with regard to their scope. For example, did you always talked about the same subject and felt we did not address your tasks with sufficient depth. (quality manual, trainee survey, our survey)	
Quality Manual was harder	1/1	Good coverage and appropriate depth	10/10
7. Have you ever felt difficulty with the framing of the questions that were made to you? (Regarding the terms used)		8. Did you always considered yourself the right person to answer our questions?	
Had difficulty	1/10	Yes	8/10
Had no difficulty (or quickly answered)	9/10	No	2/10
9. Imagine the roles were reversed, can you think of one or more ways to improve the way the questions were asked?		10. During the meetings did you ever felt, in some situation, that some of the tasks of your collaborators or steps in their tasks in reality had no reason to exist?	
Yes	1/10	Yes	3/10
No	9/10	No	7/10
11. (if yes) Can you describe one of these cases?		12. During the meetings did you ever feel, in some situation, that any tasks or steps in the tasks of your collaborators should be added, in order to make the process clearer?	
Many possible decisions that were merged into a simple process. And excess bureaucracy in the deciding of processes.	1/3		
Task approvals.	1/3	Yes	8/10
Processes with access limited by bureaucracy.	1/3	No	2/10
13. (if yes) Can you describe one of these cases?		14. How do you evaluate the workflow in the models when compared with the real operational flow of your work?	
Validation of propagation (ornamental plants) should be decided by a different person.	1/2		
Management of tractors should be added.	1/2	Corresponds to the real workflow	11/11
15. Looking at the names assigned to the transactions would you change any?		16. Which one(s) would you change?	
		Specification of intervention	1/3
Yes	3/11	Expense process	1/3
No	7/11	Records for statistical purposes	1/3

(*Continued*)

Table 1. (*Continued*)

17. Looking at the names assigned to the organizational functions would you change any?		18. Which one(s) would you change?	
		DGMI responsible	1/3
		Requester of vehicle	2/3
Yes	3/11	Names of the regional offices	1/3
No	8/11	Warehouse responsible	1/3
19. Can you identify anything produced in your organizational area that you cannot find described in the models?		20. (if yes) 16. Which one(s)?	
		HACCP Control (fishery)	2/5
		Management of lighting and air. (multi-purpose pavilion)	1/5
		Technical opinion (supply and finances)	1/5
Yes	5/11	Registration of commitment	1/5
No	6/11	Decision of the selection of budgets (supply and finances)	2/5
21. In your personal opinion, do you feel that these models can give you a concise and unambiguous notion of what goes on in the organizational area where you perform your work?		22. Before looking at the models, did you have a complete understanding of all the transactions that plays a role in the organization?	
Yes	11/11	Yes	10/11
No	0/11	No	1/11
23. Can you identify any transaction that you were not sure to be responsible for, or did not remembered that it was of your responsibility?		24. (if yes) Which?	
Yes	0/11		
No	11/11	No Answers	
25. Before looking at the models did you had a complete understanding of all transactions carried out in areas under your responsibility? (only for heads of division)		26. Can you identify a transaction that had no prior knowledge? (only for heads of division)	
Yes	1/2	Yes	0/2
No	1/2	No	2/2
27. (if yes) Which? (only for heads of division)		28. Do you agree with all transactions in the areas under your responsibility?	
No Answers		Yes	11/11
		No	0/11
30. Can you find any of your transactions, or one in an area under your responsibility that you had a different perception of the actors involved before this modeling?		31. Which one (s)?	
Yes	0/11		
No	11/11	No Answers	
32. Do you consider that the models that were produced still describe the reality of performed transactions and involved actors?		33. (if no) What has changed?	
Yes	11/11		
No	0/11	No Answers	
34. (if available for the service) How would you compare the quality manual with the DEMO models in terms of obtaining information relevant to the performing of your duties and / or the duties of your employees?		35. Can you find any reason for these models be considered an important resource in the knowledge of the organization for their own employees?	
Quality Manual is more detailed.	3/5		
Demo Models have a wider scope.	1/5	Yes	8/11
They are both good.	1/5	No	3/11

(*Continued*)

Table 1. (*Continued*)

36. Can you find any reason why these models are considered as an important resource in the knowledge of the tasks of each employee?		37. Suppose you had a new employee under your supervision and you had to explain his roles within the organization. Would you consider using any of these models as an aid in this explanation?	
Yes	10/11	Yes	9/11
No	1/11	No	2/11
38. Do you think these models give you a comprehensive and summarized vision of the organization Operation?		39. Why?	
		All is discriminated and summarized.	7/11
		Reflects the steps and processes and who performs the tasks.	1/11
Yes	11/11	Properly diagrammed.	2/11
No	0/11	Generalized view.	1/11
40. Do you believe that is useful that these models give a view abstracted of implementation (regarding people, technology, technical, implementation channels)?		41. Why?	
		Because there is constant change in those items.	9/11
Yes	10/11	Because it is more practical.	1/11
No	1/11	It makes it difficult to give credit to the proper person.	1/11
42.Do you think that the fact that these models differentiate the initiator and executor actor roles and include the acts of request, promise, execute, accept and state a transaction help understand and clarify the responsibilities of each member of the organization?		43. Why?	
		Because it clarifies responsibilities	8/11
		Needs to be well explained so that information does not get lost	1/11
Yes	9/11	Can be used for assigning blame	1/11
No	2/11	There is no need because the process is treated as a whole	1/11

Question 6 tried to establish if the collaborators felt they have explained their work enough in such way for us to understand what really happened within their departments. The responses were somehow unanimous in terms of it being the appropriate depth and coverage of the tasks taking place.

Question 7 tried to determine if the terms used by us in the interviews were of difficult understanding for the participants. By terms used, we refer to more technical words such as actor, role, and transaction, widely used in the DEMO methodology. Nine of the participants stated that either had no difficulty, or, if in fact some doubt arose, it was promptly clarified by the interviewer's explanation of the terms. One of the collaborators however stated that she had in fact difficulty during the questions as she had been "caught off guard".

Question 8 established if our questions had in fact been directed to the proper persons. Eight of the collaborators stated they were the appropriate person for all the questions asked, while two said otherwise, one stated that some questions overlapped his department and another, therefore some of the transactions we asked about, were not of his responsibility. While another, a chief of division, stated that sometimes we asked questions too in depth about the departments that she didn't know to answer and therefore the appropriate person was the head of such department.

Question 9 was intended to obtain valid input on how to better conduct our interviews in the future placing the collaborators in our place, and trying to figure better ways to conduct the interviews in order to improve the productivity. Nine of the collaborators who answered this question couldn't find any way to improve what had been done, and referred to the method as effective and appropriate. The only collaborator that answered otherwise mentioned that the process could have in fact been improved, but in their part, by having their documentation better organized and structured to facilitate gathering any needed information that was asked.

Questions 10 and 11 were to find out how the DEMO way of thinking helped the collaborators notice what tasks were not essential to their operation. Although during the meetings many collaborators realized that some of their steps had no reason to exist, during this interview only three said yes to the question. The cases mentioned were all related to excess bureaucracy in the processes, namely "unnecessary" approvals that aren't more then simply giving notice that something has been done, and in the CSD case, many of their transactions need to happen simply because the regional authorities don't give them access to the programs used in the main island.

Questions 12 and 13 followed the line of though of the previous, but in the opposite direction, finding out non existing transactions that should exist in order to improve the whole process. Although two persons answered yes, their follow up question gives us the perception that one of those answers was in fact a yes, to the previous question as it relates to an approval that had no reason to exist. The other transaction found at this point is related to an internal dispute over the vehicles used in agricultural tasks where, they are currently scheduled by the workshop and fleet department but some consider that it should be done by the DNRM who deals directly with the farmers. Leaving the dispute aside, the transaction in question did in fact exist, but it was modeled at the department currently responsible for such task.

Question 14 had the intention to validate the workflow modeled in the process step diagrams in comparison to the real workflow in order to find any flaws or changes in it. All the eleven collaborators who answered this question agreed that the workflow in the models was in fact similar to the reality of their processes agreeing to what was modeled in every step of each transaction.

Questions 15 and 16 were related to the names specified for the transactions. Although there had been already some discussion and validation on this point around one year ago, we decided to re-evaluate the appropriateness of these given names. Three of collaborators found, each, just one name that they would change in their department's models.

Questions 17 and 18 were on their turn related to the names assigned to the organizational functions. Three collaborators said they would change one or more names. In fact it was somehow surprising that one of the collaborators told us that some names were not "generic" enough. As the positions within the organization are in constant change the DAFMIM chief of division did not agree that that position was used as an organizational function, but instead suggested that we used chief of said department. In the same way, it was also suggested to change the names of the regional authorities, as they also suffer changes when another government is elected. In this case it was suggested that we changed to "regional direction with tutelage of said service". The other two suggestions by the other collaborators were related to names that are more commonly used, instead of the originally proposed ones.

Questions 19 and 20 intended to identify possible missing items that failed to be modeled originally. Five of the collaborators were able to find items that were not modeled. In the fish department the Hazard analysis and critical control points (HACCP) Control was not modeled initially as the head of the department did not find it important at the time, but as the paper print left in the process was significant both him and the chief of the division in charge now qualified it as significant. Finally on the supply and finances department there were just three new transactions proposed to be integrated in the current steps of the product acquisition process. All-in-all, the number of new items identified was quite small compared to the vast amount of transactions that kept stable during this whole year.

On question 21 we tried once again to receive input by the members of the organization on how DEMO was appropriate for the modeling, by asking them if they found the models of their departments concise and unambiguous. Every collaborator that answered to this question confirmed this quality.

Question 22, was initially purposed to give a deeper understanding of how DEMO models affected the perception of the collaborators within the organization. But unfortunately due to the time constrain it was unfordable to show all the models of the organization in every single interview, therefore they were only presented the models of their department, and for the chiefs of division, the models of all the departments they are responsible for. With this constraint it was obtained a far different answer from the expected, as the knowledge of their departments was rather clear in almost all cases. It would be interesting to evaluate their previous knowledge of the remaining departments.

The only person that answered no to this question was the responsible for the Gardens and Ornamental Plants, which admitted to be surprised by all the transactions she had in her department.

Questions 23 and 24 were meant to identify possible misunderstandings in the responsibilities within the organization, but all the collaborators interviewed had a clear perception of their tasks and responsibilities.

Question 25 is somehow similar to the twenty second but it was limited to the departments under the responsibility of each the chiefs of division. As we were only able to present the models of each of their departments to the collaborators, this twenty fifth question ended up having a wider spectrum then the previous instead of the opposite.

In this question one chief of division answered no, and admitted that he had no idea that the transaction that took place in the CSD shared a particular pattern that could be generalized and extended to all costumer related transactions that took place in that service.

Questions 26 and 27 were to validate the previous question and further extend it, but both of the chiefs of division claimed they fully knew all the transactions listed in all departments.

Questions 28 and 29 aimed at reinforcing the comprehensiveness quality while confirming if all collaborators agreed with all the listed transactions in the models deeming them as needed or even as essential.

Questions 30 and 31 were intended to find any eventual discrepancy between the collaborators' perception of reality compared to the modeled transactions. No interviewed person had a different perception of what was modeled, further reinforcing the comprehensiveness quality.

Question 32 and 33 intended to capture the validity of the work produced nearly one year before, and how it was still applicable to the current reality. Even though some collaborators and documents used changed, all eleven interviewees agreed that all models still correctly described the reality of the organization.

Question 34 we returned to the previous subject of the quality manual, and how that manual could be compared with the DEMO models. Three of the five collaborators stated that the quality manual was more detailed and therefore more helpful to the performance of their duties. While another collaborator qualified them both as equally good, and another stated that the DEMO models had a wider scope and therefore were more helpful. On this question however, we have to take in consideration that the Action Rules of the DEMO methodology were not presented, and, for their daily functions, they would be a valuable resource, maybe leading to a change in the balance of the received answers.

Question 35 aimed to get a perception on the relevance given to the DEMO models by the collaborators. The answers here varied, and although most collaborators said yes, three couldn't find any reason for the models to be relevant and another one stated that their activities were already so mechanic that the models were of little use.

Question 36 was somehow similar to the previous, but in this case the focus was on the tasks of each employee instead of the knowledge of the organization. Here most answers were positive and centered around how it offers a structured view of their work, however two of the collaborators mentioned that although it would be an interesting resource, their employees would have difficulty in properly understanding the models.

Question 37 had the intention to obtain the predisposition to use these models to explain someone who was not familiarized with the organization and their new tasks. Answers were somehow similar to the previous question. Most collaborators said yes, but one questioned the ability of someone new to understand these models, although another also mentioned that the actor transaction diagram, would be a good model to explain the procedures without complications. Still another person that also said "no" mentioned also that the models could be complicated, and it would be more profitable time wise to show them the real operations in practice.

With questions 38 and 39 we intended once more to validate one of the claimed qualities of DEMO models, and their perception by the organization's members. To do so, we asked if the models gave a comprehensive and summarized view of the organization's operation. All interviewees answered yes, and when asked why they thought like that, there was little variation in their answers. Most replies focused on how everything was discriminated and properly summarized, others stated that it was properly diagrammed and reflected their department of the organization, one also mentioned that it gave a generalized view of everything, and finally it was also stated that it clearly reflected step by step the processes of each collaborator and their respective tasks.

The objective of questions 40 and 41 were to validate the level of abstraction used in DEMO and understand to what extent this is assimilated by the collaborators of an organization. Ten out of the eleven interviewees answered that it was useful to use this level of abstraction, while one said otherwise. When asked why the responses showed the understanding of the reasons as they were mostly based on the fact that the organization is in constant change, new employees join, old employees leave and the documentation is also under constant updates, therefore this level of abstraction allowed for the models to remain correct after a long period of time, and still reflect the reality of the organization, as also demonstrated in the previous questions of the interview. Although most answers were centered in these aspects, one of the collaborators had a very different opinion that may reflect some difficulty understanding the method, as the reason used to justify the "no" was that this level of abstraction makes it difficult to give proper credit to a collaborator when its due, because no person names are used, but instead only the organizational functions.

The last question of this interview focused on determining if the collaborators found important the fact that, in the models, there was a differentiation between the initiator and executor roles as well as the specification of each transaction's main steps of request, promise, execute, state and accept. Nine of the answers were positive, eight focused on how this helped indeed to clarify the responsibilities, and how important it is to know who is responsible for what within each department. One answer had a different justification: the organization needs to be well explained so that information does not get lost, and this way of modeling did exactly prevent that. There were two collaborators on the "no" side, one stating that there was no need for this differentiation in their department because a single collaborator usually did most of the transactions as being a single process, and the other stated that clarifying the responsibilities isn't always good, as the goal of the employees is to properly do their work, as such, they normally don't do mistakes on purpose. And, that being the case, they should not look to assign blame but instead work together as a group to fix what went wrong.

6   Results Analysis and Evaluation

With these interviews we were able to validate the importance and relevance of the Ψ-theory's operation, transaction and distinction axioms and the qualities they bring about in the application of DEMO. We will now see how these axioms affected the modeling outcome and the perception the interviewees have of their organization and of its respective models, thus, validating the DEMO qualities target of research in this paper.

With these interviews we were able to validate the importance and relevance of the Ψ-theory's operation, transaction and distinction axiom, and determine how their usage in the process may have affected the outcome and perception of the collaborators interviewed.

In Table 2 we have a summary of our results analysis where we present the three DEMO qualities that we intended to validate in the research reported in this paper, along with the questions (presented in the first column) and the validating outcomes (presented in the second column). The reasoning of how the outcomes validate each quality can be found after this table.

With the first set of questions from 1 to 9 we were able to validate the approach used and how it was adequate to do this process in the eyes of the organizational members interviewed. The answers were unanimous concerning the time taking in the

Table 2. Demo qualities analysis.

DEMO Quality - Concise	
21. In your personal opinion, do you feel that these models can give you a concise and unambiguous notion of what goes on in the organizational area where you perform your work?	100% Yes
38, 39. Do you think these models give you a comprehensive and summarized vision of the organization Operation?	100% Yes
40, 41. Do you believe that is useful that these models give a view abstracted of implementation (regarding people, technology, technical, implementation channels)?	91% Yes
	9% No
DEMO Quality - Comprehensive	
7. Have you ever felt difficulty with the framing of the questions that were made to you? (Regarding the terms used)	10% Yes
	90% No
14. How do you evaluate the workflow in the models when compared with the real operational flow of your work?	100% Corresponds fully
19, 20. Can you identify anything produced in your organizational area that you cannot find described in the models? (Note: in this question, although almost half of the interviewees found missing items, the percentage of missing items in their area of responsibility varied only from 1% to 6% and the other half reported 0% of missing items)	45% Yes
	55% No
28, 29. Do you agree with all transactions in the areas under your responsibility?	100% Yes
30, 31. Can you find any of your transactions, or one in an area under your responsibility that you had a different perception of the actors involved before this modeling?	100% No
35. Can you find any reason for these models be considered an important resource in the knowledge of the organization for their own employees?	73% Yes
	27% No
37. Suppose you had a new employee under your supervision and you had to explain his roles within the organization. Would you consider using any of these models as an aid in this explanation?	82% Yes
	18% No
38, 39. Do you think these models give you a comprehensive and summarized vision of the organization's operation?	100% Yes
42. Do you think that the fact that these models differentiate the initiator and executor actor roles and include the acts of request, promise, execute, accept and state of a transaction help understand and clarify the responsibilities of each member of the organization?	82% Yes
	18% No
DEMO Quality - Stable	
14. How do you evaluate the workflow in the models when compared with the real operational flow of your work?	100% Corresponds OK
15, 16. Looking at the names assigned to the transactions would you change any?	27% Yes
	73% No
17, 18. Looking at the names assigned to the organizational functions would you change any?	27% Yes
	73% No
32, 33. Do you consider that the models that were produced still describe the reality of performed transactions and involved actors?	100% Yes

interviews being the appropriate for the information that was needed as well as the depth and coverage of the organizational processes. Most of the collaborators had no difficulties with the questions asked, showing that in a short amount of time the basic notions of DEMO can be explained and understood. This set of answers also show, that our interviews were properly prepared in terms who needed to be interviewed on each subject as the only two answers opposing to this position was a chief of division asked for his input on an area under his responsibility but not as the main source of information, and one head of a department where many processes overlap with other departments therefore making it hard to describe them in a whole. Regarding improving the interview method no valid input was give, as the only suggestion was for a better organization on the SLGA part.

Question 7 in particular was relevant to make sure that the participants were at ease with the main concepts of the DEMO approach – like actor and transaction – leading to a correct comprehension of the models. The strong positive result in this question supports the comprehensiveness quality of DEMO.

On questions 10 to 13 we evaluated if this approach had an influence in the organizational self knowledge, and how it made possible to find transactions that should exist or transactions that had no real reason for being currently executed. Although on the first round of interviews more collaborators witnessed this problem as compared with the current interview round, we were still able to witness how the some collaborators had the perception that some transactions were simply an over bureaucratization of a process and how it could be improved without those transactions. Unfortunately the decisions to simplify such procedures are not their responsibility. On the other end, concerning transactions that should exist, both of the transactions reveal the collaborators understanding of the transaction axiom, by suggesting changes in the responsibilities over currently existing transactions to better adequate to the reality of who decided. But no real example of transactions that should be implemented was given, as one was the exact opposite, a transaction that should not exist, and the other was an existing transaction of farm vehicles scheduling whose responsibility is in another department.

Question 14 was a very important question in order to demonstrate 2 points. By having a unanimous answer on how the process step diagram models reflected the proper workflow of the organization's departments, we realize the great importance of the transaction axiom. Thanks to the structuring of the many essential and common process steps in the transaction pattern, we managed to uncover some "hidden" (in the minds of the persons) transactions and, on the other hand, because the collaborators become aware that a single transaction "automatically" includes the many kinds of social interactions that can happen regarding some production, they end up evaluating the modeled process fully corresponds to their daily work. Thus, this outcome also validates DEMO's comprehensiveness. We were also able to verify that the models remained current even after multiple changes in the organization in terms of persons and documentation, demonstrating DEMO's quality of model stability, brought about thanks to the distinction axiom and its separation of the human abilities, where we normally abstract from information processing, communication and document aspects.

Questions 15 to 18 allowed us to demonstrate how the naming's determined for both organizational functions and transactions were quite adequate for the collaborators that realize the respective transactions, something that sometimes proves difficult when gathering this kind of information. There were three suggestions of transaction name changes and four organizational function changes, but taking in account the huge number of almost 500 transactions that had been modeled and the number of twice as much actor roles involved (even though many repeat themselves multiple times), the amount of change suggestions is of very small significance. This outcome strongly validates the stability of DEMO models.

Questions 19 and 20 demonstrated that, although the DEMO approach seems to be a very good approach compared to other methods, it is not infallible and, as such, these questions allowed us to detect some transactions that were not modeled on the first round of interviews around one year ago. But to give the due relevance to the amount of new transactions found in a more precise fashion, we now analyze the answers of each of the five collaborators that answered this question individually. One of the collaborators, the chief of the division DAFMIM, identified three lacking transactions while 317 transactions were modeled as being under his responsibility, that is, not much more than 1 % of missing items. The other chief of division (DNRM) identified only one missing transaction in the set of 162 transactions modeled for the areas under her responsibility. Again a percentage close to 1 %. The other three collaborators – all department heads – identified as missing transactions in their area of responsibility, respectively: 1 out of 18, 1 out of 26 and 1 out of 31, that is, percentages from 6 % to 4 %. The other 5 department heads could not find any missing item, leading to 0 % of missing items. These figures also highly contribute to validate DEMO's quality of comprehensiveness.

With question 21 we validated the conciseness quality thanks to a unanimous response on how the models really gave a concise and unambiguous notion of each of the organizational areas to their respective heads of department.

Questions 22 to 27 intended to test how these models changed the collaborators perspectives on the organization. But the responses obtained did not help us find any evidence of change, as most collaborators claimed to already know all the transactions and responsible functions for each of those transactions. These questions could perhaps have had different answers if instead of only presenting their particular department models, it had been presented the all the models of the organization, but as previously mentioned that was not a possibility due to time constraints.

Questions 28 and 29 demonstrated the importance of the operation axiom on how it allowed that interviewees' interactions within the organization were correctly modeled, by having a once again unanimous positive outcome when asked if they agreed with all the transactions under their areas of responsibility, thus also contributing to comprehensiveness.

Questions 30 and 31 helped to demonstrate the importance of the transaction axiom because all modeled transactions had all the proper responsible participants identified, thus also contributing the comprehensiveness quality.

Questions 32 and 33 demonstrated the importance of the distinction axiom by proving that all models were still up to date thanks to the separation of the ontological, infological and datalogical aspects, thus reinforcing the validation of the stability quality.

With questions 35 and 37 we tried to understand if the models could be considered useful both for existing collaborators and to help train new ones. The outcomes were not always the expected but the answers confirmed that the models were considered important for the collaborators to have, not only knowledge of their individual tasks, but also of other tasks all over the organization. Although some interviewees pointed out that some of the diagrams could be difficult to understand and thus answered negatively, the outcome is strong enough to also contribute to validate the comprehensiveness quality.

Finally, the block of questions from 38 to 43 allowed us to demonstrate the understanding by the organization's members of some key aspects of DEMO such as the focus on abstraction from implementation and the separation of the transaction steps. Although not unanimously, the majority of interviewed collaborators found these aspects important and considered them a good quality of this modeling process, as it can be seen from some of the opinions we transcribed. So 38 and 39 clearly both validate the qualities of comprehensiveness and also conciseness. 40 and 41 distinctly validate the conciseness quality, while 42 ends up also validating comprehensiveness thanks to the clarification provided by the clear identification of organizational responsibilities.

7 Conclusions

Following the tenets of Design Science Research we presented a relevant and needed contribution of an interview based qualitative validation of some of DEMO's axioms and claimed benefits – something that, to our knowledge has never been done up to now. Namely we looked at the qualities of conciseness, comprehensiveness and stability of DEMO's ontological models. This was done in the context of a large scale practical DEMO project and, to our knowledge, no publicly available study exists that practically demonstrates such qualities. And such studies like these – based in large scale projects – are essential to contribute to a more widespread and mainstream acceptance and adoption of DEMO in enterprise change projects. We interviewed 11 key departments and division heads involved in a large e-government project where around 500 ontological transactions were specified. Our research was able to demonstrate that indeed DEMO's Ψ-theory and its axioms contribute to provide a concise and comprehensive view of the essential dynamic and static aspects of an organization and that, even after a year has passed, the majority of DEMO models were still up to date and only needed to be subjected to some minor changes.

As a result of the last round of interviews, we also validated that the: (1) the duration of the interviews, (2) ability by the collaborators to answer the questions in the initial stage of the project both in the terms used by the interviewers and their knowledge of what was being asked for and (3) their opinion on the interview methodology, were all deemed adequate by the interviewees.

Our study has limitations since the DEMO approach was evaluated individually. In future studies we will apply also other modeling approaches such as simple flowcharts and/or BPMN based so that we can also evaluate them in the same dimensions of analysis target of this paper and we can compare the performance of each approach according to the perception of the organization's members. Furthermore, the number of interviewees in the research presented in this paper is not enough for a pure quantitative validation which is something that has to be done also to bring up even more solid arguments supporting DEMO's claimed qualities. We expect that, as the project advances from the pilot stage in the small island to the full fledge stage in the main island, then we will again apply DEMO for modeling further processes to be implemented in the e-government project and we will have a sample of interviewees big enough for a pure quantitative validation.

References

1. Hevner, A.R., March, S.T., Park, J., Ram, S.: Design science in information systems research. MIS Q. **28**, 75–106 (2004)
2. Hevner, A.: A three cycle view of design science research. Scand. J. Inf. Syst. **19**, 87–92 (2007)
3. Dietz, J.L.G.: Enterprise Ontology: Theory and Methodology. Springer, Heidelberg (2006)
4. Patton, M.Q.: Qualitative Research and Evaluation Methods. SAGE Publications, Thousand Oaks (2002)
5. Creswell, J.W., Plano Clark, V.L.: Designing and Conducting Mixed Methods Research. SAGE Publications, Thousand Oaks (2007)
6. Johnson, R.B., Onwuegbuzie, A.J.: Mixed methods research: a research paradigm whose time has come. Educ. Researcher **33**, 14–26 (2004)
7. Kvale, S.: InterViews: An Introduction to Qualitative Research Interviewing. SAGE Publications, Thousand Oaks (1996)
8. Dietz, J.L.G.: Is it PHI TAO PSI or Bullshit? Presented at the Methodologies for Enterprise Engineering symposium, Delft (2009)
9. Dietz, J.L.G.: On the Nature of Business Rules. Advances in Enterprise Engineering I. 1–15 (2008a)
10. Dietz, J.L.G., Albani, A.: Basic notions regarding business processes and supporting information systems. Requirements Eng. **10**, 175–183 (2005)
11. Dietz, J.L.G.: Architecture - Building strategy into design. Academic Service - Sdu Uitgevers bv (2008b)
12. Aveiro, D., Pinto, D.: A case study based new DEMO way of working and collaborative tooling. In: 2013 IEEE 15th Conference on Business Informatics (CBI), pp. 21–26 (2013a)
13. Aveiro, D., Pinto, D.: An e-Government project case study: validation of DEMO's qualities and method/tool improvements. In: Harmsen, F., Proper, H.A. (eds.) PRET 2013. LNBIP, vol. 151, pp. 1–15. Springer, Heidelberg (2013)

Knowledge Reuse in Innovation

John Favaro[✉]

Consulenza Informatica, Pisa, Italy
john@favaro.net

Abstract. Innovation is normally associated with the generation of entirely new knowledge, and many current approaches to innovation are based upon that premise. However, a number of research results in recent years have yielded new insights on the way that the mind works to produce successful innovation. What is known as "strategic intuition" involves the intelligent recombination of precedents to form new solutions to strategic innovation problems, and provides the elements of a systematic approach to innovation based upon knowledge reuse.

Keywords: Innovation · Reuse · Knowledge · Intuition · Strategy

1 Introduction

For most of the 20th century, "business efficiency" was *the* dominant corporate priority. This is no longer true. Ever since the mid-1980s, innovation has steadily risen as a priority, **overtaking** efficiency around the year 2000 and now dominating by far, as illustrated in Fig. 1.

It comes as no surprise that innovation began to acquire increased importance in the mid-1980s – it was in 1984 that Apple introduced the Macintosh. The role of Apple co-founder Steve Jobs in bringing innovation into the forefront of consciousness of the IT community in particular and the business community in general is undisputed, and his death in 2011 has only enhanced his reputation for innovation leadership.

It is also no surprise that the point at which innovation surpassed efficiency as the main corporate priority was approximately the year 2000, only three years after Clayton Christensen of Harvard published *The Innovator's Dilemma* [2], which introduced the now-ubiquitous phrase "disruptive innovation". The phrase immediately captivated the business world, and innovation firmly took center stage in corporate life.

Innovation initiatives now abound, in more or less structured forms. Many software enterprises have undertaken internal initiatives to promote innovation within the ranks of their own employees, such as the Star Search initiative reported by Gorschek *et al.* [3] Google seeks to hire co-called "smart creatives", [4] giving them unstructured time for pet projects. Studies seek to determine whether certain personality traits (e.g. introversion [5]) are indicative of more innovative people.

Even software methodologists have become involved in the discussion. The co-author of a recent book [6] on continuous delivery, an agile technique for bringing new features up and running as soon as possible, maintains that continuous delivery supports

© Springer International Publishing Switzerland 2015
A. Fred et al. (Eds.): IC3K 2014, CCIS 553, pp. 444–456, 2015.
DOI: 10.1007/978-3-319-25840-9_27

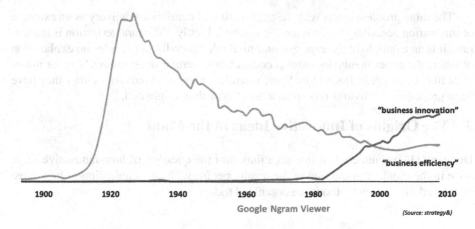

Fig. 1. Innovation has overtaken efficiency as the major business priority.

innovation as an application of the scientific method, whereby a hypothesis is proposed, feedback is obtained on that hypothesis, and the cycle begins anew. In that sense, agile methods in general are often claimed to enhance innovation.

2 Problems with Current Approaches to Innovation?

Yet, for all the excitement about innovation today, something is not working. In a recent survey of 1344 CEOs in 68 countries, the vast majority of business leaders, a full 86 %, said that they are not satisfied with the way innovation is going at their companies [7] – this in spite of an explosion of initiatives and significant invest-ment in new infrastructure such as "incubators" and the like. Carr [8] attributes at least part of this disappointment to a fundamental misunderstanding of the role of innovation in business strategy, noting that "… if innovation were a sure thing, everyone would do it equally well, and its strategic value would be neutralized". Innovation is costly and risky. In consequence, companies must carefully choose when and where to invest in innovation.

Doubts about claims that agile methods promote and enhance innovation have also surfaced, and researchers have begun to examine possibilities for inserting creativity techniques into standard agile processes [9].

Even the sacred cow of "disruptive innovation" has come under fire. Among other observations, Lepore [10] asserts that the theory of disruptive innovation relies on shaky anecdotal evidence; that the theory has not proven to have predictive value; and that the theory only provides an explanation for why businesses fail (rather than why they succeed).

Whatever the merits of these observations, the theory of disruptive innovation does indeed provide a reasonable explanation for why big and successful companies some-times are brought down by smaller upstarts instead of crushing them (or simply acquiring them). However, there is one question in particular that the theory does not answer: where the innovation comes from in the first place.

The same problem exists with the explanation of continuous delivery as an example of innovation according to the scientific method. Firstly, the characterization is inaccurate: it is an example of the *experimental* method. Secondly, it provides no explanation of where the experimental hypothesis comes from. Similar observations hold for initiatives like the aforementioned Star Search: while it manages innovations *after* they have been proposed, it provides no explanation of how they originated.

3   The Origins of Innovative Ideas in the Mind

Duggan [11] has undertaken investigations into the question of how innovative ideas arise in the mind. A discussion of the results begins with an examination of how most structured innovation methods are organized today.

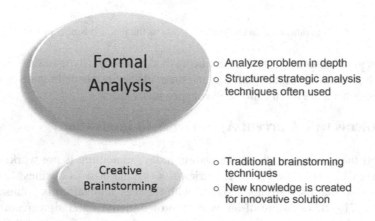

Formal Analysis
o Analyze problem in depth
o Structured strategic analysis techniques often used

Creative Brainstorming
o Traditional brainstorming techniques
o New knowledge is created for innovative solution

Fig. 2. Most innovation methods involve a two-step analytical/creative process.

As illustrated in Fig. 2, most structured methods today involve some variation on two basic steps. The first is some kind of formal strategic analysis, which can be very extensive and detailed (e.g. using Porter's Five Forces [12]). This first step is generally followed by some form of brainstorming session, whereby the assumption is that new knowledge will be generated to provide the innovative solution to the problem. The brainstorming session is often brief, as little as a single afternoon.

This approach has its roots in the early 1960s when Roger Sperry did pioneering research on split brains, showing that the left and right sides can coexist even when separated, and that each side hosts different functions [13]. He won the Nobel Prize for this work in 1981, and it led to the popular notion of a kind of duality of consciousness. In its most common expression, there are two different kinds of thinking, each controlled by a different side of the brain: when working out some form of logical problem, the analytical side of the brain is being used; when brainstorming something new, the creative side is being used (thus the two steps of Fig. 2). This notion provided a very powerful model of how the mind works, and became an important tool in business [14] and in psychology, where people were routinely classified as "right-dominant" creative or "left-dominant" analytical.

The model persists in popularity today, despite the fact that it has been demonstrated to be wrong. In fairness, the neuroscience community never did completely accept the idea that there may be "left-dominant" or "right-dominant" people. At the very least, it did not seem like an efficient use of the brain, with one side remaining idle while the other was active. A convincing demonstration was provided in the early 1990s when Seiji Ogawa started using Magnetic Resonance Imaging to determine which parts of the brain were in fact being used for specific functions. It became quickly clear that not just one side of the brain or the other was being used; rather, there was activity everywhere [15].

The work of Eric Kandel provided the basis for a better model of how the mind works. He won the Nobel Prize in 2000 for his studies of *learning and memory* [16]. Kandel determined that the brain absorbs knowledge both through sensations and analysis, and retrieves it both through unconscious and conscious searching, involving *both* analysis and intuition. More specifically, two different kinds of intuition are used to search and retrieve experiences from memory.

The first kind of intuition has become known as *expert intuition*. Herbert Simon won the Nobel Prize in economics at least in part for pioneering research in expert intuition, well before he became known in the computer science community as a researcher in artificial intelligence. Klein [17] interviewed experts to document how their intuition worked, showing that it involves the very rapid retrieval of memories from the expert's own experience, allowing him to make split-second decisions based on things he has seen before – so fast that it appears to be pure intuition rather than memory retrieval. The book that probably popularized expert intuition the most in recent years was Kahneman's *Thinking, Fast and Slow* [18].

Duggan departs from the concept of expert intuition by noting first that expert intuition counts on the solution being familiar and is therefore not likely to yield true innovation. He coined the term *strategic intuition* for a second kind of intuition that we use, which still has its roots in Kandel's model of learning and memory but is more analytical in the sense that not only are one's own experiences retrieved (as in expert intuition), but also the experiences of others (which might be known, for example, through listening or study). These are then recombined not in familiar ways but in new ways through flashes of insight. It is a much slower process, and usually happens when the mind is relaxed. Duggan proposes an analogy to Kahneman's work by way of explanation: Expert intuition is "fast", and yields combinations of familiar experiences; strategic intuition is "slow", but produces new combinations of one's own experiences and those of others.

In summary, the concept of strategic intuition is aligned with the model of how the mind works when it generates innovative ideas through the mechanism of learning-and-memory. When the mind is confronted with a new problem, it breaks it down into pieces and performs a search of memory for one's own experience and those of others that are known, and makes a new combination from those pieces.

Of specific relevance to the discussion, therefore, is the fact that strategic intuition does not focus on the generation of *new* knowledge, but rather the creative reuse (and recombination) of *existing* knowledge.

4 A Systematic Innovation Method

Although strategic intuition is the core concept that enables an understanding of how innovative ideas occur in the mind, it is only a partial solution to the overall problem. Consider the arrival of agile software development on the scene in the early 2000s. The core enabling concept was iterative and incremental development. However, the real value was created when practical methods (e.g. Scrum) were successfully developed around that core concept. Similarly, Duggan took the core insight of strategic intuition and devised a method called Creative Strategy [19] for applying it systematically to produce innovative solutions to strategic problems.

Continuing the analogy with agile software development, Larman and Basili [20] were able to trace the roots of iterative and incremental development all the way back to the origins of computer science in the 1950s. It is not surprising, therefore, that Duggan was able to trace the essence of the Creative Strategy method back to some classic works of strategy that are well known in the business world, in particular a book of Clausewitz on the strategy of war [21]. In that book, Clausewitz identifies four key elements: *examples from history*; *presence of mind*; *coup d'oeil* (or "flash of insight"); and *resolution*. Those four elements are essentially those that went into the development of the Creative Strategy method, but rendered operational with modern, updated techniques – just as agile software development was made operational with modern advances in programming languages, testing technology, continuous delivery, etc. They form the bridge from the *science* of strategic intuition to the *method* of Creative Strategy.

The first step in the method is called Rapid Appraisal, and has its origins in a method developed in the 1980s for streamlining economic development initiatives in poor countries [22]. The methods used to assess situations in the richer countries did not work well in poor countries, because information was generally unavailable or of poor quality. Rapid appraisal is a series of unstructured and semi-structured interviews on a widening circle of stakeholders. Recalling that most current innovation approaches have a long and expensive initial phase of researching the problem and writing extensive documentation, Rapid Appraisal produces no new documentation; it attempts only to arrive quickly at a workable definition of the problem and its characteristics.

Once the problem has been defined, it is broken down into its elements (analogous to the way in which top-level software requirements are broken down into sets of refined requirements), which are then put into what is called the so-called Insight Matrix.

The second step is the What-Works Scan, where the fundamental question is asked: "Has anybody, anywhere, at any time, from long ago up to the present day, ever solved any element of the problem"? This question is related to Clausewitz's "examples from history". The scan produces a set of historical *precedents* that become components of a potentially innovative solution to the problem – loosely analogous to the reusable components that might become part of a new software application.

The What-Works Scan has its origins in techniques for social policy research, where it is difficult to define a clear measure of success. Since in business it is generally also very hard to measure cleanly the value of a strategic idea, the techniques from social policy research are very relevant. The actual techniques used to search for such potential *precedents* can vary from simple interviews to sophisticated analysis.

The final step of the method is called Creative Combination, and is an adaptation of the so-called "Work-Out" technique used by General Electric at its Corporate University in the 1990s [23]. Elements are selected from the Insight Matrix produced during the What-Works Scan to produce an innovative solution.

How does one know which combination of the precedents to choose? That is the flash of insight, the *coup d'oeil* of Clausewitz. As Steve Jobs remarked: "… creative people … just saw something …", and "creativity is just connecting the dots". Furthermore, "… you can't connect the dots looking forward; you can only connect them looking backwards …" – that is, into the past, into examples from history.

The chances of having the flash of insight are enhanced through Clausewitz's *presence of mind* – that is, by freeing the mind of distractions (including "sleeping on it") in order to allow the learning-and-memory to make the long and slow search for the pieces of the solution using strategic intuition.

5   Industrializing the Method

The emergence of agile software development methods in the early 2000s represented a significant step forward, but many enhancements were needed to make them usable in industrial-strength projects. New unit test harnesses and refinements such as burn-down charts and Kanban boards were introduced. Similarly, Creative Strategy provides a well-grounded method for strategic business innovation, but it has turned out that applying it in the real world means adding enhancements.

The management consulting firm strategy & has extended the method with mechanisms intended for application in large-scale corporate environments. One observation was that although the creative combination of precedents could produce an innovative idea, it did not automatically yield an actionable strategy. However, it is only possible to bring a strategy to market, not an idea. Thus, a new step was included after Creative Combination, covering activities such as determining markets, value proposition, commercialization, etc. This step roughly corresponds to Clausewitz's *resolution* element: that is, everything necessary to carry the idea through and bring the innovation to market.

A second enhancement is particularly relevant in the context of this discussion. Recall that the What-Works Scan identifies the historical precedents to use in the Insight Matrix. Another operational enhancement that they have carried out on the method is a database of precedents that can make the searching process more effective in the amount of time available for the What-Works-Scan.

6   Implications for Knowledge Reuse in Innovation

The idea of a "database of precedents" suggests an inevitable analogy to the discipline of *software reuse* [24]. In this section, we discuss the extent to which this analogy may be applicable and the implications for knowledge reuse in the innovation process.

6.1 Relationship to Other Innovation Methods

Despite the remarkable alignment of strategic intuition with modern models of how innovative ideas occur in the mind, it remains a fair question whether that justifies building a formal method around it. After all, innovation is a highly creative act, and it is reasonable to ask whether it is compatible with the rigor of a method. In response, recall the observations of Carr [8] that innovation is hard, and if everybody could do it easily, it would not provide a competitive advantage and would just reduce to being a cost of doing business. There may be a select few who come by it naturally, but for the others a method is needed to amplify their efforts. The best that can be done is to provide something that directly supports the way innovation happens naturally, rather than ignoring it (or worse, working against it). Consider an analogy to oyster farming, which systematically mimics the natural process that occurs "in the wild", in order to system-atically produce more valuable pearls. One sets up the right conditions, and then steps back to "let it happen".

The method takes the same approach of setting up the right conditions for inno-vation in order to "let it happen": First, it is grounded in a specifically framed chal-lenge that everyone agrees is where innovation is needed (through the Rapid Appraisal phase). Second, the precedents are, *by design*, outside of one's normal field of vision and thus much more likely to bring people out of their daily rut. Third, by working with solved pieces of the puzzle, one is much more likely to come up with a workable innovation, rather than a fanciful idea. Consider the difference with some other approaches:

- *Brainstorming*. The staying power of group brainstorming as a technique for inno-vation is remarkable given that it has been under fire for over fifty years, when a study [25] unambiguously demonstrated its lack of effectiveness. Over the years, a number of books [18, 26] have described its many defects, including peer pressure, the tendency of some in the group to let the others do all the work, and the tendency of extroverts to dominate and sway opinions [5]. By seeking results in a short (usually only a few hours) amount of time, in a high-pressure group that is usually homoge-neous, it is directly opposed to the concept of strategic intuition, which involves the slow elaboration, by individuals in a relaxed state of mind, of heterogeneous expe-rience from different domains. Yet group brainstorming continues to be used despite its known defects because it is easy to plan in a brief time period – much like the proverbial drunk who searches for his car keys under the street lamp, even though he lost them on the other side of the street, "because the light is better over here".
- *SIT*. The technique of Systematic Inventive Thinking [27] involves five so-called Creativity Templates (attribute dependency, replacement, displacement, component control, division), that are used to manipulate the features of a product (e.g. a chair) in the search for a more innovative design, e.g. a new way to hold the chair up. The technique is silent, however, on the question of where the new idea comes from. This fact alone is not incompatible with Creative Strategy – on the contrary, the technique could be enhanced with strategic intuition to produce that new idea (As observed earlier, this problem is common to many methods, including for example Christensen's *Innovator's Solution* [28] – no guidance is given for how to arrive at

creative ideas, only what to do with them after they have arrived.). Nevertheless, the main problem with SIT is that the entire approach of (performing thought experiments on an existing product), rather than identifying first where innovation is needed on the product and searching for historical precedents that could provide a solution, is much less aligned with what really happens in innovation [19].

- *Design Thinking.* Although this technique has its origins in product design, it has been expanded to cover general business problems [29]. The six design methods offered in the technique (customer insights, ideation, visual thinking, prototyping, storytelling, and scenarios) suffer from both the defects discussed above: ideation is simply a form of brainstorming, which most of the others (prototyping, storytelling, scenarios) are silent about where the idea for the story, scenario, etc. comes from.

The core idea of the creative combination of historical precedents remains the closest to the way that innovation really happens. We discuss the implications for knowledge reuse in the next section.

6.2   Innovation as Creative Combination of Prior Experience

Although it is not accurate to describe continuous delivery as something other than a form of the experimental method, it is indeed accurate to describe Creative Strategy as a form of the true scientific method, as Duggan has observed [19]. Roger Bacon, the father of the scientific method, states that "... at first one should believe those who have made experiments or who have faithful testimony from others who have done so... experience follows second, and reason comes third". That is, first examine what others have done before – the examples from history of Clausewitz. The next step is described by Thomas Kuhn in The Structure of Scientific Revolutions [30]: "Scientists ... speak of .... the 'lightning flash' that inundates a previously obscure puzzle, enabling its components to be seen in a new way that for the first time permits its solution". In other words, the *coup d'oeil*, the flash of insight of Clausewitz. In summary, it is the core concept of innovation as the creative combination of examples from history that distinguishes the method.

These examples from history, or *precedents*, are therefore key elements in the method, and a closer examination of their nature is in order. Even more than the *when*, the key in searching for precedents is the *where*. With an eye on strategic intuition, it is essential to cast a wide net, going far beyond one's own personal experience into the experience of others. In particular, it is important to go outside one's own domain – and this is where the direct analogy to software reuse tends to break down. A fundamental paradigm in the field of software reuse is *domain analysis* followed by *domain engineering*, whereby the essential components of a domain are analyzed to become assets that are then recombined to build systems [31]. Creative Strategy decisively departs from that paradigm, because something different is needed.

True innovation tends to contain elements from *different domains than where the original problem lies*. There are countless illustrations of this point. For example, the Odón Device has just been enthusiastically adopted by the World Health Organization as an inexpensive solution in poor countries for dealing with difficult births. It was invented by an Argentine car mechanic who had the idea in his sleep (the relaxed "presence of mind" of Clausewitz) after watching a YouTube video about extracting the cork from a wine bottle [32].

In another example, Hollywood movie star Hedy Lamarr teamed with music composer George Antheil to invent (and patent) *frequency hopping*, which led to the spread spectrum technology that underlies today's technologies such as Bluetooth and WiFi. Their innovation was inspired by Antheil's experience in synchronizing player pianos for his composition Ballet Mécanique [33].

Design patterns [34], today one of the most fundamental concepts in software engineering, provide an example in the pure software domain. Their inspiration came not from the field of software development, but rather from the entirely different domain of architecture, specifically the work of Alexander and his colleagues on reusable patterns for designing anything from single homes to entire cities [35].

An ideal example is provided by the company co-founded by Steve Jobs. The original Apple iPod was not only the result of original design work, but rather also involved the creative combination of several successful design precedents from other sources, some of which also came from different domains. The hard drive came from Toshiba and the basic circuitry was borrowed from PortalPlayer [8]. The innovative click wheel, however, came from an entirely different place: it was based upon the design of a telephone made by Bang and Olufsen [36].

The preceding are examples of *product* innovation, to which may also be added *process* innovation [8]. Here, too, creative combination from different domains can yield innovative manufacturing or business processes. The idea for the car manufacturing assembly line process of Henry Ford came from slaughterhouses. Ford implemented the first profit-sharing business plan in America for his employees after being inspired by a book on political economy written in 1848 [37].

A particularly instructive example of business process innovation by creative combination is Netflix, started in the late 1990s with the (at the time) revolutionary concept of avoiding video rental stores like Blockbuster by sending movies by mail to the home on a subscription basis. The founder of Netflix recounted afterward [19] where the idea for the business model came from: a combination of the monthly subscriptions he knew from his local exercise gym, the online ordering service from Amazon (new at the time), and a new lightweight medium for storing movies called the DVD, replacing the heavy and bulky cassettes being used at the time and making it realistic to send them by mail.

6.3   Historical Precedents as Reusable Knowledge Components

Given the central role of historical precedents in the method, the search for precedents is the most critical step, and as much time as possible must be dedicated to this step (in a normal application of the method, the amount of time dedicated to this step can be larger than for each of the other steps by a factor of 8 to 1). Above all, this step is in the *solution space* (Fig. 3). This is very different from traditional methods, which spend most of their time in formal analysis in the *problem space*. It also reinforces the analogy to software reuse: a reusable software component is by definition in the solution space, and represents a successful prior solution to an element of a problem.

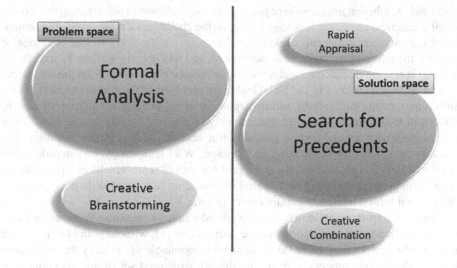

Fig. 3. Proportionally long step of search for precedents in the solution space.

The database of precedents can therefore be seen in a loose analogy to the repository of software components that the software engineer scans, searching for suitable components. The software components are combined by the engineer to fit a new application in the domain. The situation with historical precedents is analogous, but at a much higher level: these are multi-domain *knowledge components*, pieces of a solution to a puzzle, that are creatively combined in a new way.

Similar to the way in which software components can be combined into component hierarchies, the reusable knowledge of historical precedents can also be combined into innovation hierarchies. A typical example of such multiple hierarchical combination is process innovation on top of product innovation. We have seen the example of product innovation represented by the development of the iPod. On top of that product innovation, there was a step of process innovation: the innovative product (iPod) was combined with the iTunes software for organizing music and the iTunes Store for selling music to produce a new, innovative business model.

Thus, a database of precedents forms a valuable kind of corporate patrimony that can be augmented with every new Creative Strategy initiative that the company undertakes. This has a parallel with the concept of the *knowledge commons* [38] in the public domain, which Surowiecki has noted [39] as a continuously renewing resource of knowledge that forms the basis upon which new and innovative ideas are generated.

7   Conclusions: The Challenge of Knowledge Representation

At this time, the precedents in the database maintained by strategy & are not yet represented in any particularly actionable, "runnable" way. Yet in order to make the database of precedents a useful tool in performing the What-Works Scan, some augmentation of the searchability of the database in a meaningful way is needed. Given the multi-domain aspect and the inherent imprecision of precedents, the challenge of representing the semantics of a historical precedent is at least a great as the challenge of representing the semantics of requirements in systems engineering. Moreover, in the first two steps of Creative Strategy, the information available is incomplete, and it is not straightforward to define a measure of success in matching and retrieving an historical precedent from the database.

The approach adopted in the NextGenRE project [40] in the context of next generation requirements engineering and management could provide a roadmap of how to proceed in an incremental manner. Precedents will almost certainly first be recorded in an extremely unstructured format, probably free text. In order to initiate the migration to a more structured representation and storage, Wiki technology was introduced in NextGenRE. Wikis provide excellent support for textual representations with sophisticated formatting capabilities and a natural organization (e.g. "one Wiki page per precedent"), and form a relatively painless entry point.

The next step in NextGenRE was the introduction of semantic Wiki technology, which forms a point of integration into the semantic world, without going too far at one time. A first attempt may be made to identify a reasonable vocabulary for innovation (e.g. "product" and "process"), then a minimally structured set of queries containing appropriate keywords. Semantic links among Wiki pages (e.g. combinations of precedents) could be created without undue effort. Semantic wikis are rooted in formal logic, and could open up a path to the eventual development of ontologies and semantic queries as the vocabulary of precedents description is better understood with experience.

A next step in NextGenRE was the introduction of semantic template support: a facility for the generation of semantic Wiki pages with related semantic content based on predefined templates. As familiarity with the vocabulary and patterns of precedents is gained over time, it is feasible that instead of starting from the empty Wiki page, a template could be provided to help structure the description of a precedent and enhance its value in retrieval and analysis (e.g. a template for certain types of solutions to business process problems).

This approach could conceivably be adopted in this context and gradually enhanced and adjusted as experience is gained and the database is populated more fully.

There are enormous and fascinating challenges in creating support for making the knowledge in these precedents explicit, searchable, and actionable. It is an exciting goal that the knowledge representation and management community can set for itself and make a concrete contribution to the furthering of the innovation process.

Acknowledgements. The author wishes to thank Iaakov Exman for encouragement and guidance in the writing of this paper. Particular thanks are due to Ken Favaro for continual support in understanding the approach, implications, and implementation practicalities of strategic intuition and Creative Strategy.

References

1. Isaacson, W.: Steve Jobs. Simon & Schuster, New York (2011)
2. Christensen, C.: The Innovator's Dilemma. Harvard Business Review Press, Cambridge (1997)

3. Gorschek, T., Fricker, S., Palm, K., Kunsman, S.A.: A lightweight innovation process for software-intensive product development. IEEE Softw. **27**(1), 37–45 (2010)

4. Schmidt, E., Rosenberg, J.: How Google Works. Grand Central Publishing, New York (2014)

5. Cain, S.: Quiet: The Power of Introverts in a World That Can't Stop Talking. Crown, New York (2012)

6. Humble, J., Farley, D.: Continuous Delivery: Reliable Software Releases through Build, Test, and Deployment Automation. Addison-Wesley Professional, Boston (2010)

7. PwC 17th Annual Global CEO Survey (2014). http://www.pwc.com/gx/en/ceo-survey/2014/assets/pwc-17th-annual-global-ceo-survey-jan-2014.pdf

8. Carr, N.: Building Bridges. The Rough Type Press, Amazon Digital Services (2011)

9. Hollis, B., Maiden, N.: Extending agile processes with creativity techniques. IEEE Softw. **30**(5), 78–84 (2013)

10. Lepore, J.: The Disruption Machine. The New Yorker, New York (2014)

11. Duggan, W.: Strategic Intuition. Columbia University Press, New York (2007)

12. Porter, M.E.: Competitive Strategy. Free Press, New York (1980)

13. Sperry, R.W.: Cerebral organization and behavior: the split brain behaves in many respects like two separate brains, providing new research possibilities. Science **133**(3466), 1749–1757 (1961)

14. Mintzberg, H.: Planning on the left side and managing on the right. Harvard Bus. Rev. **54**, 49–58 (1976)

15. Ogawa, S., Lee, T.M., Kay, A.R., Tank, D.W.: Brain magnetic resonance imaging with contrast dependent on blood oxygenation. Proc. Nat. Acad. Sci. USA **87**(24), 9868–9872 (1990)

16. Squire, L., Kandel, E.: Memory: From Mind to Molecules. W. H. Freeman & Co., New York (1999)

17. Klein, G.: Sources of Power: How People Make Decisions. The MIT Press, Cambridge (1999)

18. Kahneman, D.: Thinking Fast and Slow. Farrar, Straus and Giroux, New York (2011)

19. Duggan, W.: Creative Strategy: A Guide for Innovation. Columbia University Press, New York (2012)

20. Larman, C., Basili, V.: Iterative and incremental development: a brief history. IEEE Comput. **36**(6), 47–56 (2003)

21. von Clausewitz, C.: On War. Random House, New York (1943)

22. Chambers, R.: Rural Development: Putting the Last First. Taylor and Francis, New York (1983)

23. Schaninger, W.S., Harris, S.G., Niebuhr, R.E.: Adapting general electric's workout for use in other organizations: a template. Manag. Develop. Forum **2**(1) (1999)

24. Favaro, J., Morisio, M. (eds.): ICSR 2013. LNCS, vol. 7925. Springer, Heidelberg (2013)

25. Dunnette, M.D., Campbell, J., Jaastad, K.: The effect of group participation on brainstorming effectiveness for 2 industrial samples. J. Appl. Psychol. **47**(1), 30–37 (1963)

26. Surowiecki, J.: The Wisdom of Crowds: Why the Many Are Smarter Than the Few and How Collective Wisdom Shapes Business, Economies, Societies and Nations. Doubleday, New York (2004)

27. Goldenberg, J., Mazursky, D.: Creativity in Product Innovation. Cambridge University Press, Cambridge (2002)

28. Christensen, C.M., Raynor, M.E.: The Innovator's Solution: Creating and Sustaining Successful Growth. Harvard Business School Press, Cambridge (2003)

29. Osterwalder, A., Pigneur, Y.: Business Model Generation: A Handbook for Visionaries, Game Changers, and Challengers. John Wiley and Sons, Hoboken (2010)

30. Kuhn, T.: The Structure of Scientific Revolutions. University of Chicago Press, Chicago (1962)
31. Neighbors, J.M.: The draco approach to constructing software from reusable components. IEEE Trans. Softw. Eng. **SE-10**(5), 564–573 (1984)
32. McNeil, D.G.: Car Mechanic Dreams Up a Tool to Ease Births. New York Times, 13 November 2013
33. Rhodes, R.: Hedy's Folly: The Life and Breakthrough Inventions of Hedy Lamarr, the Most Beautiful Woman in the World. Doubleday, New York (2011)
34. Gamma, E., Helm, R., Johnson, R., Vlissides, J.: Design Patterns: Elements of Reusable Object-Oriented Software. Addison-Wesley Professional, Boston (1994)
35. Alexander, C., Ishikawa, S., Silverstein, M., Jacobson, M., Fiksdahl-King, I., Angel, S.: A Pattern Language: Towns, Buildings, Construction. Oxford University Press, Oxford (1977)
36. Carr, A.: Apple's Inspiration For The iPod? Bang & Olufsen, Not Braun. Web, 6 November 2013. http://www.fastcodesign.com/3016910/apples-inspiration-for-the-ipod-bang-olufsen-not-dieter-rams
37. Mill, J.S.: Principles of Political Economy. Oxford University Press, Oxford (2008)
38. Hess, C., Ostrom, E. (eds.): Understanding Knowledge as a Commons. From Theory to Practice. MIT Press, Cambridge (2006)
39. Surowiecki, J.: Thinkers and Tinkerers. Foreign Affairs. Web, 27 December 2014. http://www.foreignaffairs.com/articles/142590/james-surowiecki/thinkers-and-tinkerers
40. Favaro, J., Mazzini, S., Schreiner, R., de Koning, H., Olive, X.: Next generation requirements engineering. Proc. INCOSE Int. Symp. **22**(1), 461–474 (2012)

How to Build Ontologies for Requirements Systems Engineering Projects Aiding the Quality Management Process

Anabel Fraga(✉) and Juan Llorens

Carlos III of Madrid University, Madrid, Spain
{afraga,llorens}@inf.uc3m.es

Abstract. Knowledge is centric to systems engineering, the knowledge management process must take into account that a Systems Knowledge Repository (SKR) exists as a key element for either quality improvement, traceability support and, in summary, reuse purposes. Requirements engineering in the Systems Engineering process is enhanced by using knowledge systems and quality of requirements enriched as well. The more correct, complete and consistent a requirement is, the best performance it will have and knowledge systems enable a more exhaustive and fast quality process. A knowledge management process is proposed and it is guided by a requirements domain based example using a Knowledge Management tool supporting the whole process.

Keywords: Requirements engineering · Ontologies · Knowledge · System engineering · Reuse · Quality

1 Introduction

The application of ontology engineering in systems engineering seems to be a very promising trend [4, 12]. We call it system verification based on Knowledge Management, and it deals with assisting System Engineers to get a complete and consistent set of requirements (e.g. compliance to regulation, business rules, non-redundancy of requirements…) by using Ontologies, which represent the domains of knowledge of an organization. The combination of Requirements Engineering with Knowledge Management, throughout Information Retrieval from existing sources, allows the verification process to measure quality of a set of requirements by traceability, consistency/redundancy, completeness and noise. Information retrieval enables also to verify the completeness of the ontology using a PDCA (Plan-Do-Check-Act) cycle of improvement. Requirements engineering is the first step and by traceability and Ontology based systems, similar assets of any phase of the development process used in analogous projects could be reused and adapted to a new challenge. For instance, by using a semantic approach, a requirement can be translated into a graph by means of NLP (Natural Language Processing) techniques.

In order to build a knowledge repository for managing requirements quality, the first activity must be to clearly define the typology of requirements that are going to be

© Springer International Publishing Switzerland 2015
A. Fred et al. (Eds.): IC3K 2014, CCIS 553, pp. 457–476, 2015.
DOI: 10.1007/978-3-319-25840-9_28

covered by the knowledge system, because this typology will affect the requirements structure and vocabulary. For example, an organization would be interested in managing the quality of "performance requirements". Even if it seems to be simple, the selection of the kind of requirements to formalize in a knowledge repository is not trivial. "In order to define what I want, I usually need to know what I do not want". And this is a real problem. Because in many cases "performance requirements" will NOT be considered by you as "functional requirements", but in other cases they certainly will. Figure 1 shows a requirement types taxonomy.

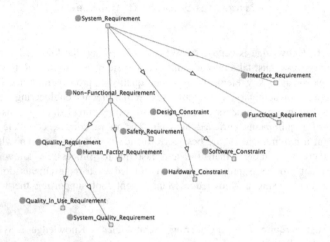

Fig. 1. Taxonomy of requirements types.

A requirement is an identifiable element of a function specification that can be validated, and against which an implementation can be verified. [1–3, 5–7, 11]

Requirements Development encompasses all of the activities involved in eliciting, analyzing, specifying and validating the requirements.

Requirements Management encompasses the activities involved in requesting changes to the requirements, performing impact analysis for the requested changes, approving or disapproving those changes, and implementing the approved changes. Furthermore, includes the activities used for ensuring that work products and project plans are kept consistent and tracking the status of the requirements as one progresses through the software development process.

There are some rules that establish how the requirements must be written and which mistakes must be avoided. The INCOSE (International Council on Systems Engineering) rule states that the requirements must be clear and concise, complete, consistent, verifiable, affordable, achievable, necessary, bounded, traceable with independent implementation [8]. Any mistake in the requirements definition phase is distributed downwards until low level requirements being almost impossible to fix. Thereby, those mistakes must be caught in the early development process.

2   Knowledge Management Process

Stage 0. Inmature Situation. An organization is in an Immature situation for the knowledge management process when there is no conscious understanding of the need of managing assets for their (re)-use internally.

Stage 1: Managing SaS. This stage is gathered as soon as the organization has implemented whatever activity for supporting an assets store. Usually, this store is based on a defined model for an assets repository (the most common is the OMG [13]).

Stage 2: Managing Terminology. This stage is reached when the organization is managing the terminology that affects the system of interest. The existence of a controlled vocabulary of terms, as concepts, allows the quality management process to produce relevant metrics about the work-products, and therefore it is a key stage to provide qualified assets as well.

In this stage we will start with our guided example, imagine a system "Car", with components like "Brakes", "Pedals", and also actions even when the pedals are "pressed", and also additional variables as "velocity" and "time of speed".

Here the vocabulary and thesaurus must be built. For instance:

The project breakdown structure (PBS) also contains useful information:

Car = System
Brake = Brake System
Pedal of the brake
Pedal States: Pressed, Released

Knowledge about the physical world also helps to understand the requirement:
Measurement unit equivalence
Speed   ---------------------------------------→ V
Acceleration ---------------------------------→ A = V'
    | Antinomy
    |_____>Deceleration

Stage 3: Managing the System Conceptual Model (SCM). When an organization supports a persistent representation of a SCM, it becomes an asset ready to be reused. The existence of a SCM allows the quality management process to produce advanced metrics about the work-products. In organizations that only want to develop the Knowledge Management Process for Knowledge Reuse, this stage represents the possibility to reuse "ground truth" knowledge at the organization level. Those organizations that do not want to develop a SCM can have a light stage, with almost no effort.

In this example the organization manages assets for indexing and retrieval. In this stage the semantic grouping of concepts must be managed, at conceptual level it is not showed but semantics will be implemented in a tool, as for instance Protégé [10], Knowledge Manager [9], and so on.

Stage 4: Managing Patterns. Patterns are developed with the intention to be matched to Systems Engineering work-products content. When an organization owns a set of patterns, it becomes possible to identify relevant structures inside the produced content.

By using patterns matching, specialized software can identify the patterns inside artifacts and be able to assign them the patterns as a means to classify them. The existence of patterns allows the quality management process to identify the typology of the work-products. On the other side, the Knowledge Management process needs patterns as the first stage to properly formalize them.

In the case of a requirement like: *"Whenever the pedal of the brake is pressed, the car shall decelerate immediately"*

A quick first syntactical analysis will present the following structures:

Whenever
The
Pedal of the brake
Is pressed
The
Car
Will
Decelerate
Immediately

Two syntactic structures have to be considered here: The compound noun "pedal of the brake" and the compound verb "is pressed". The joining of those words as a single compound phrase is performed by the Tokenization stage of the process.

We'll not produce patterns at syntactical level because we consider that "Pedal of the brake" is, in itself, a term of the domain with full meaning. This means that "Pedal of the brake" must be included in the ontology. The structure "is pressed" will be split in the two terms that form it: the verb "to be" and the verb "to press" in participle.

The patterns structure looks like (Fig. 2):

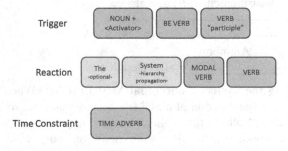

Fig. 2. Pattern structure for the guided example.

Stage 5: Managing Formalization. In stage 4 the organization owns an ontology with formalization capabilities for different work-product types. The formalization information, today, is formed by rules for producing formal representations of work-products.

The existence of Formalization information allows the quality management process to produce advanced metrics about the work-products.

Due to the existence of formalization capabilities, the supported operations in the previous stage are improved in the following way:

- Artifact Retrieve. The capability of the repository to formalize artifacts allows producing smart retrieval algorithms for semantic search, improving precision.
- Artifact Traceability. Due to the formalization capability, automatic trace policies based on similar content can be produced.

The following formalization will be used for the example we follow in this paper (Fig. 3):

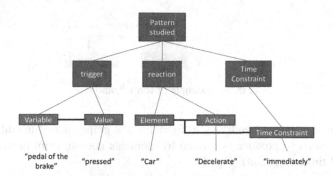

Fig. 3. Formalization of the requirement.

As it can be seen in the previous diagram, the trigger pattern produces by itself a graph (Fig. 4):

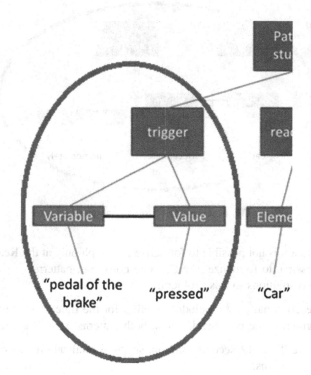

Fig. 4. Graph of the trigger.

The complete representation will be (Fig. 5):

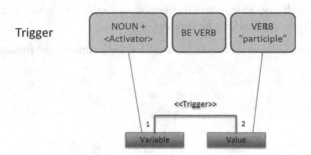

Fig. 5. Example of formalization.

But, a problem arises when trying to formalize the graphs out of the other patterns: the graph we want to produce is formed by elements coming from different patterns (reaction and time constraint) (Fig. 6):

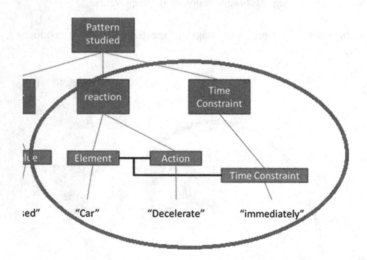

Fig. 6. Example of formalization #2.

This means that it is not possible to formalize the graph only at the Reaction pattern, and also not possible to formalize it at the time constraint pattern.

There are two solutions to this problem:

- Unify the reaction pattern to include a section for the time constraint.
- Create the graph at the next level, where both patterns exist as sub-patterns.

In our case, we´ll use the second solution, so the formalization will be produced at the next level of patterns.

This formalization process is, of course, very human dependent, and we must not be afraid of that. The result of this process must be the way the organization understands the requirement semantics. Just by defining a way, good benefits of it can be gathered (sharing same understanding, promoting the view to the supply chain etc. etc.)

For example, other Systems Engineering group could see the studied requirement as a state machine transition, and the modeling could be a different one (Fig. 7).

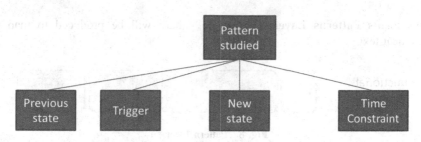

Fig. 7. Pattern structure.

Stage 6: Managing the Systems Knowledge Repository (SKR). In stage 5 the organization has included inference information to the ontology, completing the System Knowledge Base (SKB) and forming a complete System Knowledge Repository (SKR).

Inference information is necessary, as well, for the quality management process, in order to produce consistency metrics.

Usually, the inference rules for decision making purposes are based on the existence of a full developed SCM. Therefore, in some cases, the SCM is fully developed in this stage.

Due to the existence of inference capabilities, the supported operations in the previous stage are improved in the following way:

- Artifact Transformation. The existence of inference information allows generating new knowledge based on previous artifacts. For example: a rule could allow a user to state a complex query using UML, asking the repository to create new UML models with the information in the similar artifacts that is NOT existing in the query, and being notified of them by producing new artifacts.
- Artifact Traceability. Due to the inference capability, automatic trace policies based on transformations can be produced.

Once a terminology, a thesaurus, patterns and its formalization has been done, if a requirement instance of a type matching is needed, as for instance: "If the car is moving, whenever the pedal of the brake is pressed, the car will decelerate immediately". The SCM plays its role and indexing/retrieval systems are involved in the process. The knowledge system is completed and implemented in a Knowledge System as Knowledge Manager [9] that fully support this kind of process and even patterns are available for this kind of requirements.

Classification Layer of the Ontology. The <Activation> Class was needed to for defining the Trigger Pattern, so it must be included in the ontology, assigning the term "Pedal of the brake" as instance.

On the other hand, the following Relationship types were created for the graphs production:

- ≪Trigger≫
- ≪Reaction≫

Requirements Patterns Layer. One single pattern will be produced to map the requirement text:

Fig. 8. Pattern layer.

The defined sub-patterns can be found in the pattern, as well as the syntactical support elements to separate them.

As it has already been described, the formalization graph produced by this pattern will be the reaction graph, as it is affected by two sub-patterns (Fig. 9).

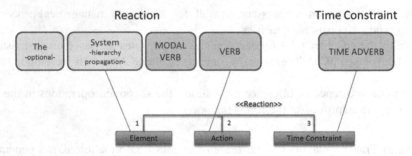

Fig. 9. Formalization of patterns in a semantic graph.

Ontology Information: Summary. Here we summarize the following operations:

- The vocabulary and thesaurus layers have been populated with:

```
Car = System
      Brake = Brake System
            Pedal of the brake  --------------→ Pedal States
                                                Pressed
                                                Released
```

Physical Environment

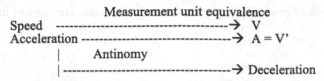

Measurement unit equivalence

Speed --→ V
Acceleration --------------------------------→ A = V'
      | Antinomy
      |------------------------------------→ Deceleration

- The Light ontology layer includes
  – Classes

    《Activator》
    Pedal of the brake
  – Relationship types
    《trigger》
    《Reaction》
- The patterns layer includes

A set of NOT INDEXABLE internal patterns with their corresponding formalizations (Fig. 10):

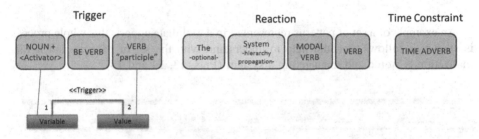

Fig. 10. Not indexable patterns.

A set of INDEXABLE requirements patterns, with their corresponding formalizations (Fig. 11):

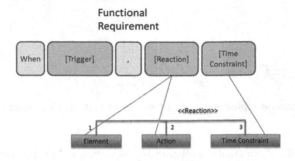

Fig. 11. Indexable patterns.

Where, the graph is indeed produced by the corresponding internal elements of the sub-patterns (Fig. 12):

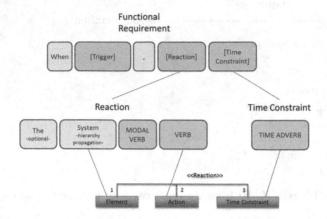

Fig. 12. Formal transformation #1.

An example of a set of patterns represented in a tool that supports this whole process is shown as follows, it could help to understand who the set of patterns and it formalization is stored and managed in a system (Figs. 13, 14 and 15).

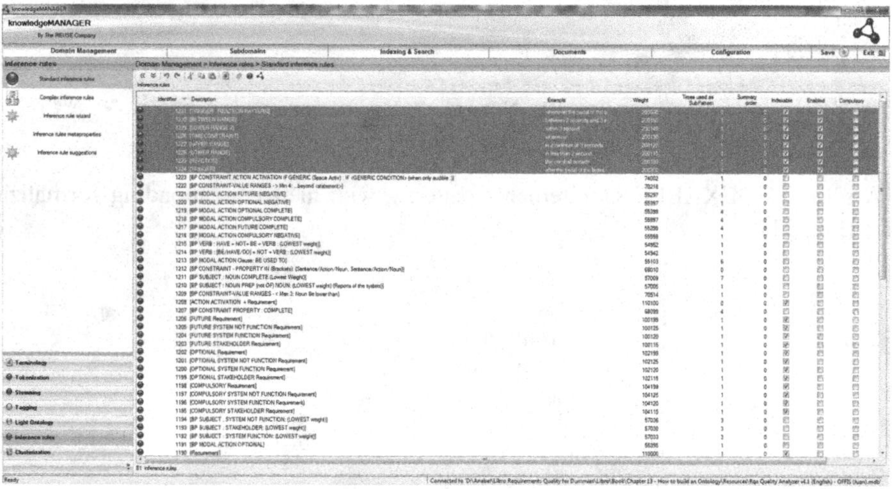

Fig. 13. Formal patterns representation in a tool: set of patterns (KnowledgeManager by The Reuse Company).

The TRIGGER pattern creates a relationship as shown in Fig. 16, using the semantic EVENT as root of the relationship. In the left side the PEDAL OF THE BRAKE (one of the terms of the thesaurus) and in the right side PRESS identified as verb and the

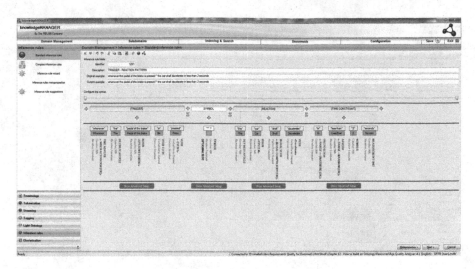

Fig. 14. Formal patterns representation in a tool: into a patter (KnowledgeManager by The Reuse Company).

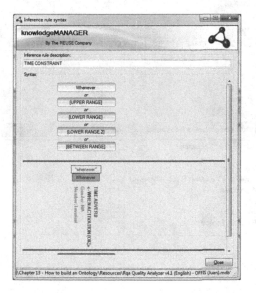

Fig. 15. Formal patterns representation in a tool: a syntactic element of one pattern (KnowledgeManager by The Reuse Company).

event done in the pedal of the brake. It is just going deeper into the implementation but iit helps to figure it out how it could be managed in a real project.

And now, what about if we index the requirement, the result is shown in Fig. 17.

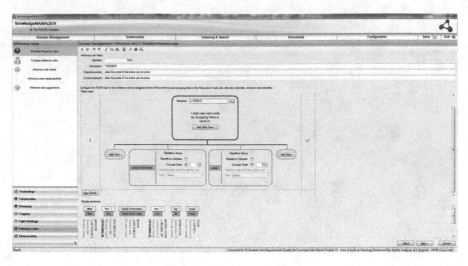

Fig. 16. Creating relationships in a pattern using semantics in KM.

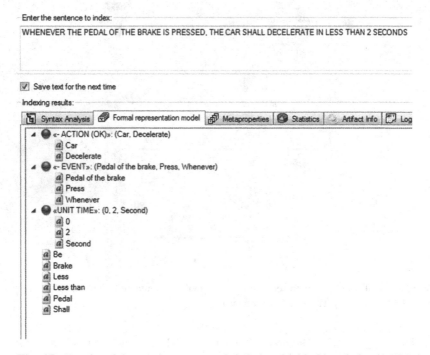

Enter the sentence to index:

WHENEVER THE PEDAL OF THE BRAKE IS PRESSED, THE CAR SHALL DECELERATE IN LESS THAN 2 SECONDS

☑ Save text for the next time

Indexing results:

| Syntax Analysis | Formal representation model | Metaproperties | Statistics | Artifact Info | Log |

▲ «- ACTION (OK)»: (Car, Decelerate)
　　a Car
　　a Decelerate
▲ «- EVENT»: (Pedal of the brake, Press, Whenever)
　　a Pedal of the brake
　　a Press
　　a Whenever
▲ «UNIT TIME»: (0, 2, Second)
　　a 0
　　a 2
　　a Second
　a Be
　a Brake
　a Less
　a Less than
　a Pedal
　a Shall

Fig. 17. Results of the requirement once it is indexed with this solution in KM.

Formalization Transformation and Post-Processing. When the input requirement arrives:

Whenever the pedal of the brake is pressed, the car will decelerate immediately

---

The following is the results after working with the formalization transformation (T2) (Fig. 18):

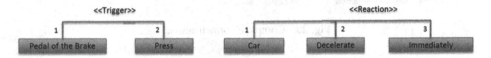

Fig. 18. Indexing results.

In order to get the expected results defined above a specific transformation must be produced to the graphs created so far (using the post- processing transformation). In order to represent:

In order to get the expected results defined above a specific transformation must be produced to the graphs created so far (using the post- processing transformation). In order to represent (Fig. 19):

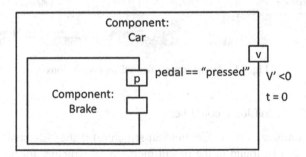

Fig. 19. System representation.

The graph should present the following structure (Fig. 20):

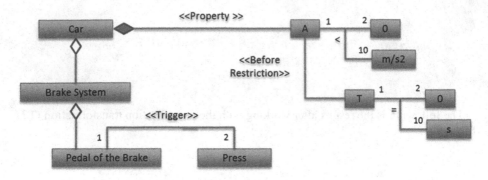

Fig. 20. Complete formalization.

Some slight differences can be found between both representations:

- The Pedal of the brake is modeled as a breakdown (an aggregation) instead of as a Property (as suggested), it means a metaproperty won't be used to represent the formalization of the pattern
- The requirement is not modeling that the car has a property named V (Speed) as it is NOT mentioned in the requirement.

And the input graph to be transformed is (Fig. 21):

Fig. 21. Transforming formal representations.

So the transformation logic could be:

1. Look in the ontology (in the hierarchical structure) if the "Element" of the [Reaction] pattern can be found in the parent hierarchical structure for the "Variable" of the [Trigger] Pattern (Figs. 22, 23).
   This restriction must be fulfilled in order to continue with the T2 transformation.
   If the restriction is fulfilled, the whole path between both terms in the ontology will be included in the graph representation of the requirement (Fig. 24):
2. If a relationship type of one relationship is of semantic type «reaction» and the node with concept order 2 in the relationship is the term Decelerate occurs substitute the node for (Fig. 25)

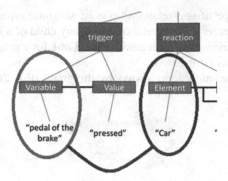

Fig. 22. Formal transformation #1.

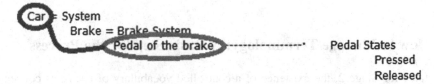

Fig. 23. Restrictions for transformations.

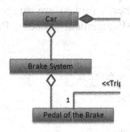

Fig. 24. Graph representation of the requirement.

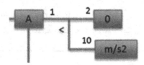

Fig. 25. Graph representation of the requirement #2.

And the relationship type to a Composition of type "Property"
3. If a relationship type of one relationship is of semantic type ⟪reaction⟫ and the node with concept order 2 in the relationship is any child of a Physical Environment that has a Measurement Unit equivalence substitute the term by the Measurement Unit equivalence.

4. If a relationship type of one relationship is of semantic type ≪reaction≫ and the node with concept order 2 in the relationship is any child of a Physical Environment that has NOT a Measurement Unit equivalence Look for a relationship of Antinomy and check that exists (if it exists)
5. Substitute the Immediately term always by the graph (Fig. 26).

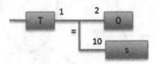

Fig. 26. Graph representation of the requirement #3.

Connected to the Physical Environment Measurement Unit equivalence, forming

3 How to Manage Terminology: A Basic Step for the Process

As stated for Stage 2, the existence of a controlled vocabulary of terms, as concepts, allows the quality management process to produce relevant metrics about the work-products, and therefore it is a key stage to provide qualified assets as well. The sequence diagram explaining the process is shown in Fig. 27. First of all, it is important to notice that diverse environments (areas) are suggested to be active in any new implantation of a Knowledge Management Process: Production, Pre-production, and

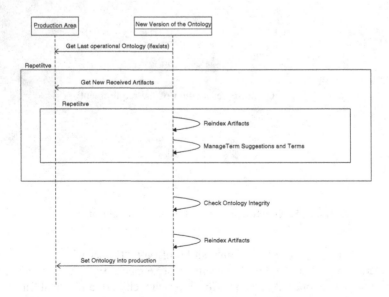

Fig. 27. Terminology process sequence diagram.

Evolution (New Version of the Ontology). In the Evolution area, the ontology is evolved based on the domain artifacts once it is ready then it is set as active and replicated in Pre-Production and Production areas. In the Pre-Production area indexing of new artifacts in an incremental mode is done based on the ontology active at that moment that maybe is not the one active in the Evolution area. In the Production area the retrieval process is done based on the indexing available and the ontology active at that point. Figure 8 shows a piece of the process between Production and Evolution areas focus on the terminology point of view.

A Terminology List is needed for standardizing and normalizing the terminology used in the custom application. The input information must/should match the controlled vocabulary. Using a glossary with different categories of terms, the ontology may store:

- Business related Terms: those terms central to the business area (domain) to be treated.
- General Language Terms: those terms related to the basic language used in the ontology (i.e. English, Spanish, German, and so on).
- Syntactically relevant phrases: those are general language terms related to the General Language Terms: Adverbs, Adjectives, etc.
- Stop Terms: those terms that could be of no relevance.
- Steps for gathering terminology.

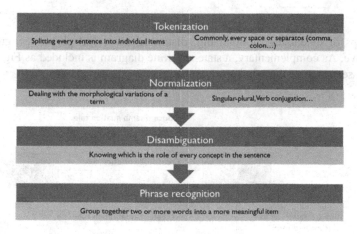

Fig. 28. Steps for gathering terminology.

Gathering Vocabulary: How It Works. The terminology of the domain must be extracted of the artifacts in the domain. For that reason, a sequence of steps must be completed, as shown in Fig. 28:

- Tokenization
- Normalization

- Disambiguation
- Phrase recognition

An example of the steps to be completed for gathering the terminology of the domain is as follows:

All Radars shan't identify the following targeting enemy objectives:

Tokenization	[All] [Radars] [shall] [not] [identify] [the] [following] [targeting] [enemy] [objectives] [:]
Normalization	[All] [Radar] [shall] [not] [identify] [the] [following/follow] [targeting/Target] [enemy] [objective] [:]
Disambiguation	[All] [Radar] [shall] [not] [identify] [the] [following] [Target] [enemy] [objective] [:]
Phrase Identification	[All] [Radar] [shall] [not] [identify] [the] [following] [Target] [enemy objective] [:]

The steps of tokenizing, normalizing, disambiguation and phrase recognition are shown above. As complementary, a state machine diagram is included as Fig. 29 as an illustrative sequence of the whole process.

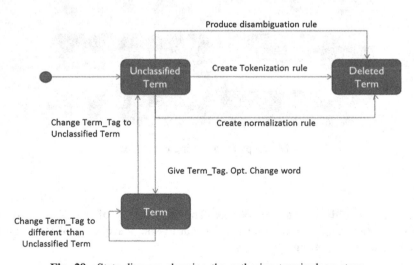

Fig. 29. State diagram showing the gathering terminology steps.

4 Ontologies Maintenance: Effort/ROI Perspective

The perception that building ontologies to improve Systems Engineering could imply a project with negative ROI is certainly real within organizations. The REUSE Company has produced a methodology that has as main goal to discourage this thinking. The fundamental aspects of this approach are:

- Define a SIMPLE schema/Meta-model for the Ontology, avoiding complex structure
- Create Semi-automatic methods to produce the ontology layers at low cost
- Define clear applications of the ontology inside the SE lifecycle, like, for instance, the requirements quality control
- Define a PDCA cycle to reduce the cost of maintenance

Finally, the experience has shown that, once the ontology is specialized to particular usages, the maintenance cost decreases very quickly once a determined level of maturity is gathered (Fig. 30).

These arguments allow producing positive ROI easily.

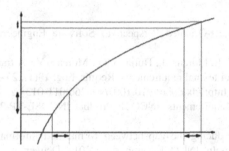

Fig. 30. ROI graphical representation.

5 Conclusion

Once the Knowledge Management Process was introduced and a guided example shown, the next step is the full definition of the example in a tool supporting all the Lifecycle of a requirement coexisting in a Knowledge System Repository. It is in progress and partially implemented at this point. The whole tool is in its process to be improved and completed for supporting the whole Knowledge Management Process for any project, company or organization.

The requirements in any project are one of the most important assets, if not the most. A bad group of requirements might have terrible implications in a developed system. For instance a requirement detailed in various parts of the requirement list using different measurement units might cause an accident or a failure during operation of any system.

Classical sequential review process of requirements is costly in terms of time consuming. Then support of tools for lexical, syntactic analysis enables to correct bad requirements writing before business of project reviews.

One of the next challenges in the Industry is to reach an ontology-assisted system engineering process to write SMART requirements at a first shot, it is available in some tools but just like an empty box to be filled by analysts without any checkup. The challenge is to check the requirements written and detect possible errors in order to improve the process from this first stage.

The use of ontologies and patterns is a reality for a better requirements engineering and knowledge reuse in any system engineering project. The process is improved in the first stages as Requirements engineering and Requirements management.

Acknowledgement. Authors thank Manuela Berg and Thomas Peikenkamp, both from OFFIS, for the requirement textual example provided for this guided definition of the process.

The Research Leading to These Results has Received Funding from The European Union's Seventh Framework Program (FP7/2007-2013) FOR CRYSTAL– CRITICAL System Engineering Acceleration Joint Undertaking under Grant Agreement N° 332830 and from Specific National Programs and/or funding authorities.

References

1. Braude, E.: An Object-Oriented Perspective. Software Engineering. Wiley, New York (2001)
2. Genova, G., Fuentes, J., Llorens, J., Hurtado, O., Moreno, V.: A framework to measure and improve the quality of textual requirements. Requir. Eng. **18**(1), 25–41 (2013). doi:10.1007/s00766-011-0134-z. http://dx.doi.org/10.1007/s00766-011-0134-z
3. Guide for Writing Requirements, INCOSE Product INCOSE-TP-2010-006-01, V.1, April 2012
4. Chale, H., et al.: Reducing the gap between formal and informal worlds in automotive safety-critical systems In: INCOSE Symposium 2011, Denver
5. Alexander, I.F., Stevens, R.: Writing Better Requirements. Addison-Wesley, London (2002)
6. Sommerville, I., Sawyer, P.: Requirements Engineering: A Good Practice Guide. Wiley, New York (1997)
7. Sommerville, I.: Software Engineering. Pearson-Addison Wesley, Boston (2005)
8. INCOSE, Systems Engineering Handbook. http://www.incose.org/ProductsPubs/products/sehandbook.aspx. Accessed 13 November 2013
9. OBSE Fundamentals, The Reuse Company, Juan Llorens (UC3 M) - José Fuentes (TRC). http://www.reusecompany.com. Accessed 13 November 2013
10. Protégé. Stanford University Development. http://protege.stanford.edu. Accessed 15 November 2013
11. Pressman, R.: Software Engineering: a practical guide, 6th edn. McGraw-Hill, New York (2006)
12. The Reuse Company (TRC). http://www.reusecompany.com. Accessed 13 November 2013
13. The Reusable Assets (RAS) - OMG. http://www.omg.org/spec/RAS/2.2. Accessed 7 January 2014

Apogee: Application Ontology Generation with Size Optimization

Iaakov Exman[✉] and Danil Iskusnov

Software Engineering Department,
The Jerusalem College of Engineering – Azrieli,
POB 3566 91035 Jerusalem, Israel
{iaakov,danilis}@jce.ac.il

Abstract. To obtain runnable knowledge – convertible into executable software – from the highest abstraction level of an application, one should start with a neat set of application ontologies. But the latter are not readily available in the literature. One needs to generate dedicated and smaller application ontologies from larger generic domain ontologies. The main problem to be solved is to optimize the size of the generated application ontologies as a trade-off between two opposing tendencies: to enlarge the selected domain ontology segments to include most relationships between relevant concepts, while reducing the same segments to exclude irrelevant terms. This work describes a chain of algorithms and a series of heuristic rules to reach the proposed solution. Finally, case studies are used to actually illustrate the whole approach.

Keywords: Application ontology · Domain ontology · Keyword extraction · Classification · Optimal application ontology size · Heuristic rules

1 Introduction

Recently [9, 24] we have been extending MDE (Model Driven Engineering), which starts software system development from a UML design model, to KDE (Knowledge Driven Engineering) which starts from the highest level of abstraction of the software hierarchy, viz. from a set of ontologies. This specific set of ontologies exclusively relevant to a kind of software system is what we call application ontologies.

Application ontologies are dedicated to the chosen kind of software system that one wishes to develop. Thus they are much smaller than the generic domain ontologies, whose set of concepts contain the respective application ontologies' concepts as subsets. There are two main reasons to work with smaller dedicated application ontologies: a- computational efficiency; b- conceptual neatness, i.e. a clear conceptualization of the software system. The latter reason assumes that conceptual neatness is essential for good software system design (see e.g. De Rosso [5]).

© Springer International Publishing Switzerland 2015
A. Fred et al. (Eds.): IC3K 2014, CCIS 553, pp. 477–492, 2015.
DOI: 10.1007/978-3-319-25840-9_29

The main goal of this paper is a systematic procedure to obtain the desired set of application ontologies, given the generic domain ontologies. Our approach is to combine a chain of existing algorithms within an integrated tool – coined Apogee – for generation of specific application ontologies.

A central issue of the referred algorithm chain is to decide the optimal size of the resulting application ontology, such that it may still be considered a legitimate ontology – including most of the relationships among relevant keywords – while excluding as far as possible terms irrelevant to the specific application.

1.1 Domain vs. Application Ontologies

One can roughly classify ontologies into specific and generic types. Specific types of ontologies have a narrowly defined purpose, and a relatively small size. Generic types can refer to whole domains, and diverse applications, and are often of large size.

A domain ontology is characterized by being comprehensive, i.e. to encompass all possible concepts that may appear in a given domain of discourse. Domain ontologies are usually of large sizes and may have a rather complex structure, containing generic concepts, their sub-types, instances and properties. Domain ontologies are commonly found in the Web.

In this paper we continue our investigation of special purpose, small ontologies. In particular, we deal with the extraction of specific/small ontologies from generic/large ones. Specific types of ontologies include among others, reputation ontologies (e.g. Chang et al. [1]) and opinion ontologies (e.g. Exman [8]).

Application ontologies as proposed in this work have a definite particular meaning. We assume that a given software system or application is conceptualized in its highest abstraction level by a set of related application ontologies.

For the sake of illustration, the development of a software application dealing with purchases in the internet – as was dealt with in ref. [24] – may be started from two ontologies: one defining the "*shopping cart*" and another one defining a "*product*".

Typical concepts in the *shopping cart* ontology are: products, items per product, tax, current price and total price. Common concepts in the *product* ontology can be: product name, product price, serial number and part number.

1.2 Related Work

General literature reviews concerning ontologies and types of ontologies are available in refs. by Guarino (July 1997) [11], Guarino (1998) [12] and Nguyen [16].

The notion of application ontology has appeared before in the literature, with a somewhat similar, but surely distinct connotation from the present work. In Guarino (July 1997) [11], an application ontology is described as a specialization of both domain ontology and task ontology. Here a task is a kind of activity like diagnosing or selling, which is clearly different from the intent of this work as a conceptualization of a software system. In Guarino (February 1997) [10], an application ontology is also described as a specialization of a domain ontology and a task (or method) ontology, which differs from this work's notion due to the introduction of the latter kinds of ontologies, in contrast to our use of scenarios as a characterization of the software system behavior.

The generation of application ontologies may involve different kinds of operations on existing ontologies and other kinds of documents, such as merging, intersection or matching of ontologies. For instance, Stumme and Madche [20], use FCA (Formal Concept Analysis) to produce, from documents described by ontologies, a conceptual lattice which is in turn used to merge the referred ontologies. Another example is found in Tijerino et al. [22], in which ontologies are generated from two-dimensional tables (like those in databases) by a few semi-automatic steps: mini-ontologies are constructed from the tables; inter-ontology mappings are found; the mini-ontologies are merged into the growing application ontology.

A very small sample of the extensive literature referring to ontology matching consists of Jean-Mary et al. [13] for ontology matching with semantic verification and Swab-Zamazal [21] about the influence of pattern-based ontology refactoring on ontology matching.

Tools of wide interest for specific operations involving ontologies include among others, Ontobraker (Decker et al. [4]) for ontology based accessing distributed information, Swoogle (Ding et al. [6]) a convenient search engine for the semantic web and ontologies that may serve as inputs to the application ontology generation, and the well-known Protégé (from Stanford University [17]) for ontology editing and graphical representation, as used in this work.

A project having some similarity to ours is ReDSeeDS, viz. Requirements-Driven Software Development System [18]. It offers a full MDE lifecycle. An example paper within this project is that by Nowakowsky et al. [15] which deals with requirements level tools to capture a software system essence.

1.3   Paper Organization

The remaining of the paper is organized as follows. We introduce the goal of our work in detail (Sect. 2), outline the solution chain of algorithms and heuristic rules in the process from domain to application ontologies (Sect. 3), overview the software design and implementation of the Apogee tool (Sect. 4), present a case study to illustrate the approach (Sect. 5) and conclude with a discussion (Sect. 6).

2   The Goal: From Domain to Application Ontologies

In this section we motivate the need for systematic extraction procedures of smaller ontologies dedicated to a desired application from generic domain ontologies.

2.1   KDE: Knowledge Driven Engineering

In our recent KDE (Knowledge Driven Engineering) approach (see e.g. [9]), we have proposed to develop software systems from a higher abstraction level than UML. The kinds of inputs to the KDE approach are:

(a) *Application Ontologies* – providing the semantics for the concepts in the software
    system;
(b) *Scenarios* – describing allowed and forbidden behaviors of the software systems,
    in terms of the concepts found in the ontologies. These should be executable by the
    application.

Concepts in the application ontologies are either candidates for UML classes, for
these classes' attributes or their methods. The classes' behaviors are obtained by
matching terms in the scenarios with the ontology concepts.

The problem is that application ontologies have been formulated manually ad
hoc, and were gradually refined as needed, when new scenarios were added to
describe the system more precisely. Despite the existence and wide availability of
domain ontologies, there was no formal starting point, nor a systematic procedure to
obtain the necessary application ontologies.

To illustrate the kinds of inputs and their possible relationships, we provide an
example for a small software system. For instance, in an internet purchase situation,
we have used an application ontology for the shopping cart as seen in Fig. 1.

A typical scenario for the same internet purchase example is shown in Fig. 2.
Note the lack of correspondence of terms in the application ontology to terms in the
scenario. Some of the terms in the ontology do not appear in the scenario and vice-
versa.

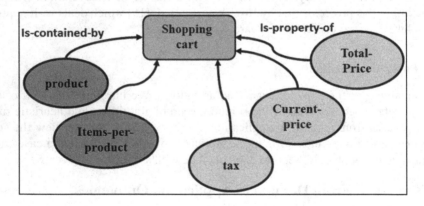

Fig. 1. Sample shopping-cart ontology for an Internet Purchase – this ontology was formulated
manually ad hoc. In the l.h.s. of the ontology {product, items-per-product} have the *Is-contained-
by* relationship. In the r.h.s. {Total-price, Current-price, tax} have the *Is-property-of* relationship
to Shopping-cart.

Internet Purchase – Sample Scenario
Pick a shopping-cart;
Put one item of product A in the shopping-cart;
Put three items of product B in the shopping-cart;
Proceed to payment.

Fig. 2. Sample Scenario for an Internet Purchase – Note that the terms {product, items-per-product, shopping-cart} appear in the shopping-cart ontology, but the term {payment} does not appear in the ontology.

2.2 Ad Hoc Formulation of Application Ontologies: Shortcomings

The manual ad hoc formulation of application ontologies has shortcomings as:

- *Lack of Systematic Criteria* – no guarantee of consistency within an application ontology and along time;
- *Lack of Knowledge Reusability* – formulation needs to be done anew even for similar systems in the same domain.

Thus, we are led to seek a systematic procedure to obtain application ontologies from domain ontologies.

3   The Solution: From Keywords to an Optimal Size Ontology

In this section we describe our solution, on how to deal with the process leading from generic domain ontologies to smaller ontologies dedicated to a desired application. We describe the chain of relevant algorithms, starting from the inputs to the process and emphasizing the main issue of an optimal size for the application ontology.

The overall idea is that a software stakeholder or developer knows the desired behavior of the software system. This behavior is expressed in a scenario as in Fig. 2. From the scenario one extracts keywords, to characterize suitable domain ontologies. Then we get into techniques to reduce the latter into smaller application ontologies.

3.1 Inputs

The inputs to the process to generate application ontologies are:

- *Bank of ontologies* – a bank of potentially useful domain ontologies;
- *Scenarios* – user provided scenarios that describe the system behavior and are a source of keywords to start the algorithm chain.

3.2   The Chain of Relevant Algorithms

The proposed chain of algorithms for the generation of application ontologies from domain ontologies is seen in Fig. 3.

Fig. 3. Algorithm chain for Application Ontologies generation – The chain is shown in a Statechart with a set of sequential states activated from left to right.

The algorithms' purposes are as follows:

- *Keyword Extractor* – selects from the input scenario (text) relevant application keywords. For the internet purchase example these could be: shopping-cart, item, product, payment.
- *Domain Classifier* – assigns keywords to a domain, in order to select potential domain ontologies from the bank of ontologies. For instance 'payment' belongs to the 'financial' domain. An important issue here is to resolve ambiguities; for instance, does 'bridge' refer to card games or to civil engineering structures?
- *Ontology Segments' Matcher* – performs matching of the keywords in the keyword list to selected domain ontologies and extract segments of those ontologies including the matched keywords.
- *Application Ontology Composer* – composes a preliminary application ontology from matched ontology segments; its size is optimized by means of heuristic rules. This is explained in more detail in the next sub-sections.
- *Application Ontology Checker* – checks the ontology for consistency and subsumption relationships among classes.

3.3   The Optimal Size of the Application Ontology

Here we focus on the main issue of optimal size of the application ontology, which occurs in the *ontology segments matcher* and the *application ontology composer* stages. We first explain the problem and then propose our specific solution.

The matching of ontology segments starts by searching, for all keywords, all the appearances of a given keyword in the relevant domain ontology. For each such appearance, one traverses the domain ontology graph upwards, until one either gets to the top (say "thing") or there is a criterion which tells one to stop earlier. All the terms before stopping are included in the current ontology segment.

The referred criteria are the significant part of the solution. These criteria are given by heuristic rules for application ontology size optimization. These criteria strive to include all the relevant terms and relationships, while excluding irrelevant terms, to achieve an optimal size of the application ontology.

3.4 Heuristic Rules for Size Optimization

We need more precise rules to improve the quality of the resulting application ontology. We therefore provide a set of heuristic rules to obtain the size optimized application ontology from a preliminary one:

- *Top Generic Concepts* – the top class concept in OWL (the Web Ontology Language, with permuted initials) is *thing*. Immediately below thing, there should appear generic terms relevant to the software system domain;
- *Eliminate Stop-words* – stop-word terms (found in stop-word lists or files [19]), which appear as independent concepts in the preliminary application ontology should be eliminated from the optimized application ontology;
- *Bottom-up Search above a Keyword* – on the bottom-up direction starting from a keyword found in a domain ontology segment, add up to k concepts above the keyword (where k is a chosen fixed small integer, say k = 1 or k = 2);
- *Category Classification Terms* – include in the application ontology categories found by the domain classifier; for instance in our ATM domain the category "*financial*";
- *Unify Identical Segments* – identical segments should be unified and not repeated;
- *Identical Terms* – if one encounters again a term found in a previous segment, one can coalesce the new segment with the previous one.

The underlying meta-rules for the above heuristics can be summarized as follows:

a. Eliminate irrelevant (stop-words) and repeated material (identical segments and terms);
b. Keep relevant terms (top generic concepts, category classification), which would reinforce the domain classification.

4   Apogee: Design and Implementation

Apogee is the tool that we have designed and implemented to generate dedicated application ontologies from larger generic domain ontologies. As far as possible, we have reused external tools by means of calls to their APIs.

4.1   Apogee Design

The modules of the Apogee tool are schematically shown in Fig. 4. It contains core internal modules and external tools called by means of APIs. The order of execution of functions within those modules is depicted in the sequence diagram in Fig. 5.

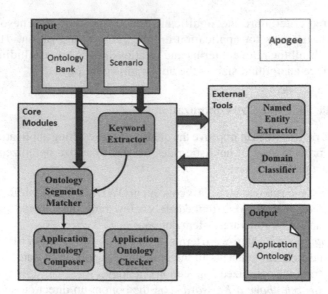

Fig. 4. *Apogee Schematic Design* – It shows: a- Input; b- External tools called by core modules; c- Core Apogee modules; d- Output. Arrows display transitions between modules.

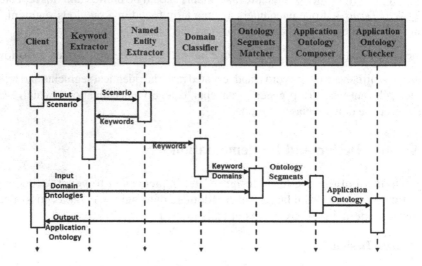

Fig. 5. *Apogee Sequence Diagram* – It shows messages (as horizontal arrows) between objects. The vertical dashed arrows are time axes from top to bottom. The diagram displays the order of the messages and object function executions.

4.2 Apogee Implementation

Apogee core modules were implemented in Java. The modules use Java libraries such as:

- OWL API – for creating, manipulating and serializing OWL ontologies;

  The external tools used were:

- dataTXT-NEX – as Named Entity Extraction [2];
- dataTXT-CL – as text classifier [3].

  The output format is textual OWL.

5 Case Studies

We have made experiments with several case studies to validate the approach. These included diverse specific variations such as providing as input to the domain classifier the whole scenario, or sentence by sentence individually.

Here we show one of the case studies, the ATM cash withdrawal, to illustrate results and difficulties.

5.1 Case Description: ATM Cash Withdrawal

In Fig. 6 one can see a sample input scenario and selected keywords. It uses a Gherkin syntax in the easily applied and understood [**Given, When, Then**] format, used by the Cucumber [23] tool, to describe the scenario.

These scenarios can be input in various "natural" languages. Note that the syntax does not demand grammatically well-formed sentences.

ATM Cash Withdrawal – Sample Scenario

Account has sufficient funds.
Given the account balance is 100.
And the Card is valid.
And the ATM contains 50.
When the Account requests cash 20.
Then the account balance should be 80.
And the card should be returned.

Fig. 6. ATM Cash Withdrawal *Sample Input scenario and selected keywords* – The scenario uses the Gherkin [**Given, When, Then**] format. Selected keywords have a green background.

It is probably preferable to input the scenario as a whole to the domain classifier, instead of sentence by sentence. For instance, let us look at the second sentence:

Given the balance is 100.

In the context of the whole scenario, a human would certainly understand this sentence, assuming that balance refers to the account. But as an individual sentence, balance is ambiguous and could have several different meanings in various domains.

5.2   Domain Classifier: ATM Cash Withdrawal

The domain classifier outcome for the above ATM Cash Withdrawal scenario is seen in Fig. 7 below.

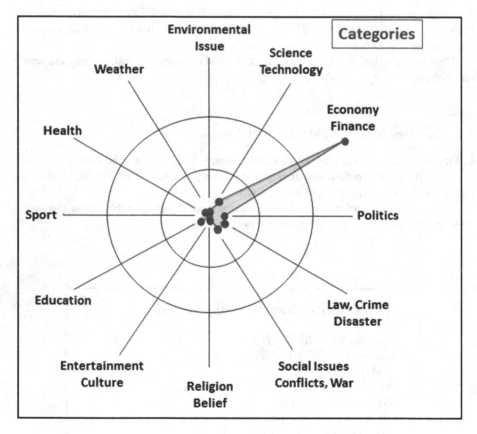

Fig. 7. ATM Cash Withdrawal *Classifier Outcome* – From the categories suggested by the classifier, the chosen category is *"Economy, Finance"* as shown by the red point farthest from the figure center.

One should note that a keyword consisting of *"balance"* alone, without account, could give a completely different result. Balance of forces has a physical meaning, fitting *"Science, Technology"*. Balance can refer to balance of power among parties, fitting *"Politics"*. Balance can be a quality of a basketball player or a characteristic of a judo competition, fitting *"Sport"*. Balance between preys and predators, fits *"Environmental Issue"*.

5.3 Preliminary Application Ontology

A preliminary class hierarchy of the application ontology for the previous values of the classifier is shown in Fig. 8. Ontology segments were obtained by going upwards from the matched keyword. The textual OWL output was inserted into Protégé [17] for display purposes. One can see roughly four segments demarcated by their indentations.

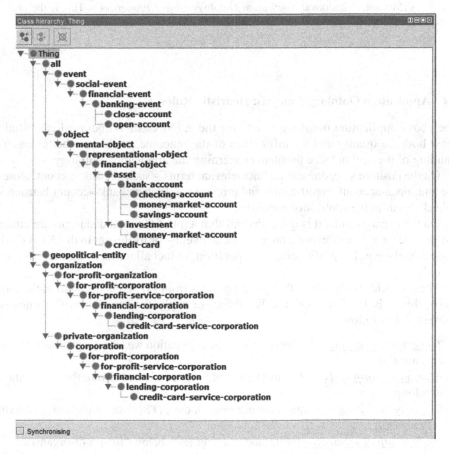

Fig. 8. ATM Cash Withdrawal preliminary *Application Ontology Classes* – This is the class hierarchy displayed by the Protégé tool.

Next in Fig. 9 one sees the corresponding object properties of the same application ontology, obtained by similar considerations.

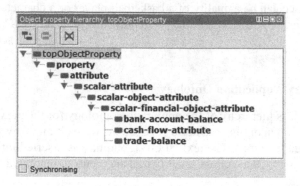

Fig. 9. ATM Cash Withdrawal *Application Ontology Object Properties* – This is the object property hierarchy, displayed by the Protégé tool.

5.4   Application Ontology Analysis: Heuristic Rules

The above application ontology results for the ATM Cash Withdrawal case study show both the quality and the difficulties of the outcome, enabling a better under-standing of the optimal size problem concerning the application ontology.

On the positive side, one can see that relevant terms – such as bank-account, close-account, open-account, credit-card – and properties – such as bank-account-balance – indeed appear in the application ontology.

On the negative side, it is quite obvious that a lot of the included terms are either too generic – e.g. social event, above financial event – or irrelevant to the ATM Cash Withdrawal – e.g. for-profit-service-corporation, in fact all terms containing 'corpo-ration'.

These considerations show that going upwards from say 'bank account' to the root class "thing" is far from optimal. We therefore apply the set of heuristic rules in Subsect. 3.4 as follows:

- *Top generic concepts* – for the current ATM application we select e.g. object, event, organization;
- *Eliminate stop-words* – for instance, "all" is eliminated from the application ontology;
- Category classification terms – for instance, in our ATM domain the category term is *"financial"*;
- *Unify identical segments* – for instance the segments below "for-profit-organization" and below "corporation" in Fig. 8 are identical, thus are unified.

5.5 Optimized Application Ontology

Usage of the above heuristic rules, in the preliminary Application Ontology (seen in Figs. 8 and 9), results in the optimized Application Ontology whose classes are displayed in Fig. 10. The rough number of apparent segments was reduced to three, corresponding to the three generic concepts below thing: event, object and organization.

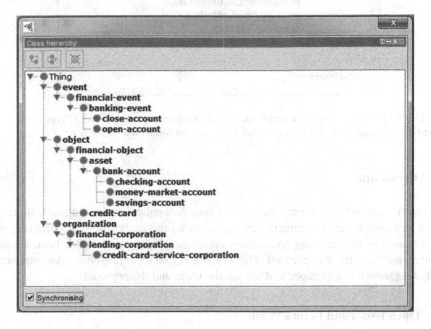

Fig. 10. ATM Cash Withdrawal optimized size *Application Ontology Classes* – This is the class hierarchy displayed by the Protégé tool. It was obtained from the preliminary Application Ontology classes in Fig. 8, using the Heuristic rules.

One can see that the optimized size Application Ontology is much smaller than the preliminary one. Thus, it keeps all the relevant keywords, while having much lesser numbers of irrelevant or too general terms, with regards the ATM cash withdrawal application.

The object property hierarchy of the optimized size Application Ontology was obtained by usage of the same Heuristic rules. It can be seen next in Fig. 11. The object property hierarchy is also smaller relative to the preliminary Application Ontology, while keeping relevant keywords.

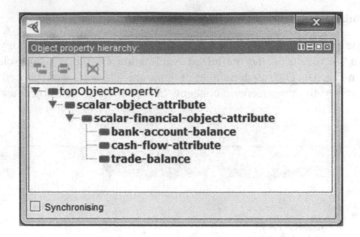

Fig. 11. ATM Cash Withdrawal optimized size *Application Ontology Object Properties* – This is the object property hierarchy, after usage of the optimization Heuristic rules.

6 Discussion

This work proposes the generation of small specific application ontologies from much larger and comprehensive domain ontologies, in a semi-automatic way, stressing size optimization of the resulting application ontologies. The approach has been actually implemented into the Apogee tool. The tool was tested by case studies. A sample case study was shown in this paper to illustrate the ideas and the approach.

6.1 Open Issues and Future Work

Several issues have been raised along this work.

Concerning the input composition and size, there is a trade-off. If one inputs a full scenario at once to Apogee, the many keywords facilitates resolution of domain ambiguity, on the other hand may introduce additional noise which is difficult to control. The alternative is to input the scenario line by line, in which case ambiguity is more prevalent, but one has less noise. The issue seems to be incompletely resolved.

Regarding the output, again one could decide a priori, by human intervention, the number of output application ontologies: one single larger application ontology or two smaller ones, etc. One would like to be able to formulate criteria to solve this issue in a more clear-cut way.

With respect to the main issue of optimal size of the application ontology, we intend to formalize the heretofore heuristic rules. For instance quantitative formulas for the criteria on how to include/exclude terms in the vicinity of the selected keywords are needed. The '*k* level above the keywords' criterion is ad-hoc and requires deeper testing. Possible quantitative formulas to eliminate irrelevant terms include those found in data mining areas, say TfIdf and the notion of Relevance within the context of the interestingness approach [7]. This issue deserves extensive investigation.

The whole line of research about special purpose small (or even nano-ontologies), is still in need of a unified approach, in particular referring to their semantic significance and their "legitimate" status as first-class citizens of the ontology world.

An interesting open issue is the demands from application ontologies in order to enable their reuse for different systems having the same requirements and purpose.

The current Apogee is an initial prototype. A second version – after comprehensive testing of the current version – could be much more flexible, e.g. to include APIs for various alternative external tools.

6.2 Main Contribution

The main contribution of this work is a systematic approach, starting from large domain ontologies, to generate software specific application ontologies, with size optimization.

References

1. Chang, E., Hussain, F.K., Dillon, T.: Reputation ontology for reputation systems. In: Meersman, R., Tari, Z. (eds.) OTM-WS 2005. LNCS, vol. 3762, pp. 957–966. Springer, Heidelberg (2005)
2. dataTXT-NEX – dandelion named entity extraction API. https://dandelion.eu/products/datatxt/nex/demo/
3. dataTXT-CL – dandelion text classification API. https://dandelion.eu/products/datatxt/cl/demo/
4. Decker, S., Erdmann, M., Fensel, D., Studer, R.: Ontobroker: ontology based access to distributed and semi-structured information. In: Proceedings of the DS-8, pp. 351–369 (1999)
5. De Rosso, S.P., Jackson, D.: What's Wrong with Git? A Conceptual Design Analysis. Onward! 2013, pp. 37–51 (2013). doi:10.1145/2509578.2509584
6. Ding, L., Finin, T., Joshi, A., Peng, Y., Cost, R.C., Sachs, J., Pan, R., Reddivari, P., Doshi, V.: Swoogle: a semantic web search and metadata engine. In: Proceedings of the CIKM 2004. ACM (2004). http://swoogle.umbc.edu/
7. Exman, I.: Interestingness – a unifying paradigm – bipolar function composition. In: Fred, A. (ed.) Proceedings of the International Conference on Knowledge Discovery and Information Retrieval, KDIR 2009, pp. 196–201 (2009)
8. Exman, I.: Opinion-ontologies: short and sharp. Accepted for Publication in 6th International Conference on Knowledge Engineering and Ontology Development, KEOD (2014)
9. Exman, I., Yagel, R.: ROM: an approach to self-consistency verification of a runnable ontology model. In: Fred, A., Dietz, J.L.G., Liu, K., Filipe, J. (eds.) IC3K 2012. CCIS, vol. 415, pp. 271–283. Springer, Heidelberg (2013)
10. Guarino, N.: Understanding, building and using ontologies. Int. J. Hum. Comput. Stud. **46**, 293–310 (1997). doi:10.1006/ijhc.1996.0091
11. Guarino, N.: Semantic matching: formal ontological distinctions for information organization, extraction and integration. In: Pazienza, M.T. (ed.) SCIE 1997. LNCS, vol. 1299, pp. 139–170. Springer, Heidelberg (1997)
12. Guarino, N.: Formal ontology in information systems. In: Guarino, N. (ed.) Formal Ontology in Information Systems, pp. 3–18. IOS Press, Amsterdam (1998)
13. Jean-Mary, Y.R., Shironoshita, E.P., Kabuka, M.R.: Ontology matching with semantic verification. Web. Semant. **7**(3), 235–251 (2009). doi:10.1016/j.websem.2009.04.001

14. Luke, S., Spector, L., Rager, D. Hendler, J.: Ontology-based web agents. In: Proceedings of the 1st International Conference on Autonomous Agents, AA 1997, pp. 59–66 (1997)

15. Nowakowski, W., Śmiałek, M., Ambroziewicz, A., Straszak, T.: Requirements-level language and tools for capturing software system essence. Comput. Sci. Inf. Syst. **10**(4), 1499–1524 (2013). doi:10.2298/CSIS121210062N

16. Nguyen, V.: Ontologies and information systems: a literature survey. Technical report DSTO-TN-1002, Australian Government, Defence Science and Technology Organization (2011). http://digext6.defence.gov.au/dspace/bitstream/1947/10144/1/DSTO-TN-1002PR.pdf

17. Protégé - A free, open-source ontology editor and framework for building intelligent systems. http://protege.stanford.edu/

18. ReDSeeDS – Requirements-Driven Software Development System (2014). http://smog.iem.pw.edu.pl/redseeds/

19. Stop-words list in English. http://xpo6.com/list-of-english-stop-words/

20. Stumme, G., Maedche, A.: FCA-MERGE: bottom-up merging of ontologies. In: Proceedings of the 17th International Joint Conference on Artificial Intelligence, IJCAI 2001, pp. 225–230 (2001)

21. Svab-Zamazal, O., Svatek, V., Meilicker, C., Stuckenschmidt, H.: Testing the impact of pattern-based ontology refactoring on ontology matching results. In: Proceedings of the Ontology Matching Workshop, hosted by ISWC 2008, 7th International Semantic Web Conference, pp. 229–233 (2008)

22. Tijerino, Y.A., Embley, D.W., Lonsdale, D.W., Ding, Y., Nagy, G.: Towards ontology generation from tables. World Wide Web: Internet Web Inf. Syst. **8**, 261–285 (2005). doi:10.1007/s11280-005-0360-8

23. Wynne, M., Hellesoy, A.: The Cucumber Book: Behaviour Driven Development for Testers and Developers, Pragmatic Programmer, New York, USA, (2012)

24. Yagel, R., Litovka, A., Exman, I.: KoDEgen: a knowledge driven engineering code generating tool. In: Proceedings of the 4th International Workshop on Software Knowledge, SKY 2013, hosted by IC3K, Vilamoura, Portugal, pp. 24–33 (2013)

Knowledge Management and Information Sharing

Knowledge Management Problems at Hospital:
An Empirical Study Using
the Grounded Theory

Helvi Nyerwanire[1], Erja Mustonen-Ollila[1(✉)], and Antti Valpas[2]

[1] Department of Innovation & Software,
Lappeenranta University of Technology, Lappeenranta, Finland
{helvi.nyerwanire,erja.mustonen-ollila}@lut.fi
[2] Department of Obstetrics and Gynaecology,
South Karelia Social and Health Care District, Lappeenranta, Finland
antti.valpas@eksote.fi

Abstract. Knowledge management describes how information communication technology systems are applied to support knowledge creation, and in capturing, organization, access, and use of an organization's intellectual capital. This paper investigates knowledge management problems in the healthcare environment. We used the Grounded Theory approach in collecting and analyzing the data. The discovered six thematic categories found were Patient, Patient Data, Physician, Midwife, ICT Systems, and Medical Equipment. We decomposed each category into multiple items by deriving them from the data and validated them with past studies. We found propositions to our categories, relationships between the categories, and discovered eleven higher levels of abstraction of statements. A conceptual framework of knowledge management categories, items, the relationships of the categories to each other, and propositions to our categories were developed by the Grounded Theory approach. The relationships between the knowledge management categories enhance the validity of the categories and expand the emerging theory.

Keywords: Knowledge management · Hospital · Empirical research · Case study · Grounded theory

1 Introduction

The concept of healthcare, as referred to in our study, includes medicine, nursing, and rehabilitation [1]. In this study the healthcare environment is referred to as a place in which medical, clinical and nursing knowledge is ingrained in practitioners [2]. Knowledge refers to the ways that information can be made useful to support a specific task or make a decision [3], personalized information [4], awareness, experience, skills and learning [5], tacit and explicit knowledge [6, 7], as well as the medical, clinical, and nursing knowledge of physicians and nurses [8]. Knowledge management is defined as a process where information communication technology (ICT) systems are applied to support the activities in organizing knowledge, expertise, skills and communication [5]. Knowledge management can be further defined as a collaborative and

© Springer International Publishing Switzerland 2015
A. Fred et al. (Eds.): IC3K 2014, CCIS 553, pp. 495–509, 2015.
DOI: 10.1007/978-3-319-25840-9_30

integrated approach to creating, capturing, organizing, access, and use of an organization's intellectual capital [9]. There exists collective knowledge in organizational networks [4], and people learn by working with each other in practice, and transfer and receive knowledge on best practices [10]. In spite of the above definitions and previous studies, there are several problems that hamper knowledge management. One reason for this communication problem can be the fact that the working communities do different things and work differently, they have different terms and vocabularies, and therefore they do not understand each other [9]. Viitanen et al. [11] argue that physicians have problems with searching for the right data. Patients' resistance to recommendations, limited time and limited personnel resources, work pressure, negative attitude or limited support from "peers" or superiors can be reasons for not following the guidelines [12]. Using the ICT system can also take a lot of physicians' work time, they have problems with accessing patient data, and they are highly critical towards ICT systems [13]. Finally, Kothari et al. [14] claim that physicians' and nurses' tacit knowledge cannot be found in ICT systems.

As can be seen in previous studies, a lot of knowledge management problems exist in healthcare. We have applied previous studies and empirical evidence to carry out a qualitative in-depth case study [15, 16] that identifies problems in knowledge management in the healthcare environment. We have analyzed the collected data with the Grounded Theory (GT) approach, and developed a conceptual framework with categories and relationships between the categories [17, 18]. Our goal has been to investigate knowledge management problems in detail. We have explored the strategies that healthcare organizations deploy while learning about their knowledge management problems, to what extent these problems are shaped by the organizational context, and how these potential problems influence both ICT system development work and patient care work in practice. We made 305 knowledge management observations supported by empirical evidence. The observations were categorized with GT analysis [17], and the analysis revealed six thematic categories: Patient, Patient Data, Midwife, Physician, ICT Systems, and Medical Equipment. We decomposed each thematic category further into multiple items by deriving them from the data and validating them with previous studies. The above thematic categories were related to each other, and we found out eleven higher levels of abstraction of statements based on the developed conceptual framework, and propositions to our categories. The rest of the paper is structured as follows. Section two describes related research, section three presents the research method, and section four outlines the data analysis. Finally, section five contains conclusions and discussion.

2   Related Research

Martikainen et al. [13] and Greig et al. [19] claim that healthcare ICT systems are expensive, regulated by political decisions, not standardized, and include complex medical data, it is difficult to save data to the systems, there exist security and trust problems, ICT systems are regulated by laws, new versions of systems are taken in use regularly, integration is not prepared for the future needs of patients, and enterprise architectural solutions are missing. Nykänen et al. [20] and Kothari et al. [14] claim that both the medical records and tacit knowledge of healthcare professionals are missing from ICT systems. Muhiya [21] has found lack of access to the medical and social history of undocumented migrants. Edwards et al. [22] state that patients' advice to physicians cause stress to the physicians. Ijäs-Kallio [23] and Misra et al. [24] list patients' resistance to their diagnoses, not telling about their social history, and patients' limited memory of their past medical illnesses and procedures. Sumanen et al. [25], Lin et al. [26] and Viitanen et al. [11] mention difficulties in keeping professional distance towards patients, lack of senior and experienced physicians, and difficulties in transferring organizational learning, such as own experiences. Junior and senior physicians have differences in their ways of working with ICT technology [27], and lack of time to use ICT systems because of clinical work [11, 28]. Abbasinazari et al. [29] claim that wrong medication caused by interruption in work can cause a critical condition to the patient. Martikainen et al. [13], Thorpe-Jamison et al. [30] and Viitanen et al. [11] state that the complexities of new medical equipment technologies and communication problems between the ICT systems, as well as the need for constant learning and studying, parallel with new medicines make physicians tired. Hebert et al. [31], and Eason and Waterson [32] mention the stress that nurses and physicians feel in their clinical work, restricted access to patient data, and work interruptions. Smith and Haque [33] mention extra work due to parallel checking of paper-based and electronic patient records. Thus, despite a growing interest in knowledge management problems in healthcare, their relationships have not been recognized in the literature. Rather, previous studies have focused on knowledge management problems in healthcare in general. Therefore, our study aims to respond to this lack of studies and to provide useful information of knowledge management in one Social and Health Care District, in its central hospital, and one central hospital department. Based on previous studies we have formulated the following research questions (RQ1 and RQ2): RQ1: What are the knowledge management problems in a hospital? RQ2: How are the knowledge management problems at a hospital related to each other?

3   Research Method

This study takes a qualitative, in-depth case study and Grounded Theory (GT) perspective to understanding the complex technological and social phenomena of knowledge management problems in the healthcare environment [15–18, 34, 35]. In this healthcare environment, two cases have been selected so that the study will either predict similar outcomes (i.e. literal replication) or produce conflicting results but for predictable reasons (i.e. theoretical replication) [36]. Theory triangulation is applied by interpreting a single data set from multiple perspectives to understand the research problems [37]. The concepts and their relationships are validated with the Grounded Theory approach [17, 18, 34, 38]. During the research, theoretical background knowledge was gained, which increased the credibility of the study [38, 39]. According to Eisenhardt [34], the combination of case study with the Grounded Theory approach has three major strengths: it produces a novel theory, the emergent theory is testable, and the resultant theory is empirically valid. In the GT approach the theory emerges from the data. According to Glaser [38], there is no need to review any literature of the studied area before entering the field. This is in line with our research, as we started collecting the data before developing our conceptual framework. Specifically, each transcript was analyzed and major emergent themes and concepts were identified in order to form similar categories [40].

In our first interview round we conducted seven tape-recorded unstructured and semi-structured interviews that investigated experiences in knowledge management problems at the research site. The seven interviews were carried out between June 2012 and November 2013. Four of the interviews were group interviews and three were individual ones. The unit of analysis and data collection site was South Karelia Social and Health Care District and its central hospital. The district has about 133,000 inhabitants, and the total number of staff employees in the district is 3843, of which 1711 work in health services [41, 42]. In the central hospital there are a total of 679 employees and 235 bed places for patients. The central hospital takes care of 17 special medical areas, such as obstetrics and gynaecology, surgery, pediatrics, neurology, etc., and it is therefore responsible for all advanced special treatment and healthcare of the whole population in the healthcare district [41].

In our second interview round we conducted nine tape-recorded unstructured and semi-structured interviews in January-February 2013. The unit of analysis and data collection site was the Obstetrics and Gynaecology department (OBD) of the central hospital. The department takes care of all childbearing and it is also responsible for advanced gynaecologic treatments and healthcare of women in the whole healthcare district [41]. The department has 8 senior physicians, 3 specializing junior physicians, 42 midwives, and 10 nurses [41, 43].

There was no access to archival material, and the secondary data included only public material of the Social and Health Care District, its central hospital and the Obstetrics and Gynaecology department [43]. All the 14 interviewees had been involved with many knowledge management issues and processes in their own fields of expertise during their working careers, which extended over a period of 10 to 30 years in different positions either in South Karelia Social and Health Care District or other healthcare environments in Finland. The interviewees were presented with the research problem and were chosen because their role was to use, create and transfer healthcare-related medical and ICT information, and translate it to knowledge relevant to the healthcare situation at hand. Table 1 contains information of the interviewees.

3.1   Data Collection and Categorization

The interviews included frequent elaboration and clarification of the meanings and terms, they were audio recorded, and the recordings were transcribed, yielding totally over 350 pages of transcripts. After transcribing the interviews, we searched for knowledge management problems in the material. On the basis of our intuition and knowledge we made a table in which the first column included a specific knowledge management problem discovered in the empirical data; the second column included knowledge management problem definition based on the empirical data; the third column included evidence from the literature based on the problem in the first column; the fourth column included the name of the literature source of the third column; and

Table 1. Interviewee details.

Interviewee number	Role of interviewee	Length of interview in minutes	Group or individual interview
1	ICT Director	45 min	Group Interview
2	Communications Manager		
2	Communications Manager	132 min	Group Interview
3	Medical Director of the Central Hospital (chief physician of the Internal Medicine and Endocrinology Department)		
4	Chief Physician of the OBD		
4	Chief Physician of the OBD	72 min	Individual Interview
4	Chief Physician of the OBD	120 min	Group Interview
5	Junior Physician of the Emergency Department		
4	Chief Physician of the OBD	100 min	Group Interview
5	Junior Physician of the Emergency Department		
1	ICT Director	90 min	Individual Interview
6	Development Manager (the National Archive of Health Information Project of the Ministry of Social Affairs and Health)	60 min	Individual interview
7	Junior Physician 1 of the OBD	89 min	Individual interview
8	Junior Physician 2 of the OBD	85 min	Individual interview
9	Senior Physician 1 of the OBD	71 min	Individual interview
10	Senior Physician 2 of the OBD	66 min	Individual interview
11	Senior Physician 3 of the OBD (chief physician of the OBD)	83 min	Individual interview
12	Midwife 1 of the OBD	79 min	Individual interview
13	Midwife 2 of the OBD	55 min	Individual interview
14	Midwife 3 of the OBD	70 min	Individual interview
15	Midwife 4 of the OBD	76 min	Individual interview

finally the fifth column included the transcript number and the interviewed person's name and occupation. By doing this we created a chain of evidence: from empirical data we derived the knowledge management problems and validated them with past studies. Archival material was studied, representing a secondary source of data, and it included public news and internal material of the development of the Social and Health Care District, the Obstetrics and Gynaecology department [43], and public news of the KanTa project of the Ministry of Social Affairs and Health [44]. Triangulation involved checking different data sources simultaneously to improve the reliability and validity of the data.

4   Analysis

After creating the chain of evidence, we used the Grounded Theory on the basis of our intuition and knowledge in fragmenting and reassembling the knowledge management problems into thematic categories [17, 18] according to the relevant terminology and past studies that were the most often refereed work of categorizing concepts in the

Table 2. Thematic category, category definition, item number (item no), item definition, and the sum total of item observations.

Thematic category	Category definition	Item No	Item	Item definition	Sum total of item observations
Patient	A patient receives care and treatment from a physician or a midwife.	1	Characteristics	Each patient brings a set of unique characteristics to the care situation.	5
Patient data	Patient's personal data, medical history, treatments, tests, examinations, diagnoses, and consultation requests.	1	Data in ICT Systems	Patient's personal data, medical history, treatments etc.in the ICT systems.	17
Physician	A physician needs knowledge of anatomy, physiology, and medical science and knowledge of how to apply this knowledge in practice.	1	Thoughts about midwives' professional distance towards patients	Midwives want to keep a professional distance towards patients, because they want to look at things from the patients' point of view by using their psychological knowledge.	1
Midwife	A midwife is a trained nurse specialized in childbearing, labour, delivery, women's special illnesses, and taking care of the baby after the birth.	1	Lack of knowledge of the patients' social issues	During the patient interview the midwives do not know the patient's social history without asking.	1
ICT systems	There are hundreds of ICT systems in use at hospitals, and the physicians and other professionals use them in their daily work with patients.	1	Communication barriers	Hospitals using different ICT systems are not able to communicate, and paper serves as a coordination tool. The barriers between healthcare and ICT are lack of knowledge and skills, and poor communication.	2
Medical equipment	Medical equipment is designed to aid in the diagnosis, monitoring or treatment of medical conditions.	1	Risks	When the system fails in the middle of a medical situation it poses a risk to patient safety.	6

studied research area. We further decomposed each thematic category into multiple items (traits) by using content analysis, derived the items from the data and validated them with past studies. An example of each thematic category, category definition, item number, item, item definition, and sum total of item observations is presented in Table 2 below.

The data table including all the items etc. is available by separate request from the second author. The six thematic categories formed in the study were Patient, Patient Data, Physician, Midwife, ICT Systems, and Medical Equipment. The Patient category had 18 observations and 8 items; the Patient Data category 51 observations and 9 items; the Physician category 121 observations and 22 items; the Midwife category 24 observations and 13 items; the ICT Systems category 83 observations and 12 items; and Medical Equipment category had 8 observations and 3 items. The sum total of empirical item observations (items) was 305, and the sum total of different items was 68. Our conceptual framework of the discovered categories (see Fig. 1 below) is grounded on empirical evidence and theories reflecting the findings in the field [17, 18]. In Fig. 1, the categories are shown as ellipses, and the dashed and bold arrows marked with the letters (A to J) describe the relationships between the categories. The small arrows with numbered circles pointing to categories are the multiple items (traits) to each category composed by content analysis. After the categories and relationships and items had been found, we determined the properties of categories and propositions (hypotheses) as to how the categories were related on the basis of the data (See Table 3). The constant comparison between the data and concepts in past studies in

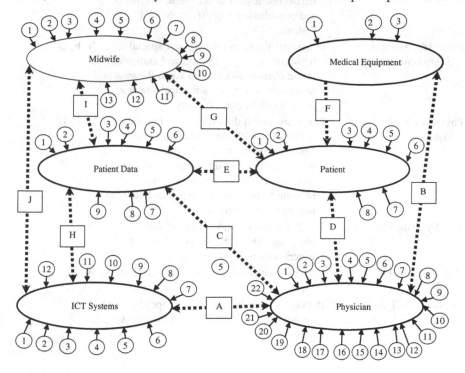

Fig. 1. Conceptual framework of categories.

Table 3. Properties of categories and propositions (hypotheses) on how the categories are related on the basis of the data.

Category/categories	Properties of categories and propositions (hypotheses) on how the categories are related (arrows marked with the letters A to J in Fig. 1) on the basis of data	Arrow marked with letters in Fig. 1
Patient, Physician	The patients do not tell about all their problems to the physician.	D
Patient, Physician	Physicians keep personal distance towards patients.	D
Physician, Patient	Patient's stress affects the physician.	D
Patient, Patient Data, ICT Systems	Patient's medical records and treatments in another hospital are not seen in the ICT systems.	E, H
Physician, Patient, Midwife	The physicians have lack of time to explain the diagnosis thoroughly to the patient and thus the midwives have to explain the diagnoses.	G
Patient, Patient Data, Midwife, Physician, ICT Systems	The midwives and physicians forget to save the patient's medical information to the systems.	E, I, C, H, J
Patient Data, ICT Systems	Patient's personal data, medical history, treatments, tests, examinations, diagnoses, and consultation requests in the ICT systems.	H
Physician, Medical Equipment, Patient	In hospital work, in primary and special healthcare services, the medical equipment is complicated to use, and when the system fails in the middle of a medical situation it poses a risk to patient safety.	B, F, D
Physician, Medical Equipment, Patient	Physicians find it difficult to read manuals and get familiar with the medical equipment. They have to get the work done on time and without errors, but the medical equipment, such as ultrasound, which is a medical imaging technique, is complicated, and the user may need a manual.	B, F, D
ICT Systems, Physician	The ICT System vendors do not ask physicians about their opinions or experiences of the IT systems.	A
ICT Systems, Physician	New ICT needs constant learning and disturbs life.	A
Physician, ICT Systems	Physicians have lack of ICT competence, and do not trust ICT systems and their data security.	A

(Continued)

Table 3. (*Continued*)

Category/categories	Properties of categories and propositions (hypotheses) on how the categories are related (arrows marked with the letters A to J in Fig. 1) on the basis of data	Arrow marked with letters in Fig. 1
Physician, ICT Systems	Physicians lack knowledge or competence on how to use a computer or ICT systems.	A
Physician, ICT Systems	Hospital ICT systems do not communicate with each other even if they are similar.	A
ICT Systems, Physician, Midwife	Too much time is spent on ICT Systems use, ICT Systems are complex, difficult to use, disturb everyday clinical work, and information is difficult to find.	J,A
Midwife, Patient	Social history of the patient must be asked from the patient herself.	G
Midwife, ICT Systems	Lack of ICT and IT computer knowledge.	J
Midwife, Patient	Midwifes work in a hurry, which causes risk to patients.	G
Midwife, ICT Systems, Patient	Midwives write patients' medical records, monitor patients' health, give consultations to other midwives, and discharge the patient.	G, I, E, J
Midwife, Patient, Physician	Acute work, work interruptions, stress due to patients, difficulties in distinguishing the differences in work between midwife and physician.	G, D
Midwife, Patient, Patient Data	Patient can deny access to her medical information outside the hospital.	G, I, E
Patient, Physician, Midwife	Patient does not remember her medical surgeries, treatments, or history and has diagnosis attitude; patient cannot describe the symptoms of medication.	D, G

order to accumulate evidence convergence on simple and well-defined categories led us to a higher level of abstraction of statements about the relationships between the categories. This theorizing is in line with Pawluch and Neiterman's [18] suggestions of creating a Grounded Theory with Glaser and Strauss's [17] approach. The higher level of abstraction of statements is presented in the conclusions and discussion section.

5 Conclusions and Discussion

This qualitative, empirical case study based on the Grounded Theory approach [17] revealed that many knowledge management problems can be found in knowledge and information-intensive environments in the healthcare domain. Based on fourteen individual and four group in-depth interviews, our study described knowledge management problems in one Finnish Social and Health Care District, in its central hospital, and in its Obstetrics and Gynaecology department. This multi-perspective point of view gave us rich data for solving our research problem.

We discovered six main thematic categories (Patient, Patient Data, Physician, Midwife, ICT Systems, and Medical Equipment), 3-22 multiple items (traits) to each thematic category, propositions to our categories, and relationships between the categories. We found eleven higher level abstractions of statements based on our conceptual framework as follows. One, physicians' tacit knowledge and experience, technological skills to use ICT systems and medical equipment, and knowledge of medical issues differ between junior and senior physicians. Two, physicians are affected by the patient's stress. Three, senior physicians can use their dominant power position over midwives, nurses and junior physicians. Four, the work of physicians, midwives and nurses is non-routine work and full of work flow interruptions, and they all have lack of time. Physicians' lack of time forces midwives and nurses to do diagnosis counseling with the patients. Five, medical and clinical decisions are influenced by patient data or missing patient data in the ICT systems. Six, new ICT systems need constant learning when medical practice issues must be updated. Seven, there exist ICT system communication barriers and use difficulties, as well as information retrieval difficulties. Eight, a patient can deny access to her files in other hospitals and there are not perfect data files of patients coming from other parts of Finland, and even more deficient files when a patient comes from abroad. Nine, there is a communication barrier between physicians and patients and nurses and patients, because patients reject their diagnoses, make their own diagnoses, and do not tell about all their problems to the physicians and midwives. Ten, primary and special healthcare have coordination problems in patient care and discharge issues. Eleven, in technology failures there are responsibility problems between physicians, medical equipment technology buyers, and firms selling medical equipment.

We also made five empirical findings which were not supported by previous studies, and we regard this data as expansion to knowledge management in healthcare. First, the same patient data must be saved several times in ICT systems. Second, internal referral notes or an internal consultation request, such as X-ray pictures must be sent on paper by post, or the patient brings them along to the physician. Third, ICT system development is monopolized to certain ICT system vendors. Fourth, patients cannot describe the symptoms of their medication to the physician. Fifth, physicians have private and secret files about patients which they do not share with anyone else.

Our study is in line with the studies of Martikainen et al. [13] and Greig et al. [19] stating that healthcare ICT systems are expensive, regulated by political decisions, not standardized, include complex medical data, it is difficult to save data to the systems, there exist security and trust problems, they are regulated by laws, new versions of systems are taken in use regularly, integration is not prepared for future needs of patients, and enterprise architectural solutions are missing; the claims of Nykänen et al. [20] and Kothari et al. [14] that both the medical records and tacit knowledge of healthcare professionals are missing from ICT systems; as well as Muhiya's [21] claim for problems due to lack of access to the medical and social history of undocumented migrants. Our study confirms the claim of Edwards et al. [22] that patients' advice to physicians cause stress to the physicians; that of Ijäs-Kallio [23] and Misra et al. [24] for patients' resistance towards their diagnoses, not telling about their social history, and their limited memory concerning their past medical illnesses and procedures; as well as the claims of Sumanen et al. [25], Lin et al. [26] and Viitanen et al. [11] for difficulties in keeping a professional distance towards patients, lack of senior and experienced physicians, and difficulties to transfer organizational learning, such as own experiences. The empirical evidence also confirmed that there are differences in ways of working between junior and senior physicians in ICT technology [27], and lack of time to use ICT systems because of clinical work [11, 28]. Furthermore, the findings of Abbasinazari et al. [29] about wrong medication due to interruption in work causing a critical condition to the patient, as well as the claims of Martikainen et al. [13], Thorpe-Jamison et al. [30] and Viitanen et al. [11] about the complexities of new medical equipment technologies and communication problems between ICT systems and their need for constant learning and studying parallel with new medicines making physicians tired were also confirmed. Our results are also in line with the findings of Hebert et al. [31] and Eason and Waterson [32] about the stress that nurses and physicians feel in their clinical work, restricted access to patient data, and work interruptions, as well as the claim of Smith and Haque [33] about extra work due to parallel check of paper-based and electronic patient records.

In our research the purpose of the Grounded Theory analysis [17] was to find out categories and relationships between the categories in the healthcare environment with the inductive research approach. The concepts were sharpened by building evidence from empirical data that described the conceptual categories which according to Glaser and Strauss [17] and Pawluch and Neiterman [18] are the building blocks of the Grounded Theory. Constant comparison between the data and concepts was made so that the accumulating evidence converged on simple and well-defined categories. Furthermore, we decomposed each thematic category into multiple items (traits) by using content analysis, derived the items from the data and validated them with past studies. After the categories were found, we defined the properties of the categories and propositions (hypotheses) of how the categories were related. In our research we took carefully into consideration beforehand who to interview, what to do next, what group to look for, and what additional data we should collect in order to develop a theory from the emerging data. Finally, a conceptual framework of the categories and items was developed, and the categories and items were grounded on empirical evidence and theories reflecting the findings in the field. The most fundamental components in this conceptual framework were its categories, items, and the relationships between the categories. The comparison with previous studies led us to eleven higher level abstractions of statements about the relationships between the categories. This theorizing was in line with Pawluch and Neiterman's [18] suggestions for creating a Grounded Theory with Glaser and Strauss's [17] approach.

In this study, a special status in the theory building was given to the focal categories, the social and healthcare district, its central hospital, and the Obstetrics and Gynaecology department. In our theory the ancillary category was the knowledge management problem. We took care of the boundary conditions in our theory creation, because the phenomenon was so atypical that it held only in this specific contextual healthcare environment. Our results validated the conceptual framework, which became the discovered theory for the phenomenon. The data which confirmed the emergent relationships enhanced confidence in the validity of the relationships. The data which disconfirmed the relationships provided an opportunity to expand and refine the emerging theory. The results which did not get support from previous studies resulted in expansion to the theory. The non-conflicting results strengthened the definitions of our categories and the conceptual framework. The previous studies with similar findings were important because they tied together the underlying similarities in phenomena not associated with each other, and stronger internal validity was achieved.

There are, however, several limitations in this study. First, we had limited knowledge of the central hospital of the social and healthcare district and one of its departments, and access to secondary sources was limited. Second, the use of only one social and healthcare district, its central hospital and one of its departments affected also our findings, and thus generalization of the results can be difficult, but not necessarily impossible. Third, our intuition and knowledge in categorizing the knowledge management problems from the data was limited. Fourth, the interviews were conducted in multiple languages, which made the interviewing, transcribing, coding and analyzing the material very demanding. Fifth, a linguistic problem was faced in our interviews because the interviewer could not speak Finnish, and some of the interviewees did not speak English well. Sixth, there was lack of medical and nursing knowledge from our side regarding the use of correct and proper medical and nursing terms in the questionnaires. We also learned that some questions in the interview protocol tool were more specific to physicians than to midwives or nurses and vice versa.

Two conclusions emerge from our study. First, the knowledge management problems influencing daily work in a health care organization cannot be controlled by a single ICT system or medical equipment, professional, communication channel, or any other issue, as the influence of the knowledge management problem can be related to another problem. Second, it is very important for healthcare professionals to understand and overcome the multiple existing knowledge management problems in their daily work, whether they come from patients, other staff or ICT systems, in order to achieve success in patient care situations.

The practical implication of this research is significant for physicians, midwives, nurses and IT managers implementing the current ICT systems, hospital systems, healthcare systems, and medical equipment in hospital settings. The results enable physicians, nurses, midwives and IT managers to forecast the knowledge management problems in daily work at the hospital and in its specific departments. By being aware of such knowledge management problems, one can be better prepared for their appearance and create strategies for overcoming them. For the researchers our study provides also evidence of the importance of the relationships between physicians, midwives, ICT systems, medical equipment, patients, and patient data. This study gives explanation of the phenomena for academics and practitioners as it has revealed the possible difficulties that the studied organization can expect when it takes care of healthcare.

In the healthcare environment in question, the cases were selected so that they would either predict similar outcomes (i.e. literal replication) or produce contracting results but for predictable reasons (i.e. theoretical replication) [36]. Theory triangulation was applied by interpreting a single data set from multiple perspectives to understand the research problems [37]. The concepts and their relationships were validated with the Grounded Theory approach [17, 34, 38]. During the research, theoretical background knowledge was gained, which increased the credibility of the study [38, 39]. According to Eisenhardt [34], the combination of case study with the Grounded Theory approach has three major strengths: it produces a novel theory, the emergent theory is testable, and the resultant theory is empirically valid. In the GT approach, the theory emerges from the data.

In the future, more data will be collected and analyzed by using the Grounded Theory approach [17], and also with quantitative analysis with Self-Organizing Maps (SOMs) [45] to find out the unique clusters and visualize the patterns. Glaser and Strauss [17] also claim for both qualitative and quantitative data in creating theory. Qualitative and quantitative data can supplement each other, and new theory can emerge from their comparison.

References

1. Koskinen, J.: Phenomenological view of health and patient empowerment with personal health record. In: Proceedings of the Well-being in the information society (WIS) conference, pp. 1–13. University of Turku, Turku (2010)
2. Räisänen, T., Oinas-Kukkonen, H., Leiviskä, K., Seppänen, M., Kallio, M.: Managing mobile healthcare knowledge: physicians' perceptions on knowledge creation and reuse. In: Olla, P., Tan, J. (eds.) Mobile Health Solutions for Biomedical Applications, pp. 111–127. IGI Global, New York (2009)
3. Stair, R.M., Reynolds, G.W.: Fundamentals of Information Systems. Thomson Course Technology, Boston (2006)
4. Alavi, M., Leidner, D.E.: Review: knowledge management and knowledge management systems: conceptual foundations and research issues. MIS Q. 25(1), 107–136 (2001)
5. Suurla, R., Markkula, M., Mustajärvi, O.: Developing and implementing knowledge management in the parliament of Finland, The committee for the future: The Parliament of Finland (2002). http://web.eduskunta.fi/dman/Document.phxdocumentId=gk11307104202716&cmd=download. 4 September 2012
6. Polanyi, M.: The Tacit Dimension London. Routledge, London (1966)
7. Nonaka, I.: A dynamic theory of organizational knowledge creation. Organ. Sci. 5(1), 14–37 (1994)
8. Hill, K.S.: Improving quality and patient safety by retaining nursing expertise. Online J. Issues Nurs. 15(3), 1 (2010)
9. Dalkir, K.: Knowledge Management in Theory and in Practice. Butterworth-Heinemann Publisher, London (2005)
10. Grover, V., Davenport, T.H.: General perspectives on knowledge management: fostering a research agenda. J. Manag. Inf. Syst. 18, 5–21 (2001)
11. Viitanen, J., Hyppönen, H., Lääveri, T., Vänskä, J., Reponen, J., Winblad, I.: Nationality questionnaire study on clinical ICT systems proofs: physicians suffer from poor usability. Int. J. Med. Inform. 80(10), 708–725 (2011)
12. Heilmann, P.: To have and to hold: personnel shortage in a Finnish healthcare organization. Scand. J. Public Health 38(5), 518–523 (2010)

13. Martikainen, S., Viitanen, J., Korpela, M., Lääveri, T.: Physicians' experiences of participation in healthcare IT development in Finland: willing but not able. Int. J. Med. Inform. **81**(2), 98–113 (2012)
14. Kothari, A., Rudman, D., Dobbins, M., Rouse, M., Sibbald, S., Edwards, N.: The use of tacit and explicit knowledge in public health: a qualitative study. Implement. Sci. **7**(1), 20 (2012)
15. Benbasat, I., Goldstein, D.K., Mead, M.: The case study research strategy in studies of information systems. MIS Q. **11**(3), 369–386 (1987)
16. Yin, R.K.: Case study research: design and methods. Sage Publications, California (2003)
17. Glaser, B., Strauss, A.L.: The Discovery of the Grounded Theory: Strategies for Qualitative Research. Aldine, Chicago (1967)
18. Pawluch, D., Neiterman, E.: What is grounded theory and where does is come from? In: Bourgeault, A., Dingwall, R., De Vries, R. (eds.) The SAGE Handbook of Qualitative Methods in Health Research, pp. 174–192. Sage Publications, London (2010)
19. Greig, G., Entwistle, V.A., Beech, N.: Addressing complex healthcare problems in diverse settings: insights from activity theory. Soc. Sci. Med. **74**(3), 305–312 (2012)
20. Nykänen, P., Kaipio, J., Kuusisto, A.: Evalution of the national nursing model and four nursing documentation systems in Finland-lessons learned and directions for the future. Int. J. Med. Inform. **81**(8), 507–520 (2012)
21. Muhiya, A.: Experiences of midwives on working with immigrants: A literature review (2013). https://www.theseus.fi/handle/10024/69293. Accessed 31 January 2014
22. Edwards, M., Wood, F., Davies, M., Edwards, A.: The development of health literacy in patients with a long-term health condition: the health literacy pathway model. BMC Public Health **12**(1), 130 (2012)
23. Ijäs-Kallio, T., Ruusuvuori, J., Peräkylä, A.: Patient resistance towards diagnosis in primary care: implications for concordance. Health **14**(5), 505–522 (2010)
24. Misra, S., Daly, B., Dunne, S., Millar, B., Packer, M., Asimakopoulou, K.: Dentist–patient communication: what do patients and dentists remember following a consultation? Implications for patient compliance. Patient Prefer. Adherence **7**, 543–549 (2013)
25. Sumanen, M., Aine, T., Halila, H., Heikkilä, T., Hyppölä, H., Kujala, S., Vänskä, J., Virjo, I., Mattila, K.: Where have all the GPs gone-where will they go? Study of finnish GPs. BMC Family Pract. **13**(1), 121 (2012)
26. Lin, C., Wu, J.-C., Yen, D.C.: Exploring barriers to knowledge flow at different knowledge management maturity stages period. Inf. Manag. **49**, 10–23 (2012)
27. Payne, K., Wharrad, H., Watts, K.: Smartphone and medical related App use among medical students and junior doctors in the United Kingdom (UK): a regional survey. BMC Med. Inform. Decis. Mak. **12**(1), 121 (2012). doi:10.1186/1472-6947-12-121
28. Chadi, N.: Medical leadership: doctors at the helm of change. McGill J. Med. **12**(1), 52 (2009)
29. Abbasinazari, M., Talasaz, A.H., Eshraghi, A., Sahraei, Z.: Detection and management of medication errors in internal wards of a teaching hospital by clinical pharmacists. Acta Medica Iranica **51**(7), 482–486 (2013)
30. Thorpe-Jamison, P.T., Culley, C.M., Perera, S., Handler, S.M.: Evaluating the impact of computer-generated rounding reports on physician workflow in the nursing home: a feasibilty time-motion study. J. Am. Med. Dir. Assoc. **14**(5), 358–362 (2013)
31. Hebert, K., Moore, H., Rooney, J.: The nurse advocate in end-of-life care. Ochsner J. **11**(4), 325–329 (2011)
32. Eason, K., Waterson, P.: The implications of e-health system delivery strategies for integrated healthcare: lessons from England. Int. J. Med. Inform. **82**(5), e96–e106 (2013)

33. Smith, C.A., Haque, S.N.: Paper versus electronic documentation in complex chronic illness: A comparison. In: AMIA Annual Symposium Proceedings, pp. 734–738. American Medical Informatics Association (2006)
34. Eisenhardt, K.M.: Building theories from case study research. Acad. Manag. Rev. **14**(4), 532–550 (1989)
35. Creswell, J.W.: Qualitative Inquiry and Research Design: Choosing Among Five Approaches. Sage Publications, California (2007)
36. Yin, R.Y.: Applications of Case Study Research, Applied Social Research Methods series, vol. 34. Sage Publications, Newbury Park (1994)
37. Denzin, N.K. (ed.): The Research Act: A Theoretical Introduction to Sociological Methods. McGraw-Hill, New York (1978)
38. Glaser, B.G.: Emergence vs. Basics of Grounded Theory. Sociology Press, Mill Valley, Forcing (1992)
39. Miles, M.B., Huberman, A.M.: Qualitative Data Analysis. Sage Publications, Thousand Oaks (1994)
40. Myers, M.D., Avison, D.E. (eds.): Qualitative Research in Information Systems: Review. Sage Publications, London (2002)
41. Timonen, T.: Key figures: statistics from the hospital, Internal Publication of South Karelia Social and Health Care District (2013)
42. Eksote: 'Etelä-Karjalan Sosiaali- ja Terveyspiiri (2013). http://www.eksote.fi. 10 November 2013
43. Raudasoja, S.: EKSOTE's Strenght is its strong customer orientation. South Karelia's Social and Healthcare District. External Publication of Eksote (2013)
44. STM.: Sosiaali- ja terveysministeriö. Tietojärjestelmähankkeet: sähköinen potilasarkisto ja sosiaalialan tiedonhallinta (2013). http://www.stm.fi/vireilla/kehittamisohjelmat_ja_hankkeet/tietojarjestelmahankkeet/kysymyksia_kanta_hankkeesta#vastaus. 5 October 2013
45. Kohonen, T.: Self-Organization and Associative Memory. Springer-Verlag, Berlin (1989)

Cooperative Knowledge Representation and Classification for Design Projects

Xinghang Dai[1(✉)], Nada Matta[1], and Guillaume Ducellier[2]

[1] Tech-CICO, University of Technology of Troyes,
12 Rue Marie Curie, Troyes 10010, France
{xinghang.dai,nada.matta}@utt.fr
[2] LASMIS, University of Technology of Troyes,
12 Rue Marie Curie, Troyes 10010, France
guillaume.ducellier@utt.fr

Abstract. Design is a knowledge-intense activity. In design projects, both domain knowledge and cooperative knowledge are produced. Present knowledge engineering methods focus on how to extract and model expert knowledge, but cooperative knowledge that is produced in cooperative activities is usually ignored. In this paper, the cooperative knowledge in design projects is studied, and a cooperative knowledge representation structure as well as a framework to classify it is proposed.

Keywords: Knowledge management · Semantic networks · Classification · Concurrent design project management · Project memory

1 Introduction

A cooperative activity is generally defined as an activity of several actors having a given goal [32]. Three dimensions must be studied in this type of activity: communication, coordination and collaborative decision-making [34]. A number of works on CSCW analyzed these dimensions and several techniques have been defined in order to give supports to cooperative activity. We mention for instance Workflow, Groupware tools [18], design-rationale approaches [3], etc.

Our study concerns knowledge management for cooperative activity in the field of design. We attempt to answer the question which kind of knowledge exists in cooperative activity and how can we represent them to reuse it.

Recent knowledge management research has proposed community of practices and story telling to enhance knowledge sharing in an organization. Experience shows that the success of these techniques depends on the dynamic of animation in these communities. Our work is based on knowledge engineering approaches. We believe that learning from experience requires two fundamental elements: reasoning strategies (also called behavior laws) [28] and context of these strategies [11]. "The learning content is context specific, and it implies discovery of what is to be done, when and how according to the specific organizations routines" [13]. These two elements are especially important for cooperative knowledge representation.

© Springer International Publishing Switzerland 2015
A. Fred et al. (Eds.): IC3K 2014, CCIS 553, pp. 510–524, 2015.
DOI: 10.1007/978-3-319-25840-9_31

This paper will begin with laying our research background through an introduction on cooperative knowledge and design project knowledge. Then the concept "project memory" will be illustrated. Finally a cooperative knowledge discovery model (CKD) will be proposed to classify knowledge rules for cooperative activities. This method will be elaborated on design project memory, followed by classification rule propositions and an example.

2 Research Background

First of all, we are going to introduce the concept cooperative knowledge in design project field to outline the characteristics of design project knowledge. Then, the concept project memory will be proposed as the general structure to represent the knowledge in design projects.

2.1 Design Project Knowledge

Design activities have gone through some major changes during the past five decades. With the use of IT tools in design projects and the more and more complex features of design product, design project tends to be multi-organizational, multidisciplinalry [12, 29]. Moreover, with the emergence of concurrent engineering design, design project no longer follows a linear management model, but a parallel one that calls for more communication, collaboration and coordination in project organization.

Design Domain Knowledge and Cooperative Knowledge. Both domain knowledge and cooperative knowledge are produced in design project. Past researches have progressed a lot on design domain knowledge management, but cooperative knowledge produced in design projects is different from design domain knowledge, and it cannot be captured or modeled through knowledge engineering methods:

- The nature of knowledge is different: The domain knowledge is related to a field and contains routines and strategies developed individually from experiences, which involve a number of experiments. The cooperative knowledge is related to several fields, i.e. several teams (of several companies) and in several disciplines collaborates to carry out a project. So there is a collective and organizational dimension to consider in cooperative knowledge. Representing domain knowledge consists in representing the problem solving (concepts and strategies) [4]. On the contrary, knowledge in cooperative activity aims at showing organization, negotiation and cooperative decision-making [8].
- Capturing of knowledge is different: The realization of a project in a company implies several actors, if not also other groups and companies. For example, in concurrent engineering, several teams of several companies from several disciplines collaborate to carry out a design project. The several teams are regarded as Co-partners who share the decision-makings during the realization of the project. This type of organization is in general dissolved at the end of the project [22]. In this type of organization, the knowledge produced during the realization of the

project has a collective dimension that is in general volatile. The documents produced in a project are not sufficient to keep track of this knowledge. In most of the cases, even the project manager cannot explain it accurately. This dynamic character of knowledge is due to the cooperative problem solving where various ideas are confronted to reach a solution. So acquisition of knowledge by interviewing experts or extraction from documents is not sufficient to show different aspects of design projects, especially negotiation [2]. Traceability and direct knowledge capturing are needed to acquire knowledge from this type of organization.

Project Memory. For the same object, people with different background can give different interpretations; concept alters according to different context. Knowledge engineering approaches based on semantic network, ontology, logic etc. has been developed for knowledge representation. As for design project, we have to focus on design rationale representation as well as its interaction with other parts of a project. In other words, a global representation of all design projects modules as well as interactions between them are needed for design project. We should represent specially:

1. The design rationale (negotiation, argumentation and cooperative decision making)
2. The organization of the project (actors, skills, roles, tasks, etc.)
3. The consequences of problem solving (evolution of the artifact)
4. The context of the project (rules, techniques, resource, etc.)

We called the structure project memory; it describes "the history of a project and the experience gained during the realization of a project" [24]. From the knowledge structure proposed by project memory, we want to focus on knowledge that is produced during cooperative activities in a design project.

3   CKD for Design Project

According to cognitive science, human can develop routines of strategies when facing to the similar problem several times, and these routines will be stocked in their long-term memory, which can be reused in the future when similar problems appear [9]. The principle of CKD method is to classify similar concept schemas of cooperative activities to identify certain repetitive ones as routines with a weight factor that indicates their importance. Classification can be defined as the process in which ideas and objects are recognized, differentiated, and understood, classification algorithms are used in biology, documentation, etc. [5]. A routine is defined as a recursive interaction schema of cooperative activity concepts. The weight factor is defined as percentage of recurrence of a routine among past similar project events. Therefore, the result of classification will be an ensemble of interactions between cooperative activity concepts. This result routine can be considered as a knowledge rule for cooperative actors to learn from, and future cooperative activities should pay attention to past knowledge rules.

A semantic network graph enable knowledge engineers to communicate with domain experts in language and notations that avoid the jargon of AI and computer science [33]. Our representation of project memory is based on a general semantic network of four modules, and then four modules are represented in sub-networks. Ontological hierarchy of concepts may be necessary for generalization. The ontological hierarchy of concept should be constructed according to a specific context, it is important to show different categories of concept as part of representation of project context.

Machine learning methods are frequently used to classify a concept automatically in a quantitative manner. However, design project interaction schemas are usually not voluminous and quite distinctive; design project information are highly structured in a computer-aided design environment. Therefore it is not necessary to use powerful machine learning algorithms for classification, detailed CKD classification method will be illustrated in Sect. 3.3.

In order to apply CDK in design projects, we have to begin with project trace from the beginning to the end of projects. Then, project trace will be conceptualized and fit into project memory structure. Finally, CDK method will be applied on certain interaction schemas to find routines (Fig. 1).

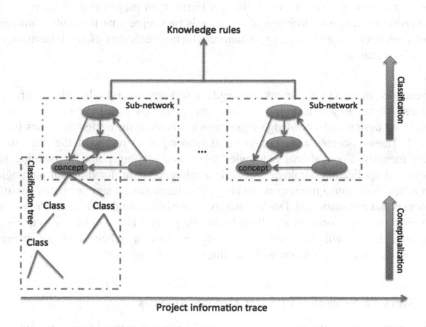

Fig. 1. CKD for design project memory.

3.1 Project Memory Structure

Section 2 has introduced "project memory" that list the four essential parts of design project. The goal of project memory is to enhance learning from expertise and past project experience [22]. Current representation approaches emphasize on organizing and structuring project information and expect users to learn from them. The problem is that human can only learn from others by matching to one's own experience, and the knowledge level or even knowledge context between expert and learner are always not the same. Traditional knowledge engineering method usually doesn't take project context into consideration (e.g. IBIS, QOC), or they neglect the interaction between different project modules (e.g. CommonKADS, DRCS). Therefore, instead of a single best classification system that suits everyone, everywhere [26], we have to come up with classification models suited within specific contexts [21].

Firstly, project memory has to be decomposed into smaller modules in order to show project memory in different perspective with different context to provide a better learning angle. The general semantic network of project memory (Fig. 3) is decomposed into 4 sub-networks:

- Decision-making process: this part represents the core activity of design project, which helps designers to learn from negotiation and decision-making experience.
- Project organization makes decision: this part represents interaction between organization and decision, which provides an organizational view of decision-making.
- Project organization realizes project: this part represents arrangement of task and project team organization, which focuses learning on project management.
- Decision-making and project realization: this part represents the mutual influence between decision and project realization, which reveals part of work environment and background.

Secondly, in each project memory module, a sub-network is built with concepts and relations. These project memory concepts are identified based on the research on engineering design and knowledge representation method for design activities [6, 20, 29, 31]. These concepts are employed and rearranged to represent the elements in project memory. Foundational ontologies serve as a starting point for building new domain and application ontologies, provide a reference point for different ontological approaches and create a framework for analyzing, harmonizing and integrating existing ontologies and metadata [25]. Dolce ontology is one of the fundamental ontologies, and the project memory concepts are aligned with the general Dolce ontology as in Fig. 2.

Lastly, CDK will be used to classify interaction schemas in or between sub-networks. The next section will introduce each sub-network.

3.2   Semantic Networks of Project Memory Modules

Based on these concepts, we are going to build our sub-networks to represent interactions between concepts in order to show the cooperative knowledge.

The first part of project memory is design rational; decision-making process is one of the most important parts in project memory. It contains negotiation process, decision and arguments that can reveal decision-making context. Concepts that are identified in a decision-making process are: issue, proposition, argument and decision. Issue is the major question or problem that we need to address, it can be about product design, organization arrangement or project realization etc.; proposition is solution proposed to solve issue by project team member; argument evaluates the proposition by supporting or objecting it, which can push proposal to evolve into another version [3, 6, 27],

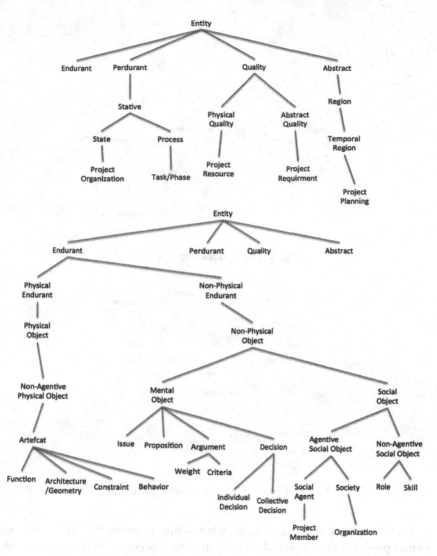

Fig. 2. Project memory concepts aligned with dolce ontology.

argument can also aims at issue which can possibly modify the specification of the issue. Propositions are considered to be possible solutions for issue, and arguments are supposed to explain the reason why. Decision is made by selecting some of the propositions for the issue and setting up a goal for next step of project realization.

One of the most important and useful knowledge that we want to represent is the context of design rationale [27]. Other project memory modules can also have mutual influences with decision-making process module. Therefore, we connect decision-making to project realization to show consequences of decision and connect decision-making to project organization to reveal an organizational influence.

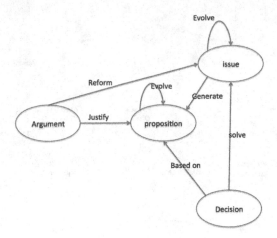

Fig. 3. Decision-making process.

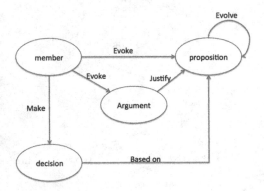

Fig. 4. Project organization making decision.

In the sub-network above (Fig. 4), we want to find a concept that serves as a bridge to connect project organization and decision-making process. So the concept "member" is introduced into decision-making sub-network to add an organizational dimension into decision-making process. Member is an important concept of project organization that links to competence, role and task.

The sub-network (Fig. 5) offers a learning perspective on project realization with an organizational dimension. Il presents us the interaction schema between task and project organization. Task is linked to two important attributes of project member: competence and role.

At last, we want to represent the triangle between task, decision and issue in order to show a mutual influence of task arrangement and decision-making process. A decision sets up a goal for a task; another issue can be evoked during a task, which initiate another decision-making process. The triangle ends by achieving the final result of a

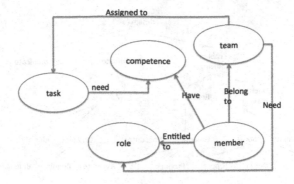

Fig. 5. Project organization realizing task.

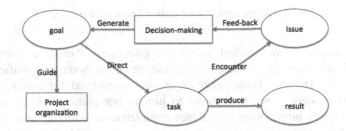

Fig. 6. Mutual influence of decision-making and project realization.

task. During a product design, the result of a task can be a new version of a product, and the version of product evolves between decision-making meeting and tasks (Fig. 6).

3.3 Design Project Cooperative Knowledge Ontology

For design projects, we have design rationale ontology [17, 30] shows us the ontology of project management and organization. Cooperative knowledge ontology should represent the interactions between design project concepts; therefore cooperative knowledge ontology concepts will be actions. Here we propose cooperative knowledge ontology, and its concepts are actions between concepts, which can be found in the four semantic networks above (Fig. 7).

3.4 Propositions of Classification Views

The ability to extract general information from example sets is a fundamental characteristic of knowledge acquisition. Machine learning technique is now a hot topic at present, it can figure out how to perform important tasks by generalizing from examples. One of the most mature and widely used algorithms is classification [10]. However, as we mentioned above, due to the particular characteristics of design project information, present machine learning techniques are not suitable for design project

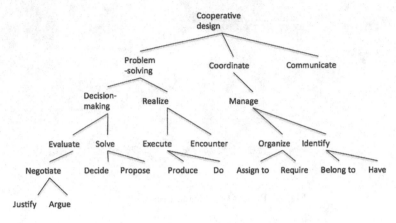

Fig. 7. Design project cooperative knowledge ontology.

memory classification. We studied four major categories of machine learning algorithms: statistical methods, decision tree, rule based methods and artificial neural network [7, 15, 18, 35]. These methods are not considered for two reasons: (1). Classification process is not transparent to human interpretation. Knowledge management for design project aims to enhance organizational learning, so it is important that knowledge classification process should be comprehensible for people with little IT background. (2). A large recursive training set is usually needed for machine learning techniques. The advantage of our classification model in project memory is that it is guided by semantic networks that indicate knowledge rules resided in interaction schemas. Therefore, according to these semantic networks, we classify interaction schemas instead of concepts. The amount of repetitive interaction schemas is significantly fewer compared to a concept; a large set of instances can be conceptualized into one class, while the probability of similar interaction schemas between concepts is much less. Additionally, the learning process will not ignore non-recursive schemas; on the contrary, they will be put aside as explorative attempts with an explanation.

Two tablet applications have been developed to capture project traces. They can register meeting information and generate XML files [24]. Project information will be structured according to a XML schema as follow (Fig. 8):

Fig. 8. XML schema of project memory structure.

```
      <xs:complexType>
        <xs:sequence>
          <xs:element name="role" type="xs:string" />
          <xs:element name="competence" type="xs:string" />
        </xs:sequence>
      </xs:complexType>
    </xs:element>
    <xs:element name="issue">
      <xs:complexType>
        <xs:sequence>
          <xs:element name="decision" type="xs:string">
          <xs:element name="proposition">
            <xs:complexType>
              <xs:sequence>
                <xs:element name="argument">
                  <xs:complexType>
                    <xs:sequence>
                  <xs:element name="criteria" type="xs:string" />
                  <xs:element name="position" type="xs:int" />
                      </xs:sequence>
              </xs:complexType>
            </xs:element>
                </xs:sequence>
              </xs:complexType>
            </xs:element>
          </xs:sequence>
      </xs:complexType>
    </xs:element>
    <xs:element name="task">
      <xs:complexType>
        <xs:sequence>
          <xs:element name="result" type="xs:string" />
        </xs:sequence>
      </xs:complexType>

    </xs:element>
```

Then project information will be classified according to different views to extract knowledge rules. Here we propose three classification views:

1. Problem-solving view: at a specific project phase, we can classify decision-making process for one particular issue. Solutions that are repetitive will be classified as essential solutions, the solutions that are distinctive will be considered as explorative attempt with its precondition as an explanation.
2. Cooperation view: an important subject that we tried to study in our model is cooperation. This classification view allows us to verify whether there are parallel tasks that involve cooperative design or regular meetings concerning whole project team. Projects that are not undertaken concurrently can lead to unsatisfactory results, e.g. solution duplication or excess of project constraint. This rule will reveal the influence of concurrent design on project result.
3. Management view: this classification view will focus on project organization influence on different project memory modules. For example, we can classify project realization with an organizational dimension to examine how project organization arrangement can influence project realization.

A weight factor that indicates recurrence rate will be attributed to each classification result to show the importance of this result. The three aspects proposed above are the most interesting and practical classification views that we find so far, however we do not exclude the possibility that more useful classification views exist. In the next section, CKD according to these three views will be applied to two example projects.

4 Example and Result

Two software design projects were undertaken by two teams in the year 2012 and 2013. The group members are students majored in computer science or mechanic design. The goal of the project is to design a tablet application, which proposes solutions for product maintenance; it should allow a technician to access and modify PLM and ERP information in order to facilitate information flow in supply chain. MMreport and MMrecord were employed to keep track of meetings from the beginning to the end of the project, they can be downloaded in APPstore for free. XML documents were generated by these two applications. We analyzed these XML documents as well as other documents (email, forum discussion and result) according to the XML schema proposed in Sect. 3.3. Next we are going to demonstrate three rules extracted by comparison between these two projects.

A problem-solving rule on the issue "function definition" can be extracted by comparing the decision-making process on this issue of both projects. We classify repetitive solutions as essential solutions for the issue function definition, and distinctive solutions as explorative cases with a precondition. The detailed classification is shown in Table 1.

Table 1. Classification on the issue "function definition".

Tablet application for product maintenance				
Year	2012		2013	
Issue	Function definition		Function definition	
Negotiation	Proposition	Argument	Proposition	Argument
	Automatic object reconnaissance	• More efficient • Help operator with little mechanical knowledge • More expensive • Technology obstacle	Manuel object search engine	• Easy to design • Require operator to have certain mechanical knowledge
	ERP and PLM connection	• Reduce data redundancy • Technology obstacle	Tablet connection with PLM and ERP	
	Tablet connection with ERP and PLM			
Decision	• Automatic object reconnaissance • Tablet connection with ERP and PLM		• Manuel object search engine • Tablet connection with ERP and PLM	

As for the issue "function definition", we have the repetitive solution "search engine" and "tablet connection with ERP and PLM". And we also have propositions that are not taken into decision but deserve to be mentioned as conditioned solution. The classification result is shown below (Table 2).

Table 2. Classification result on the issue "function definition".

Tablet application for product maintenance	
Issue	Function definition
Essential solutions	Tablet connection with PLM and ERP, object search with tablet applications
Conditional solutions	Solution	Condition
	Automatic object reconnaissance	Enough budget
	PLM and ERP connection	Feasible technology

Cooperation rules on this project can be extracted by classifying project planning, which is represented by the sub-network decision-making process and project realization. If there are tasks concern module integration and regular meetings on project specification of whole project team, then this project is undertaken concurrently. If no meetings are held with the whole group or no integration task is assigned to more than one sub-group, then this project is considered failed at concurrent design. We can see from the project information 2012, four meetings were held inside each sub-group and only one final meeting involved the entire project group, but the issue of the final meeting was "collecting each group's work", which means no integration issue was dealt with. Apparently in the project 2012, design activities were not organized concurrently, which leads to the result "database duplication" and "expensive project cost".

Linear project planning leads to bad communication between different sub-group designers, which result in poor integration design. From the management point of view, we can further this classification by adding an organizational dimension to project planning. These two classifications are shown in Table 3.

By comparing these two project organizations, we can see that in the project team 2012, competence was distributed homogenously for each group, members were divided into computer science group and mechanical design group; whereas competence was paired in the project team 2013, computer science and mechanical design both exist in each sub-group. From this classification view, we may draw the conclusion that if designers with different skills are assigned to the same task, project tends to be carried out more concurrently, which leads to a more satisfactory result.

Extraction of these rules are all guided by comparison of structured information according to different project views, rules may change as more project information will be captured. CKD classification will progress in a cumulative manner.

Table 3. Project cooperation with a organizational perspective.

Tablet application for product maintenance			
Year	2012		2013
Phase	Project realization		Project realization
Project organization	Three sub-groups for each application module (ERP,PLM, Object reconnaissance)	Competence distribution: ERP(computer science) PLM (mechanical design) Object reconnaissance (computer science)	Three sub-groups for each application module (ERP, PLM, object search engine) / ERP(computer science) PLM (computer science, mechanical design) Object search engine (computer science, mechanical design)
Project planning	• 4 working meetings inside each sub-group to validate project specification • A final meeting to simply collect each sub-group's work		• 12 work meetings of whole project team • Sub-group meetings are organized freely
Result	• Each module has its own database, the application has 3 databases in total • Automatic image recognition increase the cost drastically		• Client-server architecture that requires only one database • Centralized data management

5   Conclusion and Perspective

This paper presented our research work on cooperative knowledge, especially on how to discover cooperative knowledge in order to reuse them. A CKD method was proposed for this purpose in design project field. It is a knowledge classification guided by semantic network schemas. Instead of classifying domain expert knowledge, interaction schemas between concepts were classified; it allows us to put each important concept in its interactive context. A CKD classification is semantically expressive and comprehensible by users. Therefore, it is up to users to choose which classification view to use for knowledge extraction. We tested CKD method on two example projects, which shows that cooperative knowledge can be extracted by interaction schema classification, more importantly, the knowledge rules extracted can be quite useful for learning purpose.

No classification can be argued to be a representation of the true structure of knowledge, the design project knowledge classification showed in this paper is a application field of CKD method, class conceptualization, semantic network structure and knowledge classification views are strictly linked to design project context. In other words, a CKD classification model should be built according to application domain features. In order to enrich this application, we will try to formalize classification rules with programming languages and test our model on more complicated projects.

References

1. Bannon, L., Schmidt, K.: Taking CSCW seriously: supporting articulation work* **1**(1), 1–33 (1992)
2. Bekhti, S., Matta, N.: Project memory: an approach of modelling and reusing the context and the design rationale. In: Proceedings of IJCAI, vol. 3 (2003)
3. Buckingham Shum, S.: Representing hard-to-formalise, contextualised, multidisciplinary, organisational knowledge. In: AAI Spring Symposium on Artificial Intelligence in Knowledge Management, pp. 9–16 (1997)
4. Castillo Navetty, O., Matta, N.: Definition of a practical learning system. In: ITHET 2005: 6th International Conference on Information Technology Based Higher Education and Training, 2005, pp. 5–10 (2005)
5. Cohen, H., Lefebvre, C.: Handbook of categorization in cognitive science, Elsevier (2005)
6. Conklin, J., Begeman, M.L.: gIBIS: a hypertext tool for exploratory policy discussion. ACM Transactions on Information Systems **6**, 303–331 (1988)
7. Dietterich, T.G.: Learning at the knowledge level. Machine Learning **1**(3), 287–315 (1986). http://link.springer.com/10.1007/BF00116894
8. Djaiz, C., Matta, N.: Project situations aggregation to identify cooperative problem solving strategies. In: Gabrys, B., Howlett, R.J., Jain, L.C. (eds.) KES 2006. LNCS (LNAI), vol. 4251, pp. 687–697. Springer, Heidelberg (2006)
9. Dnecker, C., Kolmayer, E.: Element de psychologie cognitive pour les sciences de l'information ecole nationale superieure des sciences de l'information et des bibliotheques, ed., Villeurbanne
10. Domingos, P.: A few useful things to know about machine learning. Commun. ACM **55**(10), 78 (2012). http://dl.acm.org/citation.cfm?doid=2347736.2347755
11. Ducellier, G., Matta, N., Charlot, Y., Tribouillois, F.: Traceability and structuring of cooperative Knowledge in design using PLM. Int. J. Knowl. Manage. Res. Pract. **11**(4), 20 (2013)
12. Ducellier, G.: Thèse aux plateformes PLM, University Troyes, France (2008)
13. Easterby-Smith, M.P.V., Lyles, M.: The Blackwell handbook of organizational learning and knowledge management. Adm. Sci. Q. **48**, 676 (2003)
14. Fensel, D.: Ontologies : Silver Bullet for Knowledge
15. Goodman, R.M., Smyth, P.: An information theoretic approach to rule induction from databases. IEEE Trans. Knowl. Data Eng. **4**(4), 301–316 (1992)
16. Gruber, T.: Toward principles for the design of ontologies used for knowledge sharing. Int. J. Hum. Comput. Stud. **43**, 907–928 (1995). doi:10.1006/ijhc.1995.1081
17. Gruber, T.R., Russell, D.M.: Design Knowledge and Design Rationale : A Framework for Representation, Capture, and Use (1991)
18. Khoshafian, S., Buckiewicz, M.: Introduction to Groupware, Workflow, and Workgroup Computing. Wiley, New York (1995)
19. King, R.D., Feng, C., Sutherland, A.: StatLog : comparison of classification algorithms on large real-world problems. **5**, 1–70
20. Klein, M.: Capturing design rationale in concurrent engineering teams. Computer **26**, 39–47 (1993)
21. Mai, J.: Classification in context: relativity, reality, and representation. Knowl. Organ. **31**(1), 39–48 (2004)
22. Matta, N., Ducellier, G., Charlot, Y., Beldjoudi, M.R., Tribouillois, F., Hibon, E.: Traceability of design project knowledge using PLM. In: 2011 International Conference on Collaboration Technologies and Systems (CTS), pp. 233–240, 23–27 May 2011

23. Matta, N., Ducellier, G.: An approach to keep track of project knowledge in design. In: Proceeding IC3 K/KMIS, 5th International Conference on Knowledge Management and Information Sharing, p. 12 (2013)
24. Matta, N., Ducellier, G., Djaiz, C.: Traceability and structuring of cooperative knowledge in design using PLM. Knowl. Manage. Res. Prac. **11**(1), 53–61 (2013). http://dx.doi.org/10.1057/kmrp.2012.38
25. Mika, P., Oberle, D., Gangemi, A., Sabou, M.: Foundations for service ontologies: aligning OWL-S to dolce. In: WWW pp. 563–572 (2004)
26. Miksa, F.: The DDC, The Universe of Knowledge, and the Post-modern Library. Forest Press, Albany (1998)
27. Moran, T.P., Carroll, J.M.: Design Rationale: Concepts, Techniques, and Use. L. Erlbaum Associates Inc, Hillsdale (1996)
28. Newell, A.: The knowledge level. Artif. Intell. **18**(1), 87–127 (1982)
29. Pahl, G., et al.: Engineering design: a systematic approach (2007). http://www.amazon.com/Engineering-Design-Systematic-Gerhard-Pahl/dp/1846283183
30. Sathi, A, Fox, M.S., Greenberg, M.: Representation of activity knowledge for project management. IEEE Trans. Pattern Anal. Mach. Intell. **7**(5), 531–552 (1985). http://www.ncbi.nlm.nih.gov/pubmed/21869291
31. Schreiber, G., et al.: CommonKADS: a comprehensive methodology for KBS development. IEEE Expert **9**(6), 28–37 (1994)
32. Smyth, P., Goodman, R.M.: An information theoretic approach to rule induction from databases. IEEE Trans. Knowl. Data Eng. **4**(4), 301–316 (1992)
33. Sowa, J.F.: Knowledge representation: logical, philosophical, and computational foundations (1999)
34. Zacklad, M.: Communities of action: a cognitive and social approach to the design of CSCW systems. In: Proceedings of the 2003 International ACM SIGGROUP Conference on Supporting Group Work. GROUP 2003, pp. 190–197. ACM, New York, NY, USA (2003). http://doi.acm.org/10.1145/958160.958190
35. Michie, D., Speigelhalter, D.J., Taylor, C.C.: Machine learning, neural and statistical classification (1994)

Role-Driven Context-Based Decision Support: Approach, Implementation and Lessons Learned

Alexander Smirnov[1,2], Tatiana Levashova[1], and Nikolay Shilov[1(✉)]

[1] SPIIRAS, 39, 14 Line, 199178 St. Petersburg, Russia
{smir,tatiana.levashova,nick}@iias.spb.su
[2] ITMO University, 49, Kron pr., 197101 St. Petersburg, Russia

Abstract. Today, companies have to deeply transform both their product development structure and the structure of their business processes. Context-based decision support has shown its efficient applicability in this area. However, implementation of such complex changes in large companies faces many difficulties. The paper describes the methodology of context-based decision support. The context specifies domain knowledge describing the task to be solved and its situation. It is produced based on the knowledge extracted from a common ontology. The context usage is facilitated via role-based knowledge management. The major steps of the approach implementation in collaboration with an industrial partner are described. The observations made during the implementation of the approach addressing problems related to the implementation itself and generic principles that helped to overcome the problems are discussed.

Keywords: Context management · Decision support · Role · Ontology · Lesson

1 Introduction

Modern market opportunities require companies to introduce new strategic objectives and tools. They have to build strategies that provide maximum flexibility and can optimally respond to changes in their environment [1–3]. In order to cope with these requirements, companies need to deeply transform both their product development structure and the structure of their business processes. Due to the modern trends in knowledge-dominated economy from "capital-intensive business environment" to "intelligence-intensive business environment" and from "product push" strategies to a "consumer pull" management companies accumulate large volumes of knowledge usually referred to as corporate knowledge. An efficient approach was required in order to provide a mechanism, which allows for decision maker to have required knowledge "at hand" in an appropriate form for making correct and timely decisions, what in turn will make possible for a manufacturing system to quickly react on changes in its environment and to be flexible enough.

Context-based decision support has shown its efficient applicability in this area [4, 5]. It is essential in situations happening in dynamic, rapidly changing, and often

© Springer International Publishing Switzerland 2015
A. Fred et al. (Eds.): IC3K 2014, CCIS 553, pp. 525–540, 2015.
DOI: 10.1007/978-3-319-25840-9_32

unpredictable distributed environments. Efficient decision making in such situations often involves analysis of highly decentralized, up-to-date data sets coming from various information sources. The goals of context-aware support to operational decision making are to timely provide the decisions maker with up-to-date information, to assess the relevance of information & knowledge to a decision, and to gain insight in seeking and evaluating possible decision alternatives.

A number of efforts have been done in the area of sharing information and processes between applications, people and companies. However, knowledge sharing/exchange required more than this. It required information coordination and repository sharing with regard to semantics. This has led to appearance of the Corporate Knowledge Management (CKM) that can be defined as a complex set of relations between people, processes and technology bound together with the cultural norms, like mentoring and knowledge sharing.

The present research addresses methodological foundations of context-aware decision support and its implementation. The theoretical fundamentals are built around ontologies. The ontologies are a widely accepted tool for modeling context information. They provide efficient facilities to represent application knowledge, and to make objects of the dynamic environments context-aware and interoperable.

However, implementation of such complex changes in large companies faces many difficulties: business process cannot be stopped to switch between old and new workflows; old and new software systems have to be supported at the same time; the range of products, which are already in the markets, has to be maintained in parallel with new products, etc. Another problem is that it is difficult to estimate in advance which solutions and workflow would be efficient and convenient for the employees. Hence, just following existing knowledge management implementation guidelines is not possible (e.g. [6]), and this process has to be and iterative and interactive.

The paper presents lessons learned from implementing knowledge management in a collaboration experience as a result of a long-term joint work with Festo AG&Co KG, an industrial company that has more than 300 000 customers in 176 countries supported by more than 52 companies worldwide with more than 250 branch offices and authorised agencies in further 36 countries [7, 8]. Some early steps of this collaboration related to implementation of the product codification system have been reported in [9].

The paper extends previously published work [10] in the part of the context-based approach to decision support. The structure of the paper is as follows. Section 2 describes the approach itself splitting it into three subtopics (common ontology, context management and role-based knowledge management). Section 3 discusses the implementation of the approach. The lessons learned conclude the paper.

2   Approach

Efficient knowledge management assumes deriving and processing not only internal knowledge but also knowledge from various sources including (adapted from [11]):

- customer needs, perceptions, and motivations, etc.;
- expertise within and across the supply chain;

- best practices, technology intelligence and forecasting, systemic innovation, etc.;
- products in the marketplace, who is buying them and why, what prices they are selling at;
- what competitors are selling now and what they are planning to sell in the future.

Knowledge management in a global companies requires interoperability at both technical and semantic levels. The interoperability at the technical level is addressed in a number of research efforts. It is usually represented by such approaches as e.g., SOA (Service-Oriented Architecture) [12] and is based on the appropriate standards like WSDL and SOAP [13]. The semantic level of interoperability in the production network is also paid significant attention. As an example (probably the most widely known), the Semantic Web initiative is worth mentioning [14]. The Semantic Web relies on application of ontologies for knowledge and terminology description.

The approach used in the presented work [15] relies on the following principles: (a) ontological knowledge representation for its sharing via a common ontology, (b) context management techniques for identifying relevant knowledge, (c) workflow consideration based on different roles. It assumes the following steps for knowledge management implementation (Fig. 1), described in detail in the following subsections:

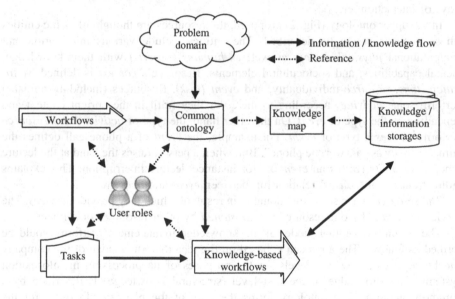

Fig. 1. Approach illustration.

1. Structural information about workflows and the problem domain is collected and described in the common ontology.
2. User roles are identified and their relevant parts of the common ontology are defined.
3. Tasks assigned to the identified roles are defined.
4. Knowledge required for performing identified tasks is defined.

5. Based on the identified roles, tasks and knowledge, new knowledge-based work-flows are defined.
6. Corresponding role-based knowledge support of the workflows is provided based on the usage of the common ontology and knowledge/information storages.

This process repeats for each particular role, with some knowledge being reused between roles.

2.1   Common Ontology for Knowledge Description

The ontology describes common entities of the company's knowledge and relation-ships between them. Besides, the dynamic nature of the company requires considering the current situation in order to provide for actual knowledge or information. For this purpose, the idea of contexts is used. Context represents additional information that helps to identify specifics of the current transaction. It defines a narrow domain that the user of the knowledge management platform works with. One more important aspect covered by the approach is the competence profiling. Profiles contain such information as the network member's capabilities and capacities, terminological specifics, preferred ways of interaction, etc.

In the upper ontology (Fig. 2) proposed, the resources are thought of as the entities whose contexts are to be described. The entities include various information and computational *physical devices* as well as *humans (experts)* with their knowledge, mental capabilities, and sociocultural elements. Resource's *context* is defined by *lo-cation, time, resource* individuality, and *event (task)*. Resources (including humans) perform some *activities* according to the *roles* they fulfil in the current context and depending on the type of *event*. On the other hand, the type of *activity* that a *resource* performs *causes* a type of *event*. For example, the *event* of a phone call defines the human *activity* as "answer the phone". But, when a person raises the hand at the lecture time, this *activity causes* an *event* as, for instance, lecture interruption. This explains bidirectionality of '*causes*' relationship between *event* and *activity*.

The resources have some functionality in result of which they provide *services*. The services *provided* by one resource are *consumed* by other resources or services.

The overall conceptual model of the knowledge management platform would be formed as follows. The approach is based on the idea that knowledge of the company can be represented by two levels for the purposes of its processing in information systems. The knowledge of the first level (structural knowledge) is described by a common ontology. The ontology forms the core of the platform. In order for the ontology to be of reasonable size, it includes only most generic common entities. Ontologies provide a common way of knowledge representation for its further pro-cessing. They have shown their usability for this type of tasks (e.g., [16–18]). The common ontology is used to solve the problem of knowledge heterogeneity and enables interoperability between heterogeneous information sources due to provision of their common semantics and terminology [19]. It describes all the products (produced and to be produced), their features (existing and possible), production processes and

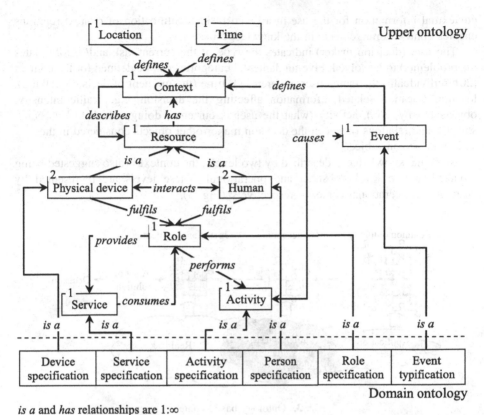

is a and *has* relationships are 1:∞
arities of other relationships are undefined

Fig. 2. Upper ontology for cyber-physical-social systems (CPSSs).

production equipment. This ontology is used in a number of different workflows. The tools are interoperable due to the usage of the common ontology and database. Knowledge map connects the ontology with different knowledge sources of the company.

Knowledge represented by the second level is an instantiation of the first level knowledge.

2.2 Context Management for Decision Support

For modern decision support systems, personalized support is important. Usually, it is based on application of the profiling technology. Each user (human or an information system) works on a particular problem or scenario represented via a context that may be characterised by a particular customer order, its time, requirements, etc.

The context knowledge accumulates up-to-date information from the environment. To take this information into account, the common ontologies has to clearly define the context and related information sources. Context management concerns organization of contextual information for the use in a given task. Identification of context relations enables context arrangements in the knowledge base.

The user (decision maker) indicates the type of the current task and/or formulates the problem(s) to be solved. Five fundamental categories can be defined for the context [20]: individuality (personal user preferences), time (the moment of decision making), location (location related information affecting the decision, e.g., traffic intensity, objects nearby, etc.), activity (what the user is currently doing, which task is being solved), and relations (between the decision maker/other objects considered in the task and other persons/objects).

Since the knowledge is described by two levels, the context is also suggested being modeled at two levels: abstract and operational. These levels are represented by abstract and operational contexts, respectively (Fig. 3).

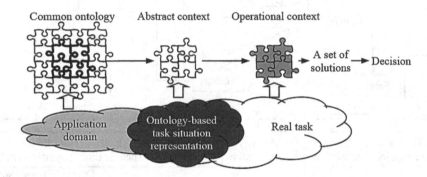

Fig. 3. Ontology-based context.

Abstract context is an ontology-based model integrating information and knowledge relevant to the current task. Such knowledge is extracted from the common ontology. The abstract context reduces the amount of knowledge represented in the common ontology to the knowledge relevant to the current task. In the common ontology this knowledge is related to the resources (services, information sources, experts, etc.) via the alignment of their descriptions and ontology elements, therefore the abstract context allows the set of resources to be reduced to the resources needed to instantiate knowledge specified in the abstract context. The reduced set of resources is referred to as contextual resources. The ontology alignment model developed by the authors is protected by USA patent US 2012/0078595 A1 [21].

Operational context is an instantiation of the domain constituent of the abstract context with data provided by the contextual resources. This context reflects any changes in environmental information, so it is a near real-time picture of the current situation for the task. The context embeds the specifications of the problems to be solved and related methods and tools. The input parameters of these methods are instantiated.

Constraint satisfaction techniques [22, 23] can be used to take into account data provided by the contextual resources and other possible constraints that might have an impact on the task solution. These techniques are naturally combined with ontology-based problem definition, and allow to set context parameters so that they would be taken it into account when the task constraints are applied. As a result, a set of feasible (alternative) solutions for the current task is produced and the decision making can be regarded as a choice between these alternatives.

The resulting knowledge representation and sharing model corresponds to the knowledge classification according to the abstraction and types proposed in [24], where the following knowledge abstraction levels are distinguished: universal, shared, specific, and individual. In the proposed approach (Fig. 4) the universal level is considered as the common knowledge representation paradigm. In order to support constraint satisfaction techniques it is based on the formalism of object-oriented constraint networks (see [25] for the detailed description) represented by means of knowledge representation language. The abstractions provided at this level are shared by the ontologies. Both shared abstraction level and specific abstraction level are considered sharable and reusable since ontologies and contexts of these levels share common representation paradigm and common vocabulary.

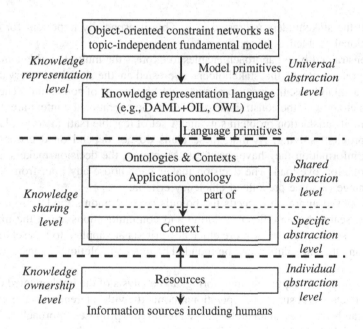

Fig. 4. Knowledge representation and sharing model.

Table 1 shows the correspondences between the steps of the described above approach and the phases of the Simon's model [26], which specifies decision making consisting of "intelligence", "design", and "choice" phases. Unlike other decision making models, Simon suggests to minimize the efforts of decision makers to evaluate consequences of the possible alternatives and his/her dependence on the multiple factors influencing the choice. Simon proposed a 'satisfactory' decision as a result of decision making. That is a decision that is not efficient or optimal, but the decision that

Table 1. Three-phase model.

Phase	Phase content	Steps of Simon's model	Proposed steps
Intelligence	Finding, identifying, and formulating the problem or situation that calls for a decision	1. Fixing goals	• Abstract context creation
		2. Setting goals	• Operational context producing
Design	Search for possible decisions	3. Designing possible alternatives	• Constraint-based generation of feasible (satisfactory) alternatives
Choice	Evaluation of alternatives and choosing one of them	4. Choice of a satisfactory decision	• Choice of a satisfactory decision

satisfies all the stakeholders interested in it. This is especially important for complex and operational problem solving.

The constrained-based approach enables to express the multiple influencing factors (e.g., the preferences of the stakeholders interested in the decision, intervals of the resources' availabilities, the resources' costs, etc.) by means of constraints. The factors' constraints along with the constraints specified in the operational context are processed as a constraint satisfaction problem solving. A set of feasible (satisfactory) plans is the result of problem solving. At that, these plans do not depend on decision makers' attentions, information they have, or stress. Moreover, the decision makers are saved from information overload. The decision maker can choose any plan from the set or take advantage of some predefined efficiency criteria.

The proposed approach exceeds the bounds of the Simon's model proposing two more steps: search for an efficient solution and communications about the implementation of this solution. If one or more efficiency criteria are applied to this set of feasible solutions, an efficient solution can be found. The efficient solution is considered as the workable decision.

In order to enable capturing, monitoring, and analysis of the implemented decisions and their effects the abstract and operational contexts with references to the respective decisions can be retained in an archive. As a result, the proposed approach can support reusable models of task solutions. Based on an analysis of the operational contexts together with the implemented decisions tacit user preferences can revealed.

2.3   Role-Based Workflow Consideration

The third idea of the approach is to consider the workflows from perspectives of different roles.

Research efforts in the area of information logistics show information and knowledge needs of a particular employee depend on his/her tasks and responsibilities

[27]. This is also confirmed in other works, e.g.: "Information demand depends on the role and tasks an entity has within a larger organization. If the role and/or the tasks change, so too will the demand" [28].

Role-based approaches have shown their efficiency in such adjacent areas as ontology modelling [29], competence modelling [30], etc.

Based on the experiences from the industrial case study, the following perspectives have been identified:

- Production engineer. Representatives of this role are responsible for designing the production process including definition of the sequence of technological operations and their grouping.
- Production manager. Representatives of this role are responsible for assigning technological operations to equipment constituting the shop floor and distributing the production program to available facilities.
- Product engineer. Representatives of this role are responsible for designing new products and defining their characteristics.
- Product manager. Representatives of this role are responsible for communications with customers, including selection of products based on customer requirements as well as analyzing customer needs and sharing these with product engineers.

It is obvious that representatives of each role require different knowledge and have different views both at workflows and knowledge used. Hence, personalization is one of the important features of efficient decision support. For this reason, the roles are described via profiles and associated with different parts (usually overlapping) of the common ontology that set certain limitations to provide only information and knowledge that is useful for a particular user in the corresponding perspective (Fig. 5). As a result, the job roles serve as additional constraints imposed on the operational context. The user is presented a part of the operational context that provides information for tasks for which the user of this particular role is allowed to make decisions. In other words, this part contains information useful (or allowed) for the particular user.

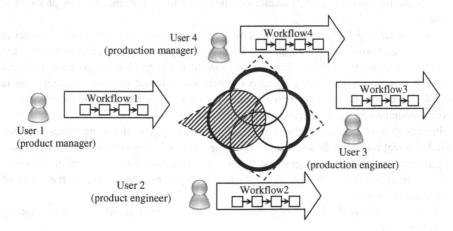

Fig. 5. Role-based perspectives of the common ontology.

A detailed description of the roles and their needs is given in the next section describing the implementation of the developed approach.

3   Implementation

The first step of the approach implementation is creation of the ontology. This operation was done automatically based on existing documents and defined rules of the model building. The resulting ontology consists of more than 1000 classes organized into a four level taxonomy, which is based on the VDMA (Verband Deutscher Maschinen – und Anlagenbau, German Engineering Federation) classification [31]. Taxonomical relationships support inheritance that makes it possible to define more common attributes for higher level classes and inherit them for lower level subclasses. The same taxonomy is used in the company's PDM and ERP systems.

For each product family (class) a set of properties (attributes) is defined, and for each property, its possible values and their codes are defined as well. The lexicon of properties is ontology-wide, and as a result, the values can be reused for different families.

Then, based on the developed ontology, the complex product modelling design and system has been implemented. Complex product description consists of two major parts: product components and rules. Complex product components can be the following: simple products, other complex products, and application data. The set of characteristics of the complex product is a union of characteristics of its components. The rules of the complex products are union of the rules of its components plus extra rules. Application data is an auxiliary component, which is used for introduction of some additional characteristics and requirements to the product (for example, operating temperatures, certification, electrical connection, etc.). They affect availability and compatibility of certain components and features via defined rules.

At the second step, the major roles, whose workflows were addressed by knowledge management implementation, have been identified. As it was mentioned earlier, the roles are product manager, product engineer, production manager, and production engineer.

Then, at steps 3 and 4, their tasks and needs are analysed. The product manager works with customers and their needs. Usually, the parameters and terminology the customer operates with differ from those, operated by product engineers. For this reason, a mapping between the customer needs and internal product requirements is needed. Based on these requirements new products, product modifications or new product systems can be engineered for future production.

For the goal of production process description the approach distinguishes between virtual and real modules. In accordance with the approach, the virtual modules are used for grouping technological operations from the production engineer's point of view. The real modules represent actual production equipment (machines) at the level of production manager.

At steps 5 and 6 the knowledge-based workflows are defined and corresponding supporting tools are built.

A system (called DESO) has been developed for a structured storage of the knowledge about data domain, and for its further processing. Depending on the particular tasks it can be supplemented by other components (tools) intended for solution of specific problems using the knowledge, contained in the common storage. In the time being the tools for the enterprise production program planning (Goal), for the production modules designing (Module), and for the industrial resources distribution and planning (Goal and Module) have been developed.

The system supporting the levels of production engineer and production manager was originally focused on the early stages of planning procedure of investment calculation and determination for the (a) derivation of production scenarios, (b) determination of investment cost, (c) assignment of locations and (d) estimation of product variable cost. The system aims at providing a knowledge platform enabling manufacturing enterprises to achieve reduced lead time and reduced cost based on customer requirements through customer satisfaction by means of improved availability, communication and quality of product information. It follows a decentralized method for intelligent knowledge and solutions access. Configuring process incorporates the following features: order-free selection, limits of resources, optimization (minimization or maximization), default values, freedom to make changes in global production network model.

This system distinguishes between virtual and real modules. In accordance with the approach, the virtual modules are used for grouping technological operations from the production engineer's point of view. The real modules stand for the real equipment used for the actual production. The production engineer sets correspondences between the technological operations of virtual modules and machines of real modules.

It also includes a tool for sequences of operations for a part production, possible alternatives of production distribution etc. This tool supports inheriting subordinate objects, what allows creating of complex hierarchical systems of objects, and using templates automating the user's work.

The main entities of the approach implementation and identified roles are presented in Fig. 6. The figure also identifies tools implemented in the first case study.

The developed so far integrated knowledge management workflow for the first case study (addressing roles of product engineer and product manager) is presented in Fig. 7 and is described in detail in [32]. At the first stage, the major product ontology is filled with generic classifications of products and their components. This is done via two tools (NOC and CONCode) since recently developed order code scheme differs from that used before. However, since multiple customers are used to operate with the old classification it has to be maintained.

At the next stage, the product managers and product engineers design new products and solutions based on existing products and components (the CONSys tool). If a new product or component is needed, its implementation can be requested from the order code structure team. Together with new products and solutions, the appropriate rules and conditions are designed as well (e.g., acceptable load, size, compatibility constraints, etc.).

When the configuration model is finished it is proposed to the customers so that they could configure required products and solutions themselves or with assistance of product managers (the CONFig tool).

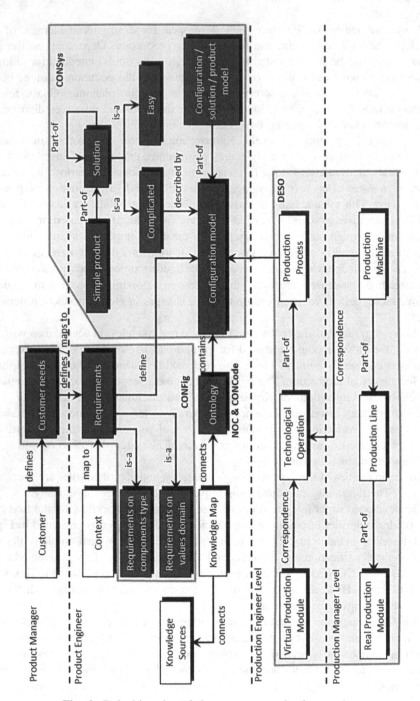

Fig. 6. Role-driven knowledge management implementation.

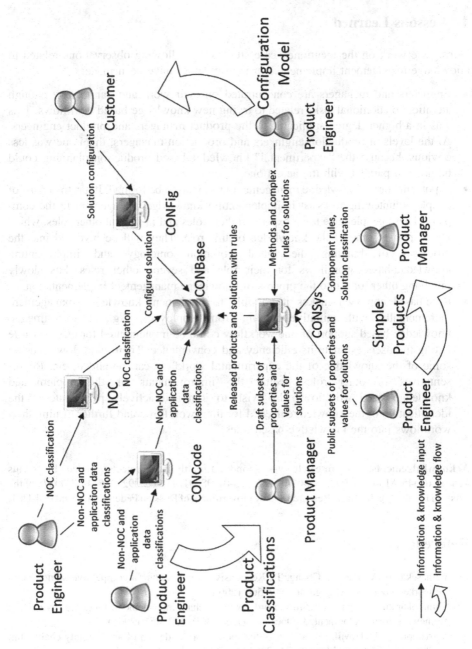

Fig. 7. Integrated knowledge-based workflow.

4 Lessons Learned

During the work on the mentioned case studies the following observations related to knowledge management implementation in companies have been made:

- Engineers and managers are concentrated on their work and cannot pay enough attention to additional tasks related to trying new knowledge-based workflows. This was in a higher degree applicable to the product managers and product engineers. At the levels of production engineers and production managers, this issue was less obvious, because the "experimental" knowledge-based production planning could be done in parallel with the actual one.
- A potential target knowledge management group has to be formed. It has to consist of people volunteering to assist in implementing knowledge management in the company. These people have to be experts in their roles and in several other roles, which would re-use some of the knowledge of this role. They will be involved into the processes of building the initial common ontology and implementing knowledge-based workflows for their role and several other roles thus slowly involving other roles into the process of knowledge management implementation.
- Role-based approach makes it possible to implement knowledge management incrementally, with initiative coming from employees. E.g., an experimental knowledge-based support of one workflow could be implemented for one user role letting the users estimate its efficiency and convenience. Then, workflows reusing some of the knowledge of the experimental workflow can be added, etc. Representatives of other roles seeing the improvements of the implemented knowledge-based workflows also wish to join and actively participate in the identification of the knowledge needed for their workflows and further turning their workflows into the knowledge-based ones.

Acknowledgements. The research was supported partly by projects funded by grants #14-07-00345, #15-07-08092, #14-07-00427, #15-07-08391, #15-07-09229, #14-07-00378 of the Russian Foundation for Basic Research and by Government of Russian Federation, Grant 074-U01.

References

1. Gunasekaran, A., Lai, K., Cheng, T.: Responsive supply chain: a competitive strategy in a networked economy. Omega **36**, 549–564 (2008)
2. Gunasekaran, A., Ngai, N.: Build-to-order supply chain management: literature review and framework for development. J. Oper. Manage. **23**(5), 423–451 (2005)
3. Christopher, M., Towill, D.: An integrated model for the design of agile supply chains. Int. J. Phys. Distrib. Oper. Manage. **31**, 235–244 (2001)
4. Blomqvist, E.: The use of semantic web technologies for decision support–a survey. Semant. Web **5**(3), 177–201 (2014)
5. Smirnov, A., Pashkin, M., Levashova, T., Chilov, N.: Fusion-based knowledge logistics for intelligent decision support in network-centric environment. In: Klir, G.J. (ed.) Int. J. Gen. Syst. **34**(6), 673–690. Taylor & Francis (2005)

6. Oluikpe, P.: Developing a corporate knowledge management strategy. J. Knowl. Manage. **16** (6), 862–878 (2012)
7. Oroszi, A., Jung, T., Smirnov, A., Shilov, N., Kashevnik, A.: Ontology-driven codification for discrete and modular products. Int. J. Prod. Dev. **8**(2), 162–177 (2009)
8. Smirnov, A., Kashevnik, A., Teslya, N., Shilov, N., Oroszi, A., Sinko, M., Humpf, M., Arneving, J.: Knowledge management for complex product development. In: Bernard, A., Rivest, L., Dutta, D. (eds.) PLM 2013. IFIP AICT, vol. 409, pp. 110–119. Springer, Heidelberg (2013)
9. Smirnov, A., Shilov, N., Kashevnik, A., Jung, T., Sinko, M., Oroszi, A.: Ontology-driven product configuration: industrial use case. In: Proceedings of International Conference on Knowledge Management and Information Sharing (KMIS 2011), pp. 38–47 (2011)
10. Smirnov, A., Shilov, N.: Role-driven knowledge management implementation: lessons learned. In: Proceedings of the International Conference on Knowledge Management and Information Sharing (KMIS 2014), pp. 36–43 (2014)
11. Botkin, J.: Smart Business: How Knowledge Communities Can Revolutionize Your Company. Free Press, New York (1999)
12. SOA: Service-Oriented Architecture (SOA) Definition. http://www.service-architecture. com/web-services/articles/service-oriented_architecture_soa_definition.html
13. Web services explained. http://www.service-architecture.com/web-services/articles/web_ services_explained.html
14. Semantic Web. http://www.semanticweb.org
15. Smirnov, A., Levashova, T., Shilov, N.: Knowledge sharing in flexible production networks: a context-based approach. In: Graves, A., Stone, G., Miemczyk, J. (eds.) Int. J. Automot. Technol. Manage. (IJATM) **9**(1), 87–109. Inderscience Publishers (2009)
16. Bradfield, D.J., Gao, J.X., Soltan, H.: A metaknowledge approach to facilitate knowledge sharing in the global product development process. Comput. Aided Des. Appl. **4**(1–4), 519–528 (2007)
17. Chan, E.C.K., Yu, K.M.: A framework of ontology-enabled product knowledge management. Int. J. Prod. Dev. **4**(3–4), 241–254 (2007)
18. Patil, L., Dutta, D., Sriram, R.: Ontology-based exchange of product data semantics. IEEE Trans. Autom. Sci. Eng. **2**(3), 213–225 (2005)
19. Uschold, M., Grüninger, M.: Ontologies: principles, methods and applications. Knowl. Eng. Rev. **11**(2), 93–155 (1996)
20. Zimmermann, A., Lorenz, A., Oppermann, R.: An operational definition of context. In: Kokinov, B., Richardson, D.C., Roth-Berghofer, T.R., Vieu, L. (eds.) CONTEXT 2007. LNCS (LNAI), vol. 4635, pp. 558–571. Springer, Heidelberg (2007)
21. Balandin, S., Boldyrev, S., Oliver, I.J., Turenko, T., Smirnov, A.V., Shilov, N.G., Kashevnik, A.M.: Method and Apparatus for Ontology Matching, US Patent 2012/0078595 A1 (2012)
22. Baumgaertel, H.: Distributed constraint processing for production logistics. IEEE Intell. Syst. **15**(1), 40–48 (2000)
23. Tsang, J.P.: Constraint propagation issues in automated design. In: Gottlob, G., Nejdl, W. (eds.) Expert Systems in Engineering Principles and Applications. LNCS, vol. 462, pp. 135–151. Springer, Heidelberg (1991)
24. Neches, R., Fikes, R., Finin, T., Gruber, T., Patil, R., Senator, T., Swartout, W.: Enabling technology for knowledge sharing. AI Mag. **12**(3), 36–56 (1991)
25. Smirnov, A., Levashova, T., Shilov, N.: Semantic-oriented support of interoperability between production information systems. Int. J. Prod. Dev. **4**(3/4), 225–240 (2007)
26. Simon, H.A.: Making management decisions: the role of intuition and emotion. Acad. Manage. Executive **1**, 57–64 (1987)

27. Lundqvist, M.: Information demand and use: improving information flow within small-scale business contexts. Licentiate thesis, Department of Computer and Information Science, Linköping University, Linköping, Sweden (2007)
28. Persson, A., Stirna, J.: PoEM 2009. LNBIP, vol. 39. Springer, Heidelberg (2009)
29. Fox, M.S., Barbuceanu, M. Gruninger, M.: An organisation ontology for enterprise modelling: preliminary concepts for linking structure and behavior. In: Proceedings of the Fourth Workshop on Enabling Technologies: Infrastructure for Collaborative Enterprises, pp. 71–81 (1995)
30. Tarasov, V., Sandkuhl, K.: On the role of competence models for business and IT alignment in network organizations. In: Abramowicz, W., Maciaszek, L., Węcel, K. (eds.) BIS Workshops 2011 and BIS 2011. LNBIP, vol. 97, pp. 208–219. Springer, Heidelberg (2011)
31. VDMA, German Engineering Federation. http://www.vdma.org/en_GB/
32. Smirnov, A., Sandkuhl, K., Shilov, N., Kashevnik, A.: "Product-process-machine" system modeling: approach and industrial case studies. In: Grabis, J., Kirikova, M., Zdravkovic, J., Stirna, J. (eds.) PoEM 2013. LNBIP, vol. 165, pp. 251–265. Springer, Heidelberg (2013)

Knowledge Capture and Information Sharing for Science and Technology

Augusta Maria Paci[✉], Cecilia Bartolucci, Cecilia Lalle, and Francesco Tampieri

National Research Council of Italy, Piazzale Aldo Moro 7, Rome, Italy
{augustamaria.paci,cecilia.lalle}@cnr.it,
cecilia.bartolucci@mlib.ic.cnr.it, f.tampieri@isac.cnr.it

Abstract. The knowledge of the intellectual capital of research organizations can benefit from knowledge capture and information sharing processes and practices in the direction of competitive advantage. The integrative perspective is described formed by communication, navigation and multicontact relationship. It supports a dynamic framework for the current 'open nature' of knowledge. Foresight project and Horizon Scanning are practices reported in this paper for knowledge capturing in science and technology. These practices have an inclusive and engaging nature that requires participatory processes and investigate cross-cutting aspects and impacts beyond technology developments. In a KMIS 2014 Conference session, the discussions related to this paper focused three overall principles for addressing competitive advantage in R&D group: team effectiveness of both performances and creativity; team collaboration and competition to increase performances; team integration, coordination and synchronization for benefits and values of successful research organizations.

Keywords: Foresight · Horizon scanning · Research organizations · Knowledge capture · Research and innovation

1 Introduction

Knowledge provides the intellectual basis in science and technology and its developments will impact on industrial change, growth and jobs. Highly specialized learned individuals in competitive fields represent the intellectual capital in the knowledge economy.

However, this knowledge is not yet captured to contribute to the forward-looking development of intellectual capital in research organizations. The value of encouraging exploration of future areas of research can be beneficial both to individuals in work processes involving a large and global community and to research organizations dealing with accumulation of knowledge. Due to the economic situation and financial contexts,

---

Note on Authorship: Cecilia Bartolucci and Francesco Tampieri are authors of Sect. 4.1, Cecilia Lalle is author of Sect. 2, Augusta Maria Paci is author of overall paper approach and content.

© Springer International Publishing Switzerland 2015
A. Fred et al. (Eds.): IC3K 2014, CCIS 553, pp. 541–555, 2015.
DOI: 10.1007/978-3-319-25840-9_33

diverging trends in public investments in research and innovation have been continuously addressed (ERIAB 2014).

Knowledge capture and information sharing can support the relationship among knowledge, social relationships and creativity and contribute to develop a "social capital" empowering scientific organizations and individuals to become reserves for the future and renewable sources for the development of the knowledge capital.

The establishment of this relationship requires conditions for stimulating researchers to openly discuss major problems arising in societal and planetary contexts, thus exchanging insights and self-preparing to a research approach, which would deliver responsible and knowledge-based solutions (McInerney and Day 2007).

Many countries and organizations in science and technology had advantages from the large accumulation of knowledge resulting from the research outcomes - a big reserve - and presently need to start knowledge capture to develop a new source for competitive advantage to create societal prosperity in the web age.

The Intellectual Capital of organizations needs to develop the knowledge capability which is defined by Alavi and Leidner (2001) as the interior capability of researchers having the potential for influencing future action. In order to facilitate researchers to develop this interior capability, knowledge capture and information sharing processes can be realized for building forward-looking innovation perspectives. Three decisive components - Communication, Navigation and Multicontact Relationship represent the current 'open nature' of knowledge in science and technology and key activities in researchers work including communities of interest debating challenging themes and interdisciplinary R&D.

Knowledge capture and information sharing processes should focus these three components - which presently are additional in daily management processes carried on within research organizations. The establishment of the above mentioned relationship requires dynamic processes supported by social activities within research organizations to enable a transition of the Intellectual Capital towards new forms of knowledge production.

These processes facilitate the development of an integrated perspective where old and new components are considered and managed within a dynamic framework with trustful and collaborative learning environments.

The dynamic framework outlines how the integrative perspective can enhance the research processes by adding knowledge management and information sharing to selected and strategic knowledge areas.

In order to develop the interior capability of researchers having the potential for influencing future action, foresight and horizon scanning are activities requiring discussions and debates among researchers on the future of societal progress empowered by R&D. These activities can help to drive forward the transformation of the intellectual capital. Foresight and horizon scanning in particular concern long-term thinking and do not interfere with the management and planning of organizations and individuals.

The experiences reported below refer to foresight and horizon scanning practices which are motivated by the need to develop, at organization level, new strategic interdisciplinary directions. They developed collaborative settings for knowledge and information sharing with the goal of developing collective knowledge for S&T (Andreta et al. 2013). Horizon

Scanning experiences are small scale practices which aim at developing a knowledge capability in small groups with selected knowledge areas. To this end, this practice targets high level researchers and activates processes for opinions exchange among pairs on key advances within selected knowledge areas.

This process is carried on with periodic meetings and the facilitator enables to point out relevant scientific areas and to discuss state of art, trends, strengths and weaknesses, personal leadership roles. Collaborative settings support the open discussion and the capture of expectations in order to prioritize potentials of future knowledge areas with individual interests.

Through the activation of practices, Intellectual Capital can represent important source for the competitive advantage for the organization.

2 State of the Art

The overall on-going shift in the relationship between science and society, according to Beck's (1996) main message, rises problems and implications for any industrialized society and requires to find new pathways for science and technology development in this new context.

According to literature in intellectual capital, the definition of knowledge meeting the research intellectual resources is to focus knowledge as a the interior capability of researchers having the potential for influencing future action (Alavi and Leidner 2001; Carlsson et al. 1996).

In Knowledge Management literature, Swan (2007) pointed out the issues in the relationship between Knowledge Management and innovation introducing the perspective of production, process and practice. These reflections studies on rethinking Knowledge Management as a plurality of techniques, methods and epistemologies are collected in McInerney and Day (2007).

Recent studies, still in draft publications, describe difficult times for science and technological organizations that should demonstrate public costs, value of results and quality of life achieved in years of scientific development (ERIAB 2014). Previously, the perspective of Science 2.0 was proposed by Burgelman et al. (2010). However, it is not yet known how to support this transition and its related processes and research runners help themselves engaging in foresight initiatives. Foresight is a well-known area that has given support to various types of intervention for transformation purposes as reported below.

Georghiou et al. (2008) demonstrate how, since the last decade of the last century, foresight has been strongly linked to public policies. During that period five generations of technology foresight were developed to respond to both stakeholders and multi-level policy, involving national, regional and local frameworks (Taylor 2004).

Debating foresight with implications for policy and decision-making at general level is considered a relevant activity by the European Commission, enabling experts to think and share technological perspectives and suggesting implications to complement potential development for economic growth (EC and JRC 2008). Various contributions find a relationship between foresight and the increase of societal participation extending the concept of stakeholders beyond private or institutional organizations to communities. The role of these communities is to develop Future-Oriented Technology Analysis (FTA) reflecting their interests in new emerging contexts and to make use of shared methodologies.

Recent literature in foresight debates on the integration of quantitative and qualitative approaches. This is addressed in detail by Haegeman et al. (2013) who propose a taxonomy for methodological combinations across current FTA practices. Of particular interest in this paper is the analysis of common 'misconceptions' such as 'subjectivity' which becomes a barrier especially in qualitative processes involving scientists.

Furthermore, the reported Epistemology-Skills-Trust cycle indicates the need to adopt mechanisms for the capturing of intellectual scientific knowledge.

The authors point out weaknesses and introduce ways to reduce the distortion, for instance through the 'legitimacy of the person making the judgment'. With regard to participation from institutional stakeholders, foresight has been considered an instrument for political participation on medium-long-term planning perspectives, capable of influencing priority-setting and political agendas (Gieseke 2012).

In the foresight studies literature, the maturation of foresight through growth processes is a result of contributions given to overall S&T development, through technological forecasting, horizon scanning and scientific/technological prospective studies including broader social and economic perspectives.

Recently, due to the economic transformation, foresights concentrate on innovation and on drivers of change STEEP (Society, Technology, Economy, Environment, Policy) and on decisive, strategic factors analyses. Multiple stakeholder classes are required for this foresight and this stresses the 'collective ability to shape the future'. In particular, it is acknowledged that the use and impact of forward-looking technology analysis for policy and decision making can be successfully led by institutional governmental organizations taking into considerations dominant criteria for success (Calof and Smith 2008). Interviewed practitioners from different countries include social and economic dimensions in technological development and innovation systems. Among the features of foresight success, Calof and Smith identify the existence of a national-local academic receptor and training capacity during the start-up phase.

Foresight is also considered supporting strategic intelligence for policy decision-making (De Smedt 2008; Montalvo et al. 2006). It was noted that to respond to innovation policies, foresight needs to meet the demand of future-looking scenarios regarding the overall changes of cultural and societal aspects (Cagnin and Keenan 2008). In order to enable decision-makers to better understand and cope with the complexities and uncertainties of the continuously changing patterns of innovation, foresight practices increased mobilization and coordination of different stakeholders as well as personalized delivery of insights and analysis to key players at group level.

More recently, Van der Gießen and Marinelli (2012) reported the multiple functions and the role of European and national policy workshops. Yet, in past literature, Eriksson and Weber (2006) suggested that adaptive foresight were developed at the crossroads of foresight and adaptive strategic planning, Innovation is seen as increasingly complex, interdependent and uncertain and therefore in need of broad and multi-disciplinary exploration and participation. establishing a close relationship between foresight, decision-making, innovation and new technologies. Particularly they highlight the capitalization of the accumulation of knowledge with the aim of delivering insights on the maturity of knowledge and planning for knowledge options in a later stage.

A relevant aspect in the literature regards the evolution of initial foresight panels involving only a few experts (De Smedt 2008). These have been superseded by large extended and interacting communities, formed by stakeholders and key players. They actively participate and progressively align their expectations to shape the future (EFP 2012; Eriksson and Weber 2006; Futman Project 2015).

All these elements mark important steps in the evolution of design of complex economic and industrial scenarios. This continuous process of collecting stakeholders' knowledge through different contexts is called 'integrative planning'. It defines the combination and mapping of available insights for future developments emerging from different groups of stakeholders and experts and mass interviews conveyed into 'packages of information' that are published as policy briefs for decision-makers, summarizing possible and alternative options and solutions.

In the same area, to bridge the gap between research and industrial demand, starting from the futuring of next generation industrial technologies in order to assist the industrial transformation, a full cycle-oriented methodology -rolling programme- was proposed to relate results from future-oriented activities such as Foresight, Roadmapping, Implementation and Monitoring (FRIM) (Paci and Chiacchio 2008). This methodology can enable participants to assess and prioritize those enabling technologies for implementation of research and development projects within the EU 7FP as proposed by Paci et al. (2013). In the last five years, foresight for research and innovation represents a relevant social process. It was intensified and supported through funded projects (covering different objectives such as Global Europe, Pashmina, NEEDS etc.) by the DGs of the European Commission and foresight studies carried on by the EU Commission Joint Research Centre (JRC 2013) in order to collectively shape a strategic frame for the EU policy strategy that envisages both industrial as well as social benefits.

In this stimulating thinking, the sketch contested science delivered by the EU foresight project VERA visions, presents a pessimistic narrative (VERA 2013).

Di Bello and Andreta (2013) provide some initial guidance for industrial competitiveness in Horizon 2020 towards an integrated planning framework for key enabling technologies. These enablers drive innovation and growth in the economy and society in a global Europe (EC 2012) in the frame of the Grand Challenges as depicted by the EU Horizon 2020 programme.

Considering digital initiatives, recent developments and the toolkit available in looking at digital futures have been described by Accordino (2013) with regard to the EU Commission web platform Futurium in order to strengthen and support the social networking aspect together with the global dimension.

3 Integrative Perspective

In the business perspective, intellectual capital is formed by four separated elements covering different roles: human capital, structural capital, innovation capital and customer capital.

In the research perspective, the intellectual capital covers multiple roles with the four elements establishing a flexible relation with the related research organization structure. It is interesting to observe the distribution of these different roles during the research process for identifying aspects useful to knowledge capture and information sharing. Currently, the activity of researchers is related to funded project and implemented in a linear chain. The following three main types of assets have been identified:

- Innovation capital: this asset is related to the accumulation of scientific and techno-logical knowledge such as scientific papers, patents and knowledge deliverables.
- Human and Structural capital: this asset represents the organization and its segmen-tation, the infrastructure and the scientific competencies.
- Customer capital: this asset refers to public-private collaboration models, grants and efforts.

In research organizations, research performance needs to achieve excellence, project results, costs and efforts in a very effective way. Efficiency does not allow flexible behaviors and implies reflections on how to support the 'open nature' of knowledge in the dynamic framework for research process by introducing the integrative perspective.

It can be useful to analyze how to move forward and which orientations are to be taken to stimulate intellectual capital to interdisciplinary and forward-looking thinking.

The dynamic framework and the integrative perspective combine the current process with three decisive components influencing today and in the future how knowledge could be shared and accessed: Communication, Navigation and Multicontact relationship (Fig. 1).

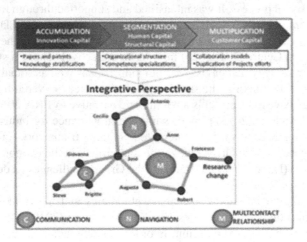

Fig. 1. Integrative perspective for a dynamic framework of research intellectual capital.

Knowledge capture and information sharing contribute to this integrative perspective and dynamic framework.

The perspective integrates organization capitals to empower the research process with the following dynamic behaviors (Fig. 2):

- Communication facilitates trusted confidence, this behavior adds value to previous Accumulation formed by scientific papers, patents and knowledge stratification and stimulates the innovation capital to develop a proactive behavior.
- Navigation with multiple channeling facilitates interdisciplinary thinking, overcoming the rigidity of the Segmentation of organizational structure, competence specializations.
- Multicontact relationship facilitates easiness sociality and impacts on the customer capital by stimulating exploitation of collaboration models and projects results.

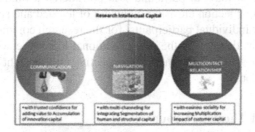

Fig. 2. Decisive components of the integrative perspective.

The components of the integrative perspective are societal driven and introduce more adaptive and flexible approaches in organizational processes.

4 Practice Perspectives

Practice perspectives for knowledge and innovation are mentioned in the literature as emerging areas to encourage learning and innovation within communities of practices (Schatzi 2001; Swan 2007). Foresight and Horizon Scanning can add insights to the understanding of the fast dynamics of knowledge, dedicating closer attention to the social activity of research communities. Foresight and Horizon Scanning support the understanding of technological trends, of complex issues, of strengths and weaknesses as well as expectation from society and companies that will drive the societal progress in next 20–30 years. The practice proposed are related to two initiatives carried on with research groups.

These practices respond to the need of establishing forward looking activities to address future emerging S&T paradigms to anticipate and increase impacts. They aim to develop the imaginative capability of researchers to envisage interdisciplinary solutions, building and implementing a societal driven vision contributing to a responsible and sustainable socio-economic development.

The objective of these practices is to accelerate the shift from project team or individual practice of accumulation of knowledge to collective practices that involve researchers to analyze strategic choices, exploit networked collaboration and build new solid and durable contacts across high-tech specialties within the organization and at global level. This in particular targets to activate dynamic processes to form a critical mass within research organizations that want to act as a key receptor in the knowledge economy.

The first practice presents the Science and Technology Foresight Project (STFP) launched in 2013 by the National Research Council of Italy (CNR) and the Trieste Area Science Park Consortium (AREA), with the support of the Ministry of Education, University and Research (MIUR) as an open bottom-up participatory process. The scope is to build a community of researchers and a networked collective knowledge to overcome disciplinary boundaries.

The practice describes the approach to new interdisciplinary S&T areas for research development of next generation technologies in the medium to long term tackling societal problems related to health, food, environment and energy challenges.

The network dimension - in particular - is considered a fundamental enabler of interaction and interconnection ensuring inclusion of learned and sound popular opinions. Benefits are for individual participants, for the research communities, and for the research organizations that covers the role of national-local receptor - as observed by Calof and Smith (2008). The first practice perspective on Foresight focus the Interdisciplinary thematic groups.

4.1   Foresight S&T Thematic Groups

The CNR Foresight thematic groups are the core engine of the project aiming to provide new insights for the collective understanding exploiting the collective intelligence of researchers through a participatory process and a bottom-up approach.

Focusing on urgent problems related to the Horizon 2020 Grand Challenges -food, health, environment, energy and transport- each thematic group highlights a specific S&T topic: Nano for sustainable food, Personalized medicine, Intelligent traceability for environment, Smart storage for future energy.

Each thematic group carries on specific, topic related activities: holds an exploratory seminar or brainstorming, improves and identifies knowledge gaps and obstacles, long term Key Enabling Technologies (KETs) areas, as well as market potential and social acceptability (HLG on KETs 2011).

The first seminars in December 2013 and May 2014 regarded the thematic group Food and the thematic group Health. In November 2014, the thematic group Environment organized a Brainstorming. All events allowed to validate the background documents and to discuss with international experts long term knowledge developments.

What primarily emerged is the necessity to develop a highly interdisciplinary approach, which allows to constantly take into consideration the strong links between sectors such as health, food and environment and evaluate how the introduction of a new technological application will impact areas apparently remote from the one of intervention.

A synergy across thematic groups was realized for a communication on food packaging, nanomaterials and environmental issues (Bartolucci et al. 2014), but further cooperation is emerging in addressing topics such as the application of nanotechnologies in agriculture, food processing and nutrition. Imitating the "magic bullets" introduced in medicine, the development of nanocapsules for the controlled release of fertilizers or pesticides could greatly reduce the environment impact and positively affect the quality of the food.

Foreseeing futures needs, the application of converging technologies could support monitoring the production along the whole food chain, providing information on environmental impact, quality, nutritional value, also contributing to waste reduction.

A further aspect is the development of personalized food in analogy to the one of personalized medicine which shouldn't necessarily address single individuals but groups of people with specific needs.

Crossing all thematic areas is the awareness that the application of new technologies, in particular of nanomaterials in the food sector, is accompanied by reluctance since it is difficult to evaluate the hazard on environment and on human health. This encourages researchers to participate in discussions on cross-cutting aspects such as appropriate analytical methods for the characterization and quantification of nanomaterials and nanoparticles within a life cycle framework, necessary for toxicological evaluations.

Social acceptance is becoming a key to the adoption of a new technology also in market driven strategies, with better information, transparency and willingness to communicate with stakeholders and the public - which is a requirement that needs to be continuously addressed.

Gathering around the table, experts with specific disciplinary skills in a highly multidisciplinary context, chosen to promote the transfer of knowledge across situations, should encourage a more holistic approach, allowing us to deal with specific issues, while keeping a broad view.

4.2 Networking

Social connectivity plays a key role in the involvement of scientists and experts from academia, government and the private sector. Knowledge management, information system and web systems are geared toward enabling users to share discussions and to stimulate scientists' knowledge exchange. The aim is to form a collective knowledge across multiple disciplinary fields of knowledge crossing boundaries (Alavi and Leidner 2001).

The CNR foresight project makes use of a web platform to enable collective knowledge sharing. The web-platform characterizes the innovative approach of the foresight project and the Sections for topics discussion, contributions, document repository, management try to respond to the need of a networking environment (http://www.foresight.cnr.it).

The web-platform plays as the operational infrastructure and supports transdisciplinary knowledge sharing ensuring connectivity and real time information exchange.

5 Horizon Scanning

The practice about Horizon Scanning (HS) represents a collaborative activity to identify emerging trends and issues in a complex future. Researchers exchange information in order to prioritize areas for competitive advantage in knowledge development.

This practice helps to overcome the segmentation currently in place in organizational structure due to competence specializations.

Horizon Scanning is an activity distinct but often related to foresight. The domain of Knowledge Management and Information Sharing is particularly important for this practice as the aim is to identify potential fields of competitive advantage. It introduces a social and collaborative activity in daily working habits in order to overcome silo-thinking in specialized fields of knowledge.

The experience was developed to accompany the social and collaborative activity of a Working Group of high-level researchers (Paci 2011) in periodic meetings. The five steps DMAIC-HS method was applied:

- Design = problem recognition. This aim to introduce the continuous development of future technologies and/or innovation actions to meet societal challenges.
- Measure = the overall appraisal. This aims to recognize the researchers' outstanding quality of S&T compliant with the trends towards the future.
- Analyze = the analysis of the state of the art. This is a qualitative analysis requiring intensive-information meetings. The activity develops synergies among participants, processes. It facilitates the update through knowledge sharing and distribute leadership across various knowledge domains. It facilitates the strategic positioning and the possibility of influencing strategic and innovation choices for industrial developments and market changes.
- Improve and Control/Monitor = the resulting analytics enable to move further the discussions, to identify forward-looking areas and to identify case studies and groups of leaders.

Another practice regarded a Working Group 'Discussion and Prevision Group'. This practice was participated by researchers in Chemical Sciences and Materials Technologies to enable discussions across the specializations of chemistry (macromolecular, bio-sensor, bio-catalysis, plastic materials, DNA,…).

Practices in Knowledge Management and Information Sharing show the need to facilitate in-house debate and to activate prioritizing processes for focusing issues and promoting interdisciplinary across several scientific/technological domains. These approaches, based on in-house expertise and team leaders, can effectively constitute important sources of competitive advantage. They also can facilitate horizontal and inter-generational learning, blurring verticality in organizational settings and linearity of thinking processes.

6   Toward the Dynamic Framework

The dynamic framework and the integrative perspective can influence the future of Communication, Navigation and Multicontact relationship in research organizations.

Knowledge Management and Information Sharing can contribute to this goal and the Conference held in Rome in October 2014 provided a good discussion floor at international level.

The session has exchanged interesting results on ongoing experiences for an effective knowledge management within organizations confirming the importance of knowledge capture as a source of competitive advantage and a factor of success of contemporary organizations.

6.1 Perspectives in Effective Knowledge Management

The theoretical contribution on social and cognitive proximities showed how Knowledge Management and Information Sharing can support clusters of knowledge flows and combination outlining the role of 'Transactive Memory Systems'. These systems will enable to highlight who knows what and the kind of problems to be solved in wide and dynamic clusters that presently lack of identitary factors (Amandine and Catherine 2014).

Another contribution proposed a unifying model for IT Knowledge Artifact (ITKA) considered "object per se" and "socially situated" seeing knowledge as a social practice, context-dependent and built on performative interactions of human actors (Cabitza and Locoro 2014).

A case study reported results based on processes for knowledge creation and technology evaluation. This paper provided a needs-driven future-oriented approach in knowledge management to foresee services and goods needed for ageing societies and related planning for knowledge and innovation actions (Saijo et al. 2014).

The practices carried on within National Research Council of Italy to knowledge capture in science and technology refer about social processes of research organizations (Paci et al. 2014). The KMIS 2014 Session C discussion was animated around three arguments:

- a new way of looking at R&D groups. In particular the focus was put on the leadership role: prior, this role was referred to a "strong leader", while presently in successful organizations the "Agile Leadership" and "Member leadership" aim to develop self-development and full participation in goal setting. The target is the team effectiveness in terms of performances and creativity of the R&D group and looks to new knowledge systems and ensure the achievement of multiple outcomes;
- the motivation factors in R&D teams and the need of showing the value of collaborations. This is part of the support that can be provided by Knowledge management and Information Sharing processes. The team collaboration and competition are essential mechanisms for increase group performances;
- the role of process management that prior was dedicated to quality aspects and today extends quality aspects to integration, coordination and synchronization of activities in successful organization to identify benefits and values.

6.2 Analysis of Behaviors

The analysis, regarding the behaviors of researchers in the practices, identified issues and barriers, resistance and polarization of opinions that may occur in knowledge and information sharing during meetings and in follow-on discussion. It provided elements for a transparent process to improve collaborative settings and to develop successful processes built on collective experiences. From the results, emerged that researchers become aware of the need to overcome barriers and silo-thinking and of the primary role of individual knowledge played in successful participatory process to build a collective knowledge.

Motivational climate, friendly relationships, team spirit and self-motivation, re-treat for meeting outside organization premises emerged as new working needs.

The analysis showed also relevant issues related to feelings of uncertainty, risk aversion, self-learning, tendency to close shops and of excessive filtering. Working together for a long and continuous period enables to overcome some of these issues as skills/capabilities are developed.

Another barrier to participation is due to peaks of work overload.

The mix of networked and personal approach in many cases supports to take care of personalized approaches and to resolve issues.

The web platform represent an important part in rapid information exchange also in case of contrasting opinions. The interactive network is capable of contributing with continuity, creativity and responsibility and the collaborative settings reinforces the individual ability to react to complexity.

7   Conclusions

An interdisciplinary approach to science and technology is core in the current industrial and societal transformation. Science and technology include other dimensions such as society and environment - driving forces for the global EU economy- and cross-cutting aspects such as awareness of ethical issues, users and societal acceptability, analysis of benefits, business planning and skill sets, assessment of risk. Research organizations are important assets for industry and society and are in a continuous process of transformation since year 2000. The proposed integrative perspective facilitates the expansion of knowledge flows. For the competitive advantage knowledge capture and information sharing processes and practices are also required.

In Italy, National Research Council's S&T Foresight Project developed a community of researchers with the aim to stimulate thinking for science and technology developments based on collective intelligence, interdisciplinary approaches and fostering societal driven future ideas and solutions.

To support the strengthening of the international dimension, European and international organizations are planning similar pathways for horizon scanning and foresight and future collaborations are envisaged to discuss and exchange experiences.

Going back to Beck's message, science and society need to overcome barriers and consider impacts of science and technology in a wider perspective. Discussions from the KMIS field of study and applications distribute, transfer and enable the adoption of practices from a variety of settings with different organizational processes.

The Conference is a major meeting point for researchers and develops a collective expertise in topic areas strategic for contemporary organizations that support industries and economies to change products, processes and organizations for a sustainable future and economic growth.

References

Accordino, F.: The futurium - a foresight platform for evidence-based and participatory policymaking. Philos. Technol. (2013). doi:10.1007/s13347-013-0108-9

Alavi, M., Leidner, D.E.: Review: knowledge management and knowledge management systems: conceptual foundations and research issues. MIS Q. **25**(1), 107–136 (2001)

Amandine, P., Catherine, T.: Transactive memory system in clusters: the knowledge management platform experience. In: Proceedings of the 6th International Conference on Knowledge Management and Information Sharing, Rome, pp 5–14 (21–24 October 2014)

Andreta, E., Paci, A.M., Taylor, S.: Participatory and stakeholder engagement: the fore-sight initiative of Italian Consiglio Nazionale delle Ricerche. In: European IFA Academic Seminar, Winterthur, Switzerland (16–19 September 2013)

Bartolucci, C., Paci, A.M., Tampieri, F.: A foresight perspective to the role of packaging in the sustainability of food and environment. In: International Conference on Eco-sustainable Food Packaging Based on Polymer Nanomaterials, Book of the Abstracts, Rome (2014)

Beck, U.: Risk Society: Towards a New Modernity. Sage, Beverly Hills (1996)

Burgelman, J.C., Osimo, D., Bogdanowicz, M.: Science2.0 (Change will happen…), in First Monday, July 5th, 2010, vol. 15:7 (2010). http://journals.uic.edu/ojs/index.php/fm/article/view/2573

Cabitza, F., Locoro, A.: Made with knowledge: disentangling the IT knowledge artifact by a qualitative literature review. In: Proceedings of the 6th International Conference on Knowledge Management and Information Sharing, Rome, pp. 64–75 (21–24 October 2014)

Cagnin, C., Keenan, M.: Positioning future-oriented technology analysis. In: Cagnin, C., Keenan, M., et al. (eds.) Future-Oriented Technology Analysis: Strategic Intelligence for an Innovative Economy. Springer, Berlin (2014)

Calof, J., Smith, J.E.: Critical Success Factors for Government Led Foresight. In: 3rd Seville Seminar on FTA: Impacts and Implications for Policy and Decision-Making (2008)

Carlsson, S., El Sawy, O.A., Eriksson, I.V., Raven, A.: Gaining Competitive Advantage Through Shared Knowledge Creation: Search of a New Design Theory for Strategic Information Systems. In: Proceedings of the Fourth European Conference on Information Systems, Lisbon, Portugal, pp. 1067–1075, 2–4 July 1996

De Smedt, P.: Strategic intelligence in decision making. In: Cagnin, C., Keenan, M., et al. (eds.) Future-Oriented Technology Analysis: Strategic Intelligence for an Innovative Economy. Springer, Berlin (2008)

Di Bello, G., Andreta, E.: Industrial competitiveness in horizon 2020: "Towards an integrated planning framework for Key Enabling Technologies (January 2013)

EC: Global Europe 2050, European Commission Directorate-General for Research and Innovation Socio-economic Sciences and Humanities (2012). http://ec.europa.eu/research/social-sciences/pdf/global-europe-2050-report_en.pdf

EC, JRC: European Commission and Joint Research Centre Institute for Prospective Technological Studies (eds.), Book of Abstracts of 3rd International Seville Seminar on FTA: Impacts and Implications for Policy and Decision-Making (2008)

EFP: The role of forward-looking activities for the governance of Grand Challenges Insights from the European Foresight Platform. ISBN 978-3-200-02811-1 (2012)

ERIAB: Placing excellence at the centre of research and innovation policy: Innovation Union and European Research Area (ERA) Stress Test. EC-ERIAB, Draft publication, February 2014 (2012). http://ec.europa.eu/research/innovation-union/pdf/expert-groups/ERA_STRESS_TEST-Placing_excellence_at_the_centre_of_research_and_innovation_policy.pdf

Eriksson, E.A., Weber, K.M.: Adaptive foresight navigating the complex landscape of policy strategies. In: Second International Seville Seminar on Future-Oriented Technology Analysis: Impact of FTA Approaches on Policy and Decision-Making, Seville, Spain (28–29 September 2006)

Geyer, A., Scapolo, F., Boden, M., Döry, T., Ducatel, K.: The Future of Manufacturing in Europe 2015–2020. The Challenge for Sustainability, European Commission, Joint Research Centre (DG JRC), Institute for Prospective Technological Studies (2003). http://foresight.jrc.ec. europa.eu/documents/eur20705en.pdf

Georghiou, L., et al. (eds.): The Handbook of Technology Foresight: Concepts and Practice. Edward Elgar Publishing, London (2008)

Gieseke, S.: FLA as a means of participation in modern democratic decision making. In: The Role of Forward-Looking Activities for the Governance of Grand Challenges, European Foresight Platform (2012)

Haegeman, K., Marinelli, E., Scapolo, F., Ricci, A., Sokolov, A.: Quantitative and qualitative approaches in future-oriented technology analysis (FTA): from combination to integration? Technol. Forecast. Soc. Chang. **80**(3), 386–397 (2013)

HLG on KETs: High level group on key enabling technologies. Final Report (2011). http:// ec.europa.eu/enterprise/sectors/ict/files/kets/hlg_report_final_en.pdf

JRC 2013: The foresight study on how will standards facilitate new production systems in the context of EU innovation and competitiveness in 2025? Joint Research Centre Final Report (November 2013)

McInerney, C.R., Day, R.E. (eds.): From Knowledge Objects to Knowledge Processes Series. Information Science and Knowledge Management, vol. 12. Springer, XII (2007)

Montalvo, C., Tang, P., Mollas-Gallart, J., Vivarelli, M., Marsilli, O., Hoogendorn, J., Leijten, J., Butter, M., Jansen, G., Braun, A.: Driving factors and challenges for EU industry and the role of R&D and innovation, ed. European Techno-Economic Policy Support Network (ETEPSAISBL), Brussels (2006)

Paci, A.M.: New trends in research and innovation: internationalization of research activities. In: 6th International Working Conference "Total Quality Management - Advanced and Intelligent Approaches", Belgrade, Serbia (2011)

Paci, A.M., Chiacchio, M.S.: Futuring the changing world for industrial technologies. In: Book of Abstracts of 3rd International Seville Seminar on Future-Oriented Technology Analysis: Impacts and Implications for Policy and Decision-Making (2008)

Paci, A.M., Bartolucci, C., Lalle, C., Tampieri, F.: Research change in transition. In: Proceedings of the 6th International Conference on Knowledge Management and Information Sharing, Rome, pp 44–53 (21–24 October 2014)

Paci, A.M., Lalle, C., Chiacchio, M.S.: Education for innovation: trends, collaborations and views in special issue on engineering education. J. Intell. Manuf. **24**, 487–493 (2013). doi:10.1007/ s10845-012-0631-z

Taylor, S.: The scan™ process and technology foresight (2004). http://www.cgee.org.br/ atividades/redirKori/118

Saijo, M., Watanabe, M., Aoshima, S., Oda, N., Matsumoto, S., Kawamoto, S: Knowledge creation in technology evaluation of 4-wheel electric power assisted bicycle for frail elderly persons - a case study of a salutogenic device in healthcare facilities in Japan. In: Proceedings of the 6th International Conference on Knowledge Management and Information Sharing, Rome, pp 87–100 (21–24 October 2014)

Schatzi, T.: Practice theory. In: Schatzki, T., Knorr Cetina, K., Von Savigny, E. (eds.) The Practice Turn in Contemporary Theory. Routledge, New York (2001)

Swan, J.: Managing knowledge for innovation. In: McInerney, R.C., Day, R.E. (eds.) From Knowledge Objects to Knowledge Processes. Information Science and Knowledge Management, vol. 12, XII, pp. 147–167. Springer, Berlin (2007)

Van der Gießen, A., Marinelli, E.: The value of FLA for strategic policy making. In: European Foresight Platform: The Role of Forward-Looking Activities for the Governance of Grand Challenges Insights from the European Foresight Platform, pp. 22–30 (2012)

VERA: VERA 2030 ERA Scenario, Short version of D3.1 ERA Scenario Report (November 2013). http://www.eravisions.eu/page/36/attach/VERA_2030_ERA_Scenarios.pdf

Ontology Methodology Building Criteria for Crowdsourcing Innovation Intermediaries

Cândida Silva[1,3(✉)] and Isabel Ramos[2,3]

[1] School of Management and Industrial Studies, Polytechnic Institute of Oporto,
Vila do Conde, Portugal
candidasilva@eseig.ipp.pt
[2] Information Systems Department, School of Engineering,
University of Minho, Guimarães, Portugal
iramos@dsi.uminho.pt
[3] Center Algoritmi, University of Minho, Guimarães, Portugal

Abstract. Crowdsourcing innovation intermediaries are organizations that mediate the communication and relationship between companies that aspire to solve some problem or to take advantage of any business opportunity with a crowd that is prone to give ideas based on their knowledge, experience and wisdom. A significant part of the activity of these intermediaries is carried out by using a web platform that takes advantage of web 2.0 tools to implement its capabilities. Thus, ontologies are presented as an appropriate strategy to represent the knowledge inherent to this activity and therefore the accomplishment of interoperability between machines and systems. In this paper we present an ontology roadmap for developing crowdsourcing innovation ontology of the intermediation process. We start making a literature review on ontology building, analyze and compare ontologies that propose the development from scratch with the ones that propose reusing other ontologies, and present the criteria for selecting the methodology. We also review enterprise and innovation ontologies known in literature. Finally, are taken some conclusions and presented the roadmap for building crowdsourcing innovation intermediary ontology.

Keywords: Ontology building methodologies · Crowdsourcing innovation · Innovation ontology · Ontology enterprise

1 Introduction

Ontologies have proliferated in the last years, mostly in Computer Science and Information Systems areas. This is essentially justified by the need of achieving a consensus in the multiple representations of reality inside computers, and therefore the accomplishment of interoperability between machines and systems [1].

Open innovation is a timely topic in innovation management. Its basic premise is open up the innovation process. The innovation process, in general sense, may be seen as the process of designing, developing and commercializing a novel product or service to improve the value added of a company.

This paradigm proposes the use of external and internal ideas, and internal and external paths to market, as means to reach advances in technology used by companies [2].

© Springer International Publishing Switzerland 2015
A. Fred et al. (Eds.): IC3K 2014, CCIS 553, pp. 556–570, 2015.
DOI: 10.1007/978-3-319-25840-9_34

The World Wide Web, the open source movement and the development of Web 2.0 tools facilitates this kind of contributions, opening space to the emergence of crowd-sourcing innovation initiatives.

Howe introduced the term crowdsourcing, in an article in Wired Magazine [3], as a way of using the Web 2.0 tools to generate new ideas through the heterogeneous knowledge available in the global network of individuals highly qualified and with easy access to information and technology. Although, this concept has been used quite a time, the creation of the Wikipedia and of many examples of free software, like Linux, are examples of crowdsourcing activity. Crowdsourcing is a form of outsourcing not directed to other companies but to the crowd by means of an open call mostly through an Internet platform. Basically, the process is trying to solve a company problem by an open call in the network. The company posts a problem and a vast amount of indi-viduals offers the solution for evaluation. The winning idea is awarded in some way and the company develops the idea. The crowd can be defined as a large set of anonymous and heterogeneous individuals, which may be composed of scientists and experts in various fields, but also of novices [4, 5].

A crowdsourcing innovation intermediary is an organization that mediates the communication and relationship between the seekers – companies that aspire to solve some problem or to take advantage of any business opportunity – with a crowd that is prone to give ideas based on their knowledge, experience and wisdom [6].

For crowdsourcing innovation intermediary the crowd is composed by groups of specialists in different areas, such as individual researchers, research team, labs, post-graduate students and highly qualified individuals.

This paper makes a literature review on ontology building, and analyzes and compares ontologies that propose the development from scratch with the ones that propose reusing other ontologies. It also review enterprise and innovation ontologies known in literature. Finally, are presented the criteria for selecting the methodology and the roadmap for building crowdsourcing innovation intermediary ontology.

To achieve this objectives we defined the following main questions, which guided the literature review: (i) What are the main concepts guiding ontologies building?; (ii) What are the existing ontologies about business and innovation?; (iii) Which methodologies should be considered to build an ontology?

To answer these questions, we started conducting an exhaustive bibliography review of the authors most relevant to the scientific area, identifying curriculum authors, books, book chapters, papers presented at conferences and published articles in scientific journals. This literature review was conducted in Scopus, Google Scholar, ISI Web of Knowledge. The documents were collected through the UM catalog, b-on; RCAAP, IEEExplore, Colcat. Then, based on the extensive bibliography retrieved, we proceeded to the identification of the most relevant papers, gathering all those whose title refers to the following combination of words: "ontologies", "ontology develop-ment"; "ontology building"; "innovation ontology"; "enterprise ontology"; and "on-tology methodologies".

This paper is organized as follows. In Sect. 2, is made a literature revision of ontology concepts such as its definition and features, classification of ontologies by different authors, application areas, and enterprise and innovation ontologies. Fol-lowing, in Sect. 3, we review literature on ontology methodologies. Finally, the conclusions of this work are presented and the roadmap for building a crowdsourcing innovation intermediary ontology.

2 State of Art on Ontologies

There are several definitions of the concept of ontology from where can be assemble that it has an informal and formal notion associated to it. Gruber [7] definition clearly shows these – "An ontology is a formal, explicit specification of a standard conceptualization".

An ontology is a conceptualization of world view with respect to a given domain. This world view is conceived by a framework as a set of concept definitions and their interrelationships, that may be implicit, existing only in someone's head or tool, or explicit which includes a vocabulary of terms and a specification of their meanings.

The specification of that world view by means of a formal and declarative representation, with semantic interconnections, and some rules of inference and logic, will perform the formal ontology. The formal representation will facilitate the interoperability between heterogeneous machines and systems.

Ontologies have been developed with the promise of providing knowledge sharing and reuse between people and systems, by building a conceptual framework of a given knowledge domain to be represented. This framework will be formalized through a specific ontology language which will clearly express a controlled vocabulary and taxonomy, as represented in Fig. 1.

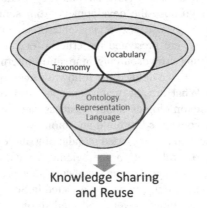

Fig. 1. Ontology building features.

The vocabulary is a list of terms or classes of objects, respective definitions and relationships between each other, provided by logical statements. They also specify rules for combining the terms and their relations to define extensions to the vocabulary.

The taxonomy or concept hierarchy is a hierarchical classification or categorization of entities in the domain of an ontology. The taxonomy should be in a machine-readable and machine-processable form in order to permit interoperability.

The full specification of an ontology domain establishes a conceptual framework, composed by the vocabulary and the taxonomy, for discussion, analysis, and information retrieval in a domain.

Ontology development requires an effective ontological analysis of the content the world view domain that it intends to represent. This analysis will reveal the terms and concepts of the domain knowledge, their relations, organization and hierarchy. Thus, they clarify the structure of domain knowledge, so, it can be called a content theory [8, p. 53].

As the objective of ontologies is to facilitate knowledge sharing and reuse between various agents, regardless of whether they are human or machines, then it can be said that ontologies are a prerequisite and a result of a consensual point of view on the world. It is a prerequisite for consensus because to have knowledge sharing agents must agree on their interpretation of a domain of the world. And it is a result of consensus because the model of meanings was built as result of a process of agreement between agents on a certain model of the world and its interpretations. Therefore, it is an essential requirement that any ontology can progress over the time [9].

Briefly, an ontology provides an explicit conceptualization that describes the semantics of the data. As Fensel [9] stated "ontology research is database research for the 21st century where data need to be shared and not always fit into a simple table".

2.1   Type of Ontologies

Over the years, researchers of this body of knowledge, tried to clarify, classify and typify the concept of ontology, in terms of its definition, components, and application areas. Table 1 present a summary of, what we considered being, the most relevant contributions.

Analyzing these table and the different views on the classification of ontologies, we can organize them in different types by the subject or issue of conceptualization, and them, each of this type can have different degrees of formality, purpose or objective, and components.

So it seems some consensus that the types of ontologies, by subject or content matter are:

- Domain or content ontology – represents the knowledge valid for a given type of domain (e.g. enterprise, medical, electronic, mechanic).
- Meta-data ontology – provide a vocabulary for describing informational content (e.g. Dublin core describes on-line information sources).
- General or common-sense ontology – provides basic notions and concepts about describing general knowledge about the world and so they are valid across several domains (e.g. time, space, state, event).
- Representational/frame ontology – ontologies that provide representational entities without stating what particular domain it represents. Do not commit to any particular domain.
- Task/method/problem solving ontology – provide terms specific for particular tasks and problem-solving methods. It defines primitives by which the problem solving

Table 1. Ontologies' classification by researcher's perspectives.

Author	Classification/Dimension
Guarino [10]	Informal conceptual system; Formal semantic account; Representation of a conceptual system with a logical theory; Vocabulary used by a logical theory; Meta-level specification of a logical theory
Mizoguchi et al. [11]	Content theory: - Object ontology, - Activity ontology, - Field ontology; Task ontology; General or common-sense ontology
Uschold and Gruninger [12]	Formality: - Informal, semi-formal, formal; Purpose: - Communication between humans, - Inter-operability among systems, - Systems engineering benefits; Subject matter: - Domain ontology, - Task/method/problem solving ontology, - Representational/meta ontology
van Heijst et al. [13]	Amount and structure of the conceptualization: - Terminological ontology, - Information ontology, - Knowledge modelling ontology; Subject of conceptualization: - Application ontology, - Domain ontology, - Generic ontology, - Representation ontology
Guarino [14]	Domain ontology; Meta-data ontology; General or common-sense ontology; Representational ontology; Method and task ontology
Lassila and McGuinness [15]	Controlled vocabulary; Glossary; Thesaurus; Informal is-a hierarchy; Formal is-a hierarchy; Formal instances; Frames; Value restrictions; General logical constraints
Benjamins and Gómez-Pérez [16]	Reusability: - Content ontologies: task, domain, representation, - Issue of the conceptualization: application, generic, representation, domain
Bullinger [17]	Subject matter: - Application, - Task, - Domain, - General, - Representation; Formality: - Informal notation, - Semi-informal/semi-formal notation, - Formal notation; Expressiveness: - Taxonomy, - Thesaurus, - Topic map, - Lightweight vs. heavyweight ontology

context can be described and domain knowledge can be put into the problem solving context.

2.2 Application Areas

Fensel [9], in his book, classifies the main broad areas where ontologies are of interesting application: knowledge management, web commerce, electronic business and enterprise application integration. Gasevic et al. [8] identified some high-level activities where the utilization of ontology technology applies perfectly, which are tasks that fall, somehow, in all these application areas. After all, those are the usual task for having knowledge share and reusability:

- Collaboration – ontologies provide a unique consensual knowledge framework that can be used as a common, shared reference to communicate and work with.
- Interoperation – ontologies enable information conversation, transfer and integration from different and heterogeneous sources. However, to permit automatic integration it is needed that all the sources recognize the same ontology.

- Education – ontologies can be a reliable and objective source of information to those who want to learn more about a specific domain, since it is expected that they result of a wide consensus of the structure of the knowledge domain they represent. So, they are also a good publication medium and source of reference.
- Modeling – the structure and hierarchy established in the ontology will represent important reusable building blocks, which many specific applications should include as predeveloped knowledge modules.
- E-commerce – Since ontologies enable interoperability between machines and systems, e-commerce can be considered an application domain for ontologies. They can be fully used in all the e-commerce tasks.
- Search engines – concepts and taxonomies from ontologies can be used to support structures, comparative, and customized searches.

2.3   Enterprise Ontologies

Enterprise ontologies are usually created to define and structure knowledge in business universe about the processes, activities, organization and strategies.

The first enterprise ontology (EO) project was developed at the University of Edinburg with the aim of promoting the common understanding between people across enterprises, as well as to serve as a communication medium between people and applications, and between different applications [18]. Its major role is to act as a communication medium, ensuring effective interchange of information and knowledge between different users, tasks and systems.

This implies that besides technical interoperability it is needed a semantic and pragmatic interoperability between applications and users [19].

The main intended uses for EO, identified by Uschold et al. [18], were:

- "enhance communication between humans, for the benefit of integration;
- serve as stable basis for understanding and specifying the requirements for end-user applications using the Tool Set which in turn leads to more flexibility in an organization;
- to achieve interoperability among disparate tools in an enterprise modeling environment using the EO as an interchange format."

To develop the EO, the authors used brainstorming technique to identify the maximum of potential terms that are relevant to enterprises. The list of terms and phrases harvested were then grouped by similar areas and established priorities to include the terms in the ontology. The resultant list of terms was categorized to identify the core and specific terms of each area and define it. The core concepts define the Meta-Ontology of EO.

The EO establish the following basic core terms in a Meta-Ontology: Entity, Relationship, Role, Attribute, State of Affairs, Achieve, Actor, Actor Role, and Potential Actor. The specific terms defined by EO were grouped into five working areas: Activity, Organization, Strategy, Marketing, and Time, as presented in Table 2.

Table 2. List of terms defined by Enterprise Ontology by working area [18].

Area	Terms
Activity	Activity Specification, Execute, Executed Activity Specification, T-Begin, T-End, Pre-Conditions, Effect, Doer, Sub-Activity, Authority, Activity Owner, Event, Plan, Sub-Plan, Planning, Process Specification, Capability, Skill, Resource, Resource Allocation, Resource Substitute
Organization	Person, Machine, Corporation, Partnership, Partner, Legal Entity, Organizational Unit, Manage, Delegate, Management Link, Legal Ownership, Non-Legal Ownership, Ownership, Owner, Asset, Stakeholder, Employment Contract, Share, Share Holder
Marketing	Purpose, Hold Purpose, Intended Purpose, Strategic Purpose, Objective, vision, Mission, Goal, Help Achieve, Strategy, Strategic Planning, Strategic Action, Decision, Assumption, Critical Assumption, Non-Critical Assumption, Influence Factor, Critical Influence Factor, Non-Critical Influence Factor, Critical Success Factor, Risk
Strategy	Sale, Potential Sale, For Sale, Sale Offer, Vendor, Actual Customer, Potential Customer, Customer, Reseller, Product, Asking Price, Sale Price, Market, Segmentation Variable, Market Segment, Market Research, Brand Image, Feature, Need, Market Need, Promotion, Competitor
Time	Time Line, Time Interval, Time Point

First, the EO was defined in an informal way, establishing its concepts in plain English and later, in the formalization phase, the terms were encoded into Ontolingua language. The Ontolingua has already adequate primitives to cover what was required to represent Enterprise Meta-Ontology, namely: objects, relations, and functions. Thus, it was evaluated the concepts that already are defined by Ontolingua and imported to EO. The formal Enterprise Meta-ontology become: Actor, Function, Set, Thing, Potential Actor, Relation and State of Affairs.

This ontology has been successfully used as a mean to achieve inter-operation through a common terminology used for specifying tasks, capabilities, and agents; and to enhance communication between humans by using terms in a consistent way.

Some of the failures of this project were the difficulty to use formal definitions and to have automatic interpretations; the lack of an interchange format to other ontologies; the fact of being too generic; and missing a graphic context for browsing the list of terms.

The TOVE (TOronto Virtual Enterprise) project, developed in the University of Toronto, came out as an enterprise ontology that solves the problems presented above.

TOVE aims to create a generic, reusable enterprise model for a company. This model must (1) provide a shared terminology; (2) defines the meanings of each term in a precise and unambiguous manner; (3) implements semantics in a set of axioms, and (4) provide a graphical context for depicting terms or concepts [20].

TOVE was implemented with two formal languages: C++ for the static part and Prolog the axioms. The ontology implementation started with a generic ontology for enterprises, but additionally, it has been created more specific ontologies covering enterprise subareas, like, business and project process, organization, logistics, transport ant store, scheduling, and information resources [21].

2.4   Innovation Ontologies

Ning et al. [22] presented a system architecture that combines ontology, inference and mediation technologies to create a semantic web of innovation knowledge, which they called Semantic Innovation Management (SIM). The framework of the system was based on metadata harvesting and RDF access technologies.

Bullinger [17], in her PhD thesis, develop OntoGate, ontology to manage idea assessment and selection of the innovation process.

Also Riedl et al. [23] proposed an ontology to represent ideas of the innovation management process. It defines the core idea concept that is enriched by other concepts like collaborative idea development, including rating, discussing, tagging, and grouping ideas.

They classified the ontology as an application ontology because it provides a description of a technical architecture to represent complex ideas evaluations along various concepts. It offers a common language to idea storage and exchange for the purpose of achieving interoperability across innovation tools.

The ontology was built following the methodology proposed by Noy and McGuinness [24], and reused other existing ontologies, as suggested.

The Table 3 presents the main classes of idea ontology, and the source of each class. When the class is new, the source ontology will be Idea Ontology. It was also reused the Enterprise Ontology to model the descriptive attributes of an idea.

Table 3. Idea ontology terms and related source ontology.

Class	Source
CoreIdea	Idea ontology
Document	Friend of a friend (FOAF)
Item	Semantically-interlinked online communities (SIOC)
Resource	Resource description framework (RDF)
Origin	Idea ontology
Rating	Rating ontology
Person	FOAF
Tagging	Tag ontology
Concept	Simple knowledge organization ontology (SKOS)

3   Ontology Building Methodologies

Ontology methodologies comprises a set of established principles, processes, practices, and activities used to design, formalize, implement, evaluate, and deploy ontologies, for which uses some development tools. These development tools include ontology representation languages, graphical ontology development environment, and ontology-learning tools.

To develop an ontology it must be first answered questions like: what is the scope of the ontology? Who is interested in it? Who will use and maintain it? Which methods and methodologies can be used to build ontologies? Which activities are performed?

Which tools gives support to the ontology development process? Which ontology language can be used to implement ontologies? Which methodology, tool and language should be used to develop and to implement an ontology for crowdsourcing innovation intermediaries?

Noy and McGuinness [24] gave some basic advices in seven steps for the process building of your first ontology, and that helps to answer these questions:

1. **Determine the Domain and Scope of the Ontology:** This step should help to define the knowledge domain covered, to limit the scope of the model, and the users and maintainer of the ontology.
2. **Consider Reusing Existing Ontologies:** Before starting to create an ontology from scratch, it is worth to check if there exist any ontology that can be refine or extend that cover our particular domain or task. Reusing ontologies specially considered if our system needs to interact with other applications that have already committed to particular ontologies or controlled vocabularies, in order to reduce the translation effort.
3. **Enumerate Important Terms:** Ontology development should start by listing all the terms we thing important or like to explain to user, and describe them briefly.
4. **Define the Classes and Their Hierarchy:** This step and the following are taken by turns.
5. **Define the Properties of the Classes:** Expresses the internal structure of concepts by explicating their extrinsic properties (name, duration, and use), intrinsic properties (weight, color, etc.), parts, and relations to other classes and individuals in those classes.
6. **Define the Characteristics Classes' Properties:** Defining things like attributes type, domain and range allowed values, cardinality, and other features.
7. **Create Instances:** Creating individual instance of classes in the hierarchy, filling in the attributes values.

Over the years, several authors have proposed some distinct methodologies, by different proposals of combining practices, activities, languages, etc., according to the project they were involved in Fernández-Pérez and Gómez-Pérez [25] and Corcho et al. [26] described some of these methodologies and compared their degree of maturity. Based on the seven steps for building an ontology [24] and in the work of [25, 26], we defined as criteria for selecting the methodology the following parameters: aim, method, phases, activities, language, tool and build cooperation capability. Therefore, the resultant list of the best-known approaches for both building from scratch and reusing ontologies analyzed are summarized in Appendix.

4 Conclusions

The ontologies are presented as a conceptual model for the systematization and formalization of consensual knowledge in a field of knowledge. This conceptualization is rendered concrete with the definition of terms and concepts from the domain of knowledge in analysis, their relationships, organization and hierarchy, and allows the sharing and reuse by different people and systems of such knowledge [1, 26–30].

Some of the difficulties of sharing knowledge and reusing ontologies are [8]: the existence of several different languages to representing ontologies, and tools may not support the language used to develop the ontology; there are many diverse ontologies that have been developed to describe the same topic or domain, resulting of using different competing methodologies and working groups. To build an ontology by combining some of them may require a lot of manual adjustments because of deep differences between them, and the resulting ontology may still inadequate to fulfill all the requirements; and difficulties on ontology maintenance, since all parts of knowledge evolve over time.

Various ontologies have emerged, particularly in the areas of business and enterprise. Ning et al. [22], Bullinger [17], and more recently Riedl et al. [31], proposed ontologies for the process of innovation management, but they represent only the component relating to the process of generating ideas. Not the best of our knowledge, the existence of any ontology that represents the entire process of creating an intermediate value of crowdsourcing innovation. Thus, an ontology of crowdsourcing innovation intermediaries will be an instrument to understand this phenomenon and thus will also be a facilitator for the emergence of such intermediaries.

The roadmap for building this ontology comprises two main phases: in the first phase is being conducted an empirical study with innovation intermediaries that rely on innovation and crowdsourcing to develop some of their tasks or solve problems. The result of this study will be a model of knowledge for innovation intermediaries with crowdsourcing. This model of knowledge will be the basis for the ontology development. The second stage will involve the development of the ontology itself. First will be developed a domain or content ontology that represent the entire taxonomy of concepts and their hierarchy of the underlying knowledge model of an innovation intermediary with crowdsourcing. After it will be developed a meta-data ontology to provide a descriptive vocabulary of this knowledge area. With these two artefacts we intend to contribute to the standardization of concepts in this area of knowledge and to enhance the emergence of such intermediaries.

The ontology development project will be performed using the NeOn methodology (cf. table in appendix), as this is a very complete methodology, which provides guidance in all phases of project development, allowing the use of ontological and not ontological resources, collaborative development, and evolution and maintenance of the ontology network. Also, the use of the Web Ontology Language, which is a suitable language for developing web ontologies, widely used, and also it provides a tool to support the development of ontology, but without being mandatory the use of this tool.

Appendix

	Aim	Method	Phases	Activities	Language	Tool	Build cooperation
Cyc KB - Knowledge Base [32]	Capture a large portion of what people normally considered consensus knowledge about the world	From scratch: bottom-up	1. Manual extraction of common sense knowledge; 2. Codification: Computer aided extraction of common sense language; 3. Computer managed extraction of common sense knowledge	Implementation; Knowledge acquisition; Documentation	CycL, an augmentation of first-order predicate calculus, with extensions to handle equality, reasoning, skolemisation, and some second-order features		No
Uschold and King [33]	Enterprise modeling processes	From scratch: middle-out	1. Identify purpose; 2. Building: Capture; Coding; Integrating 3. Evaluation; 4. Documentation	Requirements; Implementation; Knowledge acquisition; Verification and validation; Documentation	Ontolingua	Ontolingua Server	No
Gruninger and Fox [34]	Business processes and activities modeling; support design-in-large scale projects	From scratch	1. Capture of motivating scenarios; 2. Formulation of informal competency questions; 3. Specification of the terminology of the ontology within a formal language; 4. Formulation of formal competency questions using the terminology of the ontology; 5. Specification of axioms and definitions for the terms in the ontology within the formal language; 6. Establish conditions for characterizing the completeness of the ontology	Requirements; Design; Implementation; Knowledge acquisition; Verification and validation; Documentation	KIF (Knowledge Interchange Formal), first-order logic		No

(Continued)

(Continued)

	Aim	Method	Phases	Activities	Language	Tool	Build cooperation
KACTUS [35]	Complex technical systems development	From scratch: up-down; modifying other existing ontologies of application development	1. Specification of the application; 2. Preliminary design based on relevant top-level ontological categories; 3. Ontology refinement and structuring	Requirements; Design; Implementation; Maintenance	CML (Chemical Markup Language); Express; Ontolingua	KACTUS toolkit	No
METHONTOLOGY [36]	Support application development process	Re-engineering	1. Project management activities (Schedule; Quality assurance); 2. Development-oriented activities (Specification; Conceptualization; Formalization; Implementation; Maintenance); 3. Support activities (Knowledge acquisition; Integration; Evaluation; Documentation; Configuration management)	Project monitoring and control; Requirements; Design; Implementation; Maintenance; Knowledge acquisition; Verification and validation; Ontology configuration management; Documentation	OWL,DAML + OIL; RDF; XML; OCML	ODE (Ontology Design Environment) and WEB-ODE	No
On-To-Knowledge [28]	Knowledge management of heterogeneous sources in the internet	From scratch	1. Kick-off: requirements capture and specification; 2. Refinement; 3. Evaluation; 4. Maintenance	Project initiation; Project monitoring and control; Ontology quality management; Concept exploration; Requirements; Design; Implementation; Maintenance; Knowledge acquisition; Verification and validation; Ontology configuration management; Documentation	OIL (Ontology-based Inference Layer); XML; RDF	OntoEdit	No

(Continued)

(Continued)

	Aim	Method	Phases	Activities	Language	Tool	Build cooperation
SENSUS [37]	Natural language processing for developing machine translators	From scratch: up-down	1. Identify seed terms; 2. Link seed terms to SENSUS by hand; 3. Include nodes on the path to root; 4. Add some complete sub-trees	Requirements; Implementation	DL: LOOM	Ontosaurus	No
NeOn [38]	It provides guidance for all key aspects of the ontology engineering process, that is, collaborative ontology development, reuse of ontological and non-ontological resources, and the evolution and maintenance of networked ontologies, through nine scenarios.	Re-engineering	1. Initiation (Requirements specification; Scheduling; Evaluation); 2. Reuse (Non-Ontological Resource (NOR) Reuse; Search; Reuse; Statements Reuse; Evaluation); 3. Merging (Aligning; Evaluation); 4. Re-engineering (NOR Reengineering; Modularization; Evaluation); 5. Design (Conceptualization; Evolution; Localization; Evaluation); 6. Implementation (Evaluation; Maintenance; Evaluation)	Project monitoring and control Requirements Design Implementation Maintenance Knowledge acquisition Verification and validation Ontology configuration management Documentation	OWL (Web Ontology Language)	NTK – NeOn Toolkit	Yes

References

1. Hepp, M.: Ontologies: state of art, business potential, and grand challenges. In: Hepp, M., De Leenheer, P., de Moor, A., Sure, Y. (eds.) Ontology Management: Semantic Web, Semantic Web Services, and Business Applications, pp. 3–22. Springer, Berlin (2007)
2. Chesbrough, H.W., Chesbrought, H., Vanhaverbeke, W., West, J.: Open Innovation: Researching a New Paradigm. Oxford University Press, Oxford (2006)
3. Howe, J.: The rise of crowdsourcing. Wired Mag. **14**(06), 1–4 (2006)
4. Howe, J.: Crowdsourcing: How the Power of the Crowd is Driving the Future of Business. Crown Business. Random House, New York (2008)
5. Surowiecki, J.: The Wisdom of Crowds. Anchor Books, New York (2005)
6. Ramos, I., Cardoso, M., Carvalho, J.V., Graça, J.I.: An action research on open knowledge and technology transfer. In: Dhillon, G., Stahl, B.C., Baskerville, R. (eds.) CreativeSME 2009. IFIP AICT, vol. 301, pp. 211–223. Springer, Heidelberg (2009)
7. Gruber, T.R.: Toward principles for the design of ontologies used for knowledge sharing? Int. J. Hum. Comput. Stud. **43**(5–6), 907–928 (1995)
8. Gasevic, D., Djuric, D., Devedzic, V.: Model Driven Architecture and Ontology Development. Springer, Berlin (2006)
9. Fensel, D.: Ontologies: A Silves Bullet for Knowledge Management and Electronic Commerce, 2nd edn. Springer, Berlin (2004)
10. Guarino, N.: Formal ontology, conceptual analysis and knowedge representation. Int. J. Hum. Comput. Stud. **43**(5/6), 625–640 (1995)
11. Mizoguchi, R., Vanwelkenhuysen, J., Ikeda, M.: Task ontology for reuse of problem solving knowledge. In: Mars, N.J.I. (ed.) Towards Very Large Knowledge Bases: Knowledge Building & Knowledge Sharing. IOS Press, Amsterdam (1995)
12. Uschold, M., Gruninger, M.: Ontologies: principles, methods and applications. Knowl. Eng. Rev. **11**(02), 93–136 (1996)
13. van Heijst, G., Schreiber, A.T., Wielinga, B.J.: Using explicit ontologies in KBS development. Int. J. Hum. Comput. Stud. **46**(2–3), 183–292 (1997)
14. Guarino, N.: Formal ontology in information systems. In: Proceedings of FOIS 1998, Trento, Italy (1998)
15. Lassila, O., McGuinness, D.: The role of frame-based representation on the semantic web. Knowledge Systems Laboratory, Stanford, Technical report KSL-01-02 (2001)
16. Benjamins, V.R., Gómez-Pérez, A.: Knowledge-system technology: ontologies and problem-solving methods. Department of Social Science Informatics, University of Amsterdam (n.d.)
17. Bullinger, A.: Innovation and Ontologies: Structuring the Early Stages of Innovation Management. Springer, Berlin (2008)
18. Uschold, M., King, M., Moralee, S., Zorgios, Y.: The enterprise ontology. Knowl. Eng. Rev. **13**(01), 31–89 (1998)
19. Leppänen, M.: A context-based enterprise ontology. In: Abramowicz, W. (ed.) BIS 2007. LNCS, vol. 4439, pp. 273–286. Springer, Heidelberg (2007)
20. Fox, M.S., Gruninger, M.: Enterprise Modeling. AI Mag. **19**(3), 109–121 (1998)
21. Gómez-Pérez, A., Fernández-López, M., Corcho, O.: Ontological Engineering: with Examples from the Areas of Knowledge Management. E-commerce and the Semantic Web, Springer, Berlin (2004)
22. Ning, K., O'Sullivan, D., Zhu, Q., Decker, S.: Semantic innovation management across the extended enterprise. Int. J. Ind. Syst. Eng. **1**(1–2), 109–128 (2006)

23. Riedl, C., May, N., Finzen, J., Stathel, S., Leidig, T., Kaufman, V., Belecheanu, R., Krcmar, H.: Managing service innovations with an idea ontology. In: Proceedings of XIX International RESER Conference, Budapeste, Hungary, pp. 876–892 (2009)

24. Noy, N.F., McGuinness, D.L.: Ontology development 101: a guide to creating your first ontology, Standford University, Knowledge Systems Laboratory, Technical KSL-01-05 (2001)

25. Fernández-Pérez, M., Gómez-Pérez, A.: Overview and analysis of methodologies for building ontologies. Knowl. Eng. Rev. 17(02), 129–156 (2002)

26. Corcho, O., Fernández-López, M., Gómez-Pérez, A.: Methodologies, tools and languages for building ontologies. Where is their meeting point? Data Knowl. Eng. 46(1), 41–64 (2003)

27. Corcho, Ó., Fernández-López, M., Gómez-Pérez, A., Vicente, Ó.: WebODE: an integrated workbench for ontology representation, reasoning, and exchange. In: Gómez-Pérez, A., Benjamins, V. (eds.) EKAW 2002. LNCS (LNAI), vol. 2473, pp. 138–153. Springer, Heidelberg (2002)

28. Fensel, D., Van Harmelen, F., Klein, M., Akkermans, H.: On-to-knowledge: ontology-based tools for knowledge management. In: Proceedings of the Ebusiness and Ework 2000 Conference (EMMSEC 2000) (2000)

29. Smirnov, A., Pashkin, M., Chilov, N., Levashova, T.: Ontology-based knowledge repository support for healthgrids. Stud. Technol. Inform. 112, 47–56 (2005)

30. Tang, L.-a., Li, H., Qiu, B., Li, M., Wang, J., Wang, L., Zhou, B., Yang, D.-q., Tang, S.-w.: WISE: a prototype for ontology driven development of web information systems. In: Zhou, X., Li, J., Shen, H.T., Kitsuregawa, M., Zhang, Y. (eds.) APWeb 2006. LNCS, vol. 3841, pp. 1163–1167. Springer, Heidelberg (2006)

31. Riedl, C., May, N., Finzen, J., Stathel, S., Leidig, T., Kaufman, V., Belecheanu, R., Krcmar, H.: An idea ontology for innovation management. Int. J. Semant. Web Inf. Syst. 5(4), 1–18 (2009)

32. Lenat, D.B., Guha, R.V.: Building Large Knowledge-Based Systems: Representation and Inference in the Cyc Project. Addison-Wesley, Reading (1990)

33. Uschold, M., King, M.: Towards a methodology for building ontologies. In: Workshop on Basic Ontological Issues in Knowledge Sharing, held in conjunction with IJCAI-95 (1995)

34. Grüninger, M., Fox, M.S.: Methodology for the design and evaluation of ontologies. Presented at Workshop on Basic Ontological Issues in Knowledge Sharing, IJCAI-95, Montreal (1995)

35. Bernaras, A., Laresgoiti, I., Correa, J.: Building and reusing ontologies for electrical network applications. In: Proceedings of the European Conference on Artificial Intelligence (ECAI96), Budapeste, Hungary (1996)

36. Fernandéz, M., Gómez-Pérez, A., Juristo, N.: METHONTOLOGY: from ontological art towards ontological engineering. In: Proceedings of the AAAI97 Spring Symposium, pp. 33–40 (1997)

37. Swartout, B., Ramesh, P., Knight, K., Russ, T.: Toward distributed use of large-scale ontologies. In: AAAI Symposium on Ontological Engineering (1997)

38. Gómez-Pérez, A., Suárez-Figueroa, M.C.: NeOn methodology: scenarios for building networks of ontologies. In: Proceedings of the 16th International Conference on Knowledge Engineering and Knowledge Management Knowledge Patterns (EKAW 2008) (2008)

"Made *with* Knowledge": Reporting a Qualitative Literature Review on the Concept of the IT Knowledge Artifact

Federico Cabitza[✉] and Angela Locoro

Università degli Studi di Milano-Bicocca, Viale Sarca 336, 20126 Milan, Italy
{cabitza,angela.locoro}@disco.unimib.it

Abstract. Knowledge Artifact (KA) is an analytical construct denoting material objects that in organizations regard the creation, use, sharing and representation of knowledge. This paper aims to fill a gap in the existing literature by providing a conceptual framework for the interpretation of the heterogeneous scholar contributions proposed on this concept so far. After a comprehensive literature review we define one pole as "representational", grounded on the idea that knowledge can be an "object per se"; and another pole as "socially situated", where knowledge is seen as a social practice, that is a situated, context-dependent and performative interaction of human actors through and with "objects of knowing". We propose a unifying view of the dimensions of knowledge, and, in so doing, we try to shed light on the multiple ways these can inform the "reification" of knowledge into specific IT artifacts, which we call IT Knowledge Artifact (ITKA). Our model can contribute to the design of computational artifacts supporting knowledge work in organizations.

Keywords: Knowledge artifact · IT artifact · Organizational knowledge · Literature review

1 Introduction

"IT artifact" is a general expression to denote "the application of IT to enable or support some task(s) embedded within a structure(s) that itself is embedded within a context(s)" [9]. Convincingly introduced in the Information Systems literature almost 15 years ago by Orlikowski and Iacono [58], this concept has been addressed by hundreds of scholars in this time lapse from different and complementary perspectives [1] and it has been associated with a number of definitions [4] to account for the multiple manifestations and proteiform nature of software applications in organizational settings. Notwithstanding this apparent scholarly variety, recent contributions are converging towards a stronger recognition of the importance of both the semiotic and social nature of the IT artifact [43] as this is never "natural, neutral, universal, or given [but it is rather] socially created, [...] shaped by the interests, values, and assumptions of a wide variety of communities of developers, investors, users, [...] embedded in [...] a social contexts [that let][...] emerge [them] from ongoing social practices, [...][and] not static and unchanging, but in a continual evolution" [32]. In particular, it has also been recently claimed that taking the socio-technical nature of the IT artifact seriously

© Springer International Publishing Switzerland 2015
A. Fred et al. (Eds.): IC3K 2014, CCIS 553, pp. 571–585, 2015.
DOI: 10.1007/978-3-319-25840-9_35

[36, 49] is essential to promote "ethical responsibility [and] to minimize the negative consequences of information and communication technologies".

In this paper, we focus on Socio-technical IT artifacts that support knowledge, as a specific refinement of the more general concept discussed above, and we will then propose the notion of IT Knowledge Artifact (ITKA in what follows). In [15] this class of software applications has been proposed to encompass "material [IT] artifacts [which are] either designed or purposely used to enable and support knowledge-related processes within a community, [...], like idea expression and exchange, content and structure negotiation, meaning reconciliation, collective deliberation, new product and process co-design, knowledge representation at various degrees of (under)specification, problem framing and solving, mutual learning, and novice training".

Tackling this matter from a socio-technical perspective requires focusing on those IT artifacts that create and circulate new information within human practices, often on the basis of computational rules that in some way mirror domain-specific knowledge, as well as on those artifacts that enable and support knowledge-intensive activities and tasks both at human (i.e., cognitive) level and at social (i.e., community) level. A first step in this direction is to focus on the different aspects of computational support to knowledge practices, as these emerge from different research strands and scholarly works articulated around the concept of Knowledge Artifact (KA in what follows).

We believe it is time to denote these particular ITKAs in more precise and specific terms (starting from the work of [14]), in order to fill a gap in the literature on them where, to the best of our knowledge, a comprehensive review about the affinities and divergences in the use of the term KA is still missing.

The purpose of this work is then to conduct a qualitative review that would help answer some main research questions like: "what do we talk about when we talk about knowledge" (as in [23]) in the IT discourse? What are the underlying assumptions in the design of ITKAs? How these assumptions affect the design of these artifacts and, consequently, their low or high adoption and effective use by their intended users?

The phrase mentioned in the title of this contribution epitomizes in an intentionally ambiguous manner the two extremes of the whole spectrum of possible answers that can be given to the questions above mentioned; a bipole where ITKAs can be seen as either "made of knowledge" [61] or "made in virtue of knowledge" [12]. On one pole, we can recognize the tenets of the Knowledge Representation (KR) field, which assumes a realistic perspective on knowledge, i.e., a relation between the objects conceived in the mind and the apparently immutable outside forms perceived as reality. KR expresses the concrete possibility to represent things, when not to provide means for computationally reason on them, in order to capture their essence for sharing a discourse with others. This pole roots in Artificial Intelligence and is based on some principles [66] that inform the design of ITKAs as KR devices: this includes models of real objects according to a formal theory that elucidate their nature, their relations, and the computational manipulation of their instances. Knowledge, in this sense, should represent the reached consensus by a community on a description of a piece of reality, being that a domain of discourse, an application, a task, and so on, which has been disambiguated, automated and embedded in a system for managing knowledge (e.g., in the formal conceptualization of ontologies).

On the other pole, a complementary mode of knowledge, or better yet of "know-ing", draws on the distinction between the procedural "know how" and the discursive "know that" [21], but also on a dimension of interpretation, where "an individual pre-understanding is a result of experience within a (cultural) tradition" ([74], p. 74). This knowledge disposition is part of a process that is "neither subjective nor objective" ([74], p. 75) and has biological roots [52]: it emerges from patterns of interaction that couple living organisms with their internal structures and external environments, and orient their actions and changes "in many dimensions at the same time" (p. 116). In [34], Greenhalgh and Wieringa grasp such interplay of knowledge with the word "mindlines", that is "internalized guidelines"; in other words, the capacity of decisional processes by human beings to "continually being adjusted partly by grazing on written sources [...and] mainly by reflecting on experience during discussion with colleagues and opinion leaders, [...especially when they share] real stories of how they managed real cases" (p. 506).

The main contribution of our work is the proposal of a conceptual framework of key values and attributes for the analysis of the rationales and design principles at stake around the concept of (IT)KA (that is simply the computational counterpart of a KA), according to the results of an extensive qualitative review. The outcome of this analysis helped us conceive two main categories of ITKAs: the "Representational ITKA" and the "Socially situated ITKA". In general, this categorization could be a tool for the analyst and designer to interpret the peculiarities of the setting hosting ITKAs, as well as to understand the ways and goals according to which ITKAs are built and used. In addition, this analysis could be a first contribution to unravel implicit values and assumptions for ITKAs design, which more than often are worth undergoing a deconstruction process in order to reveal and possibly amend those prejudices that limit the potential "fit for use" of ITKA in different organizational settings.

The paper is structured as follows: after a brief introduction to the method adopted for our qualitative review, we show the results of the different phases of our analysis, discuss them by also providing examples of design, and then conclude the contribution with some further remarks and reflections.

2 Method, Categories and Dimensions of Analysis

2.1 Search and Selection

Our review of the literature on ITKAs relies on the methods proposed in [73, 75]. We strictly adhere to all the phases listed and described in the latter reference, whose approach corresponds to the so called Grounded Theory (GT) methodology. The "search and selection" phases are depicted in Fig. 1. In particular, for the first round of search and the retrieval of relevant articles on KAs, the Google Scholar engine was queried in Summer 2013, with the following keywords: "Knowledge Art*fact* is|are| (can be)". The top ranked articles with their abstracts were examined, and we proceeded with a selection of those works that resulted within the scope of the review and its aim to retrieve definitions of KAs. This resulted in a final selection of 21 sources, each containing a definition of KA, part of which were also used in our "design by examples" Section. A second and a third query were issued with the term: "knowledge representation", and "ontology design", respectively, in order to refine the branch of the KA literature on the representational side. A total of 40 papers were selected, and 3 papers were kept as paradigmatic of KA real applications. A fourth query with the term: "epistemic object" was finally issued in order to refine the branch of the KAs literature on the situativity side. A total of 3 papers were selected and used as above. A final round of queries were issued in the AIS eLibrary for finding theoretical sources, primary studies, and review articles based on the criteria of having being published as a journal article (e.g. MIS Quarterly Executive) or conference proceedings (e.g. ICIS and ECIS) of the AIS community. After a thorough analysis and re-read of excerpts, 15 articles were selected and exploited for the creation of our conceptual model on KAs. This selected literature spans at the intersection of theoretical approaches to organizational knowledge [13], sociological and information systems studies on IT conceptualization [39, 40, 58, 70], as well as Knowledge Management Systems reviews and Organizational Knowledge studies [2, 11, 38, 60].

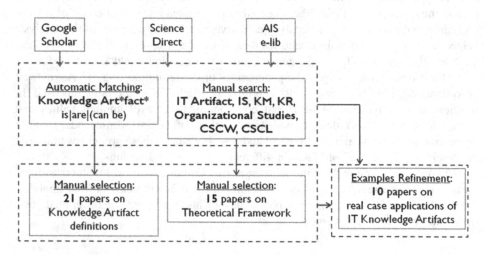

Fig. 1. Search and selection workflow.

2.2 Analysis and Presentation

The "open coding" phase was conducted on both the 21 papers containing a definition of KA, and the 15 theoretical papers regarding IT artifact and Knowledge Management conceptualizations; the "axial coding" phase was conducted by extracting conceptual dimensions from the 15 theoretical papers; to this aim, the findings of the articles were put in an output table, as in [73], with the categories of the research conceptualization as concepts. Each article could span more than one concept and attribute (see Table 1); then we classified the 21 papers on KAs definitions and the 10 real case application papers according to them (see the descriptive analysis of Sects. 3.1 and 3.2); the "selection coding" phase led to a further synthesis of our findings into the two main categories of the KA classification, and was based on all of the papers collected (see Fig. 2 for a summary of our findings).

Axial Coding: Categories and Dimensions of Knowledge Artifacts and IT Design. The two dimensions of objectivity and situativity (our input dimensions for classifying ITKA-based applications, see next paragraph) characterize the categories of our output map, depicted in Table 1, and this results in a bipolar conceptualization of the literature on KAs. This Table is a classificatory device for our review activity. The dimensions described in the Table are ideally split in two parts.

The first part of the Table is theoretically grounded and stems from the framework of Burell and Morgan [13] within the organizational studies field. We enrich their sociological paradigms of knowledge – namely, the functionalist one, which sees social structures as functions, and the interpretive one, which makes human constructs as products of their thoughts – with other perspectives: the one by Nonaka and Takeuchi [55], talking of explicit and implicit knowledge; that by Sowa [66], stating that knowledge is representable; that of Duguid [26], highlighting the cultural (and grounded in practice) side of knowledge, with a special focus on Communities of Practice.

Besides Burrell and Morgan epistemology, we have preferred to contrast the term "positivist" (objectivist) with a set of terms that should better characterize the "non-positivist" stance that is replaced by the term "costructivist", as emerging from interactions of experience and reflection. This, in our view, better expresses the epistemology of the "Socially situated KA" perspective, than the term "non-positivist". IT conceptualizations still borrow the deterministic and voluntaristic stances from Burrell and Morgan, to highlight the opposition between the unescapable automatism vs. the non-coercitive will.

The "knowledge modes and structures" [2, 40, 60, 70] describe the further opposition of "unstructured" (as in audio, video and free text) and "structured" (as in metadata, formal categories and graphics) forms of knowledge artifacts.

The second part of Table 1 focuses on the application of the principles of the first part to IT artifacts design and requirements. Orlikowski and Iacono [58] and Iivari [40] gave a classification of IT artifacts and of IT applications archetypes, respectively. These two sets allow to draw an analogy between the opposition influencing the design of informational vs. mediation roles, and computational vs. social proxy roles of ITKAs. Alavi and Leidner [2], Holslappe and Joshi [37], Massey and Montoya-Weiss [50], Iivari et al. [39], and March and Smith [48] all introduce some attributes of knowledge resources in IT that we highlight, and put into correspondence in the table according to their values (the selection and synthesis is presented in the 'attributes of knowledge in organizational settings' row). The final dimension of KM Applications by Binney [11] is a frame into which we could exemplify how the issues of passing from representational knowledge to socially situated knowledge may also fit what he calls "KM landscapes" (a representative example is that of analytical vs. innovative ITKAs).

Selective Coding: Objectivity Vs. Situativity. The two categorical dimensions of objectivity and situativity, together with their relationships with different kind of IT applications, are reported in Fig. 2.

Table 1. Conceptual framework of organizational KM and ITKAs design.

Assumptions for design	Representational ITKA	Socially situated ITKA
Paradigms [13]	Functionalist	Interpretive
Ontology [13, 26, 55, 66]	Explicit/representable (realism)	Tacit/cultural/practical/actionable (nominalism)
Epistemology [13, 34, 70]	Positivist (nomothetic)	Constructivist/interactionist/emergentist (ideographic)
IT conceptualizations (status of knowl.) [70]	Autonomous/passive (deterministic)	Integrative (action and structure not separable)/malleable/actively and interactively usable (voluntaristic)
Knowl. modes and structures [2, 26, 40, 60, 70]	Explicitly structured codifiable descriptive procedural objective essential formal rational/conceptual	Explicitly unstructured interpreted socially constructed instrumental performed comprehensive subjective flexible/conventional
IT artif. archetypes [40]	To informate/to automate	To mediate/to augment
IT artif. views [58]	Computational (algorithm, model), labor substitution, production, information processing tool	Social relation tool proxy/ensemble view
KM views [2]	Factual data oriented to be stored and manipulated condition of access to information	Personalized information Knowing and understanding Process of applying expertise Potential to influence action
Attributes of knowl. in organizational settings [38, 39, 48, 50]	Quality: validity; Main view: objective; Time: discrete, ordered; Level of certainty: decidable/specified; Usage: computational/procedural/top-down	Quality: utility (means-end oriented); Main view: subjective; Time: continuous, chaotic; Level of certainty: undecidable/incomplete; Usage: pragmatic/situational/bottom-up
KM applications [11]	Transactional/analytical/asset	Process/Developm./innovation

Situativity, for our purposes, can be epitomized in terms of the extent the KA is capable to adapt itself to the context and situation at hand, as well as of the extent it can be appropriated by its users and exploited in a given situation. Objectivity, to our aims, can be conversely considered the capability of a KA to represent true facts in an objective, crisp, and context-independent manner, as well as the extent it can be transferred among its users as an object carrying some knowledge with itself. To adopt evocative terms introduced by [31, 42], then objectivity refers to the extent a KA is "cold/immutable" (cf. Latour) and "dry" (cf. Goguen), while situativity refers to the extent an object is "warm/mutable" and "wet", respectively.

Each group of applications of Fig. 2 is associated with the research and design principles, values and assumptions of the disciplines that lay at the intersection points of the Figure. This schema aims to express how much objectivity and situativity is implied by each design input and requirement for the IT artifact as the final output. Besides the two extremes, no other group of applications contains objectivity or situativity in a pure manner, but most of them result in a differently mixed blend of the two.

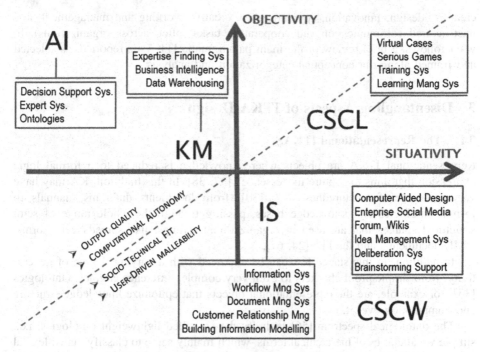

Fig. 2. Selective coding: classification of ITKA-based applications grouped by research discipline and according to the two dimensions of objectivity and situativity.

Depending on the attributes collected in our axial coding analysis (see dimensions in Table 1), we can give a brief overview of what objectivity and situativity mean in terms of the knowledge phases, structures, attributes, and knowledge-based activities that these applications and tools are supporting, while we produce a detailed description of some aspects of their design in the next Section.

Objectivity seems to better characterize all those large-scale applications that are oriented to business and enterprise activities, where the dimension of organizations requires to handle information quantitatively and in a centralized way (e.g. by collecting all the specialized knowledge distributed among the different sectors of an organization and by adopting a uniform and top-down codification). The aim of these applications is to collect, store, retrieve, and apply information across the lens of standard procedures that guarantee the control of a complex system and the rapid problem solving and decision making at a shopfloor level, as well as at a managerial level.

Information systems technology lays in the middle, is scalable, and constitutes the set of tools able to process and manage information (i.e. to structure, store and maintain the documents by indexing, organizing, classifying, filtering them, and so on). Computer Supported Cooperative Learning and Computer Supported Cooperative Work technologies support a more subjective and situated dimension in the management of the knowledge that they are called to handle. In a way, the environment of such applications is less standardised, not totally specified, and strongly oriented to the creation, design, innovation, apprenticeship, creative working and management, and unstructured communication and cooperation tasks (often across organizations). In what follows, a brief review of the main paradigmatic ITKAs is reported, as selected and framed along our conceptual categorization and analysis.

3 Disentangling Aspects of ITKA Design

3.1 The Representational ITKAS

Representational ITKA are objects where knowledge is reduced to a formal logic expression that aims to capture its "essence" [37, 38]. In this tradition, KA may have varying degree of structuredness (e.g., [30]), from documents, diagrams, manuals, to formal ontologies and knowledge bases, passing through the whole range of semi structured sources that are used in organizational settings, like spreadsheets, forms, XML documents, and the like [24, 67].

In addition, in this stance KA can be endowed with a varying degree of generativity, from simple proof checkers to even very complex inference engines. Ontologies [35], for example, are the representational objects that epitomize knowledge structure and computation over it.

The ontological spectrum [56] includes the so called lightweight ontologies, i.e., simple vocabularies of hierarchical terms, which mainly serve to classify items; lexical resources [28], which makes the local space of word meanings explicit and expandable, for tasks of advanced search and retrieval; fully axiomatized theories, which are encoded in logic languages [6] and equipped with reasoning tools. As the computational complexity tends to increase rapidly, often the model of reality that undergoes the conceptualization and axiomatization process tends to be partial and oversimplified [66].

According to the above mentioned definitions, representational KA are those stored in Knowledge Management repositories [72], in Digital Libraries [19], and the like, as structured sources of static knowledge, as well as in more sophisticated tools that computationally "activate" knowledge, like Decision Support Systems or Expert Systems [51], or any semantically enriched IT System, with its automatic or semi-automatic services for structuring, storing, extracting, retrieving, evaluating, and maintaining knowledge artifacts (e.g., [45]).

In a representational approach, some suggestions on how to couple formalisms with design requirements for a better use of KAs are devised [42]. Also [71] recognizes that "the design of knowledge artifacts includes the processes where they are applicable". KAs for enhancing and supporting highly specialized tasks in communities of experts are those depicted in [7, 8, 61]; in these works formal knowledge is collected while also keeping in mind and applying principles and methodologies that involve instances, practices, and values of specific communities and their past knowledge. In particular, the KA conceived in these researches is a case-based reasoning KA, in that it supports innovation management activities by incorporating the jargon of the technical experts for recording past experiences. A case-based reasoning mechanism for problem solving is then provided to facilitate collaboration within members of other teams.

The computational part is based on the explicit semantics added to the rather implicit system of meanings of language symbols, to provide a flexible layer of negotiation that can adhere to the new cases that need to be collaboratively examined on the basis of the past cases. New combinations of the elements represented in the KA memory support are obtained by exploiting fuzzy logic rules as computational counterpart of the qualitative variables of the specific domain and case at hand.

3.2 The Socially Situated ITKAs

Practice oriented definitions refer to something that is made during a performance, in "knowing" [22], which is seen as a "social product [emerging from the] messy, contingent, and situated outcome or group activity" [68].

In this context, KAs are not supposed to store knowledge or to be designed to "engender knowing" [63]. Rather, a KA allows its users to make apt and proper decisions or create innovation, or solve problems, and overcome breakdowns. In this stance, it acts as a support or scaffold to the expression of knowledgeable behaviors. In complex organizational settings, KAs regard also how to organize memories, report best practices, outline ideal and effective methods [18], because these representations (either textual, diagrammatic or pictorial) trigger opportunities for socialization, internalization and, by evocation and memory aid, knowledge retention and exploitation (i.e., knowledgeable behavior). As "centers of gravity for knowledge, [KAs] concentrate it, make it tangible, instrumental, effective" ([3], p. 62).

The works [20, 62] introduce the concepts of "performing-with" and "being-in-relation-with" to distinguish the relation with "epistemic objects", the classical objects of study in scientific practice, which implies a relational attitude more than a performative one. These objects are "capable of unfolding indefinitely" although "instantiated", as they are "simultaneously mutating" (cf. Suchman on the situated actions in the context of "heterogeneously enacted and intrinsically indefinite events"). An "unfolding ontology" suitable to describe them should be based on post-scriptive structures, which may include the "temporality" of things into which "epistemic objects [...] tend to [constantly point to a possibly] unreachable real whole".

The work [27] extended this notion to the visual forms through which typically both designers and engineers cooperate on a common ground. Visual forms are mutable forms that unfold "in time", as "they are not yet but might become in future iterations"; "in space, as standpoint-specific boundary objects". For their boundariness they are defined "trans-epistemic objects", i.e. "capable of traversing and permeating different epistemologies of design".

The authors of [50] investigate the process of knowledge conversion (KC), both in indirect KC (solitary work interacting with the artifact) and in direct KC (synchronization via communicative human-to-human activity). They propose a model of "personal perception of time", i.e., a one task or "monophasic" entity vs. a "polyphasic" entity or structure of events.

What is put in the foreground by these design studies are the mutable social and temporal dimensions of knowledge forms, which seem to suggest inescapable requirements for the design of KAs.

3.3   Classifying KA Definition Papers

Finally, we propose a report of our classificatory activity on the 21 papers that contain a definition of KA (depicted in Table 2).

Table 2. Papers that contain a definition of KA classified with our conceptual framework.

Objectivity	Smith [65]; Ancori et al. [5], Seiner [64], Diaz and Canals [24], Krupansky [41], Gandhi [29], Holsapple and Joshi [37, 38], Weber et al. [72], Weber and Gunawardena [71], Alavi and Leidner [2], Mangisengi and Essmayr [46], Giunchiglia and Chenu-Abente [30]
Situativity	Salazar-Torres et al. [61], Paavola et al. [59], Mödritscher and Hoffmann [53], Scrivener [63], Mansingh et al. [47], Bereiter [10], Oinas-Kukkonen [57], Tzitzikas et al. [69]

4   Discussion and Conclusions

The literature review that we have outlined above has shed light on the manifold, and sometimes even divergent, perspectives that have so far emerged about the nature of what scholars have wanted to denote with the term Knowledge Artifact in the last 15 years. Our literature review unveils the characteristic of the concept of Knowledge Artifact to be a "boundary concept" [44], that is something that allows "disparate proponents to appropriate it in consonance with the main aims and scopes of their fields". More than often, implicit assumptions are grounded and stratified in the theoretical stances that inform the fields of Knowledge Representation, Knowledge Management, Organizational studies, and so on, and the discourses upon the Knowledge Artifact are not an exception. With our review we have aimed to raise awareness of both the complementarity and tensions existing between different design stances that can be positioned at some point within the objectivity vs. situativity spectrum. This categorization is ultimately aimed at outdrawing the necessary recognition that designing for effective knowledge artifacts requires to address once again important questions on what we want human knowledge to be, or at which scale we want to support human knowledge, and accept as possible yet temporary answers contributions coming from the whole symbolic-subsymbolic range of stances that have been very briefly outlined in this work.

Wondering how the categories proposed in this conceptual framework can be reflected in distinct design principles or, even, more specific requirements, would be an inevitable follow up of this research. We hint at these principles here in a very general manner for limitations of space.

A coherent *representational* stance will require KAs to be able to: abstract and model more or less structured documents; possibly classify and store them on the basis of some domain ontology, for example to activate some semantic dependencies among the electronic resources; allow their retrieval with a system of more or less rigidly specified queries, filters, topic models and user profiles of varying complexity. The KA could also be capable of storing usable representations of facts or declarative (e.g., assertions and statements) and procedural (e.g., algorithms, process models and rules) knowledge to assist users in knowledge intensive work or give them support in decision making. An ideal KA is assessed in terms of the quality of its output (pertinency, accuracy, completeness, consistency, timeliness, etc.) and of its "autonomy" in providing answers on the basis of the available information (e.g., inputs), within acceptable range of deviations from the gold standards posed with respect to the quality dimensions mentioned above. This means that the process leading to the right output should be aware of the context, including the user, but within the variability accepted and considered in the computational model formally. In Fig. 2 these features characterize the upper Cartesian hemiplane depicted therein, where also the main classes of systems that are usually focused on a more closed and static representational treatment of knowledge are listed.

On the other hand, the situated perspective would require to design KAs that do not necessarily represent knowledge per se, as said above, but rather promote knowledge-related processes like innovation, decision making and learning: in this latter case the nature of the KA cannot be decoupled, nor generalized, from the specific setting or Community of Practice, or from the boundary between communities where the KA is supposed to play its role of knowledge facilitator and transfer medium. The first studies of communities where the creation and circulation of new knowledge is part of the practices that foster sense of belonging and identity led to characterize a specific kind of community, denoted as "knowing community" [17], which is defined as the social gathering around a KA and where actors interact also in virtue of the KA mediation. To account for this mediating capability, specific principles have been proposed for the design of situated KAs, namely: representational locality, semantic openness and reconcilement, and flexible underspecification, as discussed in [16]. Differently from the representational KA, the quality of the situated KA can be proved formally only in part, but it is rather assessed in terms of user adoption, impact, appropriation, satisfaction as support of knowledge work and collaborative practice. In short, the KA quality is estimated in terms of the extent users consider it fit to their needs, and capable to adapt to a dynamically changing context and of triggering social interactions that would allow them to create, socialize and diffuse new knowledge, which in its turn can be put in discussion again, in a continuous process [17]. In Fig. 2, we denoted this capability "socio-technical fit" (in the bottommost hemiplane) as this is certainly a major attribute that a KA does not possess independently of the social setting in which it is adopted, but rather depends on many factors that go beyond the merely computational and performance-related aspects of the artifact. As locality is important for situated KAs (by definition itself of situation), any strong assumption or strict structure hardwired in the artifact at design time could undermine, or just hinder, the processes of user appropriation [25], re-appropriation and evolutionary growth [54] the KA must somehow support a knowing community over time. For this reason, the capability of the KA to be adapted, configured, and tinkered by end users themselves to improve the above mentioned fit is the second dimension on which prospective situated KAs can be assessed (see Fig. 2, bottommost hemisphere).

Based on the literature review, future research can address several directions. First, the literature review highlights the diversity of KAs; we suggested a typology based on two dimensions, i.e. objectivity and situativity. However, other proposals, analyses and taxonomies are possible. On a more conceptual level, our categorization can also be taken as a contribution for a scholarly debate still to be developed, regarding what features should a KA exhibit, and on what kind of priorities to focus on with respect to the application domain or community. A major attention to the social practice aspects of knowledge, for instance, could motivate the design of artifacts that are made to be local, in continuous evolution, and to host necessarily incomplete, and possibly partially inconsistent and ambiguous representations.

These only seemingly paradoxical features should not then be taken as deficiencies of the tools conceived to support knowledge, but rather as features that result from a deep understanding of the semiotic nature of human representations [33] and that require a committed research agenda in the next years to come to be fully realized in running applications and knowledge artifacts.

References

1. Akhlaghpour, S., et al.: The ongoing quest for the IT artifact: looking back, moving forward. J. Inf. Technol. **28**(2), 150–166 (2013)
2. Alavi, M., Leidner, D.E.: Review: knowledge management and knowledge management systems: conceptual foundations and research issues. MIS Q. **25**, 107–136 (2001)
3. Allen, B.: Knowledge and Civilization. Cambridge University Press, Cambridge (2004)
4. Alter, S.: Work systems and IT artifacts-does the definition matter? Commun. Assoc. Inf. Syst. **17**(1), 14 (2006)
5. Ancori, B., et al.: The economics of knowledge: the debate about codification and tacit knowledge. Ind. Corp. Change **9**(2), 255–287 (2000)
6. Baader, F.: The Description Logic Handbook: Theory, Implementation, and Applications. Cambridge University Press, Cambridge (2003)
7. Bandini, S., et al.: The role of knowledge artifacts in innovation management: the case of a chemical compound designer CoP. In: Huysman, M., Wenger, E., Wulf, V. (eds.) Communities and Technologies, pp. 327–345. Springer, The Netherlands (2003)
8. Bandini, S., Manzoni, S.: Modeling core knowledge and practices in a computational approach to innovation process. In: Magnani, L., Nersessian, N.J. (eds.) Model-Based Reasoning, pp. 369–390. Springer, New York (2002)
9. Benbasat, I., Zmud, R.W.: The identity crisis within the IS discipline: defining and communicating the discipline's core properties. MIS Q. **27**, 183–194 (2003)
10. Bereiter, C.: Education and Mind in the Knowledge Age. Routledge, London (2002)
11. Binney, D.: The knowledge management spectrum–understanding the KM landscape. J. Knowl. Manage. **5**(1), 33–42 (2001)
12. Brown, J.S., Duguid, P.: Knowledge and organization: a social-practice perspective. Organ. Sci. **12**(2), 198–213 (2001)
13. Burrell, G., Morgan, G.: Sociological Paradigms and Organisational Analysis. Heinemann, London (1979)
14. Cabitza, F.: At the boundaries of communities and roles: boundary objects and knowledge artifacts as complementary resources for the design of information systems. Presented at the ItAIS, Milan (2013)
15. Cabitza, F., et al.: Knowledge artifacts within knowing communities to foster collective knowledge. In: Proceedings of the 2014 International Working Conference on Advanced Visual Interfaces. ACM (2014)

16. Cabitza, F., et al.: Leveraging underspecification in knowledge artifacts to foster collaborative activities in professional communities. Int. J. Hum Comput Stud. **71**(1), 24–45 (2013)
17. Cabitza, F., et al.: The knowledge-stream model - a comprehensive model for knowledge circulation in communities of knowledgeable practitioners. In: KMIS 2014: Proceedings of the 6th International Conference on Knowledge Management and Information Sharing. Rome, Italy. SCITEPRESS, 21–24 October 2014
18. Cabitza, F., Simone, C.: Affording mechanisms: an integrated view of coordination and knowledge management. Comput. Support. Coop. Work CSCW **21**(2–3), 227–260 (2012)
19. Candela, L., et al.: The DELOS Digital Library Reference Model. Foundations for Digital Libraries. ISTI-CNR, PISA (2007)
20. Cetina, K.K.: Epistemic Cultures: How the Sciences Make Knowledge. Harvard University Press, Cambridge (2009)
21. Collins, H., Evans, R.: Rethinking Expertise. University of Chicago Press, Chicago (2008)
22. Cook, S.D., Brown, J.S.: Bridging epistemologies: the generative dance between organizational knowledge and organizational knowing. Organ. Sci. **10**(4), 381–400 (1999)
23. Davenport, T.H., Prusak, L.: Working Knowledge: How Organizations Manage What They Know. Harvard Business Press, Cambridge (1998)
24. Diaz, A., Canals, D.: The collaborative knowledge sharing framework. Presented at the IADIS (2007)
25. Dix, A.: Designing for appropriation. In: Proceedings of the 21st British HCI Group Annual Conference on People and Computers: HCI... but not as we know it, pp. 27–30. British Computer Society, Swinton, UK (2007)
26. Duguid, P.: "The art of knowing": social and tacit dimensions of knowledge and the limits of the community of practice. Inf. Soc. **21**(2), 109–118 (2005)
27. Ewenstein, B., Whyte, J.: Knowledge practices in design: the role of visual representations asepistemic objects'. Organ. Stud. **30**(1), 07–30 (2009)
28. Fellbaum, C.: WordNet. Wiley, New York (1998)
29. Gandhi, S.: Knowledge management and reference services. J. Acad. Librarianship **30**(5), 368–381 (2004)
30. Giunchiglia, F., Chenu-Abente, R.: Scientific Knowledge Objects v.1. Technical Report, number DISI-09-006. University of Trento (2009)
31. Goguen, J.: The dry and the wet. In: The Proceedings of the IFIP TC8/WG8.1 Working Conference on Information System Concepts: Improving the Understanding, pp. 1–17. North-Holland Publishing Co., Amsterdam (1992)
32. Goldkuhl, G.: From ensemble view to ensemble artefact–an inquiry on conceptualisations of the IT artefact. Syst. Signs Actions **7**(1), 49–72 (2013)
33. Gourlay, S.: Knowing as semiosis: steps towards a reconceptualization of 'tacit knowledge'. In: Tsoukas, H., Mylonopoulos, N. (eds.) Organizations as Knowledge Systems. Knowledge, Learning and Dynamic Capabilities, pp. 86–105. Palgrave Macmillan, Basingstoke (2004)
34. Greenhalgh, T., Wieringa, S.: Is it time to drop the "knowledge translation" metaphor? A critical literature review. J. R. Soc. Med. **104**(12), 501–509 (2011)
35. Guarino, N.: Formal ontology in information systems. In: Proceedings of the First International Conference (FOIS 1998), Trento, Italy. IOS Press, 6–8 June 1998
36. Harrison, M.I., et al.: Unintended consequences of information technologies in health care—an interactive sociotechnical analysis. J. Am. Med. Inform. Assoc. **14**(5), 542–549 (2007)
37. Holsapple, C.W., Joshi, K.D.: Knowledge management: a threefold framework. Inf. Soc. **18**(1), 47–64 (2002)
38. Holsapple, C.W., Joshi, K.D.: Organizational knowledge resources. Decis. Support Syst. **31**(1), 39–54 (2001)

39. Iivari, J., et al.: A paradigmatic analysis contrasting information systems development approaches and methodologies. Inf. Syst. Res. **9**(2), 164–193 (1998)

40. Iivari, J.: A paradigmatic analysis of information systems as a design science. Scand. J. Inf. Syst. **19**(2), 5 (2007)

41. Krupansky, J.: Definition: Knowledge Artifact (2006)

42. Latour, B.: Science in Action: How to Follow Scientists and Engineers through Society. Harvard University Press, Cambridge (1987)

43. Lee, A.S., et al.: Going back to basics in design: from the IT artifact to the IS artifact. In: AMCIS 2013, The Proceedings of the 19th Americas Conference on Information Systems (2013)

44. Löwy, I.: The strength of loose concepts-boundary concepts, federative experimental strategies and disciplinary growth: the case of immunology. Hist. Sci. **30**(90), 371–396 (1990)

45. Maedche, A., et al.: Ontologies for enterprise knowledge management. IEEE Intell. Syst. **18** (2), 26–33 (2003)

46. Mangisengi, O., Essmayr, W.: State of the art in peer-to-peer knowledge management. Technical Report, number SCCH-TR-0240 (2002)

47. Mansingh, G., et al.: Issues in knowledge access, retrieval and sharing–case studies in a Caribbean health sector. Expert Syst. Appl. **36**(2), 2853–2863 (2009)

48. March, S.T., Smith, G.F.: Design and natural science research on information technology. Decis. Support Syst. **15**(4), 251–266 (1995)

49. Markus, M.L., Mentzer, K.: Foresight for a responsible future with ICT. Inf. Syst. Front. **16**(3), 1–16 (2014)

50. Massey, A.P., Montoya-Weiss, M.M.: Unraveling the temporal fabric of knowledge conversion: a model of media selection and use. MIS Q. **30**, 99–114 (2006)

51. Matook, S., Brown, S.A.: Conceptualizing the IT artifact for MIS research. In: ICIS 2008 Proceedings (2008)

52. Maturana, H.R., Varela, F.J.: The Tree of Knowledge: The Biological Roots of Human Understanding. New Science Library/Shambhala Publications, Boston (1987)

53. Mödritscher, F., et al.: Integration and semantic enrichment of explicit knowledge through a multimedia, multi-source, metadata-based knowledge artefact repository. In: Proceedings of the International Conference on Knowledge Management, Graz, pp. 365–372 (2007)

54. Mørch, A.I.: Evolutionary growth and control in user tailorable systems. In: Adaptive Evolutionary Information Systems, pp. 30–58. Idea Group Publishing, Hershey (2003)

55. Nonaka, I., Takeuchi, H.: The Knowledge-Creating Company: How Japanese Companies Create the Dynamics of Innovation. Oxford University Press, Oxford (1995)

56. Noy, N.F., Hafner, C.D.: The state of the art in ontology design: a survey and comparative review. AI Mag. **18**(3), 53 (1997)

57. Oinas-Kukkonen, H.: The 7C model for organizational knowledge sharing, learning and management. In: The Proceedings of the Fifth European Conference on Organizational Knowledge, Learning and Capabilities (OKLC 2004), Innsbruck, Austria, 2–3 April 2004

58. Orlikowski, W.J., Iacono, C.S.: Research commentary: desperately seeking the "IT" in IT research—a call to theorizing the IT artifact. Inf. Syst. Res. **12**(2), 121–134 (2001)

59. Paavola, S., et al.: Models of innovative knowledge communities and three metaphors of learning. Rev. Educ. Res. **74**(4), 557–576 (2004)

60. Rodríguez-Elias, O.M., et al.: A framework to analyze information systems as knowledge flow facilitators. Inf. Softw. Technol. **50**(6), 481–498 (2008)

61. Salazar-Torres, G., et al.: Design issues for knowledge artifacts. Knowl. Based Syst. **21**(8), 856–867 (2008)

62. Schatzki, T.R., et al.: The Practice Turn in Contemporary Theory. Psychology Press, Hove (2001)
63. Scrivener, S.: The art object does not embody a form of knowledge. Working Pap. Art Des. **2**, 25–32 (2002)
64. Seiner, R.: Metadata as a knowledge management enabler. TDAN.com KIK Consult. Serv. Data Adm. Newsl. TDAN.com. **15** (2001)
65. Smith, D.E.: Knowledge, Groupware and the Internet. Routledge, London (2000)
66. Sowa, J.F.: Knowledge Representation: Logical, Philosophical, and Computational Foundations. Brooks/Cole, Pacific Grove (1999)
67. Toro, S.G., Kulkarni, R.: A framework for knowledge management methods, practices and technologies. In: Proceedings of INDIACom-2008 (2008)
68. Turnbull, D.: Masons, Tricksters and Cartographers: Comparative Studies in the Sociology of Scientific and Indigenous Knowledge. Taylor & Francis, London (2000)
69. Tzitzikas, Y., et al.: Emergent knowledge artifacts for supporting trialogical e-learning. Int. J. Web Based Learn. Teach. Technol. IJWLTT **2**(3), 19–41 (2007)
70. De Vaujany, F.-X.: Information technology conceptualization: respective contributions of sociology and information systems. J. Inf. Technol. Impacts **5**(1), 39–58 (2005)
71. Weber, R., Gunawardena, S.: Designing multifunctional knowledge management systems. In: Proceedings of the 41st Annual Hawaii International Conference on System Sciences, pp. 368–368. IEEE (2008)
72. Weber, R.O., Morelli, M.L., Atwood, M.E., Proctor, J.M.: Designing a knowledge management approach for the CAMRA community of science. In: Reimer, U., Karagiannis, D. (eds.) PAKM 2006. LNCS (LNAI), vol. 4333, pp. 315–325. Springer, Heidelberg (2006)
73. Webster, J., Watson, R.T.: Analyzing the past to prepare for the future: writing a literature review. Manage. Inf. Syst. Q. **26**(2), 3 (2002)
74. Winograd, T., Flores, F.: Understanding Computers and Cognition: A New Foundation for Design. Intellect Books, Bristol (1986)
75. Wolfswinkel, J.F., et al.: Using grounded theory as a method for rigorously reviewing literature. Eur. J. Inf. Syst. **22**(1), 45–55 (2013)

An Information Flow Based Modeling Approach to Information Management

Carlos Alberto Malcher Bastos[(✉)], Monica Rodrigues Moreira,
Ana Cristina Martins Bruno, Sergio Mecena Filho,
and José Rodrigues de Farias Filho

Fluminense Federal University, Niterói, Brazil
cmbastos@telecom.uff.br, {monicarodriguesmoreira,
anamartinsbruno,smecena007}@gmail.com,
rodrigues@labceo.uff.br

Abstract. This paper describes an information flow based modeling methodology as a relevant part of Knowledge and Information Management Model (KMIM) based on identification and management of information assets throughout information life cycle in the organization. The methodology includes structured interviews with users and information producers as well as with top management in order to modeling information flows as they are today and how they should be in the next future, incorporating improvements. The project research was developed in a Brazilian regulatory agency, which deals with information and knowledge as its main assets for taking decisions. Information life cycle mapping and also the identification of information quality requirements for each different organization unit showed that new practices for collection, validation, processing, storage, retrieval, distribution and dissemination of information have to be developed and implemented in order to better manage the information and use it as an input to promote knowledge management.

Keywords: Information management · Knowledge management · Information asset · Information life cycle · Information flow

1 Introduction

Due to the increasing use and the easy access to information and communication technologies, information volume produced in any type of organization has increased substantially in quantity and complexity. In this scenario, organizational information and knowledge management (KM) have become essential elements in the organizational management model, especially in public organizations, in which information transparency has been increasingly a significant requirement. For any type of organization information has become important input for performing activities and tasks, and knowledge represents a valuable asset that, being intangible, is difficult to measure and manage. Information and knowledge management are viewed as valuable assets because information, in turn, becomes knowledge when it is interpreted, put into a context or when meaning is added to it by the interference of the environment and when people who manipulated it guide intelligent behavior and we can say that knowledge become wisdom [1].

© Springer International Publishing Switzerland 2015
A. Fred et al. (Eds.): IC3K 2014, CCIS 553, pp. 586–604, 2015.
DOI: 10.1007/978-3-319-25840-9_36

In this regard, it is noted that the information management (IM), is one of the concerns and challenges that are present in scientific literature in many knowledge fields.

Well-known authors who address information management [2–4] refer to information management as a set of processes that identify information needs, collect and create, classify, store, treat, provide and use information. This paper allied information management concept with recommendations provided by information science and defined by the Institute of Information Scientists [5] that clarifies that information management consists of routines and procedures for the creation, identification, collection, validation, representation, retrieval and use of information and has taken the following steps: collection, validation, processing, storage, retrieval, distribution and dissemination as procedures and routines for managing organizational information, which is understood as information life cycle.

An organization that uses in a strategic way its information does it in order to create a knowledge organization [6]. The organizational information contains multiple meanings resulting from cognitive and emotional interpretations of individuals or groups that process it in everyday life. Thus, the information management needs to create information structures and processes that are flexible and permeable. The major challenge in information management in organizations is to clearly define role of information in processes management. Hence information management implies in mapping information flows, define what information is valuable and to check its quality requirements.

Managers can not ignore how the organization uses the information, what are their main information flows, what are the information needs for each hierarchical level and also what are the required competences to administer informational resources [7]. Information management concept encourages decisions, solutions and also customer satisfaction (external customers and internal ones).

An extensive research project was designed to develop a wide knowledge and information management model (KIMM) to a Brazilian regulatory agency. One of its components is the identification, mapping and modeling of information flows in the context of information management. KIMM starts from the premise that knowledge can be generated from structured information and thus, presents a modeling approach focused on an organization's information assets. From the information flows models of each information asset, it is possible to acknowledge the organization work routine (workflow) and provide knowledge and skills that are required to manipulate information assets. It is also possible to know organization business requirements from where system requirements can be derived for information systems development, and also to develop an ontology conceptual model and glossary of terms inherent to the field of the organization studied [8–12].

The objective of this paper is to present and discuss an information flow modeling approach to information management, a part of a wider research project, conducted in technical cooperation with the Brazilian National Transportation Agency, that aims to develop an integrated knowledge and information management model (KIMM). The information flow model is used for the generation of integrated knowledge management models.

Information flow modeling as used in KIMM has its foundation in the information life cycle analysis and is oriented by information assets, having as guiding principles the quality characteristics of a good piece of information according to the operational activities of each organizational unit (departments or divisions) in the Agency organizational structure. Information flow models constitute as a fundamental tool to reveal the informational treatment of an organization's information assets and the operation of its organizational units and that is its main contribution.

The paper is organized as follows: besides this introduction, the first section briefly describes KIMM. The second presents concepts and definitions for information and knowledge field and an approach for information assets, the core component of KIMM. In the third section, methodological aspects of information flow modeling are presented and the dynamic of project execution, interviews, integration meetings and also problems encountered are reported. This section also presented the modeling performed results emphasizing problems related to information life cycle. The fourth section presents results analysis from the information flow modeling approach, bringing indicatives for improvement and KIMM implementation in the organization.

1.1 KIMM

The research project put together five different multidisciplinary teams is order to deeply studied distinct areas of knowledge and produced specific models as a component of KIMM. The areas of knowledge are: land transportation (as the organization in case is a regulatory agency for this field), information flow, business requirements, knowledge management and ontology. The methodology starts with a meeting where the five different multidisciplinary teams interact, discuss and study the organization to consolidate an integrated holistic view of the organization. Each interaction with the organization is conducted by one of the specialized team and from the information gathered their specific models are produced as a part of KIMM.

Land transportation team interaction aims to identify information assets of each organizational unit resulting in an information assets list with its main features. Other products of this team include a references and best practices list related to main business of each organizational unit in the land transportation field, resulting from research and benchmarking processes.

Information flow team interaction will be described in detail in the following section. This team is responsible for mapping the identified information assets, using information life cycle principles. In Sect. 3.1 it is presented a representation of an information flow.

Business requirements team produces business use cases models that are derived from business global information model. They are built taking into account information flows.

Knowledge management team studies knowledge, skills and professionals related to information assets and identifies gaps in knowledge flow, associated to the steps of knowledge capturing, mobilization and innovation [13] and the processes of knowledge conversion [14]. Several formal models are produced by this team: a knowledge model based on mapping skills, knowledge and professionals, related to information assets processing; knowledge-workflows, detailing required knowledge used by employees to conduct activities in processing information assets [10–12], and a knowledge tree that allows visualize in a hierarchical way all organizational knowledge, showing all expertise that can be shared [15]. Tools and practices that may support KM in the organization are proposed.

Information flows are a prerequisite for knowledge models generation and there is a strong dependency between KM Models and information flows map.

Ontology team prepared a glossary of organizational main terms and also an ontology conceptual model that formalizes the relationship between terms and can be used as input for ontology based systems development.

After the information survey and the construction of specific models by each team, results are consolidated in an integrated knowledge and information management model (KIMM).

2   Concepts and Definitions

2.1   Information Assets

In this work, to understand the generated models and the logic of its construction and how to manage and maintain them is essential to understand information assets concept and how it is used in KIMM.

Tacit knowledge cannot be formally communicated and explicit knowledge is actually information [16, 17]. Knowledge assets should also be considered as information assets. Knowledge assets are the employee's skills and expertise, the organization culture and image and companies that are able to continuously develop their knowledge assets tend to be successful [18].

In our research project we describe information assets as a set of identifiable data stored somehow, and recognized as having value to the organization, allowing the execution of their business functions, satisfying the acknowledged one or more of business requirements [19]. Several authors agree that information asset is a set of data and that has potential value to the organization [16, 17, 19–22]. An information asset is organized information that has value, so it should be easily accessible to those who need it [23].

Development of information assets requires: necessary information identification, information capture through documented processes and the construction of a structure to allow easy access by those who benefit from the information and use it. This is the information life cycle logic.

Information assets are materialized by information flows, mapped considering each step of information life cycle. This means that the processes of collection, validation, processing, storage and retrieval, distribution, dissemination and use of information can be properly formalized as the stages of the information life cycle represented in each information assets flow.

Each information asset has a specific purpose and is strategically positioned in the organizational structure, so that the established information assets architecture forms an umbrella layer that includes data, information and explicit knowledge that can be structured, communicated and transferred according to the needs of each hierarchical level [21, 22].

In KIMM, information assets identification is performed by preliminary studies, involving legal and regulatory framework of the organization under study, as well as comparisons with other institutions in the same field and other regulatory agencies. With the results of these studies, teams discuss to reach a consensus on information assets list that will be worked on over the next modeling activities. At this time, it is a preliminary list, as there was no validation by the organization. A key point in this discussion is that there are different perspectives about the same information assets identified preliminarily. For instance, transport team adopts a more business focused view of the organization with legal foundation and benchmarking research. On the other hand, business requirements team is more focused on software engineering, considering the technology of information systems. Information flow team tends to look at the information assets in a more functional way, focusing on how business functions are developed and how they relate to each other. These different views on the information assets are important to enrich the internal meetings and bring different perspectives on the main goals of information assets to the validation meetings with the organization managers. The validation step with the organization is fundamental to give continuity to modeling work so there is a conceptual alignment about what an information asset is as well as a consensus on the list of information assets that will be worked on over the modeling activities. In this validation occasion with the organization, it is common discussions about the proposed information assets preliminary list. Managers tend to mix information assets with business processes or even with services that are performed by their respective organizational units.

After the validation of the information assets list by the organization, other internal meetings are also planned in order to get a final conceptual understanding alignment by the five teams and also to be better prepared for the information survey. The result of this discussion is the final list of information assets that will be considered in the modeling work.

2.2   Information Flows and Information Life Cycle

Information flow definitions converge on three points: subsidies to decision-making processes, organizational diagnosis facilitation and improvement opportunities [24–26]. Information flow corresponds to a sequence of events from the information generation by the issuer to its uptake/assimilation/uptake receptor, generating individual and collective knowledge, supporting decision-making [24]. Information flow is the integrating element of supply chains that are precarious, originates failures enabling organizational diagnosis in the light of information management [25]. Information flow is an information process disseminator that mediates communication favoring continuous improvement initiatives [26]. Organizations have different informational environments, consisting of information flows that permeate all activities, tasks, decision-making, that is, the action of the individual in the workplace [27]. In this sense, information flows constitute fundamental elements of information environments, so there isn't an information environment without the existence of information flows and vice versa.

In the project one of the essential purposes of information flows is to equip managers with fundamental inputs to decision-making process, since flows comprise information assets considered as strategic organizational information repositories.

The absence of information management policy makes information flows throughout organizational environment without a direction, wasting relevant information to organization knowledge generation and dissemination. It would also contribute to increase in operating costs, hindering communication and interaction between individuals and organizational units, hampering understanding of the strategic role of information and scope of institutional goals [28].

Some authors cite some adversity, regarding absence of information management, which can compromise the performance and achievement of the objectives of the organizations:

(a) Redundancy of information and consequent increased cost of development; inconsistency of information obtained from various sources; fragmentation of information communicated between organizational units [29];
(b) Communication barriers, decreasing their efficiency in order to increase the effort to fulfill the organizational mission and make real use of the information; difficulty in accepting risks associated with new ways to make the information flow in the environment; attempts to increase information flow that impair their objectivity ideal [30];
(c) Disruption of flow and resulting inefficiency [26] with [31]; and
(d) Disqualified and disordered information; poor distribution of tasks between employees [32].

As recommended in several models of information management, in general, information management consists of routine and systematic procedures for the collection, validation, processing, storage and retrieval, distribution, dissemination and use of information, which can be understood as the lifecycle information [1, 2, 29, 33, 34]. From these models, the information must be managed through appropriate organization information flows.

An effective and efficient information flow has a multiplier effect with the power of mobilizing all organizational units turning into a driving force of organizational development. In this sense, the improvement of routines and procedures in the light of the information life cycle has become an organizational need meeting the challenges posed by information and knowledge management. Information life cycle used in KIMM has seven steps as shown in Fig. 1.

The collection stage is a set of activities aimed at obtaining data and information that are translated as the initial input for the generation/maintenance of an information asset. For this step, worth questions are: what I want to get in terms of information? Where should I acquire this information? What search criteria should be used?

The validation step is a set of activities designed to ensure that the information collected meet established standards for cycle information. For this step, the worth questions are: the search criteria were set correctly? These criteria were followed, that is, all information obtained meets the criteria? The format of the information is as expected? The sources are identified, and are competent to provide the information?

The treatment step is defined as a set of activities that embody the transformation of inputs for generation or maintenance of information asset. For this step, the worth questions are: partial results are directed to the end result (waste)? Partial results may be needed for use/reuse? If so, is being stored for later retrieval? The regulatory requirements are being met, regarding the form and content? The criteria for confidentiality, availability and integrity are being defined, or are already in the rules that guide the development of information asset?

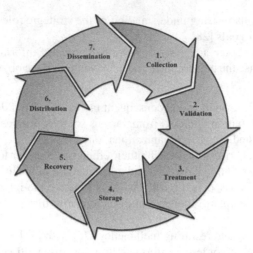

Fig. 1. Information life cycle.

An information asset, by definition, must also exist in digital format, in order to facilitate subsequent search and retrieval. In this sense, the step of storage should be concerned with issues such as: the information security requirements are being met? The information is being stored in order to be possible the recovery from parameterized queries? Those who have interest on information have the possibility to access it when need?

The recovery stage is the time when the information asset and data used in its construction, can be accessed and retrieved according to search criteria that enable the refinement of information and re-use for other purposes. In this step, the assumptions are the same as storage, since they are interdependent steps.

The distribution stage is the time when information asset is made available in appropriate format to be handle, i.e., becomes available for those interested. For this step, worth asking: what form of the distribution is appropriate to potential users? The criteria for confidentiality, availability and integrity are being met? The information asset is kept up to date throughout its life?

The dissemination step is the final stage of the life cycle, in which information asset will fulfill its goals. For this stage, worth questions are as: are there controls on the use of information assets (number of accesses, who accessed, etc.)? The target audience of information asset is being reached? Direct dissemination actions were planned, if deemed necessary?

The stages of the life cycle of information provide the structure to support the growth and development of an intelligent organization, adapted to new requirements and the environment in which it lies [35].

2.3   Characteristics of Good Information

For the effective information management it is necessary to constitute a set of policies that allow access to relevant, accurate and quality information [29]. This information must be submitted on time, with appropriate costs and easy access to stakeholders. Information to be valuable for managers must possess the following characteristics [36]: accurate, complete, economical, flexible, reliable, relevant, simple, and verifiable in time, which is consistent with other studies that describe quality dimensions of information [37].

Analyzing Table 1, it is possible to relate the characteristics of good information to the stages of information life cycle.

Table 1. Dimensions of quality information [37].

Accessibility	The extent to which information is available, or easily and quickly retrievable
Appropriate amount of information	The extent to which the volume of information is appropriate for the task at hand
Believability	The extent to which information is regarded as true and credible
Completeness	The extent to which information is not missing and is of sufficient breadth and depth for the task at hand
Concise representation	The extent to which information is compactly represented
Consistent representation	The extent to which information is presented in the same format
Easy of manipulation	The extent of which information is easy to manipulate and apply to different task
Free-of-error	The extent of which information is correct and reliable
Interpretability	The extent of which information is in appropriate languages, symbols, and units, and the definitions are clear
Objectivity	The extent of which information is unbiased, unprejudiced, and impartial
Relevancy	The extent of which information is applicable and helpful for the task at hand
Reputation	The extent of which information is highly regarded in terms of its source or content
Security	The extent of which information access to information is restricted appropriately to maintain its security
Timeliness	The extent of which information is sufficiently up-to-date for the task at hand
Understandability	The extent of which information is easily comprehended
Value-added	The extent of which information is beneficial and provides advantages from its use

For example, accurate information is the result of treatment phase actions, because the inaccuracy can be generated by the entry of incorrect data in the transformation process, while reliable information is directly related to the method of collection, i.e., its source of origin. It is important to identify sources of origin to guarantee the reliability and authenticity of the information collected. In turn, information on time is dependent on methods and tools related to the stages of storage and retrieval so that information is available when needed and in an appropriate format. For validation step, issues of format, integrity and compliance with legal and regulatory (compliance) precepts must be verified and confirmed. Confidentiality, availability and integrity are important issues to distribute information processed in previous steps as the collection, storage and retrieval.

Lack of information quality requirements in an organization can provide social and business impacts, and should be diagnosed, and efforts should be done towards its solution. Some factors that influence information quality are: information with multiple sources, use of subjective judgments, systematic errors in the production of information, information beyond storage lot [38].

As a mechanism to mitigate impacts of lack of quality information, the information flow modeling made use of information assets analysis considering the information life cycle steps and it resulted in a set of improvements propositions in order to cover all gaps identified at each stage.

3   Information Flow Based Modeling Approach to Information Management in Public Organizations – Case Study

Information management in a public organization must be supported by vision of the future, mission and institutional values, which, in case of a public organization, should be aligned to public administration principles. Policies for information management should have as premises top management vision of the future, the institutional objectives expressed in regulations and also its strategic goals defined in strategic plans of the organization. They should also provide quality information to promote assertive decisions resulting in services that meet efficiently citizens needs, achieve public confidence and eliminate resources waste. For this purpose (providing quality information) that research project was planned.

The proposed model was constructed for a regulatory agency based on its information assets, considering its organizational structure and its operation mode.

Duties and responsibilities of an organization are, in a sense, distributed by virtue of its organizational structure, which directly impacts the communication processes and organizational information management. Processing and generation of information from the perspective of information management have to minimize asymmetries, ambiguities and redundancies of information and mitigate impacts of environment weaknesses in which the organization operates, and specially developing products and services in order to meet institutional mission.

KIMM is scoped and aligned to organizational theory. The study of organizational design characterization rests on a set of four crucial variables: the first one refers to the organizational unit structure, information perceived through preliminary studies of the organization and documented in an organizational vision that is input to start modeling work. The second variable refers to the operation of each organizational unit, which can be identified through information assets analysis and its purposes in relation to the organization's business model. The third variable concerns how resources for products and services development are applied; this information is visible in information flows. And the fourth variable refers to how people are managed in relation to their skills to meet organizational characteristic, perspective studied by knowledge management modeling.

3.1   Information Modeling: Methodological Aspects

To build IM model (and KM), first, a broad and general representation of the organization is made from information obtained in interviews to highest levels of administration, and also information extracted from specific documentation readings related to the organization, relevant legislation, laws, rules, the Agency's mission and its vision for the future. The purpose of this specification is to understand and document the organization's goals, interfaces with the external environment, identifying key services involved, and especially, understand the organization's information assets. In this preparation stage, a references model is built, specific and/or internal reference models, which legitimize the organization actions, aligned to their culture. This model includes good practices in the field, selected from research to national and international agencies alike, which allows the organization to identify other possible practices to be adopted and also other information assets that could be incorporated into IM (and KM) models. Another model generated in this step is the preliminary information asset model that includes information on the purpose of these information assets, the theoretical and legal framework by which information assets are based, and also the interaction type of these information assets with the organizational units, as well as the services identified previously.

In the project, we start from the current situation understanding to then plan what will be in the future and to determine actions needed to change. First, it must known where the organization is in terms of maturity, which means perform a current situation analysis, based on the most important variables for IM (and KM). Then the desired future is identified in the medium and long term, ignoring resource constraints, for these same variables. Future scenario is supported by the information obtained thorough vision and meetings with top managers and organization's decision-makers.

On this occasion information assets previously collected are stated and prioritized. This is the main input to understand the characterization of the organizational unit under study. Through main information assets it is possible to know its roles and identify its operation model.

Organizational characterization analysis is done to identify similarities between operating units models, their points of convergence and hence procedures that could become a performance standard in the regulatory agency. It is also important to identify the characteristics of information assets that are most relevant for each type of transaction identified.

Finally, actions needed to undertake transformation in the desired direction (improvement recommendations) are identified.

In practice, initially, the current situation mapping is made and *as is* models are generated, describing how the organization works at the time, serving as a baseline for understanding information flows directly involved in its implementation. One of main results of the organization's current situation understanding is to know how the information flow is implemented and the relationship with its stakeholders. For information flows generation, meetings with actors directly involved in information assets are carried out.

And then, information flow modeling redesigns and represents all possible improvements and proposed changes, embodied by *to be* models that add IM (and KM) actions that should be deployed in the organization.

The comparative analysis of these two models, *as is* and *to be*, enables current deficiencies visualization. Gaps related to knowledge and information management processes allows identify improvements in aspects such as procedures, technology, individuals and even in the structural sense, so that the organization may be able to implement new information flows designed, giving value to their information assets. Some results of this comparative analysis (diagnostic evaluation) are presented in the next section.

For modeling and representation of information flows and other models, it was adopted Enterprise Architect [38] as a tool, using UML language and adopting stereotypes developed specifically for the project.

Flows in current situation information contain three essentia parts while flows of desired situation contain four parts, as shown in Fig. 2.

The first part represents information life cycle and explains how information is produced, showing resources that are being consumed and products generated for information asset. Information life cycle is illustrated by the hexagons in the center of the model. They are colorful and also identified by numbers and that is associated with a stage of life cycle, as can be seen in figure label.

The second part identifies actors and workers, stereotype developed for the project, which interact directly with manipulation of information depicted in the flow. It allows clarify which actors can manipulate the flow and the role played by each one. The third part represents inputs used for the information handling throughout the construction of information asset and outputs resulting from the processing of each stage of the life cycle. All legal foundation and documents are also represented as an entry in the information flow in this third part. The fourth part, presented only in desired situation (*to be*) models, represented by rectangles linked by a dotted line spread over the flow of information, indicates the propositions of improvement and in which stage of information life cycle these will impact directly, if implemented. The rectangles with a folded corner are present to help figure understanding and they are not part of the model.

Regarding the stages of information life cycle on the scope of the project, the collection step corresponds to actions to collect, acquire and seek information. The validation step includes actions to verify, compare, analyze, demonstrate and validate information. The treatment step is related to actions as clean, sort, consolidate, index, build and develop, that is, the treatment step is where the processing of information occurs. The storage step corresponds to actions to save the information processed in a particular media. In the recovery step is where the information stored is reused. The availability

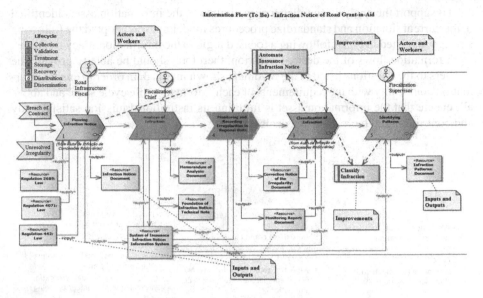

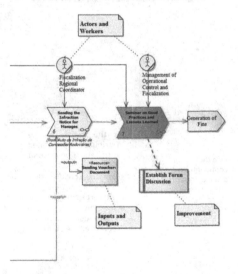

Fig. 2. Information flow model of the future situation – Part A and B (Color figure online).

of information occurs in the distribution step and it is necessary to establish appropriate mechanisms for distribution. On dissemination step is where spread occurs, in fact, the rational use of information produced and processed. This last step is where information asset fulfills its role in the organization. Dissemination strategies should take into account the characteristics of the organizational unit and the quality dimensions of information relevant to the organizational context and purpose of information asset.

To support the analysis and diagnosis activities of the information assets identified in the current situation and standardize procedures modeling, was adopted the following strategy, depicted in Fig. 4 following a focused logic to the information lifecycle stages.

Information flows of the desired situation, therefore, should be able to provide the answers to the questions listed in the roadmap, shown in Fig. 3, in order to ensure information conformity with the requirements of each information life cycle stage and above all, ensure that the information asset is fulfilling its institutional mission satisfactorily and contributing to more assertive decision.

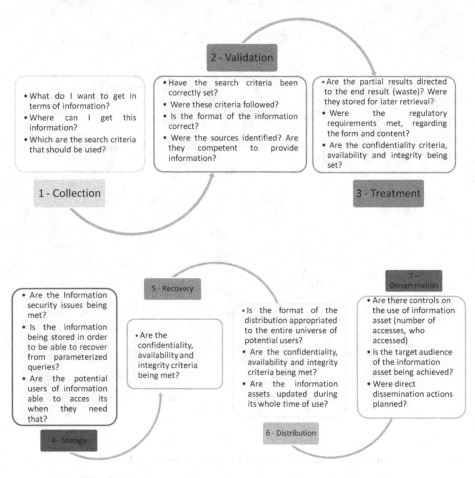

Fig. 3. Roadmap to information asset modeling – Part A and B.

The roadmap presented in Fig. 3 guides the building of desired situation models. The analyst responsible for modeling should suit the activities of each information life cycle step towards cover the gaps identified in the diagnosis performed. This adequacy can contemplate new activities, eliminating redundant or unnecessary activities, repositioning the information flow activities and even information assets creation, elimination or aggregation.

3.2 Summary of Results Obtained from the Information Modeling

Fifteen organizational units were modeled. For each of them, top management future vision was obtained and organizational characterization was identified.

Regarding future vision, three concerns were stood out, according to statements of leaders: need for information technology solutions, such as business intelligence tools implementation; automation of routines and procedures to make faster documentary proceeding, internal communication and real-time answers to questions, mainly from other government agencies; and issues inherent to intelligent surveillance, aiming organizational efficiency.

Figure 4 shows part results of organizational characterization analysis of three organizational units dealing with core activities of the regulatory body: regulation and monitoring of international and national road transport of passengers, the railway infrastructure and road infrastructure. From the analysis of their information assets were perceived similarities in their operating models, represented by three small circles on the figure, which allowed to define main characteristics of these organizational units that are responsible for the grants in the agency and, in this sense, could be characterized as managing sourcing organizations, as illustrated by the larger circle in Fig. 4. For this type of organization, free-of-error, completeness, timeliness and reputation are key information characteristics to ensuring the quality of information produced. Efforts should be made to ensure that these dimensions are present in the treatment of gaps identified in information life cycle of an information asset.

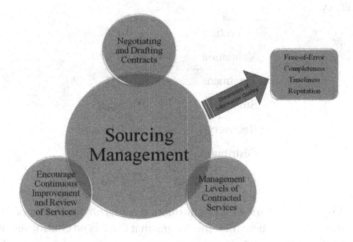

Fig. 4. Part of organizational characteristics analysis result.

Complete and error free information is essential for negotiation and drafting of contracts as well as time information is important for management and review of contracts and service levels established in the concession contracts. Maintaining a good reputation of the organization through quality information is essential to the credibility of the organization. In this sense, the stages of validation, processing, storage and retrieval of the information life cycle should include procedures and routines to ensure the presence of these features on informational treatment of organizational units information assets.

The organizational characteristics, in another perspective, can be used as an additional indicator for identifying best practices in organizational knowledge management, appropriate to each organizational unit and planning personnel training, and indicative for benchmarking best management practices.

A total of 84 information assets were identified and that led to one hundred and seven current situation information flows and 105 information flows to the desired situation. It is important to clarify that an information asset may be embodied by more than one information flow. This relationship is defined at the information collection phase and is required to better represent the information flow of the information asset and the relationship with its stakeholders. The desired situation flows were generated from the diagnosis made considering aspects of cohesion and coupling in connection with restructuring, which explains the difference in quantity between the current and desired situations.

The life cycle analysis through information flows can reveal situations as described in Table 2, where values do not directly represent data in the agency. The analysis of each information life cycle stage on each information flow aims to identify lack of activities and/or systematic procedures that can ensure compliance, accordingly the objectives of relevant step. Table 2 figures do not mean necessarily that the step is not being considered, but that could be optimized.

Table 2. Percentage of information assets that do not meet fully and systematically each step of information life cycle.

Collection	10 %
Validation	30 %
Treatment	20 %
Storage	25 %
Recovery	30 %
Distribution	15 %
Dissemination	40 %

In Table 2 is shown that for 40 % of information assets no dissemination is projected in its information life cycle activities, this means that there is no control and monitoring on the use of information assets (number of access activities, who accessed, etc.) and identification of the target audience. And yet, 30 % of all information assets do not meet validation step with fully systematized activities and so completeness of information may be compromised, as there are not always technical and/or a set of logical time organized tasks so that the information collected undergo a validation process smooth and consistent. The importance of these results should be analyzed considering the characteristics of the organizational unit.

A total of 540 improvements recommendations were listed in four predetermined categories. For information management (IM) and organizational procedures for IM categories, 411 improvement opportunities were identified. These two perspectives, automation of procedures, systems integration, defining storage activities, implementation of systems based on ontology are part of the main improvement recommendations established. Improvements were made available along with other project results through reports and those related to information lifecycle were represented in information flows in the desired position, to indicate which stage of the life cycle information would be impacted if they were implemented. Other improvements were represented in each specific model.

4 Conclusions

This paper presented the information modeling as a part of a wider research project that aims to develop a knowledge and information management model (KIMM) to Brazilian Land Transportation regulatory agency.

The modeling is based on information flows considering information life cycle and the characteristics of good information. The modeling approach proved to be an important diagnostic tool for assessing information management state in organizations focused on information.

The organization studied can be considered intensive in information and knowledge, that is, its main assets are the strategic information and the intellectual capacity of its employees, which represents a major challenge to make knowledge sharing inherent to its culture, with different employees working with creative ideas, new products and services, problems solution and forward thinking.

In this organizational context it is fundamental the ability to generate, process, analyze and disseminate information efficiently.

Information assets life cycle analysis has clarified several issues of non-compliance relating to steps.

The representation of flows as they are today and how they should be in the future showing a set of recommended improvements facilitate improvements implementation planning and the record of their impact upon implementation.

Assessing the quality of information along with the organizational characteristics also proved to be useful tool to help in prioritizing improvements to be implemented.

The EA tool was adequate for representation of flows, allowed to visually record how these flows occur and all aspects related to modeling them.

An important benefit of modeling based on information flow is the information whose production is continuing in the organization in all levels: operational, tactical and strategic, though each of these levels has specific needs and demands regarding the use and application of the information.

The modeling performed showed that, in most of the stages of the information life cycle, information assets are not systematically enforced. This implies that the implementation of KIMM involves the adoption of new practices for the collection, validation, processing, storage, retrieval, distribution and dissemination of information and mechanisms for the application of knowledge in the organization. This is a change in organizational culture and the relative strengths of the cultural factors that act as facilitators and those that act as inhibitors will determine the feasibility, robustness and speed of change in organization.

References

1. Davenport, T.H.: Information Ecology: Mastering the Information and Knowledge Environment. Oxford University Press, New York (1997)
2. Mcgee, J., Prusak, L.: Managing Information Strategically. Wiley, New York (1993)
3. Marchiori, P.Z.: A ciência e a gestão da informação: compatibilidades no espaço profissional. Ciência da Informação, Brasília, vol. 31, no. 2, maio./ago (2002)
4. Choo, C.W.: Information Management for the Intelligent Organization – the Art of Scanning the Environment, 3rd edn. Information Today, Inc., Medford (2002)
5. Institute of Information Scientists. Criteria for information science (2001). http://www.cilip.org.uk
6. Choo, C.W.: The Knowing Organization: How Organizations Use Information to Construct Meaning, Create Knowledge, and Make Decisions, 2nd edn. Oxford University Press, New York (2006)
7. Rezende, D.A., Abreu, A.F.: Tecnologia da Informação Aplicada a Sistemas de Informação Empresariais (9o.edição). Atlas, São Paulo. ISBN: 9788522475483 (2013)
8. Bastos, C.A.M., Moreira, M.R., Bruno, A.C.M., Mecena Filho, S., Farias Filho, J.R.: Information flow modeling: a tool to support the integrated management of information and knowledge. In: KMIS 2014 - International Conference on Knowledge Management and Information Sharing, Rome (2014)
9. Bastos, C.A.M., Bruno, A.C.M., Garcia, A, Rezende, L, Caldas, M.F, Sanchez, M.L.D., Mecena Filho, S.: Managing information and knowledge: a proposal methodology for building an integrated model based on information assets identification. In: 5th KMIS 2013. Portugal (2013)
10. Bastos, C.A.M., Rezende, L; Caldas, M.F., Garcia, A, Mecena Filho, S., Sanchez, M.L.D., de Castro Jr., J.L.P., Burmann, C.R.: Building up a model for management information and knowledge: the case-study for a Brazilian regulatory agency. In: Proceedings of the 2nd International Workshop on Software Knowledge - SKY/IC3 K. Paris (2011)
11. Rezende, L., Lobão, M.A., Burmann, C.R.N., de Castro Jr., J.L.P., Merino, L.A., Rocha, S.A., Bastos, C.A.M.: Modelling and knowledge management in the field of road infrastructure operation and regulation - study on the methods application in an organizational unit. In: KMIS 2012, pp. 265–268. Barcelona, Spain (2012)
12. Rezende, L., Lobão, M.A., de Castro Jr., J.L.P., Merino, L.A., Merino, R.S.A, Bastos, C.A.M.: Diagnosis and prognosis of knowledge management based on k-workflow, on conversion and knowledge flow - the case of the national land transport agency in Brazil. In: 5th KMIS. Portugal (2013)
13. Sabbag, P.: Espirais do Conhecimento - Ativando Indivíduos, Grupos e Organizações. Saraiva, São Paulo (2007)
14. Nonaka, L., Takeuchi, H.: The Knowledge Creating Company: how Japanese Companies Create the Dynamics of Innovation, p. 284. Oxford University Press, New York (1995). ISBN 978-0-19-509269-1
15. Autier, M., Lévy Pierre, P.: Les arbres de connaissances, La Découverte, Paris (1992)

16. Oppenheim, C., Stenson, J., Wilson, R.M.S.: Studies on information as an Asset I: definitions. J. Inf. Sci. **29**, 159–166 (2003)
17. Oppenheim, C., Stenson, J., Wilson, R.M.S.: Studies on information as an Asset II: repertory grid. J. Inf. Sci. **30**(2), 181–190 (2003)
18. Carlucci, D., Schiuma, G.: Knowledge assets value creation map assessing knowledge assets value drivers using AHP. Expert Syst. Appl. **32**, 814–821 (2007)
19. Higgins, S., Hebblethwaite, P., Chapman, A.: What is information architecture – white paper 1.0.0. Architecture and Standards Unit. Office of Government ICT, Department of Public Works. The State of Queensland (2006)
20. Caralli, R.H., Allen, J.H., White, D.W.: CERT® Resilience Management Model: a Maturity Model for Managing Operational Resilience. Addison-Wesley Professional, Reading (2010)
21. Davenport, T.H., Prusak, L.: Working Knowledge: how Organizations Manage What They Know. Harvard Business School Press, Boston (2000)
22. KPMG/IMPACT: The Hawley report: information as an asset: the board agenda. KPMG/IMPACT, London (1994)
23. IAD – Information Access Development. What is an Information Asset? (2012). http://www.informationassetdevelopment.com/what.html
24. Barreto, A.A.: Mudança estrutural no fluxo do conhecimento: a comunicação eletrônica. Ciência da Informação, Brasília **27**(2), 122–127 (1998)
25. Jacoski, C.A.: Peculiaridades do fluxo de informações em pequenos escritórios de projeto de edificações. In: Workshop Brasileiro de Gestão do Processo de Projeto, 5. UFSC, Florianópolis (2005)
26. Altissimo, T.L.: Cultura organizacional, fluxo de informações e gestão do conhecimento: um estudo de caso. Dissertação de Mestrado em Ciência da Informação – Universidade Federal de Santa Catarina, Florianópolis (2009)
27. Valentim, M.L.P. (Org.): Ambientes e Fluxos de Informação. Cultura Acadêmica, São Paulo (2010)
28. Greef, A.C., Freitas, M.C.D.: Fluxo enxuto de informação: um novo conceito. Perspect. ciênc. inf. **17**(1) Belo Horizonte January/March (2012)
29. Beal, A.: Gestão estratégica da informação: como transformar a informação e a tecnologia da informação em fatores de crescimento e de alto desempenho nas organizações. Atlas, São Paulo (2008)
30. Freire, I.M.: Barreiras na comunicação da informação. In: Starec, C., et al. (eds.) Gestão estratégica da informação e inteligência competitiva, pp. 33–46. Saraiva, São Paulo (2006)
31. Le Coadic, Y.-F.: A ciência da informacão. Brasília: Briquet de Lemos Livros, 1996 apud Altissimo, T. L. Cultura organizacional, fluxo de informações e gestão do conhecimento: um estudo de caso Dissertação Mestrado em Ciência da Informação. Universidade Federal Santa Catarina, Florianópolis (2009)
32. Canova, F., Picchi, F.A.: A aplicação da mentalidade enxuta no fluxo de informações de uma indústria de pré-fabricados de concreto. In: Simpósio Brasileiro De Gestão E Economia Da Construção, 6. UFSCAR, São Paulo (2009)
33. Smit, J.W., Barreto, A.A.: Ciência da Informação: base conceitual para a formação do profissional. In: Valentim, M.L. (ed.) Formação do profissional da informação, pp. 9–23. Polis. Cap.1, São Paulo (2002)
34. Lesca, H., Alameida, F.C.: Administração estratégica da informação. Revista de Administração, São Paulo **29**(3), 66–75 (1994)
35. Tarapanoff, K.: Referencial teórico: introdução. In: ____. (Org.) Inteligência Organizacional e competitiva. Brasília: Editora UnB. 343 p. (2001)

36. Stair, M.R., Reynolds, G.W.: Principles of Information Systems. ISBN-13: 978-1305108110 (2013)
37. Kahn, B.K., Strong, D.M., Wang, R.Y.: Information quality benchmarks: product and service performance. Commun. ACM **45**(4), 184–192 (2002)
38. Strong, D.M., Lee, Y.W., Wang, R.Y.: 10 potholes in the road to information quality. IEEE Comput. **18**(162), 38–46 (1997)
39. Enterprise Architect. http://www.sparxsystems.com.au/

Elucidating and Creating Working Knowledge for the Care of the Frail Elderly Through User-Centered Technology Evaluation of a 4-Wheel Electric Power Assisted Bicycle: A Case Study of a Salutogenic Device in Healthcare Facilities in Japan

Miki Saijo[1(✉)], Makiko Watanabe[2], Sanae Aoshima[3], Norihiro Oda[3], Satoshi Matsumoto[4], and Shishin Kawamoto[5]

[1] Graduate School of Innovation Management,
Tokyo Institute of Technology, Tokyo, Japan
msaijo@ryu.titech.ac.jp
[2] Graduate School of Science and Technology,
Tokyo University of Science, Chiba, Japan
makiko12317@yahoo.co.jp
[3] Kakegawa Kita Hospital, Shizuoka, Japan
sanae_ao@yahoo.co.jp, kanri-saiyou@kakegawa-kita.com
[4] Corporate Planning Division, Yamaha Motor Engineering Co., Ltd.,
Shizuoka, Japan
matsumotosat@yec.co.jp
[5] Faculty of Science, Hokkaido University, Sapporo, Japan
ssn@mail.sci.hokudai.ac.jp

Abstract. Frail elderly people suffer a variety of physical and mental weaknesses that tend to hinder their ability to make use of AT devices in the intended manner. Because of this, it is important that new AT devices undergo technology evaluation within the context in which they are to be used. In this study, frail elderly people in a Japanese daycare center and a rehabilitation hospital were given a 4-wheel, power-assisted bicycle to ride, and user-centered technology evaluations were carried out. While this vehicle was considered suitable for people age 75 and older who rarely walk, the data for the 61 people who tried out the vehicle indicate that they rode the bicycle only when a PT (physiotherapist) intervened, gave encouragement, adjusted the bicycle settings as needed for the user, and otherwise created new knowledge. In this study we summarized this new knowledge in an information DVD directed at people who work to support the frail elderly.

Keywords: Knowledge creation · Tacit knowledge · Health care for frail elderly persons · AT devices · User-centered technology evaluation

© Springer International Publishing Switzerland 2015
A. Fred et al. (Eds.): IC3K 2014, CCIS 553, pp. 605–620, 2015.
DOI: 10.1007/978-3-319-25840-9_37

1   Introduction

Our perception of health changed with the 1986 Ottawa Charter for Health Promotion. Before, it was considered the individual's responsibility to maintain a healthy lifestyle, but under the Ottawa Charter, health is something to be achieved through a collaboration of environment, community, personal skills, and public policies. Our concept of health also has gradually changed from a dichotomous to a salutogenic perspective. In the former, health and disease are treated separately and health is defined as being "low on risk factors". The latter, however, focuses on "keeping people well". From this perspective, health is viewed as an ease/dis-ease continuum [1]. The Ottawa Charter initiated a redefining and repositioning of institutions, epistemic communities and actors within the disease-health continuum [2].

Frail elderly persons, especially those aged over 75 years, who tend to be the main users of aged care services live in a dis-ease condition. Frailty is highly prevalent in old age and poses a high risk of falls, disability, hospitalization, and mortality [3]. Falls are common and often devastating among older people [4]. Walking is a risk factor for falling, and yet walking is a basic salutary factor in an individual's life. Therefore, in order to promote health among frail elderly persons, optimal approaches involving interdisciplinary and inter-agency collaboration are required to build an environment in which such people can go out without worry about falling.

While some people may have difficulty in walking, they may still be able to ride a bicycle. Already there are electric power-assisted bicycles on the market that will allow the rider to climb steep hills with ease, even if they do not have much foot power. The power-assisted bicycles were originally developed by the Yamaha Motor Company. The 4-wheel assisted bicycle is a newly developed vehicle which is safer than walking for frail elderly persons, and is actually allowed on public roads as a wheelchair. Its controls are set so that it will not exceed a speed of 6 km/h when going downhill. With this vehicle, a person who cannot walk can climb a hillside at the same pace as someone walking at normal speed. Though technically it is already available, this vehicle is not yet on the market because there remain concerns of its safety for frail elderly users. In order to find salutogenic benefits in lifestyles of the frail elderly, we undertook action research on frail elderly people's use of this 4-wheel electric power assisted bicycle in a rehabilitation hospital in the city of Kakegawa, Japan. As Antonovsky [1] said, we must start with the question, "How can this person be helped to move toward greater health?" This kind of effort must relate to all aspects of the person, and the frail elderly adult is no exception. We tackled this question through interdisciplinary and inter-agency collaboration among manufacturer, hospital, municipal government, and university.

2   Literature Review

2.1   Assessing Frailty in Elderly Persons

In order to provide appropriate care for frail elderly persons covered by long-term care insurance (LTCI) the Japanese Ministry of Health, Labour and Welfare (MHLW) [5] drew up a *Kihon Checklist*, a basic health checklist, for those aged 65 and older, to be used as a frailty index to predict the risk of requiring care under LTCI. The checklist consists of a 25-item, self-reported questionnaire, covering seven categories including physical strength, nutritional status, and oral function, as well as houseboundness, mobility, cognitive function, and depression risk [6]. Using this checklist, municipal governments classify the frail elderly persons in their communities according to their need for preventive care. Kakegawa City covers all 25 items on the questionnaire as well as medical certificates to identify the frail elderly. Table 1 is a partial view of the checklist and its screening criteria.

Municipal governments use the basic checklist as a guide to decide their own criteria and procedures for preventive care and care services, the dispatch of helpers, the lending of wheelchairs, and the need to provide rehabilitation. Care services are provided in accordance with the degree to which activities of daily life (ADL) have deteriorated. Table 2 shows the assessment levels for LTCI. Persons classified as Support Level 1 and Support Level 2 are eligible for preventive care. Table 2 was drawn up by the first author as a simplified illustration of Japanese LTCI assessment levels.

A few researchers have evaluated the validity of this checklist to predict the risk of requiring care [6, 7]. Fukutomi et al. [6], in particular, has suggested that physical strength and cognitive function are more useful indices for detecting the risk of future deterioration of ADL. However, there is no consideration of intervention to prevent future risk. Frailty is defined as a clinical syndrome in which three or more of the following criteria are present: unintentional weight loss, self-reported exhaustion, weakness (grip strength), slow walking speed, and low physical activity [3]. Geriatric interventions have been developed to improve clinical outcomes for frail older persons [4, 8]. Maki et al. [9], also evaluated the efficacy of intervention by examining a municipality-led walking program for the prevention of mental decline in the elderly aged 72.0 ± 4.0, in a randomized controlled trial. This study introduced a 90-min intervention program consisting of 30 min of exercise and 60 min of group work, and concluded that this intervention program may provide benefits for some aspects of cognition. Though their research target was not the frail elderly, from the criteria in Table 1 and the assessment level of LTCI we can easily surmise that the frail elderly requiring LTCI are unable to participate in this kind of walking intervention. We need other methodologies to assess the frailty of the elderly and to mitigate the inconveniences in their daily life.

2.2 Tacit Knowledge in Health Care

Since the 1970s, people have been given more room to take the initiative in roles where they provide expertise and participate in informing, ideating, and conceptualizing activities in the early design phases [10]. This movement is called user-centered design. For frail elderly people, there are various assisted-technology (AT) devices such as canes, walkers, and bath benches, as well as wheelchairs, that could be considered to be user-centered designs. As Mann et al. [11] notes, many elderly persons rely on these devices. NHS (the UK department of health) advocates patient-led care and urges

Table 1. Screening criteria for providing preventive care (partial view of MHLW check list).

Mobility (those with top score of 3 points are candidates for preventive care)	0 point	1 point
Q6. Can you climb stairs without holding onto a handrail or wall?	Yes	No
Q7. Can you stand up from a sitting position without holding on to anything?	Yes	No
Q8. Can you walk continuously for 15 min?	Yes	No
Q9. Have you fallen within this one year?	No	Yes
Q10. Do you worry about falling down?	No	Yes
Cognitive functions (1 point or more required)	0 point	1 point
Q18. Do people around you say you repeat the same thing and have become forgetful?	No	Yes
Q19. Do you make phone calls by yourself?	Yes	No
Q20. Do you find yourself not knowing today's date?	No	Yes
Depression (2 points or more required)	0 point	1 point
Q21. I do not feel any fulfillment in my daily life during the last two weeks	No	Yes
Q22. I cannot enjoy things I used to enjoy during the last two weeks	No	Yes
Q23. During the last two weeks, I am not willing to do what I could do easily before	No	Yes
Q24. During the last two weeks, I do not feel I am useful to anyone	No	Yes
Q25. During the last two weeks, I feel I am exhausted without any reason	No	Yes

Table 2. Japan's long-term care insurance assessment levels.

	Level	Self-sufficient	Body care	Comprehension	Behavior control	Mobile
Support	Level 1	Yes	Partial	Yes	Yes	Yes
	Level 2	Yes/partial	Partial	Yes	Yes	Yes
LTC	Level 1	Partial	Partial	Yes	Yes	Yes
	Level 2	Partial	Partial	Yes	Yes	Partial
	Level 3	No	No	Partial	Partial	Partial
	Level 4	No	No	No	No	Partial
	Level 5	No	No	No	No	No

health care teams to move from a service that does things to and for its patients to one which is patient-led and which works with patients to support them with their health needs [12]. AT devices are provided to frail elderly people on the assumption that the devices will promote their independence and lower costs [11], but there is little research on how much and what kinds of frailty are mitigated by AT devices and what kind of support is needed to make full use of such devices. As some of the studies have pointed out, an important part of health care consists of tacit knowledge.

A significant part of health-care knowledge exists in tacit form, for instance the working knowledge of health care experts [13]. Abidi et al. [13], classify tacit knowledge into (1) basic tacit knowledge or routine experiential knowledge, and (2) complex tacit knowledge or intuitive experiential knowledge. The latter is "progressively accumulated as the expert responds to atypical and high acuity clinical problems–it is deeply embedded and, hence, not easily articulated, yet manifests as the expert's intuitive judgment in challenging clinical situations." According to this classification, a health care professional's judgment of the usability of an AT device for the frail user is inherent in the professional's tacit knowledge. The questions are: How can we extract this individual knowledge and create new knowledge for patient-led care? How can we share the new knowledge among those who volunteer to support frail elderly people on their outings?

2.3   Knowledge Creation and Technology Evaluation of AT Devices

Knowledge creation starts with socialization, which is the process of converting new tacit knowledge through shared experiences in day-to-day social interaction [14]. Nonaka and Toyama also state that "knowledge creation is a synthesizing process through which an organization interacts with individuals, transcending emerging contradictions that the organization faces", and "one can share the tacit knowledge of others through shared experience" [14]. In order to transform tacit knowledge into shared new knowledge, socialization and efforts to transcend contradictions are needed. In professional health care work places, medical doctors, nurses, physiotherapists, and occupational therapists work together. In this sense their routines consist of an inter-agency, interdisciplinary collaborative experience [15]. Also the main contradictions in elderly health care are inherent in the care itself. All human beings must eventually die, but the care professional's work is to challenge this destiny. The need to transcend emerging contradictions and shared experience are embedded in their routines. The problem is how to elucidate their tacit knowledge and reconstruct it to create new knowledge. In this study, we describe this process through a technology evaluation of a newly developed 4-wheel electric power-assisted bicycle, called a Life Walker or LW vehicle, for frail elderly persons. This requires, however, that the health care professional collaborate with outsiders, engineers and researchers, to undertake the technology assessment of the newly developed AT device. Collaboration with outsiders requires socialization, both mental and behavioral. We assume that this shared new experience will extract their tacit knowledge to create new knowledge in the methodology of caring for frail elderly persons.

Technological evaluations of manual and powered wheelchairs have already been made in the field of anthropometry, the measurement of physical characteristics and abilities of people [16, 17], with the objective of acquiring information that is essential for the appropriate design of a wheelchair. There is little study, however, of the usability of a wheelchair, or how it is actually used in real-world health care circumstances. And there is no user-centered technology evaluation of newly developed AT devices for supporting the frail elderly on outings in real-life circumstances with inter-agency and interdisciplinary cooperation.

McNamara and Kirakowski [18, 19] propose three aspects that need to be considered from the viewpoint of user-centered technology evaluation, namely, functionality, usability, and experience. They state that these are unique but independent aspects of usage. Functionality focuses on the product and is evaluated by answering the question, "What will the product do?" Usability is defined by the ISO 9241-11 definition of usability as, "the extent to which a product can be used by specified users to achieve specified goals with effectiveness, efficiency and satisfaction in a specified context of use" [20]. Experience is the individual's personal experience of using the device. The question asked here should perhaps be, "How do I relate to this product?"

In this study of the Life Walker (LW), we apply these aspects of technology evaluation to the framework of knowledge creation. The aim of this study is to elucidate the process of knowledge creation among inter-agency and interdisciplinary health care professionals, engineers, and researchers in evaluating AT technology meant to promote patient-led care among frail elderly persons.

3   Research Methodology

3.1   Research Questions

In introducing the Life Walker (LW), a microcomputer-controlled, 4-wheel electric power-assisted bicycle, to frail elderly persons certified as requiring care in a rehabilitation hospital and at a daycare center for the elderly, we began with two questions: (1) What conditions are necessary to have an LW used in a hospital or daycare setting in such a way that there will be knowledge creation, and (2) How can the functionality, usability, and experience of the LW be measured quantitatively and qualitatively? The three categories in the second question were further broken down into additional questions as follows:

Functionality. What are the distinguishing characteristics of the LW and what kind of elderly person is it best suited for?

Usability. What kind of elderly person will ride the vehicle and for how long?

Experience. What do the elderly persons and care professionals experience through riding the LW?

3.2   Case Study Method

We loaned the LW for two months to a rehabilitation hospital and a daycare center for the elderly. Figure 1 shows the flow of data collection for this study.

Period. Nov 5, 2012 to Jan 30, 2013.

Targets. (1) Frail elderly people (hereafter "facility users") in the Kakegawa Kita Hospital and the Kakegawa City Sayanoie daycare facility for the elderly. Facility users were rated at Support Level 1 through LTC level 4; and (2) Hospital and daycare care professionals: physiotherapists (PT), occupational therapists (OT), care managers (CM), care workers (CW), and social workers (SW).

Methods. Questionnaire survey, test riding of LW, and interviews of care professionals.

The questionnaire for this study included the following kinds of information. The authors prepared a form that was filled out by the care professionals (as it turned out, only PTs in the rehabilitation hospital filled out the form). Also we interviewed care professionals at both the hospital and the daycare center.

Personal Portfolio. Age, sex, care rating, dates at facility, experience riding LW, type of rehabilitation, continuing or interruption of LW experience: PT's predictions of whether each facility user will become a continuing rider or not.

Riding Report. Rider's basic checklist (if there were multiple test rides, only changes from the first ride were recorded), reason for test rides, evaluation of driving skills.

Recording of Interview. The authors carried out focus group interviews (FGI) of the care professionals just before or after the loan of the LWs and during the loan period.

The sub-categories of the research questions were handled as follows.

Functionality. Functionality is a concept related to the technical aspects of a product, and involves answering the question, "What does the product do?" [18, 19].

The specifications of the LW vehicle used in this study are as follows.

- Overall length × Overall width × Overall height: 1.190 mm × 655 mm × 990 mm
- Dry weight (with battery): 36 kg (38 kg)
- Tires: Sponge-type; do not blow out
- Drive system: Dual electric-assisted run and full-electric run
- Control system: Microcomputer control with motor controlled automatic brake
- Max forward speed: 6.0 km/h
- Max reverse speed: 0.5 km/h

When the rider is a healthy adult, the functions of the vehicle are dependent on its specifications; while there may be small divergences, the functions of the vehicle do not change drastically among different riders. The functionality of this vehicle as judged by its specifications would be described thus: With the electric-assist function, the vehicle can be made to move forward with just light pressure on the pedals. The vehicle can be stopped with handbrakes or by simply removing one's feet from the pedals. Reversing is also possible. The vehicle is pre-set so that it cannot go forward at a speed greater than 6 km/h or reverse any faster than 0.5 km/h.

However, the users who are the subjects of this study are all certified as requiring some level of nursing care in their ADLs, as shown in Table 2, and the functionality of the LW in the case of a facility user greatly depends on the physical and mental condition of the rider. Therefore, the question needs to be revised to, "What kind of rider can move this vehicle?" In FGIs carried out prior to the test riding of the LW, we asked the health care professionals routinely caring for the facility users what kind of person they felt was suitable for this vehicle. During the loan period of LW vehicles, we asked for the care professionals who would be willing to encourage facility users to ride the LW. We then asked them to predict whether each rider would continue to ride the LW or not. Later, we included these predictions in the personal portfolio.

Usability. Usability, according to ISO is defined as "the extent to which a product can be used by specified users to achieve specified goals with effectiveness, efficiency, and satisfaction in a specified context of use." In this study, we created a histogram and accumulated data chart comparing the riding rates of continuous riders and those who discontinued riding, a histogram of riding rates (number of rides/days at facility) by attributes (sex, age cohort, care level) to clarify what kind of people rode the LW at what frequency. The effectiveness of the rides should also be measured in terms of physical and mental effects, but that is not within the scope of this study.

Experience. According to McNamara and Kirakowski [18, 19], judging experience requires answering the question, "How do I relate to this product?" The facility users who tried out the LW required assistance in their ADLs, but they nevertheless made the decision to ride the LW, tried it out several times, and then decided to either continue or discontinue riding during the period in which the vehicle was on loan to the facility. By keeping a record of their reasons (for deciding to continue or discontinue use), it was possible to learn what they found appealing, or conversely, what they found to be a problem, about the LW. Likewise, it was possible to find out how they were influenced to try out the experience of riding the LW. This data was collected by the authors from the riding survey tables and through interviews carried out after the vehicle loan period.

This study was carried out only after it had been reviewed and approved by the Tokyo Institute of Technology research ethics committee (authorization NBR: 2012021).

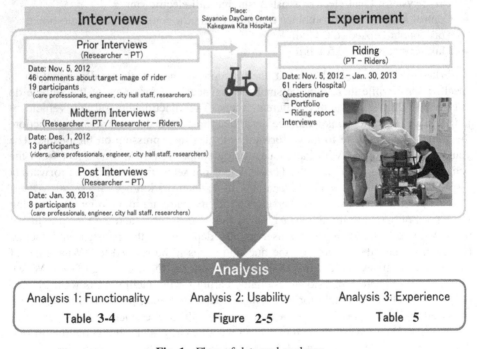

Fig. 1. Flow of data and analyses.

4   Results

4.1   Circumstances Which Elucidate Knowledge Creation

In this experiment, not one facility user at the daycare center attempted to ride the LW. At the rehabilitation hospital, there were 61 users who rode the vehicle. This result indicates that, without exception, there was no knowledge creation generated by the loan of the new AT device to the daycare center. The daycare center is a place where elderly persons certified as requiring care can spend time during the day. They do not come to the center with a specific purpose in mind, such as those who go to the rehabilitation hospital, and there are no specified activities for each individual. Still, just like at the rehabilitation hospital, the daycare center PTs and OTs did demonstrate the vehicle, riding the LW themselves and inviting the facility users to do the same. No one took them up on the invitation, however. This indicates how important it is to have someone try out the vehicle first in order to disseminate use of the new AT device.

4.2   Functionality

The Type of Frail Elderly Person for Whom the LW is Best Suited. Functionality of the AT device is decided by the interaction between the device's characteristics and the user's characteristics. In this sense, what kind of elderly person is best suited to LW is an important question. Table 3 shows the characteristics that emerged from the prior interview that best describe the type of elderly person who would be able to make use of the LW. As was shown in Fig. 1, we collected 46 comments from interviews with care professionals. Comments that were basically the same were combined into one.

Table 3. Target image of frail elderly persons who are best suited to use the LW vehicle.

Occupation	ID	Target image
CM	S	Forward-looking person
CM	S	Candidate for secondary preventive care
CM	S	Person requiring preventive care or support as opposed to a regular daycare user
CM	S	Person who is being taken care of by a general support center (preventive care)
CM	S	Person who wants to undergo rehabilitation
CM	S	Person who can walk using a cane
CM	S	Person who thinks they can still ride a bicycle or who uses a bicycle like a cane, but who others do not believe is actually capable of riding a bicycle
CM	S	A relatively healthy elderly person who still farms (such as in mountain valley areas or in tea fields)
CM	S	Person who has not yet ridden a mobility scooter

(Continued)

Table 3. (*Continued*)

Occupation	ID	Target image
CM	T	Person with relatively strong legs who can pedal (including those with dementia)
PT	A	Person who is mobile near the home using a cane or walker but who has difficulty going further afield
PT	SW	Person who uses machines or bicycles for physical training and rehabilitation
PT	A	Person who thinks they will be able to go outside if they have this kind of vehicle
CW	M	Anyone who shows interest; anyone who likes vehicles
SW	N	Person who is still riding bicycles but is losing confidence in their ability to do so
Facility director	IN	Person who doesn't want to depend totally on a mobility scooter

While no quantitative analysis has been made, the comments shown in Table 3 contrast the target image held by PTs and care managers. The PTs speak of the facility user's behavior and psychological factors while the care managers tend to base their target image on external markers such as the facility user's level of care, the AT device, and even the user's occupation. Nevertheless, as will be demonstrated in the next section on usability, these kinds of external markers are not effective measures of a user's suitability for riding the LW.

PT's Predictions of Whether a User Will Continue or Discontinue Riding. Table 4 shows the PT's predictions and actual results of facility users' continuity of riding. The rate of accuracy of the PT's predictions was around 50 %. In Table 4, the "Neither" category for prediction indicates those who rode only sporadically and for whom the PT had difficulty deciding which category (continue or discontinue) they belonged to.

Table 4. Rate of accuracy of PT's prediction.

	Result		
Prediction	Continued	Discontinued	Total
Continue	31.1 %	19.7 %	50.8 %
Neither	9.8 %	14.8 %	24.6 %
Discontinue	6.6 %	18.0 %	24.6 %
Total	47.5 %	52.5 %	100.0 %

4.3   Usability

The breakdown of the users who tried out the LW vehicle in the rehabilitation hospital is as follows: Age: 65–74: 22; 75-above: 39. Men: 36; Women: 25. Care level: Support Level 1– LTC Level 4. The number of days that a user came to the facility varied according to the individual's care level and medical condition. In order to determine what kind of persons rode the LW vehicle and to what extent, it was necessary to calculate the rate of riding by dividing the number of rides by the number of days at the facility. Figure 2 compares the riding rates and accumulated riding rates of those who continued to ride against those who discontinued. Those who continued where persons who may have taken several breaks, but who nevertheless continued to ride the LW vehicle throughout the period it was on loan to the rehabilitation hospital. Those who discontinued were persons who tried out the vehicle a few times but then stopped riding. There were only 4 riders in the "Neither" category so we traced their record of riding and re-categorized them in "continued" or "discontinued".

Looking at Fig. 2, you can see that most of those who discontinued riding generally did so after they had reached a riding rate of around 20 % (4 times). This indicates that the users were able to judge by their fourth ride whether or not the LW vehicle suited them. Those who continued to ride are distributed between 20 % to 90 %, and half exceeded a riding rate of 50 %, indicating that they rode the LW vehicle at least half of the days that they came to the hospital.

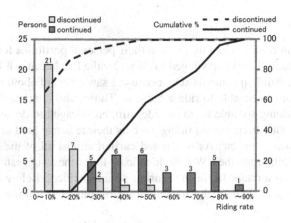

Fig. 2. LW riding rate and accumulated riding rate by continued/discontinued category.

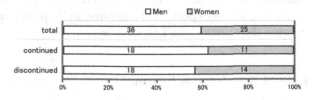

Fig. 3. Rates by sex.

Figures 3, 4 and 5 indicate that more men than women, and those older than 75 as compared to those age 65 to 75, were likely to continue riding. In terms of care levels, the greatest number of continuing riders spans LTC Levels 1, 2 and 3, while most of those at the less demanding care levels of Support Levels 1 and 2 did not attempt to ride the vehicle.

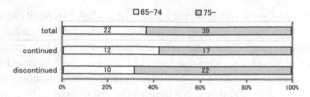

Fig. 4. Rates by age cohort.

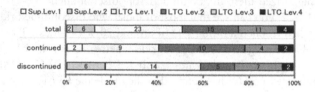

Fig. 5. Rates by care level.

4.4   Experience

The most common reasons cited by users in their personal portfolios for riding the LW vehicle were: because I was encouraged to do so by the PT, because it looked like fun, because I want to build up some muscle, because I saw or heard about someone riding, and because I want to be able to ride a bicycle. Those who continued riding gave as their reasons for doing so: able to go outside, fun, encouraged to do so, condition was good, etc. Those who discontinued riding gave as their reasons: hard to ride, tiring, too slow, condition was poor, etc. As was noted earlier, almost all of the 61 riders at the rehabilitation hospital rode the LW vehicle at least four times, suggesting that the PT's encouragement was a major factor in securing so many riders. Below are some of the comments made in an FGI of the rehabilitation hospital PTs during the experiment (Table 5).

What these comments tell us is that the rehabilitation hospital assigned staff specifically to help with the trial rides, repeatedly encouraged users to try out the vehicle, made adjustments as necessary to ensure a smoother ride, called out a beat to give the rider a rhythm by which to pedal, and otherwise made proactive efforts to act as an interface between the rider and the LW vehicle.

4.5   Sharing Created Knowledge as Information

We summarized the knowledge which PTs created through the experience of test riding of LW in a manual DVD title, "How to assist the elderly in riding the LW". The volunteers who support frail persons are the target of the manual. A 90-year old female user and a female PT at the Kakwgawakita hospital appear in this DVD to show how to ride and how to support someone riding the LW. The DVD consists of 6 parts: (1) introduction, (2) communication before riding, (3) support in getting on the LW, (4) support while driving, (5) support in getting off the LW, and (6) communication

Table 5. PTs' comments in midterm interview.

Topics	Comments
Trial rides	Start by gathering everyone together, saying "Would you like to ride it? Let's try it." Have a person get on the vehicle, show them how to operate it and walk by them as they go a few rounds. On the first ride, the rider goes a few turns and tries out all the vehicle's functions. Each individual is asked if they want to give it a try, and a staff person accompanies each person who decides to try riding. Once one person is done, the next person is brought forward
	Those who want to try a second ride are contacted by the facility staff
	Those who ride two or three times tend to continue riding
	A common characteristic of those who continue riding is that they are self-sufficient at home but have difficulty going out on their own
	These were mostly persons at support level 1 and nursing care level 2
	There were some care level 3 people who tried out the vehicle, but they generally needed assistance in controlling the vehicle, and some had to have help to keep their feet on the pedals
	Those who had disabilities of their hands or feet had difficulty controlling the vehicle
Encouragement from hospital staff during the trial ride	Progress is uneven during the trial rides, but we tried to encourage the rider by helping them with rhythm, for example, calling out, "One, two, one, two" each time they stopped pedalling. The usual staff worked with the riders [so they had people they were familiar with helping them out]
	After the rider got on the vehicle, the seat was adjusted (height, etc.) to ensure a comfortable ride. It was slightly different from adjusting a bicycle seat, but the staff kept communicating with the rider, asking "How does it feel?" and so on, and making minor changes as needed

(2012 Dec 1, Midterm FGI interview, Kakegawa Kita Hospital)

after riding. The narration describes how to assist an elderly person to ride the LW. The introduction addresses the issue of functionality: "What does the product do?" The communication before riding addresses the question: "What kind of rider can move this vehicle?" The narration for this part of the DVD suggests that observation and communication are key to judging whether a frail elderly person will be suited to riding the vehicle. Communication after the ride corresponds to experience. The narration suggests the rider should be asked to express his/her impressions and concerns about driving the LW. The parts in which the forms of assistance given are described were based on the PT's tangible arrangements and words of encouragement during the test ride. These include the following:

- If you have paralysis, get on the vehicle from the side of your body that is not paralyzed.
- If you are shaky, the assistant will hold the handles or the arm in back to help you stabilize.
- Check to make sure the rider is not making an excessive effort.
- As the rider gets used to the vehicle, the tendency is to pedal too hard. Remind the rider to "pedal lightly".
- The rider is most likely to pedal too hard right at the very start. Remind the rider, "Just a light push of the pedals will start the vehicle moving".
- Check that the rider is relaxed enough to look around.
- When the rider is doing well, say something encouraging, like, "That's the way!"
- When the rider does not do well, put out your hands to help drive the vehicle, and let them know they haven't failed by saying something like, "Don't worry. That happens sometimes".

5   Discussions

A new AT device, we found that the LW vehicle could be fun and highly satisfactory to frail elderly users who had difficulty walking on their own or going outside. But this was an experience that could not have been achieved without the interaction between the PTs and the users. The PTs provided encouragement, helped in adjusting the vehicle settings, made judgments as to who might be suited to ride the vehicle, and otherwise created new knowledge to guide frail elderly persons to using the vehicle. In order for this kind of knowledge creation to take place, it is imperative that the organization provide an environment suitable for the use of the device, that the user be proactive, and that dedicated staff be appropriately assigned to assist as necessary.

In the setting of the rehabilitation hospital where PT supervision was available, it was found that the LW vehicle was best suited to users at LTC Levels 1 to 3 who were over the age of 75. At the same time, there was considerable discrepancy among individuals as to whether they would continue or discontinue use of the vehicle. It was also found that it was difficult for PTs to predict who would or would not continue riding, and in this regard there is a need for new knowledge creation.

The primary focus of the salutogenic perspective on health is the individual's ability, right up to the time of death, to adapt to his or her own condition as necessary to stay in good health. Today, there are a variety of devices that support this ability. The more devices that are made available to frail elderly persons, the easier it will be to realize the kind of health that is the focus of this perspective. Unfortunately, major manufacturers are reluctant to develop new products for the over-75 market because of the fear of accidents and the probability of being sued as a result. Still, the market for devices for the over-75 age cohort is certain to expand and there is ample room for innovation based on new knowledge creation. As this study has shown, having a new device evaluated within the context of a care facility serves as an impetus to transform the tacit knowledge of professional caregivers to explicit knowledge. This requires, however, close collaboration among the device maker, researchers, and caregivers. In the current study, the city hall staff also played an important role as intermediaries bringing together diverse professionals and the staff of the care facilities.

There are four topics that this study must undertake in the future.

1. Accurate measurement of effectiveness which is a critical index for judging usability. We need to elucidate the psychological and physical effects of continuous use of the LW vehicle.
2. We need to provide a method, such as a rehabilitation menu, by which care managers, who are not as knowledgeable as PTs about nursing care, can judge who is best suited to use the LW vehicle.
3. We need to consider what kind of knowledge creation is needed to enable users to ride the LW vehicle outside of the facilities.
4. We need to consider to what extent the PTs' knowledge shown in the DVD can be transferred to the volunteers who support frail elderly persons on their outings.

References

1. Antonovsky, A.: The salutogenic model as a theory to guide health promotion. Health Promot. Int. **11**(1), 11–18 (1996)
2. Kickbusch, I.: The contribution of the world health organization to a new health and health promotion. Am. J. Public Health **93**, 383–388 (2003)
3. Fried, L.P., Tangen, C.M., Walston, J., Newman, A.B., Hirsch, C., Gottdiener, J., Seeman, T., Tracy, R., Kop, W.J., Burke, G., McBurnie, M.A., Cardiovascular Health Study Collaborative Research Group: Frailty in older adults: evidence for a phenotype. J. Gerontol. Med. Sci. **56A**(3), M146–M156 (2001)
4. Rubenstein, L.Z.: Falls in older people: epidemiology, risk factors and strategies for prevention. Age Ageing **35**(S2), 37–41 (2006)
5. Ministry of Health, Labour and Welfare: Act for partial revision of the long-term care insurance act, etc., in order to strengthen long-term care service infrastructure (2011). http://www.mhlw.go.jp/english/policy/care-welfare/care-welfare-elderly/dl/en_tp01.pdf
6. Fukutomi, E., Okumiya, K., Wada, T., Sakamoto, R., Ishimoto, Y., Kimura, Y., Kasahara, Y., Chen, W., Imai, H., Fujisawa, M., Otuka, K., Matsubayashi, K.: Importance of cognitive assessment as part of the Kihon checklist developed by the Japanese ministry of health, labour and welfare for prediction of frailty at 2-year follow up. Geriatr. Gerontol. Int. **13**(3), 654–662 (2013)
7. Tomata, Y., Hozawa, A., Ohmori-Matsuda, K., Nagai, M., Sugawara, Y., Nitta, A., Kuriyama, S., Tsuji, I.: Validation of the Kihon checklist for predicting the risk of 1-year incident long-term care insurance certification: the Ohsaki cohort 2006 study. Nippon Koshu Eisei Zasshi **58**(1), 3–13 (2011) (In Japanese)
8. Applegate, W.B., Miller, S.T., Graney, M.J., Elam, J.T., Burns, R., Akins, D.E.: A randomized, controlled trial of a geriatric assessment unit in a community rehabilitation hospital. New J. Med. **322**, 1572–1578 (1990)

9. Maki, Y., Ura, C., Yamaguchi, T., Murai, T., Isahai, M., Kaiho, A., Yamagumi, T., Tanaka, S., Miyamae, F., Sugiyama, M., Awata, S., Takahashi, R., Yamaguchi, H.: Effects of intervention using a community-based walking program for prevention of mental decline: a randomized controlled trial. J. Am. Geriatr. Soc. **60**(3), 505–510 (2012)
10. Sanders, E.B.N., Stappers, P.J.: Co-creation and the new landscapes of design. CoDesign **4** (1), 5–18 (2008)
11. Mann, W.C., Ottenbacher, K.J., Fraas, L., Tomita, M., Granger, C.V.: Effectiveness of assistive technology and environmental interventions in maintaining independence and reducing home care costs for the frail elderly: a randomized controlled trial. Arch. Fam. Med. **8**(3), 210–217 (1999)
12. Pickles, J., Hide, E., Maher, L.: Experience based design: a practical method of working with patients to redesign services. Clin. Gov. Int. J. **13**(1), 51–58 (2008)
13. Abidi, S.S.R., Cheah, Y.N., Curran, J.: A knowledge creation info-structure to acquire and crystallize the tacit knowledge of health-care experts. IEEE Trans. Inf Technol. Biomed. **9** (2), 193–204 (2005)
14. Nonaka, I., Toyama, R.: The knowledge-creating theory revisited: knowledge creation as a synthesizing process. Knowl. Manage. Res. Prac. **1**(1), 2–10 (2003)
15. Saijo, M., Suzuki, T., Watanabe, M., Kawamoto, S.: An analysis of multi-disciplinary and inter-agency collaboration process: case study of a Japanese community care access center. KMIS 2013 **66**, 470–475 (2013)
16. Paquet, V., Feathers, D.: An anthropometric study of manual and powered wheelchair users. Int. J. Ind. Ergon. **33**(3), 191–204 (2004)
17. Das, B., Kozey, J.W.: Structural anthropometric measurements for wheelchair mobile adults. Appl. Ergon. **30**(5), 385–390 (1999)
18. McNamara, N., Kirakowski, J.: Defining usability: quality of use or quality of experience? In: International Professional Communication Conference Proceedings, pp. 200–204. IEEE (2005)
19. McNamara, N., Kirakowski, J.: Functionality, usability, and user experience: three areas of concern. Interactions **13**(6), 26–28 (2006)
20. ISO 9241: Ergonomic requirements for office work with visual display terminals: part 11: guidance on usability (1998)

Author Index

Printed in the United States
By Bookmasters